ALLE ZEIT WACH 1842

Viviana Colombari (Editor)

Reliability Data Collection and Use in Risk and Availability Assessment

Proceedings of the
6th EuReDatA Conference
Siena, Italy, March 15 - 17, 1989

Springer-Verlag Berlin Heidelberg NewYork
London Paris Tokyo

Dr. Viviana Colombari
Tema S.p.A.
Via Medici del Vascello 26
20138 Milano
Italy

ISBN-13:978-3-642-83723-4 e-ISBN-13:978-3-642-83721-0
DOI: 10.1007/978-3-642-83721-0

Softcover reprint of the hardcover 1st edition 1989

2161/3020 543210

The conference was organized for EuReDatA
by Tema S.p.a.

EuReDatA Programme Committee

J.R. Aupied, EdF (F)
G.C. Bello, ICARO (I)
A. Bertrand, IFP (F)
P. Bockholts, TNO (NL)
V. Colombari, TEMA (I)
M. Drury, ICL (UK)
K. Ekberg, Vattenfall (S)
V. Ionescu, Motor Columbus (CH)
A.Z. Keller, University of Bradford (UK)
A. Lyytikainen, VTT (SF)
G. Mancini, JRC (CEC)
T.R. Moss, RM Consultants (UK)
U. Samdal, Norsk Hydro (N)
H.J. Wingender, NUKEM (D)

National Organizing Committee

S. Bagnara
G.C. Bello
V. Colombari
G. Gavelli
G. Mancini
M. Melis

Foreword

Reliability data collection and its use in risk and availability assessment is a subject of increasing importance. The founders of EuReDatA, and in particular, Arne Ullman, the originator and first Chairman of the Association, recognized the need for a body capable of acting as a catalyst and providing a unified approach to this subject.

It is therefore a prevailing objective of the European Reliability Databank Association to initiate and support contact between experts, companies and institutions active in reliability engineers and research.

Although the first and principle interest of Euredata is reliability data and data banks, the Association is aware that these are to establish and maintain reliability and safety.

It is with this objective that EuReDatA regularly holds conferences and seminars covering a range of reliability topics.

The rising interest in the matter and the growing awareness of its importance have consequently led to such a frequency of expert meetings, that an engineer can attend only a few of the more important event. Thus, according to the objectives of the Association, the information communicated at EuReDatA conferences has to be made publicly available. The success of the first attempt three years ago let us expect a simular interest in this volume.

H.J. Wingender
EuReDatA Chairman

V. Colombari
Organiser and Editor

Contents

Session 3: Common Cause Data

Session 4: Uncertainties - Source and Propagation

Session 5: Computer Aided Risk Analysis

Session 6: Reliability and Incidents Data Acquisition and Processing

Session 7: Human Reliability

Session 8: Probabilistic Safety and Availability Assessment

Session 9: Feedback of Reliability into System Design, Operation and Management

Session 10: Data Fusion

Session 11: Reliability Modeling and Techniques

Session 15: Electronic Reliability

Session 16: Reliability Tests

Closing Session

* Contributions marked * were not available at the time of printing

Error Analysis in Man-Machine Systems

S. BAGNARA(1), A. RIZZO(2), and F. STABLUM(3)

(1) Dept. of Philosophy and Social Sciences
University of Siena
(2) Institute of Psychology,
CNR, Roma
(3) Dept. of Psychology,
University of Padova

Summary

The study to be reported was aimed at evaluating the reliability of MMS in an automated process control system. The methodology was based on error analysis and took into consideration: a) the errors occurred, b) the error detections, and c) the recoveries from error. It was adopted to evaluate both a prototype and the implementation of the MMS. The two evaluations had different aims. The first, related to the "design" phase, was aimed at establishing whether the designed tasks were compatible with the cognitive abilities and competence of the operators. The objective of the second evaluation, related to the "re-design" phase, was to identify where and how to improve th MMS.

INTRODUCTION

The analysis of how human errors take place in a given Man-Machine System (MMS) can provide useful information on the human knowledge and skills required to interact with the system, and, consequently, on how to design it in order to minimize both the occurrence of errors and their effects (1). Moreover, many studies both in cognitive psychology (2,3,4,5) and in analyses of industrial accidents (see, e.g.,6) have shown that human errors can not be completely eliminated because of their unpredictable variety. It has been also claimed (7) that errors should not be completely prevented because of their heuristic value in

the adaptation to never-before-encountered situations. Therefore, error analysis can also provide much information in designing error-tolerant MMS.
The present paper will report a study devoted to the evaluation of the reliability of an automated process control system of a hot strip mill in a steelworks. The methodology of evaluation was based on the analysis of: a) the errors occurred, b) the error detections, and c) the recoveries from errors.

Human error analysis

There is a general consensus (1,8) that human errors depend on user's Knowledge organization and cognitive control and on characteristics of the environment where the user's performance takes place. Therefore, error analysis can yield information about the human cognitive processes, the effectiveness of human-machine interaction, and the system itself. In order to get and properly exploit these informations a classification of errors based on the factors which are involved in their production is needed.

Error	**Error-Shaping Factors**
Slip Intention correct, action not carried out as planned	1. Recency and frequency of previous use 2. Environmental control signals 3. Shared schema properties 4. Concurrent plans
Rule-Based Mistake Faulty application of well-known rules or procedures both in identifying a situation or/ and in adopting a plan of action	1. Mind set (It's always worked before) 2. Availability(First come, best preferred) 3. Matching bias (like relates to like) 4. Over-simplification 5. Over-confidence
Knowledge-Based Mistake Selected plans do not fit with the problem to be solved	1. Bounded rationality: Selectivity Working memory overload Out of sight out of mind 2. Thematic 'vagabonding' and 'encysting' 3. Memory cueing/reasoning by analogy 4. Matching bias revisited 5. Incomplete/incorrect mental model

Table 1. Major error shaping factors for each type of error (adapted from Reason, 1987).

The Generic-Error Modeling System (GEMS), put forward by James Reason (8), provides a simple, generic (i.e., context-free) framework within which to locate the principal cognitive limitations and biases that give rise to the more predictable varieties of human error. It allows to classify errors as a function of the level of control of human performance and of intended or not intended activity and to determine the role played by cognitive and situational factors in the occurrence of an error. The GEMS identifies three types of error (slip, rule-based mistakes, and knowledge-based mistakes) together with their major shaping factors. Both error types and shaping factors are reported in Tab.1.

Error detection

The feedback-based evaluation of the results of actions is of crucial importance in performing properly an intended activity. How the evaluation takes place depends on the level of attentional control devoted to the current activity. It takes place from time to time, and requires specification of the adopted intention and a change in the level of attentional control when the actions are mediated by skills. It is continuous and does not require changes both in the specification of intention and in the level of cognitive control when hypotheses of solutions are tested as it happens when knowledge-based behavior is exhibited.

Evaluation of results is also critical for error detection. According to Norman (9), there are essentially only two ways by which error are detected:

a) **self-monitoring**. The action could be monitored and compared with expectations. If the action was not as planned, then the error would be detected.

b) **deleterious outcome.** If the result of an error is to cause some deleterious outcome in the next activity, then the error may be detected.

Both ways rest upon feedback from the environment and are instrumental, though varying in strength, for the detection of every type of error (see, 10). Nevertheless, the availability of feedback by itself is not sufficient for a reliable evaluation of results and for error detection. As for both slips and mistakes detection, mental workload and the cognitive distance between the level at which the displayed feedback can be

captured and the level where attention is directed in the control of the ongoing activity (either goal setting, or intention specification, planning, action execution) play a crucial role (10,11,12).
In fact, when a slip occurs the feedback is provided at the level of the executed action while the user's attention can be directed at a higher level of specification of the intention, then the comparison between the feedback informations and the expectations can be difficult, perhaps even impossible (9,11). This means that there is a cognitive distance, within a schema of action, between the specification of the intention and the activated and executed actions.
On the other hand, when a mistake occurs, if the terms of the feedback are not those of the user's intention, the user will be required to translate the feedback into terms that are compatible with his/her intention in order to make the evaluation. This means that there is a cognitive distance between intentions and schemata of actions.

Error detection analysis can allow to evaluate the level of effectiveness of the exchange of information between the operator and the system, if it takes into consideration whether errors are detected, the events that trigger the detection, the cognitive distances between the feedback and the level of cognitive control and the activity planning structure.

In the present study, this analysis has been conducted by identifying and classifying the detection episodes as proposed by Allwood (13). He observed that subjects consistently exhibited the following behavioural patterns when detecting errors:
a) **Direct Error Hypothesis Behaviour** (DEH).It is shown by the users when they suddenly detect a real or suspected error. This pattern is not necessarily time-locked to the occurrence of an error.
b) **Error Suspicious Behaviour** (ES). The users exhibit perplexity about the results obtained, even though no error actually occurred. Again, this pattern is not time-locked to an error.
c) **Standard Check Behaviuor** (SC). It consists of a routinary control on current or next-to-be-performed actions.

The events that trigger error detection are either mismatches between feedback and explicit or implicit expectations, or outcomes of actions that, when available, fit with the expectations, but are found not consistent with the goals of the activity. As for their appearance, these events can be classified on the basis of the type of feedback they are linked with:

a) **Forcing Function** (FF). The system does not allow the user to perform or to continue in executing an action (9).

b) **External Feedback** (EF). The system makes available information, not necessarily in close relation to the actions under execution, which turns out not to fit with either the expectations or the goals of the activity.

c) **Inner Feedback** (IF). It is a possible outcome of mental simulation of activities and of their expected consequences and results. The information active in the user's working memory shows that an expectation is going to fail or has failed.

Error recovery

Error recovery can be analyzed in depth by adopting the framework of the "Norm Theory" developed by Kahneman and Miller (14). By using this framework(15), error recovery cognitive processes can be distinguished as a function of whether a backward or a forward causal analysis is carried out or a repertoire of correct solutions are scanned, without any causal analysis about the source of the detected error. The causal analysis can be either automatically or consciously conducted.

In the present study this detailed analysis of error recovery has not been adopted. The analysis was centered on whether the users were able to correctly diagnose the situation they entered when an error occurred, and to overcome the mismatch.

THE EVALUATION TESTS

The study was carried out in the hot strip mill area of a steelworks. The automated process control system (see Figure 1) to be evaluated aims at optimizing the integration between the production planning and the plant conditions. As a function of the market demands, the Planning and Production Department establishes, day by day, the Daily Schedule (DS) of the activity. The DS is sent to the Hot Strip Mill (HSM) Area Coordinator when he starts working. On the basis of the DS, the HSM Coordinator has to develop the Operation Schedule (OS) (i.e., the detailed daily activity in the area) while considering the plant conditions and the following aims/constraints: a) To exploit to the utmost the Hot Strip Mill; b) To make the most of the space in the Slab Stock Yard.

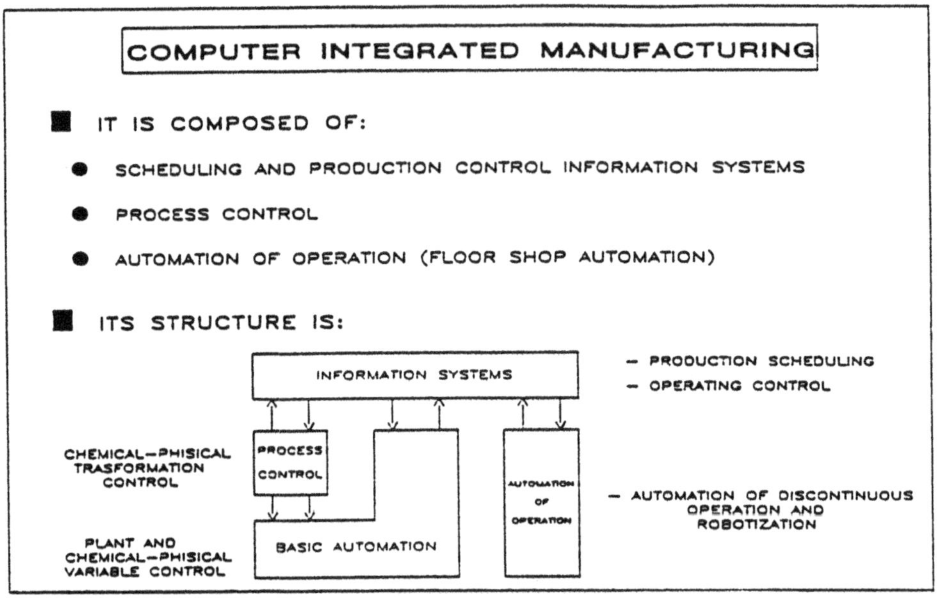

Figure 1. The production control system in CIM.

The prototype evaluation test

This test was carried out by using a prototype of the system which presents the operator with tasks assessed as "critical" by a previously performed task analysis (16).

Subjects.- Eight operators took part in the simulation.Their length of service in the Rolling Mill area ranged 13-17 years.

Training.- The operators were trained to become acquainted with: a) The functional architecture of the system; b) The procedures necessary to accomplish the task. At the end of the training, a test was given to evaluate the degree of competence of the operators. They were asked to fill in a 22-item questionnaire about the procedures to be used and the structure of the system. An additional training session was given to the operators to practice them in "concurrent verbalization" (17).

Task.-The operators were asked to plan the Hot Strip Mill production, having the plant in standard conditions and the aim to achieve a given quantitative and qualitative outcome.

Data collection and analysis.- The operators' activity was video-recorded and afterwards analyzed by two human factors specialists, one trainer, one of the designers of the system, and one expert from the Planning and Production Department. The operators were explicitly asked for verbalizing what "kind" of mismatch triggered the detection process. The classification of errors, type of mismatches, error detection behavioural patterns, and overcomes of mismatch were based on the video-recorded material, verbal reports, and operators' observations on their own behaviour made soon after the task had been completed.

The implementation evaluation test

This test was carried out when the system was already implemented. Four operators took part in the evaluation. Their activity consisted of all the operations to be performed in a whole shift. At the beginning of the test, the operators were given a questionnaire to assess their knowledge about structure and functionalities of the systems. The data collection and analysis were the same as in the prototype evaluation test.

RESULTS

The results from the two tests will be first reported separately and then compared.

Prototype evaluation Test (PT).- The operators took about 90 minutes (on average) to complete the the evaluation test. Overall, they made 95 errors and detected 74 of them. Nine episodes were left out from the analysis due to controversial classification. The relationships between types of error and error detection behavioural patterns are shown in Table 2. The Slips were detected mainly when DEH behavior was exhibited ($p. < .01$), while both R-B and K-B Mistakes were detected when operators showed an ES behaviour ($p. < .01$). The difference between DEH and ES behaviour for R-B Mistakes did not reach significance. The SC behavior accompanied the detection of Slips. The highest percentage of undetected errors belongs to R-B Mistakes.

	Slip	R-B Mistake	K-B Mistake	Tot.
SC	8 (11%)	3 (4%)	-	11 (15%)
DEH	26 (35%)	6 (8%)	1 (1%)	33 (44%)
ES	6 (8%)	11(15%)	13 (18%)	30 (41%)
Tot.Det.	40 (44%)	20 (27%)	14 (19%)	74 (100%)
Not Det.	5	13	3	21
TOT.	45	33	17	95

Table 2. **Prototype test**.- Types of errors as a function of error detection behavioural patterns.

There was no use of IF. The FF accompanied the DEH pattern for Slip, and, very much less effectively R-B Mistake detection. The EF was used for detecting all the types of error, and accompanied the DEH pattern when Slips were detected and the ES pattern when operators detected Mistakes. The most of Slips were recovered, while the most of Mistakes were not.

	Slip				R-B Mistake				K-B Mistake			
	FF	EF	IF	Tot	FF	EF	IF	Tot	FF	EF	IF	Tot
SC		8		8		3		3				
DEH	14	12		26	4	2		6		1		1
ES	3	3		6	4	7		11	4	9		13
Tot det	17	23		40	8	12		20	4	10		14
Recovery	16	19		35	3	4		7		2		2

Table 3. **Prototype test**.- Types of error as a function of types of feedback, error detection behavioural patterns, and recoveries.

Implementation evaluation Test (IT).- In this test, the operators made 86 errors, and detected all of them. Only one error, though detected, was not recovered. Only, one episode was left out from the analysis because of controversial classification. No K-B Mistake occurred. As shown in Table 4, the most of the error were Slips. Both Slips and R-B Mistakes were detected mainly through DEH.

	Slip	R-B Mistake	K-B Mistake	Tot.
SC				
DEH	66 (77%)	14 (16%)		80 (93%)
ES	3 (3.5%)	3 (3.5%)		6 (7%)
Tot.Det.	69 (80.5%)	17 (19.5%)		86 (100%)
Not Det.				
TOT.	69	17		86

Table 4. **Implementation test**.- Types of errors as a function of error detection behavioural patterns.

As reported in Table 5, the detection of Slips was mainly mediated by External Feedback, whereas External Feedback and Forcing Function feedback played about the same role in the detection of R-B Mistakes. There were two detections of R-B Mistake mediated by Inner Feedback.

	Slip				R-B Mistake				K-B Mistake			
	FF	EF	IF	Tot	FF	EF	IF	Tot	FF	EF	IF	Tot
SC												
DEH	2	64		66	6	6	2	14				
ES	2	1		3	1	2		3				
Tot det	4	65		69	7	8	2	17				
Recovery	4	64		68	7	8	2	17				

Table 5. **Implementation test**.- Types of error as a function of types of feedback,error detection behavioural patterns, and recoveries.

Comparison between Prototype and Implementation tests.- The comparison between the results from the Prototype and Implementation tests shows that:

a) There were very much less errors in the IT. It should be reminded that, while the PT lasted about 90 min on average, the IT comprised a whole shift, i.e., six hours, and many more tasks and real "critical" events.

b) At variance with the PT, there was no K-B Mistake in the IT. Furthermore, no error went undetected in the latter test.

c) In the IT, the percentage of Slips was very much increased.

d) The Standard Check behavioural pattern, though exhibited by the operators, was of no use in detecting errors in the IT.

e) In the IT, the DEH pattern was much more instrumental in detecting errors than in the PT. The reverse was true for the ES pattern.

f) The Forcing Function feedback was related to Slip detection in PT, whereas this type of feedback played a greater role with A-B Mistakes in IT.

g) There was no use of the Inner Feedback in PT, while there was some indications of it in IT.

h) All but one detected errors were recovered in IT, whereas about 40% of them were not recovered in the PT.

DISCUSSION

The discussion will deal with the results from the two tests separately, by taking into account the number and type of errors, the detection and recovery performances.

Prototype Test.-

The high number of Knowledge-Based Mistakes clearly indicates that the operators did not possess an adequate mental model of the system and/or they underwent to mental overload in dealing with the system. The qualitative analysis of the video-recorded material and the verbal protocols confirm both these interpretations.

The operators maintained that in many cases they knew what information was required to properly handle the situations encountered, but were unable to keep all the information under control. This indicates that the most of the possessed knowledge was not yet automatized, so that the operators easily experienced mental workload.

On the other hand, the operators ran frequently into the "false analogy" bias (8). When a situation, for which there was no proper procedure at hand, had to be faced with, frequently they adopted inappropriate analogies derived from the mental model of the previous control system.

The incompleteness of the mental model of the system is also confirmed by how the operators made the R-B mistakes. They were produced these errors because of "mind set" and "availability" biases. The operators were well-practised with a small number of procedures only. So, they made use of the familiar procedures (because easily available) even when those procedures did not fit with the actual situation, or stuck to them once adopted, because unable to exit from them and to find alternatives.

As for the detection performance, it depended heavily on deleterious outcomes, accompanied by Error Suspicious behavioural patterns. The data were confirmed by the qualitative analysis which showed that there usually was a very large distance, both in the cognitive and temporal dimensions, between the detected discrepancy and the error. The operators were not equipped for carrying out a causal analysis, neither they possessed a repertoire of procedures to cover such a distance. In this endeavor, the system was not of much help.
The fact that the DEH pattern was mainly instrumental in detecting the Slips indicates that the operators were forced to direct the attentional control to the surface of interaction (on how to execute a procedure rather in planning activity). This is also corroborated by the type of errors more likely to go undetected: R-B Mistakes. The detection of this type of errors requires attention be allocated to control the timing of actions and the planning rather than individual actions. Given the level of practice, the faulty mental model, and the characteristics of the interaction required by the system, attention was instead dedicated to how to carry out individual operations.

The data obtained about the type of feedback which triggered the detection and on recovery performances confirm the previous conclusions. The detection never took place as a result of a mental simulation: It depended either by a forcing function or by external feedback. Both cases were related with deleterious outcomes. As noted before, this implies that the temporal and cognitive distances between a detected discrepancy and its source is often very large, so rendering the recovery difficult if not impossible, as demonstrated by the present results: About 40 % of the detected errors were not recovered. These results cannot be attributed only to the fact that the operators lacked in experience. Of course, this is at least partially true, because it is well known that experienced operators in many cases need not to undertake a causal analysis to correct an error. They can rely either on a repertoire of "escape routes", or on methods for overcoming a mismatch, or on the knowledge of "peculiarities", rather than of general rules, which allow to identify very rapidly the source of a mismatch (18). As a matter of fact, the system did not allow the operator to trace back the operations done and did not provide any help or advice for recovery. Moreover, the operator was not given on-line feedback about the possible consequences of an action.

Implementation test.-
After the re-design, the evaluation test of the implemented system showed that the quality of interaction was very much improved. There was no K-B Mistake, and all the errors were detected and all but one recovered. The mental model was enough complete to allow to face successfully with all the drawbacks which take place in the real work setting. This is also confirmed by the fact the DEH pattern accompanied the detection of more than 90% of errors. The cognitive distances between mismatches and their sources were reduced. The operators had not to wait for a Forcing Function to become aware that an error had been made. They were also able to mental simulate the activity and to detect an error through Inner Feedback. They made also a much more active use of the External Feedback. Information could be evaluated well before a deleterious outcome appeared. This improvement can not be attributed only to the operators' longer training and experience with the system, since the system itself had been very much improved (as for it flexibility, feedback provided, and help tools) as a consequence of the early evaluation of the prototype.

It has to be noted that the quantitative and qualitative analyses of the data obtained through the evaluation test still showed some deficiencies. Some recoveries of error were not straightforward: There were four cases where the diagnosis was faulty, followed by recovery actions which brought about new and more complicated errors. At the very end, the operators were able to find out the correct solution. Furthermore, Error Suspicious pattern was still exhibited, even though in relation to low level errors. All this indicates that some improvements are still needed. Both types of deficiencies were related to "unforeseen" situations, such as the combination of two errors or of one error and an external interference.In both cases, it was very difficult to tell apart the sources of variation, mainly because the feedback was designed as providing information about individual disturbances rather than about the interweaving of two or more of them.

CONCLUSION

The results of both evaluation tests showed the usefulness of the proposed methodology based on error analysis. When adopted in a design phase it allowed to establish whether the designed tasks were compatible with the cognitive abilities and competence of the operators and to evaluate the training. So, this evaluation could be used both for deciding whether the design has to be changed as for its aims and main characteristics, and for reconsidering the adopted training methods. When the analysis of error was used in the implementation, it allowed to evaluate the Man-Machine interaction and to identify what had to be still improved.

REFERENCES

1. Norman, D. A. : Design rules based on analysis of human error. Communications of the ACM. 26 (1983) 254-258.
2. Reason, J. T. : Actions not as planned: The prize of automatization. In G. Underwood and R. Stevens (eds.) Aspects of consciousness. Vol.1. London: Academic Press 1979.
3. Reason, J.T. Lapses of attention. In R. Parasuraman and D. Davies (eds.) Varieties of attention. New York: Academic Press 1984.
4. Norman, D. A.: Categorization of action slips. Psychological Review. 88 (1981) 1-15.
5. Nisbett, R. & Ross, L. Human Inference: Strategies and shortcomings in social judgement. Englewood Cliffs, N.J.: Prentice-Hall 1980.
6. Rasmussen, J., Duncan, K., and Leplat, J. (eds.) New technology and human error. London: Wiley 1987.
7. Rasmussen, J. Information processing and human machine interaction. New York: North Holland 1986.
8. Reason, J.T. : Generic error modelling system (GEMS): A cognitive framework for locating common human error forms. In J. Rasmussen, K. Duncan, and J. Leplat (eds.) New technology and human error. London: Wiley 1987.

9. Norman, D.A. Working papers on errors and error detection. San Diego: University of California 1984.
10. Rizzo, A., Bagnara, S., and Visciola, M.: Human error detection processes. International Journal of Man-Machine Studies. 27 (1987) 555-570.
11. Lewis, C., and Norman, D.A.: Designing for error. In D.A. Norman and S. W. Draper (eds.) User centered system design. Hillsdale, N.J.: Erlbaum 1986.
12. Hutchins, E.L., Holland, J.D., and Norman, D.A.: Direct manipulation interfaces. Human-Computer Interaction.1 (1985) 311-338.
13. Allwood, C. M.: Error detection processes in statistical problem solving. Cognitive Science. 8 (1984) 413-437.
14. Kahneman D., and Miller, D.T.: Norm Theory: Comparing reality to its alternatives. Psychological Review.93 (1986) 136-152.
15. Bagnara, S., Ferrante, D., Rizzo, A., and Stablum, F.: Causal analysis in human computer interaction: When does it occur? Cambridge: Fourth European Conference on Cognitive Ergonomics 1988.
16. Bagnara, S., Rizzo, A., Stablum, F., and Visciola, M.: "Generics"in human decision making. In H.J. Bullinger and B. Shackel (eds.) Human-computer interaction - INTERACT '87. Amsterdam: North Holland 1987.
17. Ericsson, K.A., and Simon, H.A.:Protocol analysis. Cambridge: MIT Press 1984.
18. Partridge, D., and Paap, K.: An introduction to learning. Artificial Intelligence Review. 2 (1988) 79-101.

Environmental Impact Assessment and Risk Analysis. Theoretical Background of the Computerized Methodology "LIVIA"

V. COLOMBARI (**), G.C. BELLO (*) , E. GALATOLA (*)

(**) TEMA SpA , ACOAM Division, Milano (Italy)
(*) consultant

SUMMARY

Industrial facilities exchange materials and energy with the surrounding area either in normal operating conditions or in exceptional accidental situations. The total effect is then due to the sum of a continuous deterministic component (impact) and of random stochastic components (risks). A method has been developed to identify and evaluate the impacts (using a logical approach based on the coaxial matrix theory) and to merge their effects with those caused by accidental events , in order to map the likely total interaction between the facility and the environment. The computerized version of the method , named LIVIA , has been tested in various applications in the refining and petrochemical fields.

THE LOGICAL APPROACH

The identification of the impacts (either deterministic or stochastic) is done by logical vectors composed by five fields , i.e.

1. Activity --> 2. Primary Cause --> 3. Environmental Component Affected --> 4. Impact --> 5. Receptor

Each field accepts up to three levels of detail. Therefore a typical impact vector could appear as

1. Operation, Refinery , Topping plant
2. Exchange of materials , Emission of gases , Emiss.of SO2
3. Atmosphere , Troposphere , Lower levels
4. Modification of chemical composition , Increase of SO2
5. Human population , Residents in XYZ

It appears that each vector starts with the Activity causing the impact and ends up with the Receptor of the effects.

The creation of the set of vectors is realized by means of four coaxial matrixes (see fig. 1) , i.e.

- Activity vs Primary Cause
- Primary Cause vs Environmental Component
- Environmental Component vs Impact type
- Impact Type vs Receptor

Default values for each axis of the matrixes are stored in computerized files and the user can either select stored options or introduce new fields.

Each vector can have "branching points" or generate "secondary vectors".

A Branching Point exists whenever more than one point is specified on an axis . As an example the above mentioned Topping Plant can release products other than SO_2 (NO_x , liquid effluents , etc.) , or it can exists receptors of the effects other than Human Population (Agricultural activities, etc.). At each branching point LIVIA will automatically create a new vector (brother vector).

A Secondary Vector exists whenever the primary impact generates a chain of secondary impacts. As an example the above mentioned primary vector generates a secondary vector where the Environmental Component affected is the Rain Water (increase of pH , due to increase of SO_2 in the atmosphere) ; this secondary vector generates a tertiary vector where the Environmental Component affected is the Soil (alteration of the surface quality due to acid rains).

LIVIA can generate complex structures with as many branching points and secondary vectors , as appropriate.

At the end of the vector creation phase , the user has exaustively mapped the interactions between the Activity and the Receptors , in a qualitative way.

QUANTIFICATION OF VECTORS

The impact vectors are quantified by LIVIA in two different ways , depending upon the nature of the vector itself (a risk vector or a deterministic impact vector). The risk vectors are quantified the usual way (expected frequency and likely damages caused). The deterministic impact vectors are quantified attributing a "weight" and an "intensity".

The weight is calculated as a function of

- the renewability of the receptor
- the rarity of the receptor
- the economic importance of the receptor
- the persistence of the impact
- the reversibility of the effects
- the dimensions of the area affected

The weight then represents the "importance" of the vector in the environmental scenario considered and has values ranging from 1 to 12. The higher values of weight identify the vectors that are of primary importance , either because of the relevance of the targets affected , or because of the persistence and irreversibility of the effects.

The intensity is calculated as a function of the values of the impact parameter (i.e. SO_2 concentration in air)

- before the proposed activity (Iante)
- after the proposed activity (Ipost)
- the best acheivable value (Ibest)
- the worst acceptable value (Iworst)

The values of intensity range in the field (-100/+100) , where the value zero occurs whenever Iante=Ipost (no effect). (see fig. 2).

Ibest and Iworst have to be agreed before the application of LIVIA , while Iante is measured and Ipost is calculated by using effect simulation codes (i.e. codes for studying dispersion of SO_2 in the air).

The product of the weight and the intensity (ranging from -1200 to +1200) is an index of the impact caused by each vector.

The sum of the above mentioned products divided by the sum the weights represents the average weighted impact intensity , ranging from -100 (completely unacceptable impact) to +100 (extremely positive impact).

The average weighted impact intensity is then an index of the total impact caused by the proposed activity in the environment ; positive values indicate an improvement of the overall quality of the environment , while negative values pinpoint situations of decreasing environmental quality.

ANALYSIS OF SUB-SYSTEMS

If used for broad scope analysis LIVIA considers not only the environmental components strictly defined (chemical , physical and biological components) , but also other factors affected by the activity , such as

- economical components
- social and occupational effects
- alteration of the landscape
- alteration of the cultural scenario

This causes two orders of problems to be tackled :

1. Some of the components are not suitable for quantitative evaluations based on mathematical algorithms (i.e. the beauty of the landscape)
2. The merging of the effects in a unique impact index could produce highly misleading results.

To tackle the first problem LIVIA accepts "expert judgments" for components unmanageable by math techniques.

To tackle the second one , it is advisable to process the vectors generated not as a unique sample , but as separated homogeneous sub-samples , obtaining results such as :

- the total averaged impact on atmosphere
- as above but on the water and soil system
- as above but on the social-economic scenario
- as above but with ref. to the human health , etc.

Using this approach one can produce a serie of indexes of impact , dealing with the relevant environmental components and prepare a table suitable for decision makers.

MERGING OF IMPACTS AND RISKS

The results of impact assessment and risk analysis are quite different , i.e. the first one is an index of alteration of the quality of the environment that will last at least for the duration of the activity (and may be longer than that if irreversible effects are caused) , while the second one is expressed as expected frequency and likely effects of random events.

There is no way to merge them in an overall index of deterministic and stochastic impact , unless we proceed to a dynamic representation (impact vs time).

The strategy proposed by LIVIA is that of calculating the Total Loss of Environmental Quality , caused by both the determistic permanent loss of quality due to the continuous normal operational exchanges between the activity and the environment and by the random loss of quality due to stochastic accidental events.

The environmental quality depends upon various polluting agents and a set of indexes has been developed for this purpose. For instance the air quality can be described by a PSI (Pollutant Standard Index) , taking account of the concentration of SO2 , NOx , CO , particulates , etc. Similarly we developed an index for surface water (IQA) and for the soil. Considering the index developed by the air , this will assume the values 100 for top quality air and of zero for very polluted air.

Let's assume (see fig. 3) that the air quality index for a certain area is equal to 70 , before a new proposed activity goes into operation , and is reduced to 50 due to the continuous emissions of pollutants by that activity.

Let's also assume that random events (accidents) with a frequency of occurrence 'lambda' cause a further transient decrease of the air quality index. The final value of the air quality index depends upon both the continuous emissions and the accidental releases (this final value in some case cannot reach the initial one , because of any irreversible effect).

The Total Loss of Environmental Quality is calculated as the integral in time of lost environmental quality.

This integral represents the total price that the environment has to pay to the new activity , taking account of both the normal operating conditions and of the accidents. The Total Loss of Environmental Quality can be easily converted into a Total Expected Damage , at least from a theoretical standpoint.

The link between the two indexes are the Vulnerability Models and in particular

- the Chronic Exposure Vulnerability Models (for permanent deterministic impacts) , and
- the Acute Exposure Vulnerability Models (for accidental impacts).

Let's assume (see fig. 4) that the rate of chronic diseases in the area is equal to 10 before the proposed activity enters into operation and reaches the value of 30 after the plant start-up. Accidental events can cause transient increases in the desease rate. The Total Expected Damage is equal to the integral in time of the increase of deseases , due to both normal operation and accidents caused by the proposed activity.

In order to actually perform this convertion an effort shall be made to implement Vulnerability Models much more comprehensive and reliable than those presently available.

CONCLUSIONS

Major conclusions of our work can be summarized ad follows:

1- Environmental Impact Assessments need to be supported by a coherent logical background , tackling exaustiveness and explicitation of the logical paths between the activity generating the impacts and the final receptors.

2- Quantification of the impacts is a mandatory goal if the intention is to produce results suitable as a tool for decision-making.

3- Results of the analysis should be processed by homogeneous sub-samples , to avoid misleading conclusions.

4- Merging of continuous impacts and random risks can be done only through a dynamic representation.

5- Total expected environmental damage can be in principle calculated , provided that better vulnerability models are developed.

Aknowledgements

Many thanks to Dr.Ing. Tania M. Amorim for her assistance in developing intensity evaluation criteria ; to Dr. Luisa Folli for her contribution on the biological aspects ; to P.I. Luca Moroni for the informatic implementation of LIVIA; to Alfredo Verna , Maurizio Gilioli , Antonio Salvatore , Federico Micheli and to Prof.s Mercedes Bresso and Severino Zanelli for their contribution of ideas and suggestions .

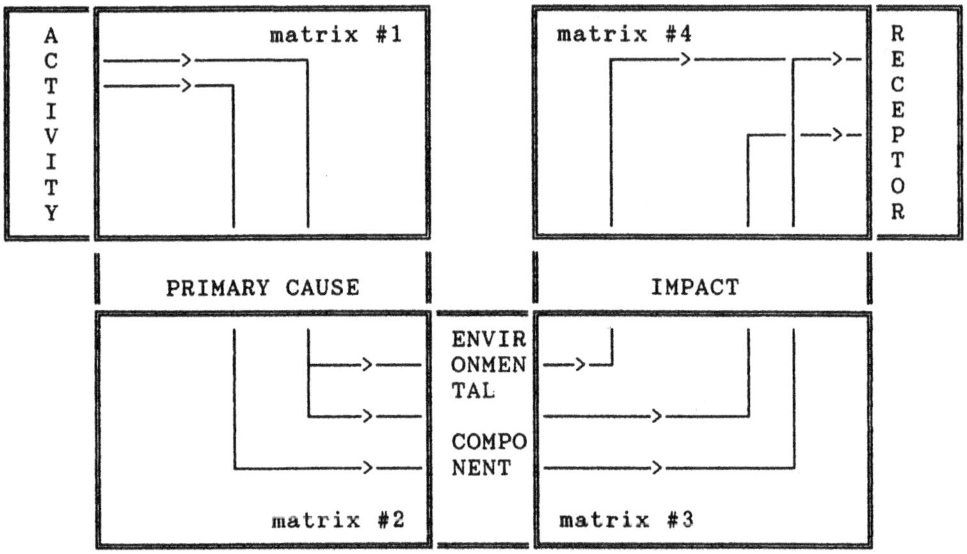

FIG.1 - COAXIAL MATRIXES FOR IMPACT VECTOR IDENTIFICATION

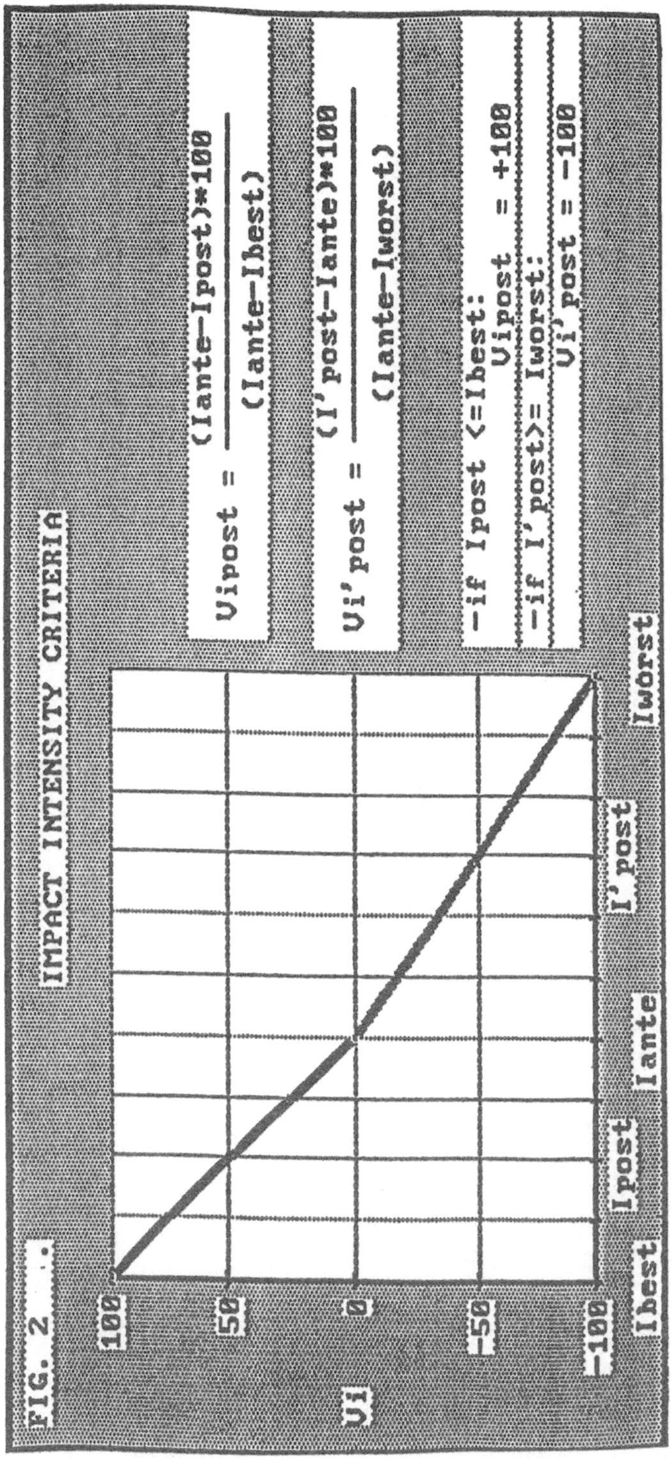
FIG. 2 ..
IMPACT INTENSITY CRITERIA
Ui
100
50
0
-50
-100
Ibest
Ipost
Iante
I'post
Iworst
Uipost = (Iante-Ipost)*100 / (Iante-Ibest)
Ui'post = (I'post-Iante)*100 / (Iante-Iworst)
-if Ipost <=Ibest:
Uipost = +100
-if I'post>= Iworst:
Ui'post = -100

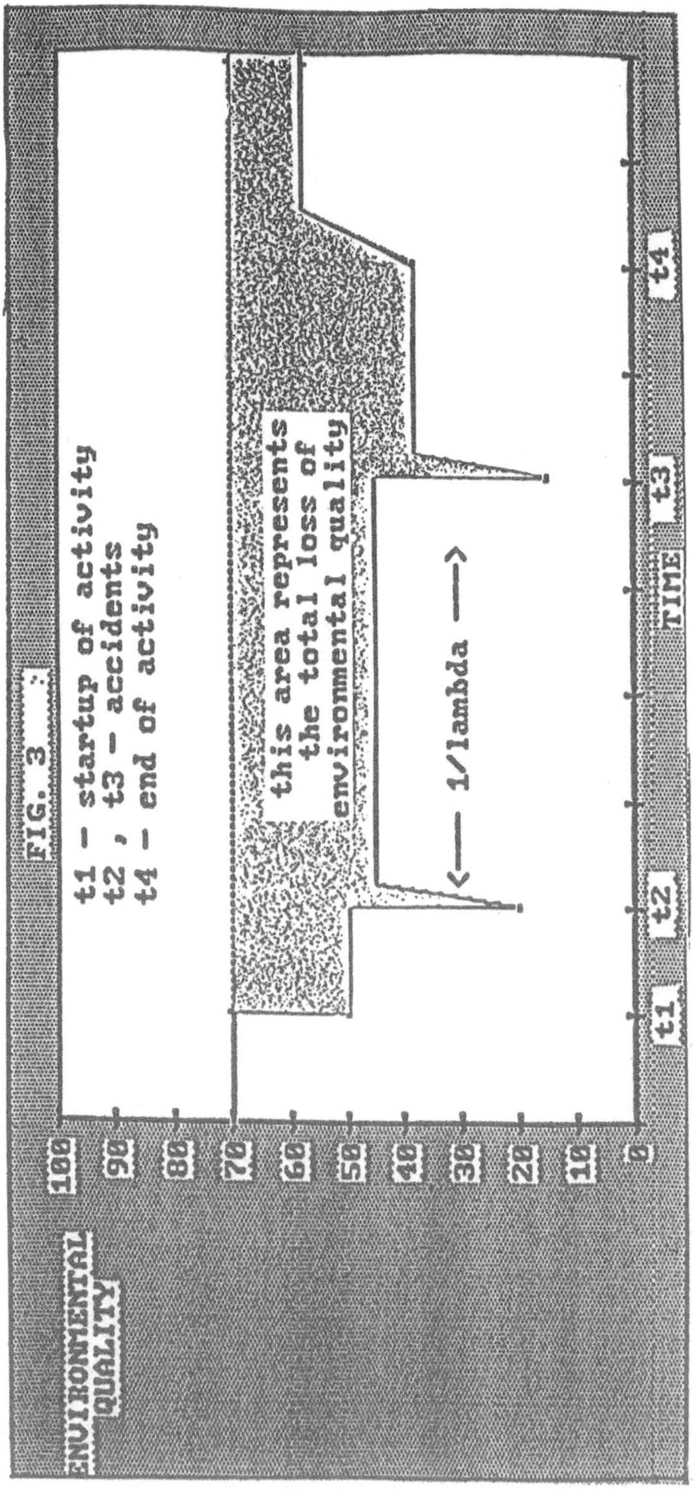
FIG. 3 :
t1 - startup of activity
t2 , t3 - accidents
t4 - end of activity
this area represents
the total loss of
environmental quality
<— 1/lambda —>
ENVIRONMENTAL
QUALITY
100
90
80
70
60
50
40
30
20
10
0
t1
t2
t3
t4
TIME

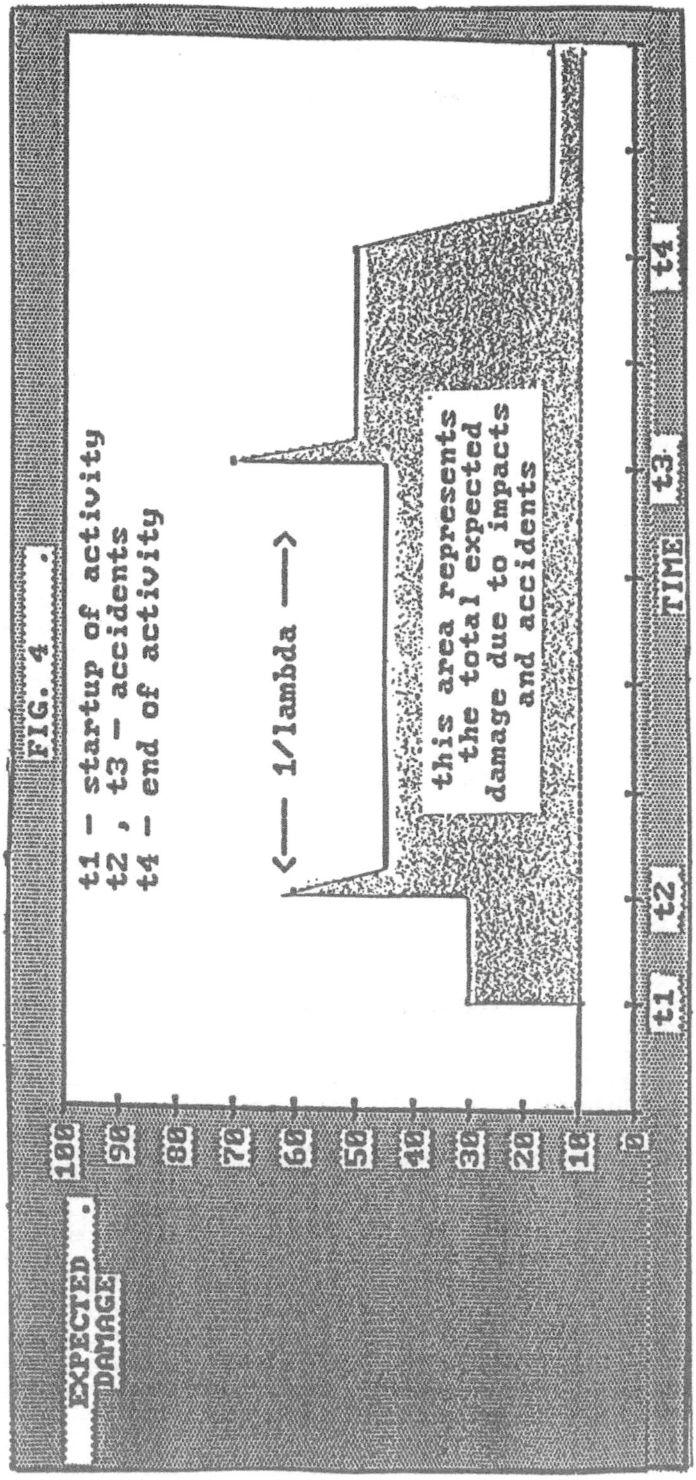
FIG. 4
EXPECTED DAMAGE
100
90
80
70
60
50
40
30
20
10
0
t1 - startup of activity
t2 , t3 - accidents
t4 - end of activity
<— 1/lambda —>
this area represents
the total expected
damage due to impacts
and accidents
t1
t2
TIME
t3
t4

Addressing the Problem of the Relevance of Reliability Data to Varied Applications

P.J. McIntyre*, I.K. Gibson+, H.H. Witt#

* Central Electricity Generating Board, Sizewell 'B' PMT, Knutsford, Cheshire, WA16 8QG, England.
\+ United Kingdom Atomic Energy Authority, Safety & Reliability Directorate, Culcheth, WA3 4NE, England.
\# Australian Nuclear Science & Technology Organisation, work performed on attachment to +.

SUMMARY

Reliability data is collected for many reasons on a wide range of components and applications. Sometimes data is collected for a specific purpose whilst in other situations data may be collected simply to provide an available pool of historical data. Data can also be extracted from information that was gathered without recognition that it could be adapted for use as reliability data at a later stage. It is not surprising that there should be significant differences in the strengths and weaknesses of data obtained in such different circumstances. This paper describes work undertaken to investigate how to make best use of available data to provide specific and reliable predictions of valve reliability for nuclear power station applications.

1.0 INTRODUCTION

Reliability data is collected for many reasons on a wide range of components and applications. Sometimes data is collected for a specific purpose whilst in other situations data may be collected simply to provide an available pool of historical data. Data can be extracted from information that was gathered without recognition that it could be adapted for use as reliability data at a later stage. It is not surprising that when faced with a new reliability study the analyst often finds that directly applicable data is not available and that the data that is available has varying relevance to the analyst's needs. To some extent this is the position found by the CEGB in considering data needs for the P.R.A. of its first series of PWR power stations. Amongst the steps taken to provide adequate data was the collaborative study with UKAEA Safety and Reliability Directorate that is discussed in this paper.

To restate the problem, each new application has characteristics that may differ from those used in constructing the historical database. Furthermore, the component chosen for an application may differ in type or detailed design from those represented in the database. As each characteristic of a component design or an application may influence reliability there is usually an attempt to identify these influences in order to select the most appropriate subset of historical data for prediction in a particular case.

For example, for the Sizewell 'B' PWR Pre- Construction Safety Report valve data has been gathered from PWR applications and split by actuation method. Thus broadly similar duties will be present on generally similar plants to Sizewell B but a judgement has been made that actuation method is the dominant influence on reliability. There is much engineering common sense in these assumptions. However, the possible effects on reliability of for example fluid pressure or of aggressive fluids such as silt bearing cooling water are not explicitly modelled. Unfortunately, whereas there may be a large population of valves with a particular actuation method as these populations are sub-divided, using factors such as those mentioned above, the resulting more relevant populations are smaller. Statistical and other uncertainties reduce confidence in prediction as populations reduce, counteracting the benefit of more relevant data.

A somewhat different approach to the use of historical data has been attempted from time to time where, from a thorough understanding of the mechanisms of failure in an application, a direct prediction of failure rate is made based on the values of physical or chemical parameters judged important in influencing failure. An example would be the use of values of creep strength, stress, temperature and time to predict the probability of failure by creep in a component.

Such a prediction is very specific to the application and relies on a thorough engineering understanding of the failure processes acting. Although an attractive concept, such an approach is likely to be unattainable in most industrial applications. It is possible to envisage approaches lying between the above data driven and engineering theory driven approaches. Such an approach is described below.

2.0 APPROACH

The approach taken has been to attempt to utilise engineering knowledge to direct the analysis of historical data. Essentially the approach is similar to that of the scientific method. Initially a theory of failure and what causes or accelerates it is postulated. An appropriate expert proposes the relationship between fundamental aspects of the component design and application, and the modes of failure of the component.

The second step is to conduct an experiment to test the experts postulates. For component reliability the results of "experiments" are historical data gathered on many types of component in many applications. Data must be selected that has the quality, completeness and extent of information to provide a reliable testing ground for the engineering postulates.

Finally, the data must be examined statistically to determine if the engineering postulate is confirmed. If not then further iteration will be required to review the engineering model and to further explore the quality of the data. The analysis methods used may depend on the form of the data, or at least require the careful review of historical data in order to match the requirements of the method chosen. Ultimately the aim is to quantify the effect of the variables identified in order to be able to predict the reliability in new, differing applications. The following 3 sections consider the stages of engineering input, data and analysis. It is emphasised that the strength of the approach is dependent on strength in each of these three stages, weakness in one will significantly weaken the whole. In the initial study reported here the component studied is the gate valve.

3.0 ENGINEERING INPUT

In order to represent the relationship between failure modes, symptoms and causes a failure diagram was developed as outlined schematically in figure 1. This concept was developed to provide the valve engineer with a framework on which to describe failure systematically.

FIGURE 1 SCHEMATIC EXAMPLE OF A GATE VALVE FAILURE DIAGRAM FORM

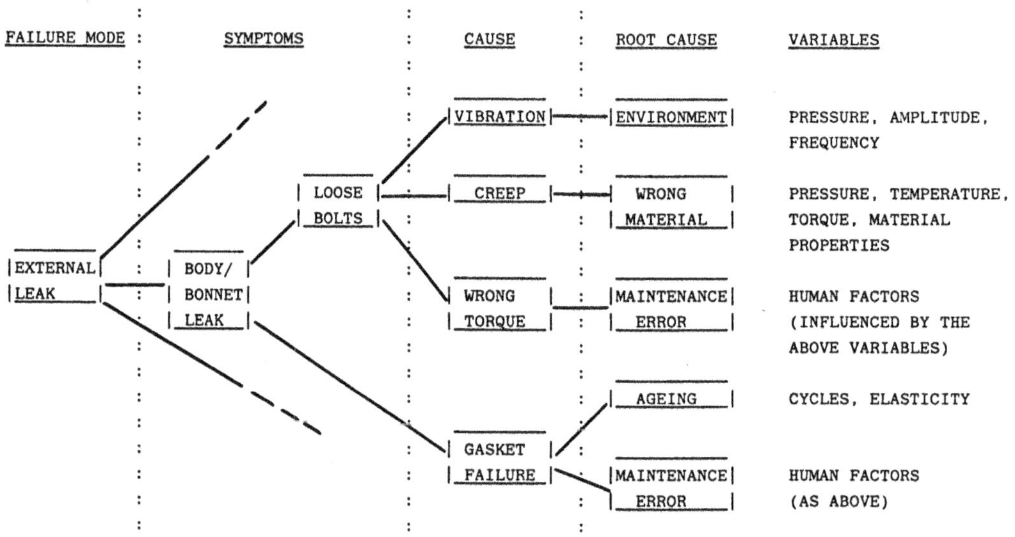

The diagram is split into five sections with headings to aid the valve engineer. The sections have the following characteristics:-

a) Failure Mode - a description of the failure identifying the fundamental design requirement no longer being met. Usually of greatest concern to the system designer and fault tree analyst.

b) Symptoms - identification of the sub-components involved in the failure and description of the damage to them.

c) Causes - the aspects of the component design or application that has brought about the symptoms and failure.

d) Root Causes - the underlying reasons why a professionally designed component failed prematurely, ie. why the (immediate) cause had not been successfully avoided within the design life of the valve.

e) Variables - measurable variables of the design and application the value of which is alleged to affect the probability of failure. Where possible the engineer should define whether a linear or other dependence on the parameters is applicable to aid analysis. A binary representation of effect/no effect may also be used.

Sections (a) and (b) are directly observable in the field. Section (c) can often be inferred. Thus the failure diagram can be used to guide the recording of information during data collection, assuming that the diagram is reasonably complete. The relative frequency of the failure modes and symptoms in sections (a) and (b) should be derivable from a comprehensive survey of field experience of failures.

The root cause of the failure may not be as easily determined in the field although the availability of the failure diagram may help. The underlying reason for failure is often not immediately clear.

Many failures have human performance shortcomings as a root cause. Although premature failures driven purely by physical phenomena probably occur (due to the overlap of the distribution of operating conditions with the distribution of component resilience dictated by economic considerations), many failures appear to be accelerated by human performance related reasons eg misapplication, manufacturing inadequacy or maintenance error. This is discouraging due to the likely difficulty in parameterising human factors' contributions to failure.

In order to make progress an assumption has been made that human failings often lead to an earlier effect by those same physical variables that would have produced on eventual component wear-out failure. This is demonstrated in figure 1 where the effects of "wrong torque" can be modelled using the same variables as the other causes of "loose bolts".

There remain certain other difficulties. Severity and location of a problem (eg of external leakage) will affect the likelihood of a failure being reported and speed of repair. This will also be plant system dependent. Additionally, different failure mechanisms may lead to differing severity of symptoms eg to seepage or gross leakage. Yet it is often difficult for a reporter to define how severe a symptom is. These difficulties will be reflected as uncertainties in data collected in the field. Such uncertainties will affect the confidence in any output of an analysis.

4.0 DATA SOURCES

Potential valve reliability data sources were surveyed to identify data of sufficient breadth of application and quality of information to justify a discriminating analysis. Three types of data source could be identified.

Firstly, there are several event databases that provide descriptions of component failures reported as being "significant" against some criteria, eg Licensee Event Reports (LER's) but many failures (and successes) go unreported.

Secondly and conversely, many reliability data manuals contain little engineering information but present failure rates for more or less homogenised and known populations (eg IEEE 500). The lack of engineering information precludes more detailed analysis for these data summaries. Useful data sources for further analysis require both an appreciation of the prevalence of failure in the population and also the engineering details of component, application and failure event. Much of this information is lost (or was never present) in compiling the data summaries.

Thirdly a limited number of data sources have been identified that contain sufficient information on the valve applications, population and failure details to allow some exploratory data analysis. One of these was the product of a data collection exercise at a power station from which was identified a useful population of 448 gate valves in 47 groups spread over the range of applications on the station.

Component failures, repairs and planned maintenance had been logged by the data collectors using standard forms. Some information on the application was included (e.g. pressure, temperature, medium etc.) but aspects of the valve duty such as normal operating state and frequency of demands were not.

As a first step the data were reviewed to identify and resolve apparent errors and shortcomings. A coding scheme was evolved to describe the histories of the valves to be described by Gibson et al [1]. The coding scheme was imposed on the scrutinised data, and histories were produced for each valve. A great deal of effort was required in this data scrutiny and coding stage before quantitative examination and analysis of the data could be performed.

5.0 DATA ANALYSIS

Two routes were taken in analysing the data in this study. The first was the straightforward, yet always valuable review and summarising of data by means of tabular or graphical representations. Many of the characteristics of the data can be appreciated without any of the assumptions needed to use more sophisticated techniques. Later the data was examined using the proportional hazards model (PHM) in order to attempt to quantify the effects of the variables postulated as likely to be influencing failure.

PHM is discussed in Kalbfleish and Prentice [2] and Cox and Oakes [3] and there have been several application papers in the literature recently e.g. Baxter et al [4] to which the reader is referred for further explanation of PHM, related models and their estimation procedures.

The principle of the model is that the hazard function h(t) may be represented as:

$$h(t) = ho(t) \exp (B_1 z_1 + B_2 z_2 + \ldots + B_n z_n)$$

where z_1 to z_n are the assumed independent variables influencing failure; B_1 to B_n are parameters weighting the effect of these variables and ho(t) is a baseline hazard function equivalent to the hazard function when the effect of all variables is zero (i.e. all $B_i = 0$). Although a standard distribution may be forced onto ho(t) we used a distribution free approach in this initial study.

The crucial input to a PHM analysis are the variables whose effect is to be explored. In addition to identifying the variables it is necessary to determine how they will be represented in the model. The model can be used with continuous or binary (0 or 1) variables. In some cases it is clear that a binary approach is appropriate e.g. if the nature of a fluid medium is considered relevant then water and steam can be allocated values of 0 and 1. Where a third fluid is present in the data then dummy binary variables can be used to distinguish between the fluids in the following binary coded manner.

Medium	a	b	c
Variable 1	0	0	1
Variable 2	0	1	0

Clearly, if the process is well understood then the fluid could be characterised by continuous variables such as viscosity, density etc. It must be an engineering judgement that a binary representation is reasonable in all relevant applications.

Within the limits of the underlying assumptions, PHM analysis of the gate valve data has been undertaken. The value of this analysis is seen as being in the ability of the technique not only to identify a variable that affects the failure rate, but also in quantifying the effect of the variable. A well fitting model describes the data analysed and the parameters and variables allow for a degree of predictability beyond the applications in the data. Increased confidence in the quality of data used, in the understanding of the failure mechanisms and in reflecting this understanding in the assumptions of the PHM analysis should produce increased confidence in predictions arising from such an analysis.

6.0 RESULTS OF DATA ANALYSIS

The characteristics of the data as found by the simple data review are described below followed by the results of a preliminary PHM analysis.

6.1 Preliminary Data Review

To simplify the review effort was initially concentrated on valves of one manufacturer, named dataset 1 consisting of 448 gate valves for which 742 events were recorded. 152 of these events were classified as failures, 492 as maintenance and 98 as other events (such as 2 job orders for one repair - for which appropriate corrections had to be made).

Figure 2 shows the distribution of the numbers of failures per valve. Over 80% of the valves in dataset 1 were recorded as having no failures within the data collection period. Only 7 valves had four or more failures. Figure 3 shows a similar distribution for all valve events with over half of the valves having no recorded events in the period yet one valve had 18 events. Thus, we are dealing with a population which predominantly passed through the approximately 30,000 hour collection period without record of maintenance or failures.

TABLE 1
GATE VALVE DATASET 1- FAILURES BY SYSTEM

SYSTEM	(valve hours) Operating Time	All Failures: No. of Failures	All Failures: Mean Failure Rate	External Leak: No. of Failures	External Leak: Mean Failure Rate
High Pressure Steam	55843	57	42.5	51	38.1
Low Pressure Steam	117262	30	10.7	13	4.6
Reheated Steam	82764	17	8.6	14	7.0
Demineralised Water	57336	1	0.7	0	0
Feedwater	68831	2	1.2	0	0
High Pressure Feed	72488	23	13.2	22	12.6
Low Pressure Feed	113310	22	8.1	18	6.6
TOTALS	567834	152	11.1	118	8.8

Mean Failure rate in 10^{-6} failures/valve hour

Figure 2

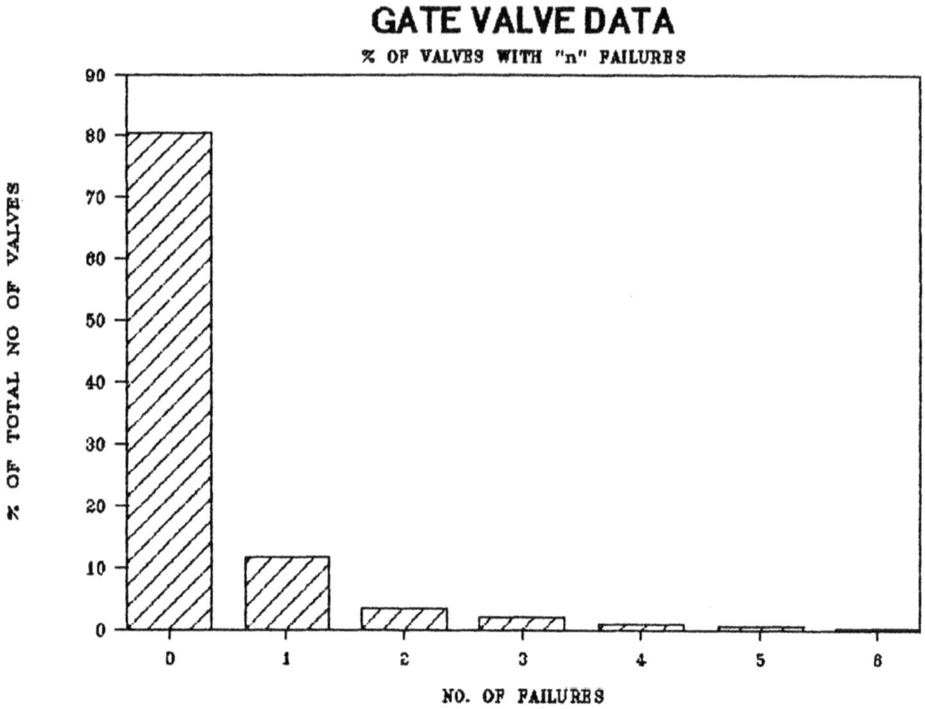

Figure 3

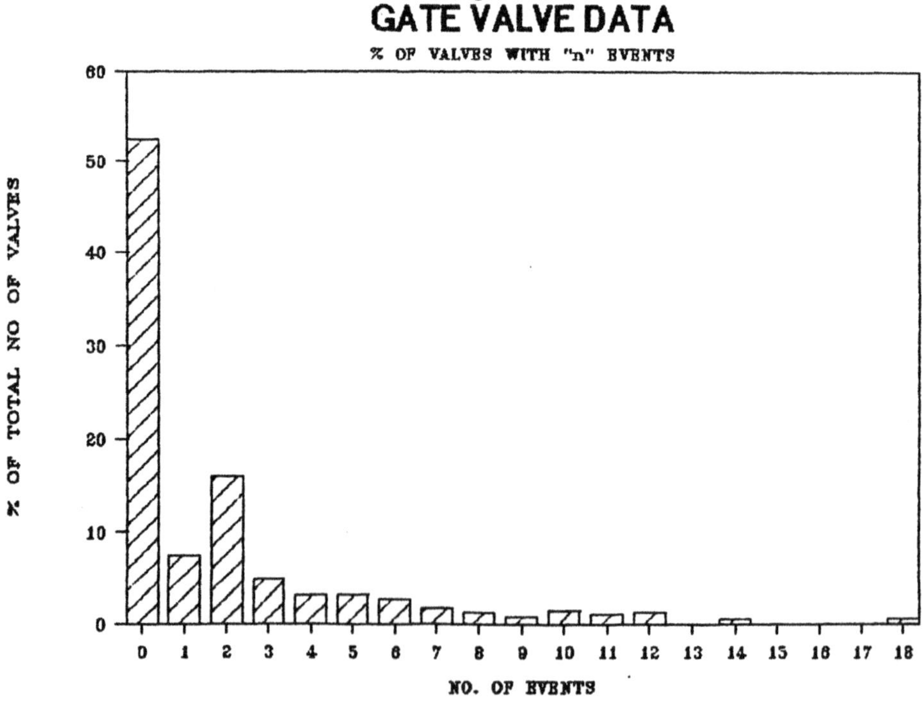

Dataset 1 was spread across 7 plant systems. Table 1 displays the distribution of failures between these systems and gives notional failure rates. Steam systems appear less reliable than water systems with High Pressure apparently being more important than Low Pressure in increasing in the failure rate. However, the results are dominated by external leaks (and in particular by gland leakage). Dependence on variables such as pressure and medium is less clear for other failures. It should be noted that external leakage is more easily revealed in the plant and may therefore remain unobserved for a shorter time than other failures.

The review also covered time variations, effect of bore, model and other potentially useful explanatory variables. From this work the following conclusions were drawn as important inputs to the PHM analysis.

(i) there appears to be some temporal distortion in the data collected

(ii) most valves did not fail in collection period, the data being highly censored

(iii) most failures are external leaks via the glands

(iv) steam and high pressure systems appear to be predisposed to external leakage, particularly via the gland

(v) there are large variations in individual valve reliabilities and a lesser variation between boiler units

(vi) medium and pressure appear important variables

(vii) valve function and actuator effects cannot be used as variables due to uncertainties about operational considerations

(viii) model and manufacturer are important but could not be included in the analysis

6.2 Proportional Hazard Analysis Results

The following variables have been used so far in the PHM analysis of the gate valve data: Pressure, Temperature, Bore, Fluid Medium. Most attention has been placed on the analysis of the data from one manufacturer looking for the effect of pressure, temperature, bore and medium on external leakage. To explore the technique, runs were performed for several combinations of the variables to gauge the consistency of results and the effect on the model of missing out a 'significant' variable. The variables were entered in the following way in these tests:-

Pressure :	binary - defined by station as high or low
Temperature:	continuous - °C (may have been more appropriate to use °K)
Diameter :	continuous - inches
Medium :	binary - steam or water

TABLE 2

TYPICAL PHM RUNS - GATE VALVE DATASET 1

RUN	INPUT VARIABLES	LIKELIHOOD RATIO STATISTIC (Model CH1-2)	B ESTIMATE	STANDARD ERROR
1	MEDIUM		1.54	0.21
	PRESSURE		1.68	0.24
	DIAMETER	74.5	-	-
	TEMPERATURE		-	-
2	PRESSURE	38.7	1.30	0.24
	TEMPERATURE		2.6×10^{-3}	$<10^{-3}$
3	PRESSURE		1.23	0.24
	DIAMETER	33.7	9.1×10^{-2}	2.4×10^{-2}
4	MEDIUM		1.77	0.53
	TEMPERATURE	24.4	-4×10^{-3}	2×10^{-3}
5	MEDIUM		0.79	0.20
	DIAMETER	17.3	-	-
6	MEDIUM		1.54	0.21
	PRESSURE	74.5	1.68	0.24

The results are presented in Table 2. The Likelihood Ratio Statistic is a measure of the total fit of the model when used in conjunction with a CHI-squared distribution with degrees of freedom equal to the number of significant variables. The parameter as derived is given along with a calculated standard error. From these an estimate has been made of the probability of the calculated value being statistically insignificant from zero at a 95% level to determine which variables to remove from the analysis (shown by dashes).

Medium and pressure are identified as the 'significant' variables in the first run which carries all four variables. An increase in pressure and/or the passage of steam are seen to increase the likelihood of an external leakage fault. These seem reasonable results from an engineering viewpoint as does the elimination of bore and temperature. Temperature of itself is not an obvious driver of external leakage although there is obviously a link with pressure in determining the medium.

The assumption of proportionality, explicit in modelling by PHM, appears to have been reasonable but further work is needed to clarify the validity of the assumptions made in the analysis. Several other runs were performed to test the effect of differing representations of the variables. These demonstrated the care with which the representations have to be chosen to ensure that engineering knowledge drives the model.

7.0 DISCUSSION

The work reported in this paper is intended to be developed further. The engineering aspects in Section 3 are considered to be a useful framework. The issue of modelling human performance, as it affects failure, has been side-stepped. Further consideration of this topic may be worthwhile in conjunction with other studies into human performance and error. A potential feed into dependent failure studies could also be pursued. Indeed the failure diagram may be of greatest benefit in providing a framework for root cause analysis in these areas and in aiding the data collector as to what to expect when looking at component failures.

On data we intend to consider the analysis of data from larger databases of wider applications, covering a range of components. The effort on improving the quality of databases up to a standard worthy of analysis should not be underestimated and is essential to avoid flawed analysis results.

With respect to the PHM analysis the work to date has essentially been restricted to exploration of the characteristics of the model and the packages available to perform it. This exploration has thrown up further questions: the use of the graphical and other methods of testing the applicability and fit of the proportional hazards model requires further study as do the implications of the calculational route for data with much censoring and ties.

The choice of variables and how they are represented in the model requires further consideration. In particular the analysis should be extended to other variables such as component type and model, to other failure modes and to the effect of representing human factors as variables. The use of other models or methods of analysis also deserves exploration.

8.0 CONCLUSION

A method of describing component failure has been produced to assist in identifying variables that may affect failure rate and has been applied to gate valves. A coding scheme has been developed to aid in screening and coding valve data for analysis. Preliminary analysis of power station gate valve data following encoding has led to the identification of pressure and medium (steam or water) as significant variables in driving gate valve external leak failures.

Proportional Hazards Modelling has quantified the relative effects of these variables but there are reservations to the applicability of the model used on the test dataset. The results of the work have been sufficient to encourage pursuit of further, more extensive studies.

9.0 ACKNOWLEDGEMENT

This paper is published with the permission of CEGB Sizewell 'B' Project Management Team and UKAEA Safety and Reliability Directorate. The contribution of students and colleagues in both organisations is gratefully acknowledged.

10.0 REFERENCES

1. Gibson I.K., McIntyre P.J. and Witt H.H.: Aspects of a model to improve the reliability estimates of engineering components. To be published at Reliability 89, Brighton, June 1989.

2. Kalbfleisch J.D. and Prentice R.L.: The statistical analysis of failure time data. Wiley - Interscience 1980.

3. Cox D.R. and Oakes D.: Analysis of survival data. Chapman & Hall 1984.

4. Baxter M.J., Bendell A., Manning P.T. and Ryan S.G.: Proportional hazards modelling of transmission equipment failures. Proc. Reliability 87 Vol 2, 4B/1-11.

Planning and Management of the Offshore Reliability Data Project (OREDA)

Magne Toerhaug

Veritas Offshore Technology and Services A/S
Hoevik, Norway

Summary

The OREDA project was established by seven oil companies to collect data on failures on selected offshore equipment. Data were collected from maintenance systems of six companies operating offshore oil and gas fields, and pooled into an event and inventory data bank. The paper summarizes how the project was organized and managed and presents the resulting data base.

1. INTRODUCTION

The Offshore Reliability Data Project (OREDA) has been established by seven oil companies to collect data on selected offshore equipment and technical failures experienced with this equipment. The seven oil companies, hereafter denoted the Participants, are

- BP Petroleum Development Ltd. Norway
- Norsk Agip
- A/S Norske Shell
- Norsk Hydro a.s
- Den norske stats oljeselskap a.s (Statoil)
- Saga Petroleum a.s
- Total Oil Marine PLC

Veritas Offshore Technology and Services A/S (VERITEC) was selected to be Main Contractor for the Project. Subcontractors to VERITEC were CREA srl (Italy), R. M. Consultants (England) and Sintef (Norway).

The present OREDA Project is based partly on the results of three previous projects:

- A prestudy examining the feasibility of establishing reliability data from maintenance records of offshore operators - undertaken in the period 1980-82 with funding from the Royal Ministry of Local Government and Labour in Norway and NIFO.

- The first OREDA project - collection of reliability data from the maintenance records of seven operators in the Norwegian and UK sectors of the North Sea and from the Adriatic Sea. This was undertaken in the period 1983-85 with funding from the companies that are also Participants in the current OREDA project and Elf Aquitaine Norge A/S. This project resulted in the Offshore Reliability Data Handbook, published early 1985.

- An intermediate OREDA project, undertaken mainly in 1986 - for planning continuation of the OREDA work, funded by the participants of the previous OREDA project. In this project it was decided to put more emphasis on how environment, maintenance and operational characteristics affect failure rates and on collecting data of importance to planning of maintenance.

In the present OREDA Project, six of the seven Participants contributed with data:

- BP, Shell and Total from their operations in the UK sector of the North Sea,
- Agip from their operations in the Adriatic Sea,
- Norsk Hydro and Statoil from their operations in the Norwegian sector of the North Sea.

All data come from the operation of fixed production platforms, with the exception of data from the operation of one production ship.

In the intermediate project it was decided to collect data on a few classes of equipment only, namely (no priority indicated):

- pumps
- compressors
- gas turbines

- heat exchangers
- separators
- valves
- various subsea equipment

2. ORGANIZATION OF OREDA PHASE II

The parties to the OREDA work were:

- The seven participating oil companies,
- their operating companies, providing the data,
- main contractor and
- the three subcontractors.

The project organization is shown in figure 2.1.

Figure 2.1: Organization of OREDA

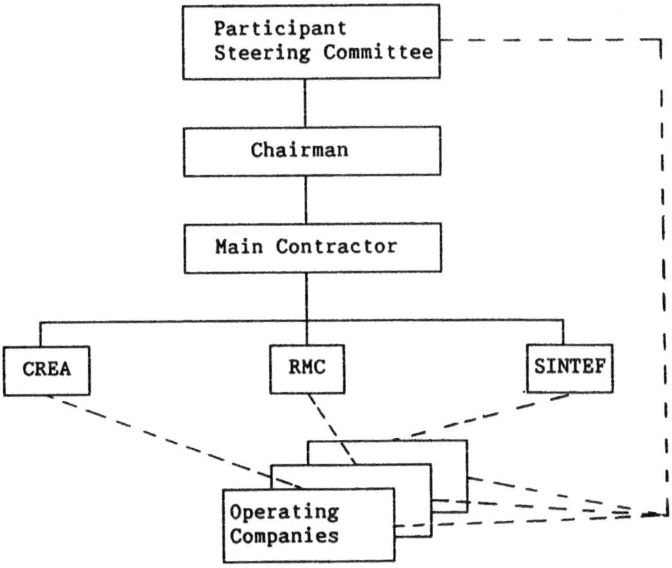

______: Lines of management and communication

------: Lines of communication

The **Participant Steering Committee (SC)** consisted of one representative from each participant. The SC had authority to

a) Approve or disapprove work plans and budgets including allocation of key personnel of the Project.
b) Decide on changes in the Project and budget.
c) Review the progress of the Project.
d) Approve or disapprove of the results of the Project.

The Steering Committee elected its Chairman and Deputy Chairman among its members.

The Steering Committee Chairman had the power to approve on behalf of the SC

Progress reports, timesheets and travel, etc.,
refinements and adjustments of work within limits defined by the SC,
corresponding reallocation of budgets, within the total budget limits stated or as defined by the SC,
technical matters.

The **Main Contractor** conducted the overall management of the project, including specification of work orders, progress reporting to the SC, cost and quality control. In the data collection and treatment process, MC was responsible for:

- definition of data collection specification,
- quality control - and -assurance of subcontractor work,
- anonymization, final quality control and pooling of the collected data,
- programming of data collection software,
- programming of final database software.

The **Subcontractors** had the responsibility for:

- Data collection with selected operator's companies as follows:

  - C.R.E.A srl. with office in Ravenna, Italy, collected data with Agip from Agip's operation of platforms in the Adriatic Sea.
  - R.M. Consultants Ltd. with office in Abingdon, England, collected data with BP, Shell and Total Oil Marine from their operations of platforms in the UK.

- Sintef with office in Trondheim, Norway, collected data from Norsk Hydro operations of a production ship and from Statoil's Statfjord Division.

  In addition, Sintef assisted R.M. Consultants in collecting data on Subsea systems from BP and Shell.

The Subcontractors communicated directly with the operating companies responsible for the data collection. These operating companies played a vital part in the project and provided direct support to the Subcontractor's data collection teams. In some cases this support was several man months of work.

4. OREDA PROJECT EXECUTION

The project was executed in 16 activities as follows:

Activity 1: Nomination of Subcontractors.
Each Participant was free to select the subcontractor for data collection on his facilities. This was the start of the project, in April 1987.

Activity 2: Preliminary Site Visits.
In order to examine the amount of work necessary to collect the data, to identify sources, meet contact persons and discover potential quality problems, a visit to all operating sites with source data was undertaken. The main contractor and the responsible subcontractor participated. This resulted in a basis for detailed planning of the data collection.

Activity 3: Selection of equipment classes and
Activity 4: Plan and budget agreement.
Each subcontractor was required to submit plans and budgets for the data collection. On the basis of this as well as on the total project budget, main contractor prepared overall plans and budgets for the data collection. This was presented to the SC for discussion and approval.

Activity 5: Data Collection.
Main contractor prepared detailed data collection guidelines and a data collection software for use in the project. This was sent to the

subcontractors for use in their work. The software was based on dBase III and all data submitted to the project from the subcontractors were on floppy disks. This greatly reduced main contractors work load and eased later control and editing of data.

The Subcontractor undertook data collection and internal quality control in accordance with predefined principles. Also meetings between main contractor, subcontractors and members of the SC were conducted to secure consistent interpretations of the source data and the data collection guidelines.

Main contractor conducted spot checks on data collected. Criteria for rejection of data samples had been established, but no rejections were necessary. Main contractor also run a computerized data analysis of the entire data base in order to control the consistency of the coding. In this so called "frequency control", the coding of each data field was counted and printed. For example, in the data field "manufacturer", there could be 14 records reading "General Electric" and 7 reading "GEC" and so on. This analysis was used as basis for unification of data and also to compare results with the data of the OREDA Handbook.

The final data were submitted from the Subcontractors in September 1988.

(**Activity 7:** Undertaken by Participants.)
Activity 8 and 9: Unification, anonymization and pooling of data.
This was undertaken by main contractor and a final quality control of the data was carried out simultaneously with anonymization of the records. Main contractor also programmed the software for handling the data (see next chapter).

Other activities were related to management of the entire project and will not be covered here. The project was completed in December 1988.

5. PROJECT RESULTS

In brief, the project results are

- A database with detailed inventory and failure event data. The base covers inventories, as detailed in table 5.1 below. The total size of the database (in dBase III+) is 8.5 Mbyte. The software uses a packing software, reducing this to 1.6 mbyte.

- A report with detailed definitions of the data in the database.
- Software for handling the database, including basic statistical analyses, selection of subsets of data, handling of files and establishment of new subfiles, editing and later extentions to new data.
- Software for data collection.
- Data collection guidelines.
- Software users manuals.

The attached tables show examples from the data base. Please note that the format has not been entirely finalised when this paper was written.

Table 5.1: Amount of data collected.

System	No. of inventories	No. of failures
Pumps	221	3 055
Compressors	50	1 639
Gas turbines	115	2 611
Vessels	331	438
Heat exchangers	170	118
Valves	646	427
Subsea systems	41	39
Overall	1 574	8 327

Table 5.2 Inventory data (Example, Heat Exchanger)

```
Inventory  Taxonomy :  HEPL                 Report no.:   260001
Report     Item name:                        Database  : HEDRAFT
--------------------------------------------------------------------------------
Function              : OP
Manufacturer          : Alfa Laval
Manuf.of control syst:
Model                 : A15 BFG
Redundant subsystem 1:              Number of red. systems:
Redundant subsystem 2:              Number of red. systems:
Redundant subsystem 3:              Number of red. systems:

Operational time      :   13000
Calendar    time      :   13000     No of demands          :   0
Installation date     : 01/09/86    Surv. start date       :  / /
Major replacements    :    / /      / /          / /

Installation name     :                        Source      :
Tag no                :                        State of data:
Sub-tag nos           :
Unique nos            : Serial no: 30100.66654
Reported by           :                           Date      :  / /
Checked  by           :                           Date      :  / /


Main fluid hot side     :  Crude oil
Main fluid cold side    :  Sea water
Rated heat transfer(kW) :    1346.7
Pressure hot side (barg):         2
Pressure cold side(barg):         2
Temp hot side,inlet     :        56
Temp cold side,outlet   :        38
Size diameter (m)       :         0
Size length   (m)       :         0
Number of tubes/plates  :        94
Material of tubes       :  Titanium
Material of body        :  See comment
Design code             :  N.A.


Environment: Location :                     Climate   :
             Temp(C)  :   0                 Vibration :

             Other    :

Maintenance: Type     :       Freq :        Condition  :
                        SCM          6      Inspection
                        SCM         12      Inspection
                                     0
                                     0

Additional text:

Full name of item: Crude cooler E-120-A No. 1 (F)

Surface area: 758 sq.ft. Max allowable w.p.: 6.9 bars.
Dimensions: - Length: 750 mm
            - Breadth: 720 mm
            - Height: 2075 mm

Material specification:  - Gaskets: Nitril
                         - Port lining: Titanium
                         - Plates: Titanium
                         - Covers: Carbon steel
```

Table 5.3 Failure data (Example, Heat Exchanger)

Failure report

Inventory no. :	260001	Ser.no.:	0	Failure no :	1

Failure mode	: OVH	Severity	: I		
Subsystem(s) failed	: FW side				
Failure mode(s),subsystem(s)	: Contamination ("grønske")				
Items repaired	: None				
Repair activity	: Sample sent for analysis				
Failure detected date	: 20/05/87	Restoration :	Mechanical:	0.0	
Failure detected time	: 0.0	:	Electrical:	0.0	
Failure repair time	: 0.0	:	Instrument:	0.0	
Downtime	: 0.0	:	Other :	4.0	
		:	Total :	4.0	
Observation method	: PPMT		State:		
Reported by	:		Date :	/ /	
Checked by	:		Date :	/ /	

Additional text:

Table 5.4 Statistical analysis output (Examples)

Standard calculation output format.

	Total	Mean
No of inventories:	170	
Operational time :	996000	5858.8
Calendar time :	3265560	19209.2
No of failures :	118	0.7
No of demands :	0	0.0
Confidence level :	0.95000	

Sever-ity	Fail mode	No. of failures	Failure rate (10^{6} hours) Lower	Mean	Upper	Repair time Min	Mean	Max
C	OVH	1	0.025	1.004	5.594	0.0	0.0	0.0
C	SIL	34	23.641	34.137	47.702	0.0	0.0	1.0
C	SPT	1	0.025	1.004	5.594	4.0	4.0	4.0
C	EXL	18	10.711	18.072	28.562	0.0	7.3	20.0
C	INL	6	2.211	6.024	13.112	0.0	1.0	3.0
C	SEL	3	0.621	3.012	8.802	0.0	3.7	6.0
C	UNK	2	0.243	2.008	7.254	0.0	2.0	4.0
C	COR	1	0.025	1.004	5.594	24.0	24.0	24.0
D	IHT	1	0.025	1.004	5.594	0.0	0.0	0.0
D	INL	2	0.243	2.008	7.254	0.0	0.0	0.0
D	UNK	1	0.025	1.004	5.594	0.0	0.0	0.0
D	EOP	1	0.025	1.004	5.594	0.0	0.0	0.0
D	SEL	1	0.025	1.004	5.594	0.0	0.0	0.0
I	EXL	20	12.266	20.080	31.012	0.0	0.4	4.0
I	COR	4	1.094	4.016	10.283	0.0	0.0	0.0
I	L	1	0.025	1.004	5.594	0.0	0.0	0.0
I	INL	2	0.243	2.008	7.254	0.0	0.0	0.0
I	IHT	1	0.025	1.004	5.594	0.0	0.0	0.0
I	OTH	1	0.025	1.004	5.594	0.0	0.0	0.0
I	UNK	4	1.094	4.016	10.283	0.0	3.3	10.0
I	EOP	8	3.468	8.032	15.826	0.0	0.3	1.0
I	OVH	4	1.094	4.016	10.283	0.0	0.0	0.0
U	UNK	1	0.025	1.004	5.594	0.0	0.0	0.0
TOTAL	ALL	118	98.064	118.474	141.879		2.0	

6. LESSONS LEARNT

6.1 ORGANIZATION AND COOPERATION

In general this has worked satisfactory seen from main contractor point of view.

The subcontractor work was organised with a strict work order and progress reporting system, which also worked well. It is our opinion that the current project is close to what the subcontractors will have to compile for their internal follow up anyway.

Execution was greatly eased by

- a good "work climate" among the group of SC, SC Chairman, subcontractors and Participants,
- direct assistance from personnel within operating companies of Participants.

6.2 PLANNING

Positive experience has definitely been made with formalizing the data collection process, including
- precise definitions of terms,
- specification of data to be collected and format of data,
- work procedures,
- procedures for quality control.

We do not believe that a data collection of this kind can be carried through within the budgets used in OREDA without such formalism. The above does not mean that there is no room for improvement in the current system.

Problems encountered: There should have been more time and efforts devoted to the site visit work, including a full trial data collection. This was not done due to cost, but is recommended for future work with unknown data sources.

Also, the large amount of data asked for caused problems. These problems were solved in most cases, and the contribution from personnel from the operator's own organisations has been most valuable.

6.3 DATA QUALITY

The data problems to be expected were:

- lack of data
- erratic data
- inconsistencies in data from different sources, Subcontractors or different operators entering data.

Generally the data quality is good. There have, however, been problems of finding the data asked for on the fields of instrumentation and maintenance.

OREDA Phase II Data Collection Experience

T R Moss RM Consultants Ltd, Abingdon UK
H Sandtorv SINTEF, Trondheim, Norway
C Pratella CREA, Ravenna, Italy

SUMMARY

This paper describes data collection activities in Phase II of the OREDA (Offshore Reliability Data) project. Data collection involved the collection and coding of engineering, functional and failure data for over 1600 equipment inventories and 8000 failure events. Problems encountered in acquiring data from maintenance records and maintaining satisfactory quality control for such a large and geographically dispersed project are also discussed.

1. INTRODUCTION

OREDA - the Offshore Reliability Data project started in the early 1980's. Phase I involved the collection and processing of failure information from the maintenance records of a number of offshore platforms to generate global failure rates for a wide range of safety and production equipment. About 150 different types of equipment were surveyed and the results published in the OREDA Handbook.

Phase II was a more ambitious project. The objective here was to create a robust data base of inventory and failure-event information for a range of vital topside and subsea equipment. The availability of this detailed information was seen as a significant step forward from the parametric reliability data base generated in Phase I. It provides the participating companies with the facility to select populations of equipment within a well-defined envelope of manufacturer, design and functional parameters, to calculate failure rates for the different failure modes and to review individual reports of the failure/repair activities.

This paper reviews the experience gained during data collection for Phase II. It is based mainly on experience from a typical computerised maintenance information system and highlights some general problems associated with extracting reliability data from maintenance records.

The need for comprehensive quality assurance in such a large, diverse project is also addressed. Some details of the extent of the quality assurance to ensure traceability of documentation and consistency in coding and data input to the data base are discussed in the paper.

2. DATA COLLECTION

Details of equipment classes, number of platforms, inventory and failure records, etc are given elsewhere (1). In total over 1600 Inventory and 8000 Failure reports were generated by the data collectors. The information required for the Inventory files was divided into General Data (as shown in Fig.1) and Equipment - Specific Data. Fig.2 shows the Equipment - Specific Data required for Pumps. The failure data requirement, was applicable to all equipment, and is shown in Fig.3. Together these records require a data base of about 40Mbytes, however, there were some missing data as discussed later.

The approach adopted by each sub-contractor was slightly different and based on the sub-contractor experience and the flexibility of the maintenance information system (MIS). In some cases the data were partitioned into discrete sets and then transferred in blocks onto the data collection pc where it was subsequently reviewed and processed. For the majority of cases, however, the MIS was programmed to provide hardcopy output which was subsequently reviewed, the relevant data extracted onto intermediate data forms - similar to those shown in Figs.1 to 3 - and then input to the database by professional data processing staff.

This latter procedure was partly dictated by the volume of non relevant information output in addition to the OREDA data required from the MIS. The process is obviously time-consuming but nevertheless has a number of advantages which may not be immediately apparent.

These include:

1. It allows experienced process engineers (who frequently lack keyboard skills) to be employed for reviewing the data.
2. Professional data processing staff are used for inputing the data ensuring low data-transfer error rates.
3. Complete manual records are available which ensure full traceability of information. These can subsequently be used for quality checking and the addition of data not available during the first pass.

3. MISSING DATA

In most cases all the required inventory data were not available from the MIS and recourse needed to be made to other information. This included accessing microfiche records, Piping and Instrumentation Diagrams (PID's), engineering drawings and maintenance schedules. Most of the deficiencies in the failure records were associated with the lack of information in the MIS on the cause of the failure and its effects on the system. It is important to realise that repair histories are not generally designed to provide reliability data. The job cards are completed by maintenance personnel who faithfully record details of the work carried out. From this information it is necessary to deduce the cause of the failure and its likely criticality.

The situation varied from company to company and sometimes even from platform to platform but generally the data which could only be collected from site and required frequent contact with Maintenance Department personnel were:

Condition-monitoring information
Instrumentation details
Active or standby redundancy
Run-times
Equipment installation dates
Actual operating pressures and temperatures

It is worth noting that considerable effort was put-in by the maintenance departments on each site to provide the missing data. Without their enthusiastic support many of the uncertainties could not have been resolved. Clearly an exercise of this magnitude cannot be successful without the full co-operation of the site management and staff.

4. QUALITY ASSURANCE

Formal quality assurance procedures were introduced by each sub-contractor at the start and actively pursued during the course of the project. Basically this involved the submission of a detailed quality plan to Main Contractor and the establishment of a Document Control Centre (DCC) by the Sub-Contractors into which all project documentation was stored and recorded.

During the course of the project RMC - who were responsible for over 60% of the total data collection - recorded over 400 transmittals. These ranged from general monthly financial and technical progress reports made to Main Contractor down to internal transmittals to RMC staff concerning assumptions made on equipment boundaries etc. An example of a document transmittal is given in Fig.4.

The Data Collection Guidelines issued by Main Contractor required self-check and verification by the sub-contractor. This was agreed to involve sampling the various stages of data collection and recording to ensure accuracy of data transcription and interpretation. Ten percent sampling was the norm but in instances where the number of recorded failures was small 100% sampling was employed.

Samples of data recorded on the data collection forms were checked against the marked-up hardcopy output from the MIS in the initial stages. Subsequently checks were made between the OREDA program output and the data collection forms. The emphasis was on coded fields since mispellings in the free-text field were generally self-evident and left uncorrected. An example of a completed Self-Check and Verification form is shown in Fig.5.

A final quality audit of each data collection exercise was carried out by the sub-contractors QA Director and the Project Supervising Officer. One completed QA Audit Report is shown in Fig.6.

5. PROBLEM AREAS

The main problem during inventory data collection was in identifying the information required on instrumentation, maintenance and, equipment redundancy. Defining equipment boundaries was also a problem because of the differences between companies as well as differences between company-defined and OREDA boundaries. The problem can be illustrated by noting that for one company a gas-turbine driven centrifugal compressor is described by no less than 550 sub-tab numbers. From this set of numbers those items within the OREDA boundary needed to be identified so that only failures of the relevant sub-tags were recorded in the failure data base.

Problems with failure data hinged on interpreting the historical records of repair actions in terms of the failure mode and severity definitions specified in the OREDA guidelines. The equipment history listings rarely yielded any information on whether a system failure had occurred. The Condition For Work statement generally showed whether and what kind of outage was necessary for the work. However, it gave no indication of whether the work was done at a convenient outage opportunity or whether the system had to be taken out of use to deal with the problem immediately. It was shown by experience that the most efficient approach was to consider the sub-system failure first and then deduce the system effect. The procedure adopted was:

- Identify the failed sub-system, ie the sub-system which contained the failed item.
- Select the sub-system failure mode from the Euredata list.

- Consider the effect on system operation and whether system failure could result. Record the appropriate system failure mode.
- Decide and record the severity of the related system failure.

In this way the failure modes specified in the guidelines were employed throughout the data collection phase. Inevitably however differences between analysts did arise. The assumptions and interpretations made by individual analysts were thus discussed within the team and then recorded on file.

In general the problems encountered were relatively few given that reliability data were being derived from maintenance records remote from the platforms. What was certainly under-estimated was the length of time required to extract inventory information and, to a lesser degree, to interpret the historical repair data. It is clearly important that experienced process engineers are employed for extracting such data. Even then active and sympathetic participation by the local maintenance planning department is essential if complete and consistent data are to be recorded on the data base.

6. CONCLUSIONS

Extraction of high-quality reliability data from maintenance records is possible given active collaboration between the company maintenance planning department and the data collector.

With the enthusiastic and professional support afforded to the data collection teams by the companies in this exercise a comprehensive failure experience data base for the major items of equipment used offshore has been established for OREDA participants.

Even so reliability data collection and processing based on historical maintenance records has limitations. On-going data recording based on specially-designed interface programs linking the Maintenance Information System and the reliability data base have a number of advantages and in the end are likely to be significantly less expensive.

References.

1. Planning and Management of the Offshore Reliability Data Project - Torhang M, Sixth Eure Data Conference, Siena, 15-17 March 1989.

INVENTORY REPORT GENERAL INFORMATION		
*Report No.:		
*Reported by:		
*Checked by:		
*Source:		
*Installation:		
*Item name:		
*Company Tag No.:		
*Company Sub-tag Nos.:		
Unique Nos. :		
*Taxonomy Code :		
*Function:		
*Manufacturer:		
Manufacturer of Control System:		
*Model/Type:		
*Redundant Subsyst.:		
Operational Time (hours):		
*Calendar Time (hours):		
No. of Demands/starts:		
Date Installed:		
Dates of Major Replacements:		
Note: Alphanumerical fields vill be free format text.		

Notes 1. Starred items indicate a manditory requirement.
2. Certain items have predefined values

FIGURE 1 - GENERAL INVENTORY DATA FORM

INVENTORY REPORT PUMP - SPECIFIC DATA		
*Type of Driver:		
*Fluid Handled:		
Fluid Characteristics:		
*Power:		
Utilization of Capacity:		
*Suction Pressure:		
*Discharge Pressure:		
Speed:		
Number of Stages:		
Body Type:		
Shaft Orientation:		
Shaft Sealing:		
Transmission type:		
Pump Coupling:		
*Environment:		
*Maintenance Program:		
*Instrumentation:		
Pump Cooling:		
Bearing:		
Bearing Support:		
Additional Info.:		
Note: Alphanumerical fields will be free format text.		

FIGURE 2 - EQUIPMENT-SPECIFIC INVENTORY DATA FORM

FAILURE EVENT REPORT (Also to be recorded on this form: all overhauls)		
*Report No.:		
*Inventory Report No.:		
*Reported by:		
*Checked by:		
Source:		
*Taxonomy Code:		
*Failure Mode, System Level:		
Subsystem Failed:		
*Failure Mode, Subsystem(s):		
Severity Class:		
Items Repaired:		
Repair Activity:		
*Failure Detected:		
Repair Time:		
Restoration Manhours		
Method of Observation:		
Additional Info.:		

Note: Alphanumerical fields will be free format text.

FIGURE 3 - FAILURE DATA FORM

RM Consultants Ltd
Suite 7 Hitching Court
Abingdon Business Park
Abingdon
Oxford OX14 1DY

DRAWING/DOCUMENT TRANSMITTAL NOTE

To T R MOSS A O F VENTON J M MORGAN A B RITCHIE J P STEAD DOCUMENT CONTROL CENTRE	Date 5/1/88 Job No. J1161 Ref. No. OR/GEN/04
Project OREDA PHASE II STUDY	

No. of copies	Reference No.	Rev.	Title
1	OR/GEN/04	1	Self check & verification guidance notes

Notes

Purpose of issue For use as guide

FROM [signature]

RM CONSULTANTS LTD ABINGDON

Self check and verification guidance notes

These notes are based on a detailed examination of the Red Book and the draft Guidelines for data collection. The aim was to establish the most economical way of satisfying Veritec's requirements. A need to keep a record of assumptions is seen. It would provide a reference to assist in achieving consistency between participants and a convenient basis for self check confirmation that assumptions are relevant and consistent.

Self Check

Percentage of items prepared *
All items +

1. Task and resources - before start

   Check time and manhours allowed for each item. Is it sufficient?

2. Data check - before finalisation

Reference	*)	Check consistency of source
Consistency	*)	pattern
Interpretations	*	Interpretations must be recorded with asumptions. + Check all for consistency at recording time.
Soundness	*	Is it complete, is coverage adequate, is is sensible.

3. Assumptions

Relevance	*	Assumptions must be centrally recorded. + Check all for
Consistency	*	consistency and relevance at recording time.

4. Calculations

Arithmetic	+	When completed
	*	before finalisation.

5. Computer records and reports

Proof reading	+
Figures	+
Graphic reliability	+

Self check should be applied to any results to be delivered.
Self check at end of each part of paper phase.
Self check at end of each part of computer entry phase.

<u>Internal Verification</u>

All deliveries except progress report, invoices, minutes of meetings.

Focus on - main conclusions
basic methods
verification of self checks.

Deliverables - Report for each equipment class.
Data bases on disk.

1. At beginning of an individuals work

<u>Check approach</u>

Source documents
Method of data selection
Data marking up
Data recording
Identification of missing data
Action to record, obtain missing data
Assumptions
Recording and checking record.

2. When first set of data forms are complete.

<u>Interpretation</u>

Check forms in association with individual to confirm.
They follow guidelines
Data sources are appropriate
Calculations, assessments and assumptions are correct.
Asumptions are recorded correctly.

3. When first platform complete
Check three sets of data forms to confirm interpretation.
Verify reports amd computerised records.

PLATFORM: XX SUBCONTRACTOR: RMC
OPERATOR: XYZ OFFSHORE CO. SUBJ. PROJ. MAN.: T R MOSS

	Item to be verified. GAS EXPORT COMPRESSORS Planned date. 9/2/88	Checked/Verified Date	Sign.
Self Check: - Tasks and resources	RESOURCES ARE ACCEPTABLE. IT IS NOTED THAT THE PLATFORM POSSESSES ONLY 2 COMPRESSORS AND NOT 4 AS ORIGINALLY PROPOSED.	9/2/88	[signature]
- Data	ALL FAILURE REPORTING FROM GT/COMPRESSOR UNIT RUN REPORTS. ALL INVENTORY REPORTING FROM DATA COLLECTED IN ABERDEEN.	9/2/88	[signature]
- Assumptions	ASSUMPTIONS WHERE NEEDED ARE RECORDED ON SOURCE MATERIAL AGAINST APPROPRIATE ITEM.	9/2/88	[signature]
- Calculations	CALCULATIONS - MINS/DECIMAL USING CONVERSION TABLE OF OR/GEN/28. YRS/HRS USING 1YR = 8760 HRS.	9/2/88	[signature]
- Report/Notes	REPORTING ONTO RMC/VERITEC STANDARD FORMS WITH SPECIAL ATTENTION PAID TO STARRED DATA FIELDS.	9/2/88	[signature]

Internal Verification

- Main Approach

The main approach is satisfactory.

- Reports/Notes

Data entry to the data base discs is generally satisfactory. However, the accuracy of data entry needs systematic checking.

[signature] 9/2/88

FIGURE 5 - SELF CHECK AND VERIFICATION EXAMPLE

R M CONSULTANTS | Report No.....1988.

Sheet 1 of 3

AUDIT REPORT - XX EXERCISE

Auditors	Names	Signatures	Audit Date
T R Moss A B Ritchie			23/8/88

General Comments:-

Very satisfactory audit, minor error found in data transfer from forms to database should be eliminated in final pass.

Reallocation of responsibilities transmitted verbally should be formally transmitted.

QA Director: T R Moss (for J M Morgan)

QA Manager: A B Ritchie (for J Deane)

OREDA

QA AUDIT - CHECK LIST

No.	Question	Satisfactory Yes	Satisfactory No	Remarks
	PLANNING AND CONTRACT CONTROL			
1.	Ask to see a sample of client specification (CTR's) for work undertaken. Check that the specification has been adequately reviewed to identify and provide for special or unusual requirements.	✓		
2.	Where required, has a quality plan been produced? Sample quality plan.	✓		
3.	Has Project Officer been appointed to be responsible for the work?	✓		Reallocation of Project Officer responsibilities not promulagated formally since JMM transfer to other work.

FIGURE 6 - QA AUDIT REPORT EXAMPLE

4.	Where work is shared between consultants have the interfaces been defined?	√	

REVIEW PROJECT FILES

5.	Is the work adequately documented, with documents correctly dated and identified in accordance with RMC P1 and P2?	√	
6.	Have the correct project control procedures been followed in accordance with RMC P3 and P4?	√	Special spreadsheet program developed and employed to meet clients requirements. Essentially same information is required in P4.
7.	Have progress meetings been held and minuted?	√	
8.	Where required, have project reviews been held?	√	
9.	Is there evidence of project planning? Sample planning charts. Are the plans kept up to date?	√	
10.	Are data collection records and reports dated and signed by the originator and where required checked and approved?	√	Some inventory reports not dated but not required on. Veritec form. Date included in database.
11.	Are Veritec Guidelines to Data Collection available? Ask to see copies. Sample self-check and verification forms.	√	
12.	Are data source references noted on Inventory and failure reports? Sample reports.	√	
13.	Are boundaries defined in accordance with guideline document? Ask to see copies of marked-up Boundary specifications.	√	
14.	Are assumptions made in applying Inventory and Failure Report definitions noted? Ask to see list of assumptions for each equipment class.	√	Original assumption noted on file. Current assumptions maintained on noticeboard.

15.	Are recommended failure mode and severity classes being strictly adhered to? Sample hard copy forms for each equipment class. Check on completeness of data recorded.	√	Some missing data - particularly trade manhours although total manhours available.
16.	Has data been transferred correctly from I and F reports to diskette? Sample diskette data and check against hardcopy.	√	Original diskette (compressors) inventory sampled. Pignone spelt wrongly. 1 error in Additional Information - further samples no errors - corrections now being done.
17.	For main contractor diskettes check that commercially sensitive data has been deleted. Sample diskette data.	x	No longer required.

REPORTS

18.	Are project reports to the agreed specification? Are reports signed by the originator and checked and approved? Sample reports.	√	Draft report to spec.

Failure Rate Estimation Based on Data from Different Environments and with Varying Quality

S. LYDERSEN and M. RAUSAND

SINTEF, Division of Safety and Reliability
N-7034 TRONDHEIM

Summary

Data to be included in reliability databases are typically collected from different sources. The "true" reliability of a component may vary from source to source, due to factors such as different manufacturers, design, dimensions, materials, and operational and environmental conditions. The quality of the data may vary in completeness and level of detail, due to one or more reasons such as data registration methods, company boundary specifications, subjectiveness and skill of the data collector, and time since the failure events occurred.

The paper discusses reliability estimation based on data with the characteristics mentioned above. Special problems and uncertainties are highlighted. The discussion is exemplified with problems encountered during data collection projects such as OREDA.

Introduction

Component reliability data is a necessary input to practically all types of reliability analyses. The sources for such data are, however, often hampered with inconsistencies and other quality problems. When seeking data for a specific component, the analyst often faces one or more of the following problems: The data sources have varying levels of detail and quality. The failure modes and boundary specifications for the component may be ambigously defined. The relevance of the data may be questionable and may also be hampered with confidentiality restrictions.

These problems are described and discussed in this paper. The ideas and results are presented in terms of a constant failure rate in continuous time. With obvious modifications, they also apply for "failure on demand" probabilities.

Some of the topics discussed in this paper are presented by the authors in a wider context in [1].

Data from Different Sources

The "true" reliability of a component will generally be dependent on a number of external and internal factors.

- Component specific factors (design, make, finish etc)
- Environmental factors (humidity, pressure, temperature etc)
- Operation and maintenance factors.

During the data analysis for the OREDA Handbook [2], the variations between the samples were often very significant. The variations in failure rates for a specific drilling equipment is shown in Figure 1. Estimates, together with 90% confidence intervals are presented in Figure 1 for each of the samples. As seen from this figure, the failure rates show significant variations between the samples.

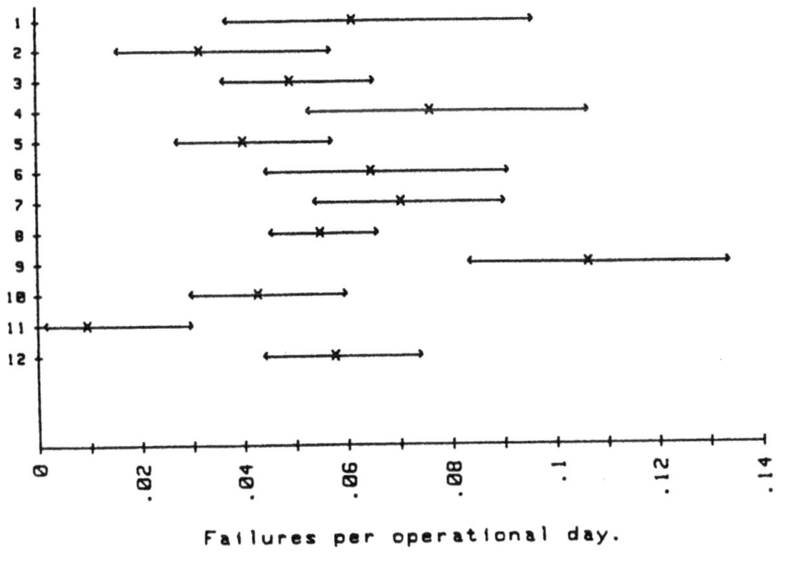

Figure 1. Failure rates and 90% confidence intervals for drilling equipment on 12 different drilling rigs. Plot from the computer program ANEX [3].

In reliability analyses, we usually need a single failure rate estimate for each component and a confidence interval for this estimate. A straightforward approach, although very doubtful, is to pool all the data into one single sample and estimate the failure rate and the confidence interval according to the general formulas for the exponential/Poisson model. This approach will not take the variation between the samples into account. The failure rate estimate will therefore normally be biased and the estimated confidence interval unrealistic narrow.

During the OREDA project, the author M. Rausand suggested a new multi-sample estimator. The estimator has been called the "OREDA estimator". It is based on a Bayesian point of view. The failure rate is regarded as a stochastic variable Λ with one realization per sample. For example, in sample number i, all times to failure are assumed independent, exponentially distributed with failure rate λ_i. The stochastic variable Λ has some probability density $\pi(\lambda)$, with mean

$$\theta = E(\lambda) = \int_0^\infty \lambda \, \pi(\lambda) \, d\lambda, \tag{1}$$

and variance

$$\sigma_\lambda^2 = \mathrm{Var}(\lambda) = \int_0^\infty (\lambda-\theta)^2 \, \pi(\lambda) \, d\lambda. \tag{2}$$

Figure 2 shows a possible prior distribution probability density for the situation in Figure 1.

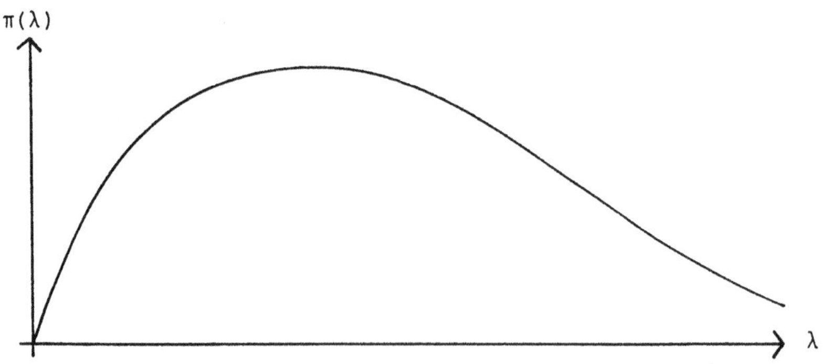

Figure 2. Prior probability density for the failure rate Λ in a sample.

No specific distribution class, such as the gamma distribution, is assumed for the prior density $\pi(\lambda)$ in the OREDA estimator.

Let X_i denote the number of failures, and t_i the total time in service in sample number i; $i=1,2,\ldots,k$.

An initial estimate for θ is found by pooling the data:

$$\hat{\theta} = \frac{\sum_{i=1}^{k} X_i}{\sum_{i=1}^{k} t_i} \tag{3}$$

An estimate of the variation between the samples is given by

$$\hat{\sigma}_\lambda^2 = \frac{V - (k-1)\hat{\theta}}{s_1^2 - s_2} \cdot s_1 \tag{4}$$

when this is greater than 0, else 0, where

$$s_1 = \sum_{i=1}^{k} t_i \tag{5}$$

$$s_2 = \sum_{i=1}^{k} t_i^2 \tag{6}$$

$$V = \sum_{i=1}^{k} \frac{(X_i - \hat{\theta} t_i)^2}{t_i} = \sum_{i=1}^{k} \frac{X_i^2}{t_i} - \hat{\theta}^2 s_1 \tag{7}$$

An estimate for the mean failure rate θ is calculated by:

$$\theta^* = \frac{1}{\sum_{i=1}^{k} \frac{1}{\frac{\hat{\theta}}{t_i} + \hat{\sigma}_\lambda^2}} \cdot \sum_{i=1}^{k} \frac{1}{\frac{\hat{\theta}}{t_i} + \hat{\sigma}_\lambda^2} \cdot \frac{X_i}{t_i} \tag{8}$$

An approximate 90% confidence interval for θ is calculated by the formula:

$$\theta^* + \frac{u_{0.05}^2}{2s_1} \pm u_{0.05} \left[\frac{\hat{\theta}}{s_1} + \frac{s_2}{s_1^2} \hat{\sigma}_\lambda^2 + \frac{u_{0.05}^2}{4\, s_1^2} \right]^{1/2} \tag{9}$$

and u_α is the upper 100α% percentile of the standard normal distribution, $u_{0.05} = 1.645$.

Note that this gives a confidence interval for the mean of the prior distribution. It is not a confidence interval for a predicted value of the failure rate. An approximate confidence interval for the failure rate of a component from this item class under similar conditions, is given by

$$\theta^* \pm u_{0.05} \sqrt{\hat{\sigma}_{\theta^*}^2 + \hat{\sigma}_\lambda^2} \; . \tag{10}$$

The OREDA estimator has been thoroughly studied by Spjøtvoll [4], who concluded that the estimator seems to be better than most alternatives. The estimator was used in the OREDA project and is also implemented in the commercially available PC program ANEX [3].

It should be noted that the data illustrated in Figure 1 all originate from drilling rigs in the North Sea with rather similar operational and environmental conditions. The failure rates are significantly different, but the confidence intervals at least cover values of the same size of order. For a number of items in OREDA [2], estimated failure rates showed much more significant discreapancies between the samples.

Unfortunately, few reliability data handbooks or databases include characteristica on the underlying distribution $\pi(\lambda)$. One exception, besides OREDA, is T-boken [5], which contains estimated gamma distribution parameters for $\pi(\lambda)$.

Varying Level of Detail and Data Quality

Practical data collection shows that the level of detail varies between data sources. The main source of reliability information is normally the maintenance recording system and work order forms. The maintenance recording system is primarily designed for follow up and planning of maintenance resources/activities. It is normally not designed to give reliability information. Failure modes are often poorly defined. False alarms and spurious activations in a safety system seldom result in a work order in the maintenance files. Hence, these failure modes are under-represented in several data sources, such as the OREDA Handbook [2].

Boundary specifications often vary from source to source. If we, for example, want to estimate the reliability of a specific type of compressors, problems with the boundaries may arise. For example, the gear box may be part of the compressor's TAG number in some sources. In other sources it may have a separate TAG number, or be included in the TAG number of a more complex unit, such as the motor/turbine. These problems are even more complicated when looking at the control and monitoring system associated to the compressor. A thorough system knowledge, and often some "detective's job", is required to obtain unique boundaries for a specified unit.

Another problem is to obtain detailed inventory data, that is, the number of items of the different kinds/makes, testing schedules, time in operation or standby, modifications etc. This type of information is often poorly recorded and difficult to trace in written sources.

In some sources it may be difficult even to find the number of failures encountered on the item, while other sources may provide detailed failure histories for each labeled component. The normal situation is in between these extremes: The number of failures and a reasonably good failure mode classification is available. Further, the amount of environmental and operational data may vary.

In essence, this means that for some samples, parts of the required data are missing. Typically, some of the environmental/operational data are missing, and some samples will hardly have any failure mode classification. Hence, estimation methods allowing for "missing observations" are needed. This is not a straightforward problem. Typically, in the samples where a specific set of data are missing, these data may have quite different values, compared to samples where these data are available. For a general discussion on this subject, see for example [6].

A data collector may experience everything from neatly updated maintenance files to an operator who states "I don't think we have had more than a few failures on this component for several years". That is, the data may be more or less accurate with respect to operational time, actual number of failures encountered, etc.

The data collector who bases his finding e.g. on maintenance files or interviews, has to use a certain amount of common sense or engineering judgement in the data classification. The subjectiveness and skill of the data collector is a factor which may be difficult to include in the database.

In the IEEE Std 500 Handbook [7], each supplier of data gave a "confidence factor" to account for his evaluation of the data quality.

Established failure event databases are, naturally, never completely error-free. This was also one of the experiences gained during the EuReDatA Benchmark Exercise on Data Analysis performed in 1988 - 1989. It was pointed out that derivation of reliability parameters based on automatic data processing may provide erroneous results due to codification and data inconsistencies, incompleteness of data, and data errors. A sound data analysis should start with a "manual" procedure mainly based on engineering judgement [8].

Varying Data Relevance

Not all the failure history data on an item class may have the same relevance. In many cases, the reliability will have a trend in time. "Old" reliability performance data are less relevant than the more recent.

Varying data relevance also arise for other reasons, which may be illustrated through an example: Consider a user who wishes to estimate the reliability of a given component under specified environmental/operational conditions. Little, if any, data are available on the given item and conditions. To improve the estimation accuracy or feasibility, data on similar items/conditions could be taken into account. If the desired component/conditions is a 4 1/2" XMV (automatic master valve) for a gas production tree, with well head pressure 100 bar, similar items/conditions may be:

- Other well head pressures
- XMV, oil production
- XMV with other dimensions
- Similar gate valves, such as AFV (automatic flow wing valves)

The best way to include data from similar conditions, is by setting up a model for how reliability depends on the relevant factors, such as pressure, GOR (Gas/Oil ratio), size etc. This involves the use of experts in one or more engineering disciplines. In principle, this yields stressor-dependent models. Estimation of parameters in such models is briefly discussed in [1], which also gives further references.

However, in many cases, such modelling is not feasible. The analyst must confine to having little or no data for the desired item/conditions, and some data for similar items/conditions. Some sort of moving average or window estimation technique may prove useful in this situation. This is illustrated in Figure 3 in terms of one factor.

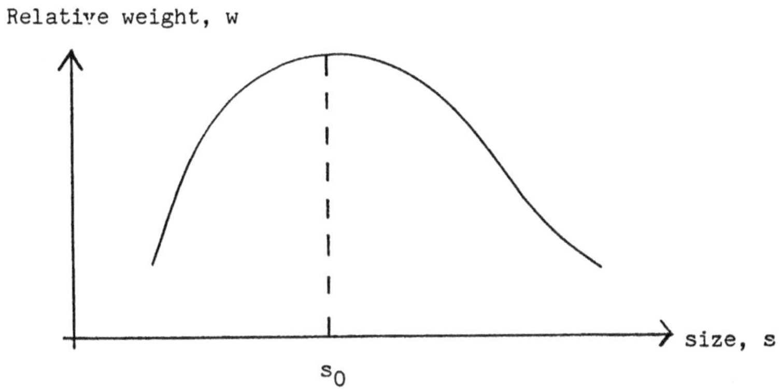

Figure 3. Example on reliability estimation with weights corresponding to data relevance. A reliability estimate is required for the size s_0. Relative weight w is applied for the data during estimation.

The concept of estimation based on **debatable evidence** provides an alternative way of treating data with varying relevance. It is best explained in terms of an example given by Apostolakis [9]: The problem is to assess the probability of failure to insert the control rods into a nuclear reactor core (to "scram") and, thus, to terminate the neutron chain reaction. A debate went on between the Nuclear Regulatory Commission (NRC) and the Electric Power Research Institute (EPRI) in the United States, as to what the actual field performance data are. This was before the Tsjernobyl incident. The debatable issues were:

- Only one event that could qualify as a scram failure had occured (the Kahl reactor, Germany, 1963). After this incident, scram system modifications have been made, and EPRI claims that it should not be counted.

- Should the number of years of experience include naval reactors and reactors producing plutonium and tritium for use in nuclear weapons, or only commercial and army reactors (as claimed by the NRC)?

- The number of tests per reactor year is a matter of disagreement. The EPRI analysis assumes at least 38 test per year, while NRC is willing to accept only 12 tests per year.

Thus, the statistical evidence was of the form {k events in n tests}, where k is 0 or 1, and n is 7908, 39212, or 114332, depending on one's point of view.

Generally, estimation of the reliability parameter(s) shall be based on statistical evidence E, which is one of the events E_1, E_2, ..., E_r. For example, E_1 = {1 event in 7908 tests}, ..., E_r = {0 events in 114332 tests}. The idea presented by Apostolakis [9] is that the analyst ought to assign probabilities to E_1, ..., E_r, such that $P(E_i)$ represents his belief that E_i is the "true" evidence. It may be argued against this approach that it is, to some extent, based on the subjective belief of the analyst. However, a "conventional" approach, where the analyst states that E_j is the only evidence in which believes, and bases estimation on E_j alone, may be said to be even more subjective. Apostolakis [9] presents a Bayesian estimation method for this situation.

Let θ be the parameter to be estimated, and let $\pi(\theta)$ be its prior distribution. If little is known about θ a priori, a non-informative prior distribution could be used. Bayesian estimation of θ may now be based on the "posterior distribution"

$$\pi(\theta|\underline{E},\underline{P}) = \sum_{i=1}^{r} \pi(\theta|E_i)\, P(E_i), \tag{11}$$

where

$$\underline{E} = (E_1, \ldots, E_r),$$

$$\underline{P} = (P(E_1), \ldots, P(E_r)),$$

$$\sum_{i=1}^{r} P(E_i) = 1,$$

and

$$\pi(\theta|E_i)$$

is the posterior distribution given the data E_i.

Varying Confidentiality

A client who has supplied data to the database, is allowed to read his own data in full detail. Further, he may be allowed to see an "anonymized" version of the rest of the data. For example, the FACTS incident database operated by TNO in the Netherlands contains both restricted and completely accessible information. If the information requested by a client is restricted, it is summarized and anonymized before delivery to the client [10].

Stressor Modelling

Most available reliability data handbooks and databases give little data on how reliability depends on operational and environmental conditions. MIL-HDBK-217E [11] is an exception, where component failure rates are tabled as functions of stressors such as temperature, applied voltage, application, etc. The Nonelectronic Parts Reliability Handbook [12] groups the data according to application, such as "ground mobile", "ground fixed", etc.

Information on reliability dependence on stressors is frequently needed in reliability engineering.

In more advanced reliability databases, it should be possible to estimate the reliability as a function of the environmental and operational conditions. During a comprehensive reliability study of Surface Controlled Subsurface Safety Valves (SCSSV's) performed by SINTEF, a detailed and comprehensive failure event database for such valves was established. To some degree, it was possible to estimate reliability as function of some environmental and operational conditions [13], [14] using proportional hazards modelling (Cox models) [15].

However, the lack of environmental/operational data has made such estimation difficult or impossible in many practical cases.

Reliability Data Dossier

In many practical situations where limited data are available, a reliability data dossier provides a good documentation of the available information. This implies that for each component type included, the analyst must

- Extract reliability data from the sources available to the analyst.

- Write down the extracted data in a systematic way.

- Give a recommended failure rate (or other reliability measures) for the foreseen application and operational and environmental conditions. The recommended failure rate is based on the analyst's judgement of all the available sources.

This approach does not necessarily require sophisticated mathematical methods, but rather good knowledge of the relevant components and applications.

References

1. Lydersen, S.; Sandtorv, H.; Rausand, M.: Processing and Application of Reliability Data. SINTEF Report STF75 A87034, 1987.

2. Offshore Reliability Data Handbook (OREDA). OREDA Participants, Veritec/ PennWell Books, Høvik, Norway, 1985.

3. ANEX ("Analysis of Exponential Life Data"). Computer program for IBM AT or PS/2 developed by SINTEF, Division of Safety and Reliability.

4. Spjøtvoll, E.: Estimation of Failure Rates from Reliability Data Bases. Presentation at the SRE Symposium 1985, Trondheim, September 30 - October 2, 1985.

5. T-boken. Tilförlitlighetsdata för komponenter i svenska kraftreaktorer (Reliability Data for Components in Swedish Power Reactors). RKS - Rådet för Kärnkraftsäkerhet (The Swedish Council for Nuclear Power Safety), 1985.

6. Bayarri, M. J.; de Groot, M. H.: A Bayesian View of Weighted Distributions and Selection Models. In Clarotti, C. A. & Lindley, D. V. (eds.) Proceedings from the course Accelerated Life Testing and Experts' Opinions in Reliability, Lerici, Italy, July 28th - August 1st, 1986. North Holland Publishing Company/Elsevier, Amsterdam, 1988.

7. IEEE Std 500-1984. IEEE Guide to the Collection and Presentation of Electrical, Electronic, Sensing Component, and Mechanical Equipment Reliability Data for Nuclear Power Generating Stations. The Institute of Electrical and Electronic Engineers / Wiley, New York, 1984.

8. Pamme: Contibution to the EuReDatA Benchmark Exercise on Data Analysis. Interatom Report no 70.04440.7, 22 September 1988.

9. Apostolakis, G.: Expert Judgment in Probabilistic Safety Assessment. In Clarotti, C. A. & Lindley, D. V. (eds.) Proceedings from the course "Accelerated Life Testing and Experts' Opinions in Reliability", Lerici, Italy, July 28th - August 1st, 1986. North Holland Publishing Company/ Elsevier, Amsterdam, 1988.

10. Bockholts, P.: Collection and Application of Incident data. In Aniello, A. & Keller, A. Z. (eds.) Reliability Data Bases. Proceedings of the ISPRA Course held at the Joint Research Centre, Ispra, Italy, 21-25 October 1985 in collaboration with EuReDatA. D. Reidel Publishing Company, Dordrecht, Holland, 1987.

11. MIL-HDBK-217E. Military Handbook for Reliability Prediction of Electronic Equipment. Department of Defence, Washington DC, 1986.

12. NPRD-2. Nonelectronic Parts Reliability Data - 2. Rome Air Development Center, New York, 1981.

13. Molnes, E.; Rausand, M.; Lindqvist, B.: Reliability of Surface Controlled Subsurface Safety Valves. Phase II - Main Report. SINTEF Report STF75 A86024, 1986.

14. Tjelmeland, H.: Regresjonsmodeller for sensurerte levetidsdata, med anvendelse på feildata for sikkerhetsventiler i olje/gass produksjonsbrønner (Regression Models for Censored Lifetime Data, with Application on Failure Data for Safety Valves in Oil/gas Production Wells). In Norwegian. M.Sc. thesis at the Norwegian Institute of Technology, Trondheim, Norway, 1988.

15. Cox, D. R.: Regression Models and Life-Tables (with discussion). J. R. Stat. Soc. B. 34, 1972, 187 - 220.

Data Selection for Probabilistics Safety Assessment

B. Tomic, L. Lederman

International Atomic Energy Agency
Division of Nuclear Safety
Reliability and Risk Assessment
Vienna, Austria

SUMMARY

The IAEA has, in the framework of its PSA activities, compiled from the available literature a component reliability data base. This data base contains more than 1,000 records drawn from 21 different sources. In order to assess the actual ranges of data a number of graphs containing failure rates or probabilities on similar components and failure modes were plotted. It was recognized that order of magnitude differences exist among the sources. The influence of the data set used for the final PSA result was assessed by use of a plant PSA model. The original data set for 8 components with the highest importance measure was replaced by consistently high and low data drawn from generic sources. The results show an increase by a factor of 15 in the normalized core melt frequency. The exercise was repeated for the 23 most important components and a factor of 46 in core melt frequency was observed.

INTRODUCTION

Numerical results of probabilistic safety assessments (PSA) depend on a large extent upon the data used to calculate the failure probabilities of various plant equipment.

Component reliability parameters come from a variety of sources, namely:

- Generic published information
- Abnormal operational events reporting system
- Specific field data collection campaigns
- Laboratory testing
- Expert opinion

It is a common practice in PSA to calculate system failure probability using fault tree models with average values for the unavailability of the basic events.

Core melt scenarios are further developed based on the anticipated plant response to groups of selected accident initiating events. The core melt frequency is calculated combining the frequency of initiating events and the probabilities of the various modes of system failures as expressed by fault tree minimal cut sets.

The uncertainties in data are represented by probability density functions. These are propagated in accident sequences minimal cut sets and displayed in connection with the core melt frequency.

Considering the importance of data in PSA, problems with data availability and the difficulties in data selection, the International Atomic Energy Agency (IAEA) in the framework of its Probabilistic Safety Assessment (PSA) Programme has compiled from the publicly available literature a bank of generic component reliability data. Twenty-one sources have been surveyed providing a total of 1,010 records. All component and failure modes usually considered in PSAs are included in the IAEA data bank. The bank is stored on an IBM compatible personal computer in dBASE III files for convenient retrieval of information. An IAEA Technical Document describing all the data sources, which also includes a complete listing of the data bank, was published [Ref. 1].

OVERVIEW OF DATA RANGES

A comprehensive overview of data on selected components and failure modes derived from the IAEA data bank is presented in Reference [2] in the form of plots. Some aspects of this work are reviewed in this section.

A close observation of the data compiled indicates order of magnitude differences in the component failure rates among the various sources. These differences are both in terms of point values (means or medians) and of confidence intervals.

One of the reasons for the variation found in the failure rate is the improper (or inconsistent) definition of component boundaries and failure modes.

Other factors influencing failure rates are component design, operating mode and environment, testing/maintenance practices and component age. Although the exact contribution of each of these factors is hard to assess, maintenance practices are often considered to have a substantial effect on component reliability.

The component age factor could also play an important role. The failure rates are assumed to be in the flat portion of the so-called "bath-tub" curve used to characterize time dependent changes in failure rates. However, the continuous aging of operating plants make it questionable whether the failure rates for equipment in use for many years are still in the flat portion of the curve.

The system of data collection adds to the problems which influence the determination of failure rates. Determination of component failure rates from event reporting systems, which are not designed for that purpose, are often incomplete.

Source classification

To minimize the influence on the component failure rates from some of the factors mentioned, sources considered in the IAEA data base compilation have been conveniently grouped into three categories, namely:

- Plant specific sources
- Updated sources
- Generic sources

Each of these three exhibits some unique characteristics and in principle should be considered separately.

The plant specific category contains reliability data based on operating experience of a single plant or a group of similar plants. The updated category includes the sources where the failure data were obtained by updating generic data with the plant specific operating experience. The third category is termed generic and it contains all the remaining sources from the IAEA Data Base. This category includes the entire range of ultimate data sources from expert opinion (single or aggregate) to collection of operating experience of a number of plants.

In the data overview plots generic sources have been separated from the other two sources. The plant specific and updated categories commonly have few data points which are, in addition, easily identifiable, and therefore plotted together.

Operating mode

The operating mode of the component is also one of the factors which influence its failure rate. The operating mode is more important for components that perform their function by continuously moving (e.g. by rotating, such as pumps) and less important for 'open-closed' components (such as valves) or for electrical and electronic components that do not change state on a macroscopic level.

Information on component operating modes is usually sparse and rarely found in the sources. Owing to data availability, the only component for which plots for different operating modes was possible was the motor driven pump.

Failure mechanisms

Another problem which is encountered when generic data are used for specific analysis is the treatment of failures which are revealed when a component is demanded. The failure mechanism generally includes elements of a time dependent process and of a demand related one. The time based failure model is based on the assumption that the failures of a component follow an exponential distribution during the standby period and become apparent when the component is demanded. The demand failure probability model is characterized by a probability that the standby component will fail at the moment when it is demanded. It is therefore assumed that the failure is caused by the immediate influences caused by the demand. The fractional influence of each is, however, difficult to access [Ref. 3]. Moreover, if this dual aspect is introduced in the modelling, unique data are needed. This type of data are rarely available and none of the sources in the IAEA Data Base provide data of this type. Instead it is assumed that both contributions are included in either demand or time related failure data. In this situation care should be taken when using per-demand failure probabilities (q), given in data bases.

Often these values are the average unavailability ($\bar{U}$) calculated over a particular test interval (T) based on a standby failure rate λ_s, as:

$$\bar{U} = \lambda_s T/2$$

However, if this average unavailability is utilized as the failure probability for components which have significantly different demand intervals, this could result in either an underestimation or an overestimation of the component reliability.

For the purpose of plotting the data, and considering that the majority of sources provided failure data for the failure modes in question as demand related, average unavailabilities have been calculated to represent the failure probabilities of components for which data was given in terms of standby failure rates.

Display of data ranges

The component reliability data for the data ranges overview plots of Reference [2] have been grouped in terms of source category, operating mode, component type, and failure mode.

Another characteristic of the plots is that sometimes different types of data are plotted together. Some of the sources provide point estimates as the mean value of a distribution; others provide medians; while some others recommend values without a formal distribution, terming them either best estimates or recommended values. Owing to the limited number of available data points, it was not possible to plot separately means, medians or best estimates. All these data are therefore plotted in the same graph as 'centre points'. The same applies to the upper and lower bounds, which are sometimes 95% and 5% of log-normal distributions, sometimes 90% and 10%, and occasionally best estimate maximum and minimum values. Information about the center point or the upper and lower bounds are provided in the data tables accompanying each plot.

Illustrative examples of these plots are given in Figures 1 and 2.

INFLUENCE OF DATA VARIATION IN CORE MELT FREQUENCY

It is recognized that changes in component failure data of selected components can lead to significant changes in core melt frequency in PSA studies. To demonstrate the magnitude of this, an example was developed using a realistic plant model and realistic data. The plant will be further referred to as plant X.

The frequency of core melt for plant X was calculated using a reference data set and plant components have been ranked based on their importance to core melt according to the Fussel-Vesely importance measure (I^{FV}), [Ref. 4] defined as:

$$I^{FV} = \frac{\partial\, CMF}{\partial\, a} \cdot a$$

CMF = core melt frequency
a = component failure probability

For the purpose of this example two data sets have been derived from Ref. [2] for eight components ranked highest in accordance with I^{FV}. Both data sets are based only on generic sources found in the IAEA Data Base. Generic sources were considered as most representative for a situation in which no (or very few) plant specific data is available and the analyst is establishing a data base for a particular problem. The first data set was established by consistently selecting the highest mean value for each particular component and failure mode. The second data set was based on the consistently lowest mean values.

Table 1 depicts the mean values for the component failure probabilities for the highest and lowest data points found in the sources and the reference data set. Table 2 indicates the effect of using each one of the two data sets in terms of the core melt frequency normalized to the base case. A factor of 3.7 in the CMF as compared to the base case was obtained when the set of consistently highest event probabilities was used. When lowest event probabilities were used the factor was 0.24. Therefore a variation in CMF by a factor of 15 represents the effect of using one or the other data set. This factor increased to 46 when data for the 23 most important components have been modified in the same manner.

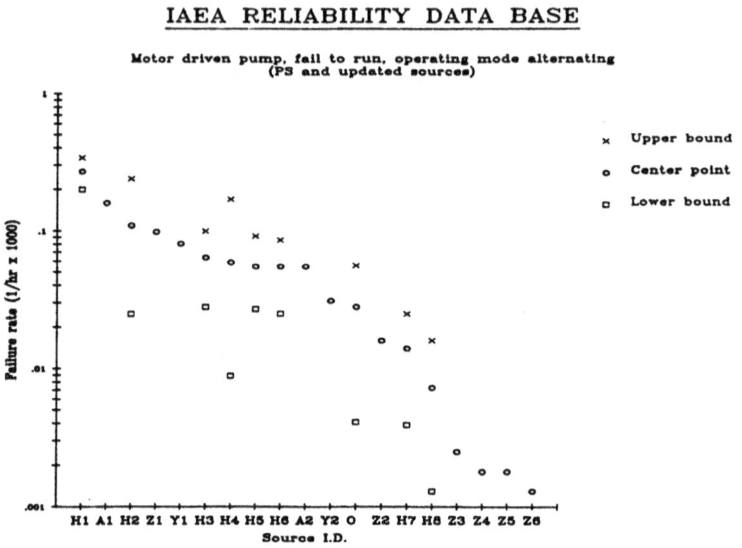

Figure 1

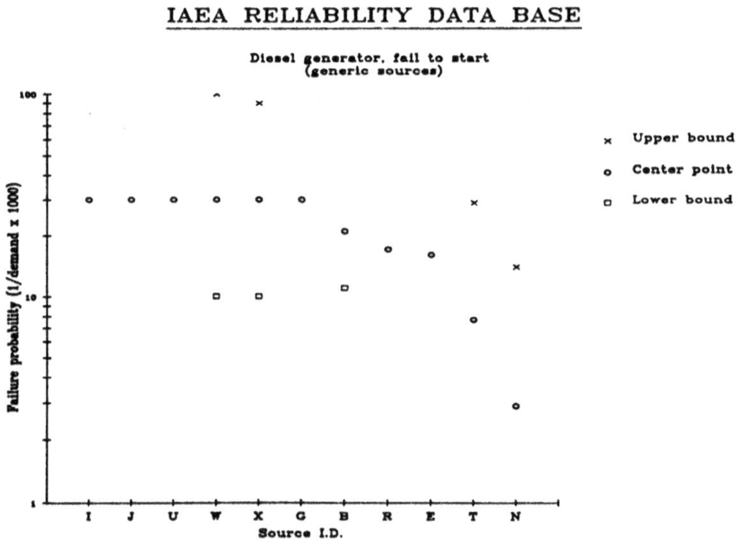

Figure 2

It should be noted that components listed in Table 1 are actually "super-components" and that their failure probabilities have been individually evaluated based on a detailed sub-event description. In this example only changes in component failures have been considered. Unavailabilities due to test and maintenance and the effect that changes in component failure probability may have in the probability of common cause failure have not been considered. Therefore it is to be expected that the actual effect of changes in the data set in the core melt frequency could be substantially higher than the values in this example.

Table 1

Range of Component Failure Probabilities (mean values)

Components Ranked by Risk Reduction Importance	High Generic Event Probability	Low Generic Event Probability	Reference Data Set
1. Pressure Safety Relief Valve S1--fails to reclose	3.5E-2	3.6E-3	2.3E-2
2. Pressurizer Safety Refief Valve S2--fails to reclose	3.5E-2	3.6E-3	2.3E-2
3. Diesel Generator 1	1.8E-1	8.4E-3	3.9E-2
4. Diesel Generator 2	1.8E-1	8.4E-3	3.9E-2
5. Chilled Water Unit B (ECS)	2.3E-2	7.0E-3	1.9E-2
6. Chilled Water Unit A (ECS)	2.3E-2	7.0E-3	1.9E-2
7. Motor-Operated Valve MOV1 (SWS)	1.4E-2	1.9E-3	1.0E-2
8. Motor-Operated Valve MOV2 (SWS)	1.4E-2	1.9E-3	1.0E-2

Table 2

Core Melt Frequency Variation

Data Set	Normalized Core Melt	
	8 Components Changed	23 Components Changed
High Generic	3.7	6.0
Low Generic	0.24	0.13

CONCLUSION

As indicated reliability data play an important role in calculating the core melt frequency. It is therefore necessary to establish requirements to be imposed on data, particularly when PSA results are used to demonstrate compliance with probabilistic safety criteria extablished at plant level (e.g. core melt frequency).

The establishment of a consistent generic data set for PSA requires a thorough evaluation of data from various sources considering their particular characteristics and tracing the origin of their ultimate source to identify dependencies between them.

The indiscriminate selection of generic data from various sources can pose a serious problem of credibility in the final results of PSA and inhibit the full utilization of invaluable PSA insights which are often less dependent of bottom line results.

In this framework, it is suggested that IAEA work in this field [Refs. 1 and 2] can be most useful. Particularly the use of data range plots to locate a proposed data value considering the of information available in the literature and therefore warning the analyst about any bias towards selecting consistently higher or lower data values.

A more ambitious application is, however, the establishment of criteria for the pooling of data and acceptable ranges for component failure probabilities in PSA. Work in this area is being carried out under the framework of an IAEA Coordinated Research Programme on Data Collection and Analysis for PSA.

REFERENCES

[1] IAEA TECDOC 487 "Component Reliability Data for Use in Probabilistic Safety Assessment", IAEA, Vienna, 1988.

[2] IAEA TECDOC "Overview of Ranges of Component Reliability Data for Use in Probabilistic Safety Assessment, to be published.

[3] Winfield, D.J., "Long Term Reliability of Standby Diesel Generators", Reliability Engineering and System Safety", Vol. 21 No. 4, 1988.

[4] Vesely, W.E.; Davis, T.C., "Evaluations and Utilizations of Risk Importances", NUREG/CR-4377, August 1985, U.S. Nuclear Regulatory Commission.

A Reliability Data Bank for the Natural Gas Distribution Industry

M. SCARRONE, N. PICCININI

Dipartimento di Scienza dei Materiali ed Ingegneria Chimica
Politecnico di Torino, Corso Duca degli Abruzzi n.24,
10129 Torino, Italy

C. MASSOBRIO

Italgas S.p.A., Corso Palermo n.2, 10100 Torino, Italy

Summary

The present study is an update and extension of those already reported. It illustrates the results obtained from ordinary maintenance operations on six large sets of urban mains with a view to optimasing the establishment of a reliability data bank. The study undescores the extreme individuality of gas distribution plants and the need for a sector data bank. A prototype form for the collection of pressure Regulating Installations (PRI) faults and breakdowns is also presented.

Introduction

Previous studies on the availability of a Pressure Regulating Installation (PRI) on natural gas (NG) distribution networks have stressed the difficulties involved in quantitative evaluation owing to the scarcity of trustworthy reliability data (1-3). Attention has been particularly drawn to the extreme individuality of city mains and hence the need for a sector data bank operating in such a way to ensure constant updating or its reliability data. Further work has since been done in cooperation with Società Italiana per il Gas (Italgas) which has displayed great interest in the updating of its own informatic facilities (E.g. in view of collecting data, since 1986, the Italgas modified the rules and the recording form for maintenance operations) (4).
The present study is an update and extension of those already reported. It illustrates the results obtained from ordinary maintenance operations on six large sets of urban mains with a view to optimasing the establishment of a reliability data

bank. A prototype form for the collection of PRI faults and breakdowns is also presented. It is, in fact, convenient to commence with a reliability bank dealing with this type of installation, which can subsequently be extended to embrance intermediate and initial reduction plants.

The networks and installation examined

As already mentioned, the study covered six sets or city mains. These are all located in N.Italy and are assigned the letters: B, C, E, F, G, H. Mains B and C are wholly or partly coincident with those describes in references (1) and (2). As on previous occasions, data were collected from ordinary maintenance operations carried out on the PRIs. A total of about 400 installation were investigated in the period 1986 to 1988. This corresponds to about 6 10^6 hours of operation.
The PRI layout and operating schemes were presented in (1) and (2). For the sake of convenience, as will appear later, the layout for a branch unit is shown in Fig. 1. As is well known, in this case there are two structurally identical pressure reduction lines, called the service and emergency line respectively. An ininterrupted supply is ensured by the automatic entry into operation of the emergency line if the service line breaks down.

Causes of PRI faults and breakdowns

In view of the considerable documentation available, namely all tha maintenance operations performed over the two-year period, identification of PRI faults and breakdowns was based solely on the historical evidence, since a failure mode and effect analysis (FMEA) had already performed (1, 2).
A complete list of causes of failure, the total number of events recorded, and the corresponding numerical failure rates (λ) are shown in Table 1. It will be noted that when processing the data to determine the failure rates no account was taken of

distinctions between:

- monitor and reductions unit pilots,
- reduction units on service and emergency lines,
- monitors on service and emergency lines.

In addition, the small number of fault events recorded for most items made it impossible to attempt an assessment of their statistical distribution. The failure rates in the table were thus taken as a means for the two-year period. The present study offers further evidence of the extreme individuality of NG mains. Critical examination of the data, in fact, revealed that some types of fault were confined to particular mains, or linked to their prior history and changes in gas consumption during the course of the year.

Some dissagregated data are presented by way of example in Figs. 2 and 3. The seasonal event "Main filter clogged" (Fig. 2) occured in different periods of the year, due to changes in gas consumption pattern throughout the year. On mains H, on the other hand, the most frequent main filter event was "Presence of hydrocarbons", compared with "Clogging by dirt" on the other mains. This difference springs from the fact that the NG in mains H comes from another source. Another example, about the 70% of the failure "Service regulator blanking disk fails" (Table 1) comes from main B.

Availability of branch mains PRI

The most critical situation on a city NG mains is that applicabile to users served by branch PRI.

The availability of this type of installation must thus be very high. It was therefore decided to evaluate, by way of example, the probability of the Top Event (TE) known as "Abnormal increase of downstream pressure", as previously studied (2), in the light of the new failure rates obtained. The corresponding fault tree is set out in Figs. 4-5. The list of primary events with their failure rates, the test intervals and the repair rates used for the numerical resolution of the fault tree (5) can be seen in Table 2. The probability of occurence of this TE was $\lambda = 1.62\ 10^{-3}$ occ/y with the data set out in Table 5 of refe- rence (2), which partly included numerical data from chemical firm's data bank. This figure is now reduced to 1.58 10^{-5} occ/y, which is a much more reliable result. As far as the failure rates previously calculated are concerned, full corroboration was obtained of the numerical values. For example, events E212/E213 "Faulty operation of monitor pilot in open position" have a failure rate of 1.1 10^{-6} occ/h and 1.3 10^{-6} occ/h in the previous work and this study respectively.

Conclusion

Work on the establishment of a reliabilty data bank for the gas industry has been greatly aided by Italgas's practice of recording all ordinary maintenance operations on its installations on special forms.

This means, amongst other things, that maintenance teams should not find it difficult to fill in a form that is only slightly more comprehensive and sophisticated than the one they are accustomed to. Extraordinary operations, on the other, are handled in a different way and as matters now stand retrivial of the relevant information is a more complicated process.

As already mentioned, the data shown in Table 1, and Figs. 2-3 were obtained from the historical data recorded during the two-year period (1986-1988). Examination of the records makes it clear that the number of causes of faults is often very small. Doubts may thus be cast on the significance of the numerical values thus obtained. This may certainly be true with respect to this study, which has obviously analysed only a part of the data available, themselves not collected for the direct stocking on a reliability data bank. Since such a bank is designed to evaluate components with an undoubtedly very high reliability, the data must obviously be much more numerous and collected over long periods. In this connection, it will be noted that what makes the establishment of a reliability data bank particularly attractive to the gas industry is the possibility of keeping all components of its mains under observation, as oppo-

sed to just a sample.
Bearing in mind all the PRI component fault and breakdown modalities so far identified, a data sheet for use as a form in the establishment of a reliability data bank is attached to this paper. It has been devised according to the general Euredata criteria (6) for use with programmed maintenance data and is still in the process of optimisation. Consideration is, in fact, being given to the use of a different form for the extra-ordinary maintenance records. The presence of different forms would undoubtedly complicate matters. On the other hand, it would allow the clearer distinction and analysis of data linked to these two types of intervention. Not only, but the extreme individuality of gas distribution plants needs to maintein separately the source of informations from the individuals mains so that each of them had its own data bank.

Aknowledgements

The authors would like to thank M. Zeppegno and G. Pozzi, of Italgas, who approved the development of this work. The authors are grateful to R.Didio and P.Rosti (Italgas) for the cooperation and reviewing of the manuscript.
Acknowledgements is made to USL 1-23 Regione Piemonte and Ministero della Pubblica Istruzione for the financial support of this research.

References

1. N.Piccinini, M.Scarrone, E.Malinverni, P.Rosti, "Availability study of a pressure regulating installation on a natural gas distribution mains, EUREDATA Sem. on Case Studies on Availability Assessment, Girona (Sp), 8 April 1987

2. N.Piccinini, M.Scarrone, E.Malinverni, P.Rosti, "Availability of a pressure regulating installation on a city mains, Reliability Eng.Syst.Safety, 20, 303-321 (1988)

3. R.Drusiani and P.L.Vivarelli, "Valutazione dell'affidabilità in sistemi di alimentazione con gruppi di decompressione di 2° salto, 4° Convegno Naz. su Sicurezza dell'uso del Gas, Brescia (Italy), 15-16 November 1984

4. R.Didio, G.Vacirca, " Progetto di gestione tecnica integrata a mezzo di sistemi informatici", 1° Conv.ATIG su Metano Tecnologia e Sicurezza, Taormina 12-13 Maggio 1988

5. S.Contini, "SALP-PC a fault tree analysis package on personal computer user manual", Technical Note N°I.87.165 PER1427 /87 JRC Ispra Establishment, 1987

6. EUREDATA Project Report N°3 "Guide to reliability data collection and managment", S.P./I.05E3.86.20 JRC Ispra Establishment, 1986

TABLE 1 - Faults Recorded in Ordinary Maintenance

Faults	Events (n)	Overall failure rate λ (h^{-1})
Service regulator diaphragm fails	4	6.4 10^{-7}
Service regulator blanking disk fails	21	3.3 10^{-6}
Defective service regulator setting	1	1.6 10^{-7}
Foreign bodies in seat of service regulator	9	1.4 10^{-6}
Poor service regulator seal	4	6.4 10^{-7}
Dirt in service regulator	1	1.6 10^{-7}
Monitor regulator diaphragm fails	*	-
Monitor regulator blanking disk fails	2	3.2 10^{-7}
Defective monitor regulator setting	1	1.6 10^{-7}
Foreign bodies in seat of monitor regulator	*	-
Poor monitor regulator seal	*	-
Dirt in monitor regulator	*	-
Defective monitor regulator calibration	*	-
Defective monitor regulator fast opening valve setting	*	-
Monitor regulator fast opening valve diaphragm fails	*	-
Monitor regulator fast opening valve blanking disk fails	*	-
Pilot diaphragm fails	*	-
Pilot blanking disk fails	1	1.6 10^{-7}
Defective pilot setting	15	2.4 10^{-6}
Poor pilot seal	1	1.6 10^{-7}
Dirt in pilot	1	1.6 10^{-7}
Water in pilot	1	1.6 10^{-7}

* Fault suggested from operators'experience.

Following of **TABLE 1**

Faults	Events (n)	Overall failure rate λ (h^{-1})
Membrane of sensitive member of slam shut valve fails	*	-
Poor slam shut valve seal	22	$3.5\ 10^{-6}$
Dirt in slam shut valve	6	$9.6\ 10^{-7}$
Defective slam shut valve calibration	1	$1.6\ 10^{-7}$
Slam shut valve fails to close	4	$6.4\ 10^{-7}$
Foreign bodies in seat of built-in slam shut valve seal	1	$1.6\ 10^{-7}$
Dirt in built-in slam shut valve	*	-
Defective built-in slam shut valve calibration	1	$1.6\ 10^{-7}$
Built-in slam shut valve fails to close	10	$1.6\ 10^{-6}$
Vent valve diaphragm breaks	1	$1.6\ 10^{-7}$
Poor vent valve seal	4	$6.4\ 10^{-7}$
Dirt in vent valve	*	-
Water in vent valve	4	$6.4\ 10^{-7}$
Poor cut-off valve up-stream seal	*	-
Poor cut-off valve down-stream seal	1	$1.6\ 10^{-7}$
Poor by-pass cut-off valve seal	*	-
By-pass cut-off valve locked in closed position	*	-
Poor by-pass manual regulation valve seal	*	-
By-pass manual regulation valve locked in closed position	*	-
Main filter clogged	124	$2.4\ 10^{-5}$
Water in main filter	1	$1.6\ 10^{-7}$
Hydrocarbons in main filter	23	$3.7\ 10^{-6}$
Pilot filter clogged	*	-
Water in pilot filter	*	-
Hydrocarbons in pilot filter	*	-
Preregulator pilot filter clogged	28	$4.9\ 10^{-6}$
Water in preregulator pilot filter	*	-
Hydrocarbons in preregulator pilot filter	2	$3.2\ 10^{-7}$
Water downstream from assembly	2	$3.2\ 10^{-7}$
Defective setting with fluctuating downstream pressure	15	$2.4\ 10^{-6}$
Pilot pressure switch impulse intake defective	3	$4.8\ 10^{-7}$

* Fault suggested from operators'experience.

TABLE 2 - Primary Events for the TE "Abnormal Downstream Pressure"

Events / Protections	Failure Rate $\lambda(h^{-1})$	Test Interval $\theta(h)$	Repair Rate $\mu(h^{-1})$
E209 Faulty operation of regula- E309 tor pilot in open position	$1.3\ 10^{-6}$	-	$1.2\ 10^{-2}$
E210 Pilots low pressure pulse E310 inlet broken	$4.8\ 10^{-7}$	-	$1.2\ 10^{-2}$
E211 Faulty operation of regula- E311 tor in open position	$4.8\ 10^{-6}$	-	$1.2\ 10^{-2}$
E212 Faulty operation of monitor E312 pilot in open position	$1.3\ 10^{-6}$	-	$1.2\ 10^{-2}$
E213 Faulty operation of monitor E313 in open position	$3.4\ 10^{-7}$	-	$1.2\ 10^{-2}$
E201 High pressure from mains	* $1\ 10^{-7}$	-	-
P202 Faulty operation of regula- P302 tor pilot in open position	$1.3\ 10^{-6}$	$1.68\ 10^{2}$	-
P203 Pilots low pressure pulse P303 inlet broken	$4.8\ 10^{-7}$	$1.68\ 10^{2}$	-
P204 Faulty operation of regula- P304 tor in open position	$4.8\ 10^{-6}$	$1.68\ 10^{2}$	-
P205 Faulty operation of monitor P305 pilot in open position	$1.3\ 10^{-6}$	$1.68\ 10^{2}$	-
P206 Faulty operation of monitor E306 in open position	$3.4\ 10^{-7}$	$1.68\ 10^{2}$	-
P207 P307 Vent valve fails to open	$1.6\ 10^{-7}$	$7.20\ 10^{2}$	-
P208 Flow rate beyond capacity P308 of vent valve	•$2.5\ 10^{-1}$	-	-
P214 P314 Faulty slam shut valve seal	$3.5\ 10^{-6}$	$7.20\ 10^{2}$	-
P215 Slam shut valve fails to P315 close	$6.4\ 10^{-7}$	$7.20\ 10^{2}$	-

• Initial unavailability Q(y); * hypotetical values; θ, μ values calculated from Maintenance Italgas Manuals

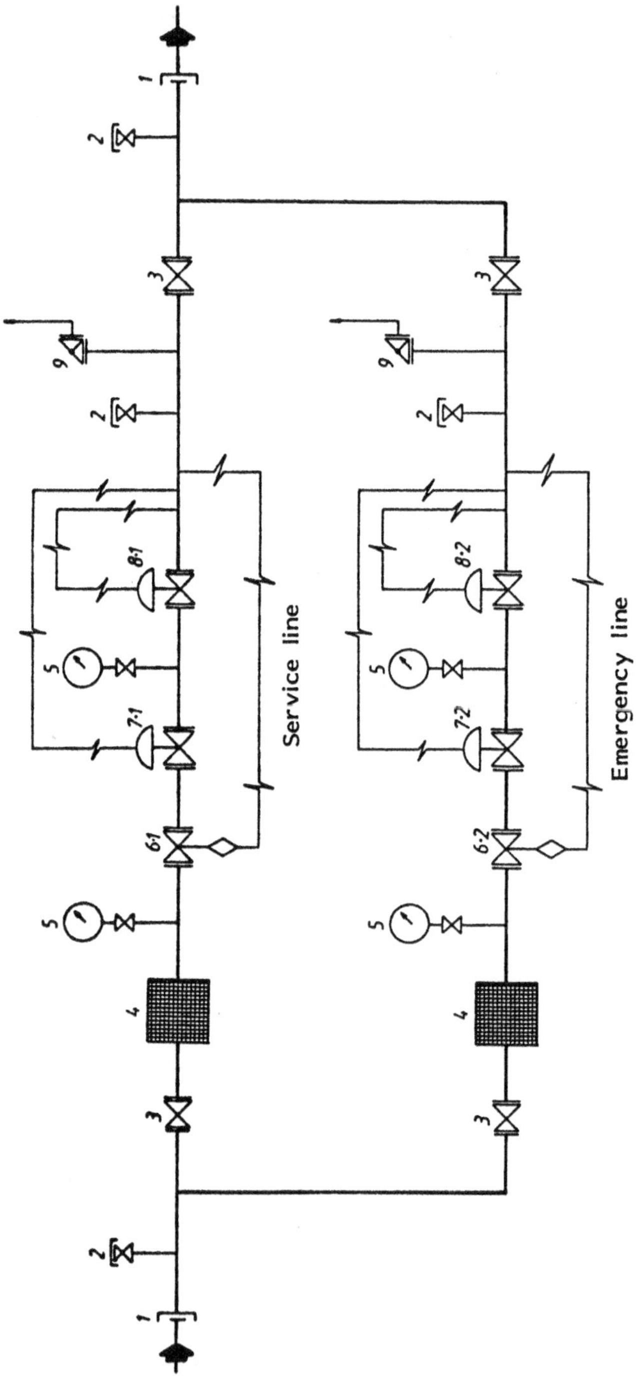

Fig. 1 - Flow Sheet of Unit for Branch Mains

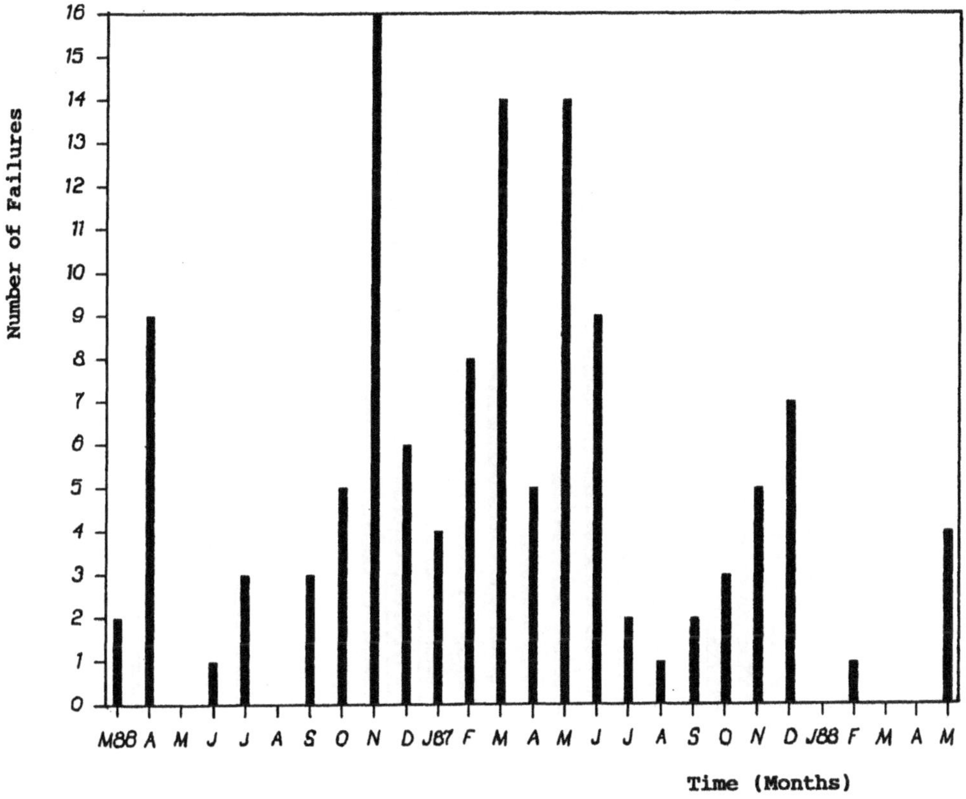

Fig. 2 - Distribution of the Total Number of the Failure "Main Filter Clogged" versus time

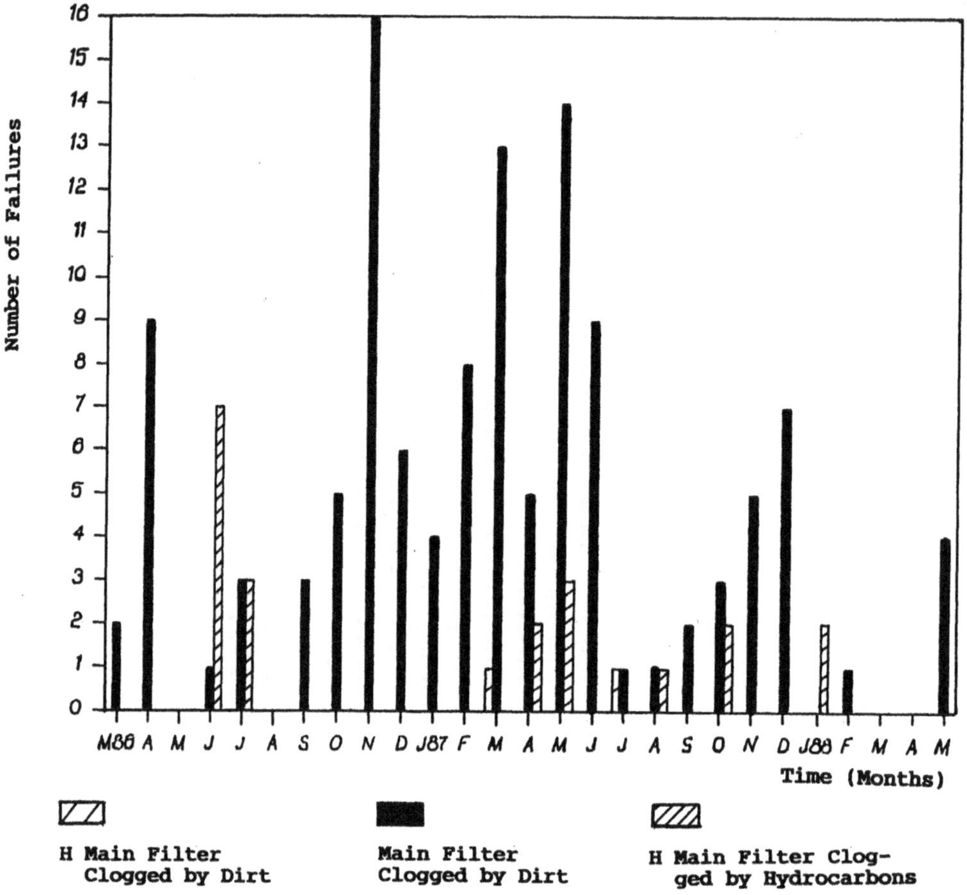

Fig. 3 - Comparison between Mains C, E, F, G and Main H for the Failure "Main Filter Clogged". For Main H the Failure is Divided in: "Main Filter Clogged by Dirt" and "Main Filter Clogged by Hydrocarbons"

Fig. 4 - Fault Tree for
a Branch Mains PRI

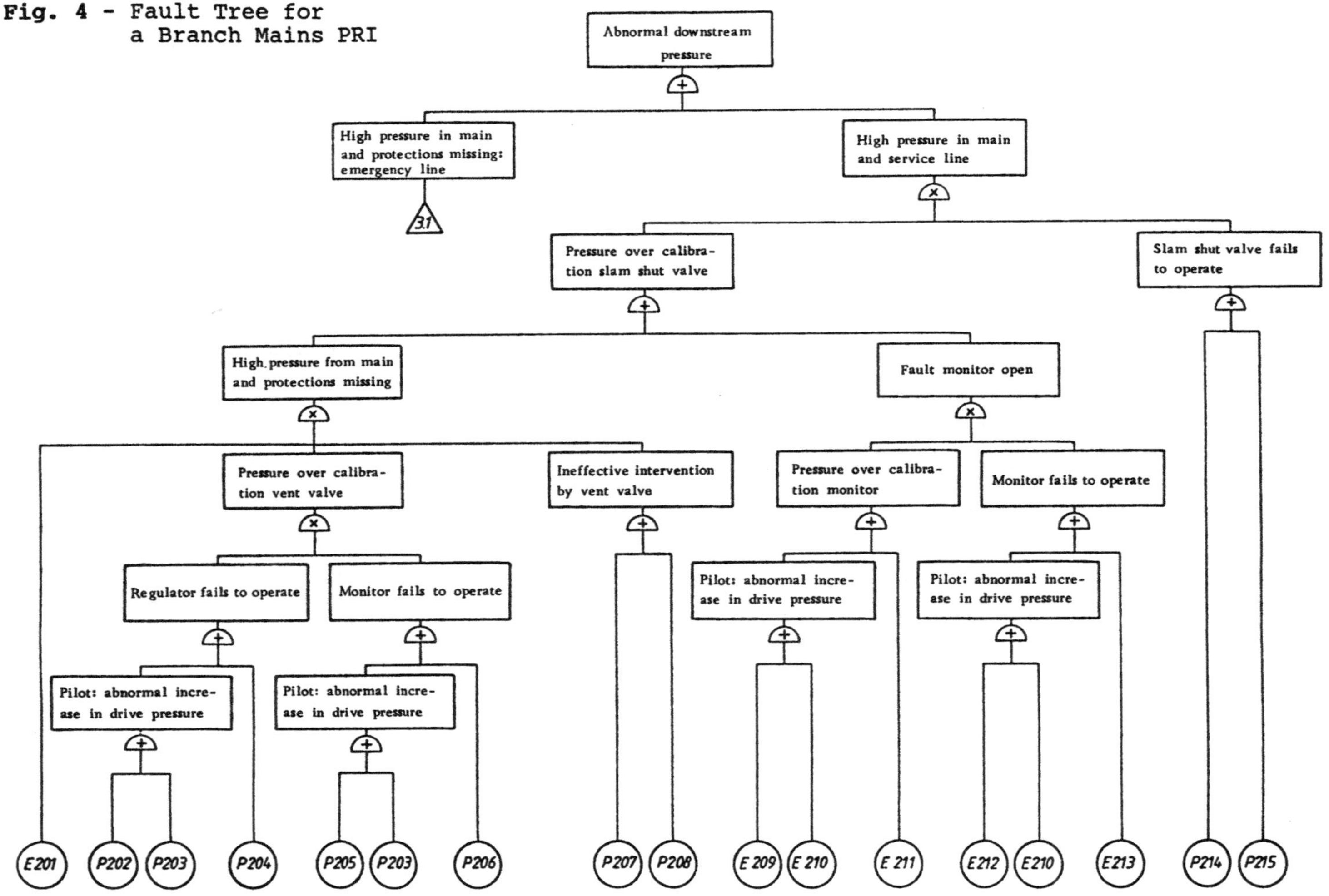

Fig. 5 - Following of the Fault Tree Reported in Fig. 4

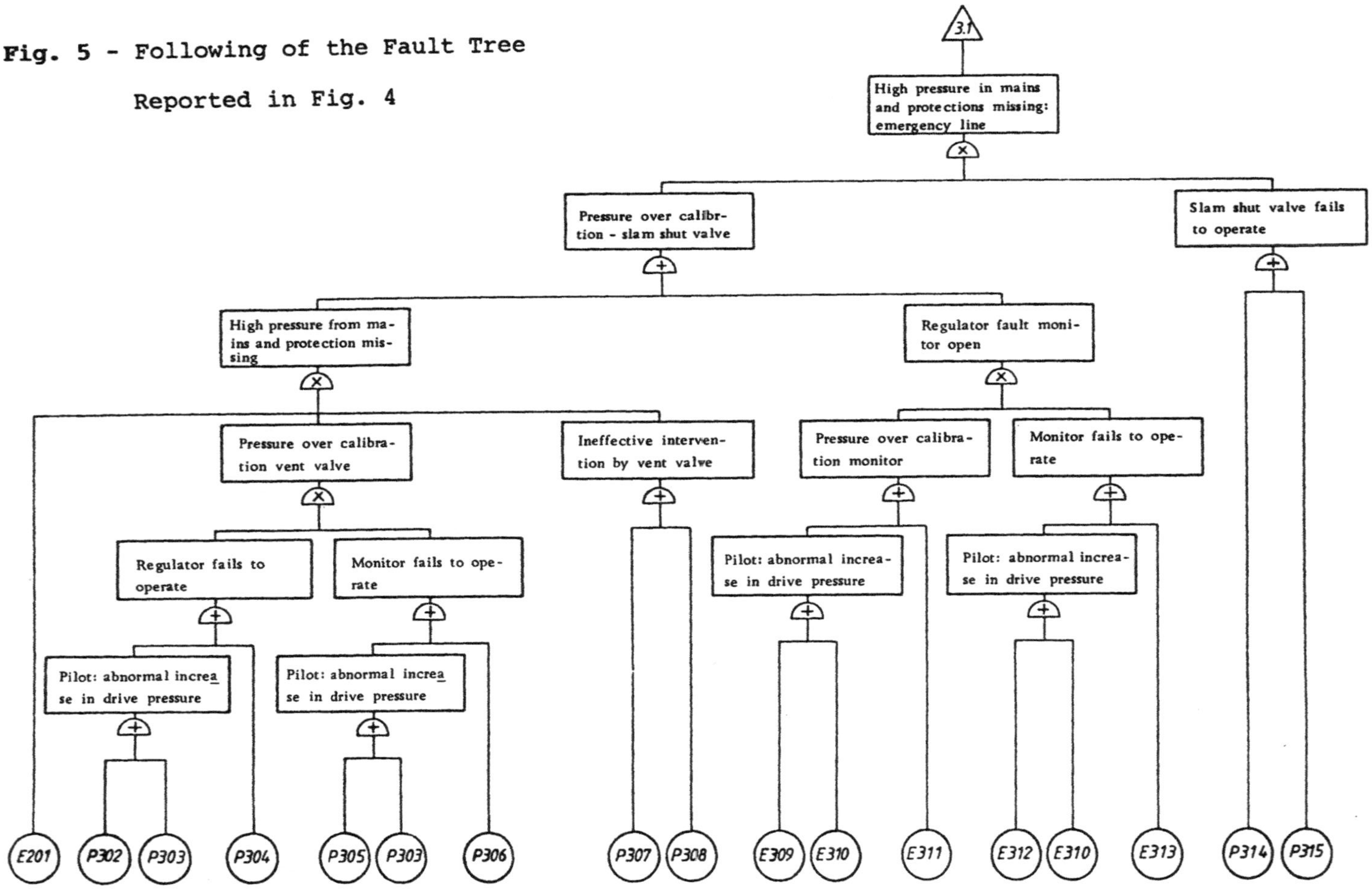

ITALGAS - BANCA DATI AFFIDABILITA'- Modulo per Manutenzione Programmata

Rapporto N°

CODICE COMUNE di ESERCIZIO	CODICE IMPIANTO	DENOMINAZIONE UBICAZIONE	INTERVENTO		
			DATA	TIPO	DURATA
* * *	* * *	* * *	***	***	***

COSTRUTTORE IMPIANTO ....

DATA INTERVENTO ....

ORE EFFETTIVE INTERVENTO ....

COSTO MATERIALE £ ....

DATA ULTIMO CONTROLLO PERIODICO ....

TIPO DI INTERVENTO

[] Sostituz.Comp. Difettoso

[] Riparaz. Comp. Difettoso

[] Regolaz. Comp. Difettoso

[] Altro

DIFETTO RISCONTRATO IN

[] Controllo Periodico A1

[] Controllo Periodico A3

[] Altro

TIPO DI GUASTO

[] Giovinezza Comp.

[] Giovinezza Rete

[] Usura Normale

[] Usura Anormale

[] Casuale

[] Incidente

[] Cattiva Manut.

[] Montaggio Difet.

SEVERITA' DEL GUASTO

[] Nessuna Conseguenza

[] Degradate Prestazioni senza Interruzione Servizio

[] Interruzione Forzata del Servizio

[] ....

....

....

....

*** RIDUTTORE ESERCIZIO

[] 6-3RS Presenza acqua
[] 22-7RS Usura membrana
[] 28-8RS Usura pastiglia
[] 32-9RS Corpi estranei sede otturatore

*** PILOTA RIDUTTORE ESER.

[] 10-4PS Intervento fuori limiti taratura
[] 23-7PS Usura membrana/e
[] 29-8PS Usura pastiglia/e

*** RIDUTTORE MONITOR

[] 36-3RM Presenza acqua
[] 13-5RM Mancato intervento
[] 19-7RM Usura membrana
[] 26-8RM Usura pastiglia
[] 31-9RM Corpi estranei sede otturatore

*** PILOTA RIDUTTORE MONITOR

[] 9-4PM Intervento fuori limiti taratura
[] 21-7PM Usura membrana/e
[] 27-8PM Usura pastiglia/e

*** PRERIDUTTORE PILOTI

[] 18-7PP Usura membrana
[] 27-8PP Usura pastiglia

*** VALV.ACCEL.MONITOR

[] 40-4VA Intervento fuori limiti taratura
[] 46-7VA Usura membrana
[] 50-8VA Usura pastiglia

*** BLOCCO PRESSOSTATICO

[] 7-4PS Intervento fuori limiti taratura
[] 12-5BP Mancato intervento
[] 16-6BP Scarsa tenuta
[] 17-7BP Usura membrana elemento sensibile
[] 48-8BP Usura pastiglia
[] 52-9BP Corpi estranei sede otturatore

*** BLOCCO INCORPORATO

[] 8-4BI Intervento fuori limiti taratura
[] 14-5BI Mancato intervento
[] 20-7BI Usura membrana elemento sensibile
[] 49-8BI Usura pastiglia
[] 53-9BP Corpi estranei sede otturatore

*** VALVOLA DI SFIORO

[] 11-4VS Intervento fuori limiti taratura
[] 15-5VS Mancato intervento
[] 24-7VS Usura membrana
[] 30-8VS Usura pastiglia

*** VALVOLA DI INTERCETTAZIONE

[] 42-6VM Scarsa tenuta a monte
[] 43-6VM Scarsa tenuta a valle

*** VALVOLE LINEA BY-PASS

[] 44- 6VB Scarsa tenuta valv.intercett.
[] 54-10VB Valv.intercettazione bloccata
[] 45- 6VB Scarsa tenuta valv.regolazio.
[] 55-10VB Valv.regolazione bloccata

*** FILTRO PRINCIPALE

[] 2-2FP Intasato
[] 4-3FP Presenza acqua

*** FILTRO PRERID.PILOTI

[] 3-2PP Intasato
[] 5-3PP Presenza acqua

*** FILTRO PILOTA RID.ESERC.

[] 35-2PS Intasato
[] 39-3PS Presenza acqua

*** FILTRO PILOTA RID.MONIT.

[] 33-2PM Intasato
[] 37-3PM Presenza acqua

*** VARIE

[] 1-1A Difettosa regolazione oscillaz. press.regolata

*** ALTRO

....

....

....

....

COMPILATO DA

FIRMA

* Codice prestampato

Guidelines for Process Equipment Reliability Data by CCPS

Gary R. Van Sciver

Rohm and Haas Company

Abstract

The Center for Chemical Process Safety (CCPS) of the American Institute of Chemical Engineers (AIChE) will release a book on reliability data in 1989. The book is titled *Guidelines for Process Equipment Reliability Data* and is part of a series of safety guidelines published by CCPS. The book was written and published by a task force of AIChE members from industry with the help of Science Applications International Corporation.

In the last few years there has been increasing emphasis on risk reduction in the chemical process industry. This has resulted in an increased use of tools such as quantitative risk analysis which require component reliability. This new CCPS publication will partially fulfill that requirement. Also, the book is intended to act as a resource for any future data collection exercises in the chemical process industry.

The book contains generic reliability data from a wide variety of sources. There are descriptions of those sources (and others) for further reference. Also, there is a section explaining how to collect data and how to convert it to the same format as the guidelines data.

Introduction

The Center for Chemical Process Safety (CCPS) of the American Institute of Chemical Engineers (AIChE) will release a book on reliability data in 1989. The book is titled *Guidelines for Process Equipment Reliability Data* and is part of a series of safety guidelines published by CCPS. The book was written and published by a task force of CCPS members from industry with the help of Science Applications International Corporation (SAIC).

The book contains generic reliability data from a wide variety of sources. Also, there are descriptions of those sources (and others) for further reference. Finally, there is a section explaining how to collect data and how to convert it to the same format as the guidelines data.

<u>Background</u>

CCPS was established in 1985 by AIChE to develop and to disseminate technical information for use in the prevention of major chemical accidents. The Center is supported by over 60 industrial sponsors in the chemical process industry (CPI), who provide funding and professional guidance to its technical committees. Since its founding, CCPS has published several books in its guidelines series.

Past CCPS Publications

- *Guidelines for Hazard Evaluation Procedures* addresses methods of identifying, assessing, and reducing hazards.

- *Guidelines for Use of Vapor Cloud Dispersion Models* surveys the current vapor cloud dispersion models, shows how to use them, and discusses their strengths and weaknesses.

- *Guidelines for Safe Storage and Handling of High Toxic Hazard Materials* discusses techniques that are used to minimize releases of high toxic hazard vapors.

- *Guidelines for Vapor Release Mitigation* discusses the techniques that are used to mitigate vapor releases from venting, equipment failure, etc.

Future CCPS Publications

- *Guidelines for Chemical Process Quantitative Risk Analysis* (CPQRA) builds on the earlier books to show the engineer how to make quantitative risk estimates of the identified hazards.

- *Guidelines for Process Equipment Reliability Data* complements CPQRA Guidelines (above) by providing information on equipment reliability needed for reliability analysis and quantitative risk assessment. These two books are being released together.

The book was written and published by a CCPS task force with the help of SAIC. The task force members were:

S. Barry Gibson (Chairman)	DuPont
Harold W Thomas	Air Products
William H. Ciolek	Amoco
Joseph Sweeney	ARCO
Brian D. Berkey	Hercules
Gary R. Van Sciver	Rohm and Haas
William K. Lutz	Union Carbide

Thomas W. Carmody and Lester H. Wittenberg of CCPS were responsible for the overall administration and coordination of the project. Joseph R. Fragola, General Manager, and Erin P. Collins, Staff Scientist of SAIC, provided technical support and reliability data.

Objectives

In the last few years, there has been increasing emphasis on risk reduction in the CPI. Fires, explosions and toxic releases are a growing concern. This has resulted in an increased use of risk analysis techniques which require component reliability data.

When using reliability data in an analysis, the ideal situation is to have historical data from the identical equipment in the same application. However, this "plant-specific" data is usually unavailable and "generic data" must be used. Generic data is compiled from information taken from equipment used in other plants and other processes which share common reliability characteristics with the equipment in question. The use of data from a wider population of equipment substantially increases the confidence bounds of the data.

The purpose of *Guidelines for Equipment Reliability Data* is to provide the practicing engineer and risk analyst with an understanding of the usefulness and limitations of generic reliability data, to provide tables of generic data, to provide a summary of pertinent data sources, and to provide guidance for the collection of plant data.

Contents

Briefly, the contents of the book are as follows:

Chapter 1 - *Introduction*, includes a broad description of the background, scope, limitations and use of the book.

Chapter 2 - *Taxonomy*, describes the process equipment taxonomy (classification scheme), the rationale and process for its development, and the factors considered in its construction.

Chapter 3 - *Data*, includes the data base and a discussion of the selection and treatment of the data. The data is presented in the same order as the equipment listed in the taxonomy.

Chapter 4 - *Data Sources*, summarizes and characterizes data sources available in the CPI. A discussion of the search and selection process used to establish the list of sources and a discussion of the presentation format are included.

Chapter 5 - *Collection of Plant Data*, describes the type and amount of data required to make plant-specific data suitable for inclusion in this book or for internal use. Potential sources of plant data and the treatment of such data to make it conform to the requirements of this generic data base are addressed.

Other sections of the book include a bibliography, taxonomy listing, equipment index, glossary and acronyms.

Taxonomy

The taxonomy is the organization of the equipment failure rate data. The data is grouped by equipment categories. Factors which define these categories include function, size, speed, service, failure mode, etc. The taxonomy is the hierarchy, based on these factors, which is used for organizing the multitude of process equipment failure rates into data cells.

A data cell is the "compartment" in which data is stored. Each data cell houses data for the specifically defined category of equipment. Each data cell has a unique taxonomy number which is determined by its level in the taxonomy. For example, the taxonomy number 3.1.7.3.2.2 specifies the third element of the first level, the first element of the second level, the seventh element of the third level, etc. A boundary figure is also included with each data cell to further define the equipment associated with that cell.

The hierarchical structure of the taxonomy is summarized below:

I. Equipment Description (2 to 8 levels)

II. Service Description
  A. Operating Mode (Alternating, Running or Standby)
  B. Process Severity (Clean, General Industry, Moderately Severe or Severe)

III. Failure Description
  A. Equipment Type (Active or Passive)
  B. Failure Mode (Catastrophic, Degraded or Incipient)

The development of an effective taxonomy was essential, but also very difficult. The taxonomy that has been developed for this book is one of the most useful results of the effort. The first step in creating the taxonomy was to develop a list of important CPI equipment to be included. The equipment was then categorized by generic type into the upper levels of the taxonomy:

1.0 Electrical Equipment
2.0 Instrumentation
3.0 Process Equipment
 3.1 Heat Transfer Devices
 3.2 Piping
 3.3 Rotating Equipment
 3.4 Solids Handling
 3.5 Valves
 3.6 Vessels and Accumulators
 3.7 Miscellaneous
4.0 Protection Systems
5.0 Utilities

Each type of equipment was further reviewed in order to establish the lower levels of the taxonomy. Factors that were determined to be important distinguishers of reliability were used to define these lower levels.

Many of the data cells defined by the taxonomy are not presented in the data base because the specific type of data needed was not available. It is hoped that this book will promote the development of new data to fill these empty data cells. As new data is collected, it is likely to bring about modifications to the present taxonomy structure. The new data should provide better answers to the following questions.

1) What equipment should be included in the taxonomy?
2) What are the most important distinguishers of reliability?

The taxonomy is the foundation for data collection and storage. Although the taxonomy presented is not perfect, it is hoped that it will be used as a starting point for future process equipment failure rate data development.

Data

Chapter 3 contains generic equipment reliability data for some of the CPI equipment types listed in the taxonomy. Data was taken from some of the sources described in Chapter 4. These sources were reviewed to extract equipment reliability data. Data values that were considered to be statistically sound and applicable to the taxonomy level were entered into the data base. When more than one data point was available for a given taxonomy level, the data was statistically combined.

The equipment reliability data is presented in tables. An example data cell table is shown in Figure 1. Data cell tables are presented in the book only where data existed at a particular level in the taxonomy. The data cell tables are organized by taxonomy number.

Data Sources

The selection of data sources was a three-step process:

Step 1 - *Search*, literature searches were conducted both by CCPS and by SAIC. CCPS concentrated on CPI data sources while SAIC concentrated on nuclear power industry data sources. At the same time, a questionnaire was sent out to the CPI. The survey requested information on the data sources used by the companies (including plant-specific data which could be used by CCPS). Also, members of the CCPS compiled lists of data sources with which they were familiar and that they had used for reliability or risk analysis. As a result, an extensive list of data sources was assembled.

The effort to collect data sources is considered by CCPS to be an ongoing project. Users of this book are encouraged to assist in this effort by recommending additional sources which can be used in subsequent editions of the book.

Step 2 - *Selection*, the list of potential data sources was reviewed by CCPS to identify those which might contain original data. Copies of these were then obtained and further reviewed for final selection.

Step 3 - *Summary*, "useful data" was defined as information that was publicly available and statistically sound. It also had to be pertinent for use in reliability or risk analysis. The list of sources that were rejected should help to prevent duplication of data source review when the anticipated future editions of this book are developed. In total, 66 sources were accepted and about 200 sources were rejected.

As the data sources were reviewed, six categories were identified. These categories were used to organize the sources:

1. CPI Equipment Data Bases (primarily data)
2. CPI Equipment Data References (primarily text)
3. CPI Quantitative Risk Assessments
4. Non-CPI Equipment Data Bases
5. Non-CPI Equipment Data References
6. Non-CPI Quantitative Risk Assessments

Collection and Conversion of Data

In most chemical plants, there is a wealth of reliability data residing within the various plant records. Seldom however, is it organized and filed in a fashion to be readily usable in reliability or risk analysis studies. Chapter 5 presents the steps necessary to identify the sources of data normally available as well as the necessary treatment of this data to add it to the failure rate data base. This overview is intended to guide and assist the reader in the development of record keeping systems that will help in the establishment of the data base.

Conclusion

The purpose of this book is to help fulfill a growing need within the CPI for failure rate data to be used in reliability studies and quantitative risk analysis. This edition is our first attempt at meeting that need.

It is recognized that this first attempt is by no means perfect. Additional work is needed in such areas as 1) data scope, 2) data organization and 3) data collection and conversion. It is hoped that the feedback that the CCPS receives from this first edition will help to make future editions more effective.

DATA ON SELECTED PROCESS SYSTEMS AND EQUIPMENT			
Taxonomy No.	1.3.1.1	**Equipment Description**	DIESEL GENERATOR
Operating Mode	STANDBY	**Process Severity**	

Population	**Samples**	**Aggregated time in service (10^6 hrs)**		**No. of Demands**
		Calendar time	**Operational time**	

Failure mode	**Failures (per 10^6 hrs)**			**Failures (per 10^3 demands)**		
	Lower	**Mean**	**Upper**	**Lower**	**Mean**	**Upper**
FAILS TO START				795.0	17600.0	54500.0
FAILS TO RUN	13.8	2250.0	7710.0			

Component Boundary

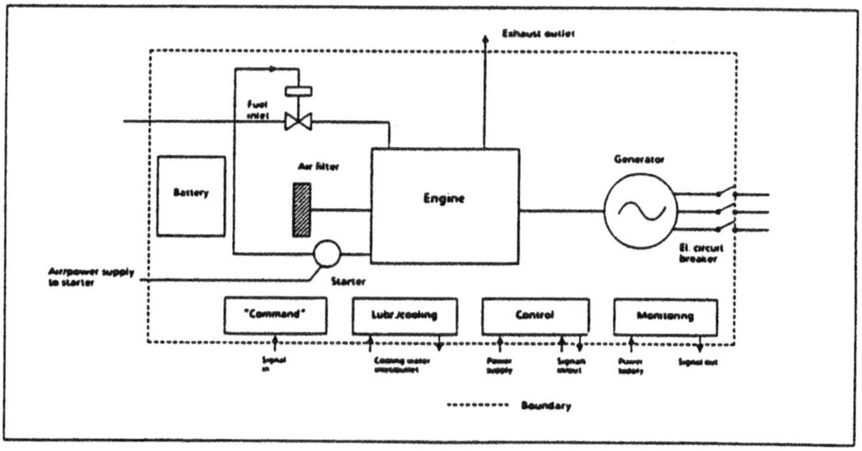

Comment: Based on nuclear power plant emergency diesel service.

Data Reference No. (Table 3.1): 8, 8.1, 8.2, 8.3, 8.8, 8.9, 8.10, 8.11, 8.13, 8.14

Figure 1

Example Data Table

IFP Databanks on Offshore Accidents

A. BERTRAND AND L. ESCOFFIER

Division of Planning and Information
Institut Français du Pétrole
B.P. 311
92506 Rueil-Malmaison Cedex, France

Summary

IFP has created two databanks concerning offshore accidents. The first, called **TANKER**, concerns accidents to carrier vessels having caused an offshore oil spill of at least 500 metric tons. The second, called **PLATFORM**, concerns accidents to drilling vessels or offshore oil platforms (whether used for exploration, development, production or as living quarters) having caused a shutdown of activity for at least 24 hours.

This paper describes how the banks were created. The format chosen presents the data in two forms:

- a series of synthesis data sheets, based on a press brief containing all the articles, reports and documents available on the accident, which clearly describes how each accident occurred and its consequences;
- a computerized databank giving, for each accident, about 30 parameters selected according to criteria making them suitable for use in counting or statistics.

The problems encountered are mentioned, and a number of practical applications are given.

1. Introduction

The Institut Français du Pétrole has compiled two relatively complete and homogeneous databanks to obtain valid counts and statistics on offshore accidents having resulted in a sizeable financial loss. We have given the databanks very specific definitions beyond the general goal of studying both offshore oil pollution and the safety of offshore petroleum activities.

The **TANKER** databank concerns accidents since 1955 to carrier vessels (tankers, freighters, barges, mixed carriers, passenger liners, etc.) having caused an offshore oil spill of at least 500 metric tons of a petroleum product (crude oil, fuel, gasoline, lubricants, naphtha, bunker fuel, etc.). The **PLATFORM** databank concerns accidents since 1955 to all kinds of offshore drilling, production or accommodation units, fixed or mobile (jackups, semi-submersibles, submersibles, drillships, barges, etc.) having caused an unplanned and unforeseen work stoppage of at least 24 hours. Work stoppages for maintenance or normal refitting, for instance after a hurricane, have not been taken into account.

Defining these parameters has the double advantage of making the goal of each databank clear. It also eliminates cases that could easily render the databanks unhomogeneous and unexploitable, such as minor incidents that occur quite frequently but about which little detail is available, or for instance various related accidents that may be severe in terms of loss of life (such as helicopter accidents) but otherwise do not fit our specific criteria.

Although their structure is quite similar, the two databanks have some basic differences. TANKER is specifically oriented toward the study of offshore petroleum pollution, while PLATFORM is oriented toward research on safety (sequence of events, preliminary factors, personnel evacuation, etc.) In addition, all the accidents in the tanker databank have by definition an aggravating factor (i.e. a large oil spill), which is not the case for the PLATFORM databank. This factor makes data gathering for TANKER much more difficult. Only 5% of all naval accidents reported in the technical press have been retained for the TANKER databank, whereas 50% of platform accidents reported have been for PLATFORM. This strict selection is nevertheless often preceded by a long search to ascertain whether the critical criterion of 500 tons of pollutant spilled has in fact been reached.

We will now explain the method chosen for compiling our databanks, and then summarize the problems encountered.

2. Compiling the Databanks

Each of our databanks is in fact made up of an **information** databank providing all the data available on the accident and a **computerized** databank for counting and statistics.

The information databank is a **press book** containing every mention that we have ever found on the accident (computer listings, articles, reports, books, etc.) as well as on the unit involved (registers, etc.). This databank is obviously never completely finished, since new information keeps coming in. The press book plays a dual role. It helps us to decide whether an accident fits the criteria set for our databanks, and if so, it enables us to write a **synthesis data sheet** for the accident. The synthesis data sheet starts with some simple facts (date of the accident, type of unit, construction date, etc.) and then clearly details the sequence of events of the accident as found in the press book. In case of conflicting reports, the interpretation of the accident judged to be the most probable is given.

The computerized databank contains data collected in a strictly-defined language of codes and abbreviations, which reflect only part of the information summarized by the synthesis data sheet. About 30 parameters have been chosen for computerizing, according to the following criteria:

- that the information be generally available, avoiding the danger of creating a databank full of blanks (if some types of accidents, for instance blowouts, call for more detailed or specific data to be collected, it has been found that it is preferable to create a special subsidiary databank just for that kind of accident);

- that the information be useful for statistical purposes;

- that the information be of interest to potential clients.

The main difficulty in compiling a computerized databank is to make it as accurate and detailed as possible without letting it get too large and unwieldy or too full of blanks. For instance, we decided to keep the parameter "hemisphere" as one of the codes for the location of the accident because it enabled us to perform statistical suveys on the effect of seasons. On the other hand, the names of tugboats involved in an accident, although often reported in the press, were deemed to be useless for statistical purposes and so were not incorporated into the computerized databank. This latter kind of information, however, is not lost because it is carefully recorded in the accident synthesis data sheet, which can include the most unexpected kind of facts.

One concrete example would be the accident reported as follows in the press: "Barge XYZ ferried a Soviet army tank out to platform ABC in the Caspian Sea, where it successfully extinguished a fire and gas blowout by demolishing the platform." In this example the computerized databank will only report the approximate location of the accident with two codes (Caspian Sea as geographical zone, U.S.S.R. as country), the name of the production platform, and three key codes to indicate the nature of the accident: gas blowout, fire and collapse of the unit. Since it cannot report either the name of the carrier barge or the fact that the blowout was extinguished by an army tank, the computerized databank obviously yields less information than the information databank.

In another real example, the report reads, "During a fire on tanker LMN, ten crew members jumped overboard and were killed by venomous sea snakes." It is obvious that the computerized databank cannot report this detail without going well beyond its original aim of studying oil slicks due to tanker accidents. Here again the computerized databank proves to be less detailed than the information databank.

A final example is the following sentence from the report of a production platform accident in the Gulf of Mexico: "Unable to approach the scene of a blowout any closer than 60 meters, a sharpshooter succeeded in shooting a hole through a twisted flowline running into the water to see whether oil was flowing through it." This detail is indeed difficult to computerize.

On the other hand the computerized databank is clearly superior for making lists and performing statistics. The correct choice of parameters is thus very important. In addition to an exact description of the sequence of events describing the accident, we have chosen codes that cover: (1) the activity of the unit at the time of the accident, (2) preliminary factors (bad weather, rupture of a towline, etc.) having favored the beginning of the accident, (3) the exact geographic location of the accident (port, lake, river, delta, strait, etc.) and (4) the consequences of the accident, including human casualties or pollution for example.

3. Main Obstacles to the Development of these Databanks

During the creation of our databanks we discovered several sometimes unexpected obstacles that had to be overcome.

3.1. The first is the diversity of sources of information reflecting the wide variety of data required, such as the characteristics of the ships or platforms involved, the circumstances of the accident, or the ecological, economic or other consequences of the accident. A press book necessarily grows in time and may also be made up of books or long reports that are too detailed to be able to incorporate all the elements proposed at once. As such the synthesis data sheet must often be written in several installments, requiring much time and effort, a good memory and the ability to summarize without repetition.

3.2. The second obstacle is a restriction that is at the core of all calculation or chain of reasoning, i.e. the accuracy of a result depends on the accuracy of the data used to obtain it. This is also known as the GIGO principle, or "garbage in, garbage out." In our case the danger is even greater because of the reverence attributed to data processing in general. As such we must be very careful to meticulously check our data before entering them. This entails surprising and often underestimated difficulties.

The truth is that accurate data do not come running into formation like obedient soldiers at roll call. They often act more like recalcitrant children at bath-time, who need to be chased down from all corners of the house. Ships frequently change names (one tanker known as *Mariner* changed names 6 times in 16 years). Others are hard to identify. There are, for instance, 15 tankers named *Showa Maru*, 12 named *Polaris*, and 5 named *Hosei Maru 8*. The names of geographic locations can change according to the atlas used (the Cap des Trois Fourches in Morocco has been found under the name of Cap de Tres Forcas). They are sometimes ambiguous. We have had to deal with seven cities of Portsmouth and ten cities of Freeport. Even the deadweight tonnage of a ship may be ambiguous if its unit of measure is not specified (metric or long ton) or if it is confused with the ship's gross tonnage.

One irritating consequence, which we have not completely eliminated in spite of careful entering of data and repeated checking, is the possibility of entering "doubles," i.e. the same accident entered twice in the bank under different dates or names (especially because salvaged ships and rigs tend to change names immediately after they are repaired, or if they are sold to a new owner).

3.3. A third obstacle is that of fluctuating information. This does not interfere too much when consideration is given to such generally definitive data as the type of unit involved or the year of the unit's construction. But it is very apparent concerning the circumstances and consequences of an accident, which may change for several reasons.

The first is a question of **economics**. For instance, some units that have been declared total losses by the insurance company involved, and that normally should not be heard from again, may suddenly reappear on the market several years later, often with a different name, having been salvaged, repaired or converted by a third party at the last minute. On the other hand, other damaged units, at first judged to be salvageable, and sent to a shipyard for repairs, may eventually be given up as a total loss as the situation evolves.

Another reason that information fluctuates is due to **politics**. We can cite several examples. (1) The total loss of jackup *Pohai 2* in 1979 was at first reported by the Chinese authorities to be the result of a gas blowout heroically fought by crew members. Later, after the fall of the Gang of Four, it was denounced to be the result of a towing accident involving serious negligence. (2) The 1982 accident to the Soviet drillship *Valentin Shashin* in the Barents Sea is still surrounded by mystery. Our synthesis data sheet summarizes the two official versions proposed respectively by Norway and by the U.S.S.R. (3) Some ships suddenly change names for political reasons, such as the Brazilian tanker *Presidente Joao Goulart* which became the *Presidente Campo Salles* shortly after President Goulart's resignation.

The synthesis data sheet must necessarily make an interpretation of this fluctuating information. To fully exploit the information available, we have been led to divide the accidents into three categories according to whether they have definitively been acccepted into the databank, definitively been refused because they do not meet the defining criteria, or are pending until supplementary information has been obtained so that a final decision can be made. The necessity of keeping an ongoing file for pending accidents is a continual constraint.

3.5. There is a host of minor policy problems that need to be solved along the way. For example:

- Should we include in our databank those accidents that are due to acts of war or deliberate scuttling?

- How do we assign a name to anonymous barges? Do we use the name of the tugboat, or that of the operator, or do we simply mention the barge's flag and the year of the accident? The same question arises for production platforms to which we generally assign the name of the block where they are located.

- How do we define a tanker that is carrying coal, or a freighter carrying barrels of asphalt? How do we classify a mobile drilling rig (jackup, semi-submersible, etc.) when it is being used for production activities?

- How do we calculate the number of names possessed by a ship which has sometimes returned to a name she had previously?

- How can we define a production platform with a concrete base and a steel superstructure? Must we create a new category called "fixed-mixed" (steel/concrete) or give preference to one construction material over the other?

. Is it right to take into account accidents to fixed platforms that are being constructed on site whereas we have decided to eliminate those to all rigs being built in naval shipyards? And, if not, can we justify taking into account the frequent accidents to mobile rigs jacking down onto a new drilling location?

It would be easy to find a solution to each of these minor problems individually, but the fact that there are so many of them makes it a very time-consuming task. They call for a certain amount of creative thinking which has led us to "shade" our databanks by defining:

. several different kinds of collisions and blowouts;

. several different dangerous conditions of the seafloor (fragile, slippery, sloping, etc.);

. the concept of "type" of accident as being the most characteristic element in a chain of events;

. the concept of "activity at the time of the accident" (we eliminate cases of long-range inactivity, but keep cases where the unit is inactive pending the onset of a hurricane, or while waiting to be moved to a new work location).

3.4. In order to fully exploit any databank, very detailed statistical files must be kept for reference purposes. For the TANKER databank, for instance, we need to know the entire world fleet by flag, by deadweight tonnage, by age, etc., and the movement of maritime traffic along major shipping lanes. It is not enough to simply know the percentage of Liberian tankers having had accidents. To fully exploit this statistic, we must also be able to compare it to the percentage of Liberian tankers in the entire world fleet. Likewise, to fully exploit the percentage of jackup rigs having had accidents, we need to be able to compare it to the percentage of jackups in the entire fleet of mobile offshore drilling platforms.

These statistical files make up a reference databank that is often long and complicated to put together.

4. Statistics from the Offshore Accidents Databank

The quality of the results obtained from these databanks on offshore accidents obviously depends on the accuracy and the amount of data entered. We must accept the fact that this type of databank is, on all levels, a superb "fuzzy set," and take this into account.

For example, in many cases data on weather conditions at the time of the accident are reported categorically in terms of 1 - 2 - or 3 (good weather, bad weather, storm) even though weather conditions themselves often change during the course of any significant accident. Similarly, trying to apply a 12-degree Beaufort scale to describe sea conditions soon proves to be impossible simply because the scale itself is much too precise.

Yet the 1 - 2 - 3 rule (minor accident - major accident - catastrophe) is often found in our databanks, yielding statistical results that are often generalized or reflect all too obvious facts. Nevertheless, these results have the undeniable advantage of enabling us to assign a figure to obvious generalities, even if it is sometimes approximate.

For instance, to say that the Brittany coast or the Strait of Malacca is prone to oil slicks due to ships running aground is one thing. To be able to add that the two regions have respectively suffered 35% and 10% of all oil pollution due to tanker grounding accidents is obviously an invaluable asset.

Similarly, to be able to determine the approximate age at which a tanker becomes a high accident risk (20-23 years) is obviously of great interest to safety researchers, as is the fact that lightning plays a role in 2 out of 1000 tanker accidents.

The examples that we give to conclude this paper are:

- for the PLATFORM databank:

  . the percentage of accidents to mobile platforms and to jackups according to type of accident;
  . the percentage by type of accident for fixed platforms;
  . the percentage of accidents by type of mobile platform;

- for the TANKER databank:

  . the causes of accidents to loaded and empty tankers (note the preponderance of explosions in the latter case);
  . the frequency of accidents as a function of flag;
  . classification by flag of the number of accidents and amount of oil spilled considering the dwt of the petroleum fleets (note the changes in classification as soon as we take into consideration the relative size of each flag's fleet);
  . the frequency of accidents according to age group for small tankers (dwt less than 30,000).

In the last case we note a critical age group in which tankers are more prone to accidents.

Table 1. Percentage of accidents to mobile platforms and to jackups by type of accident.

type of accident	mobile platforms (%)	jackups (%)
blowout	24.4	19.6
leg and mat damage	12.5	26.8
weather damage	8.5	8.4
capsize	8.3	7.6
collision	7.1	5.6
drift	6.9	4.4
leak	4.9	4.8
fire	5.6	4.4
structural damage	3.8	4.0
cracks	2.9	3.2
grounding	2.9	2.8

Table 2. Percentage by type of accident for fixed platforms.

type of accident	fixed platform (%)
blowout	36.8
fire	19.2
collision	10.0
collapse	6.8
explosion	4.8
act of war	4.8
weather damage	4.0
collapse	3.6

Table 3. Percentage of accidents by type of mobile platform.

type of platform	% accident
jackup	54.7
semi-submersible	22.0
drillship	12.6
submersible	10.7

Table 4. Percentage according to the primary accident cause of the number of accidents (N) and the amount of oil spilled (Q) for loaded tankers and unloaded tankers

Type of accident	loaded tankers		unloaded tankers	
	N(%)	Q(%)	N(%)	Q(%)
Grounding	37.4	31.5	28.7	17.9
Collision	27.1	25.6	12.9	16.3
Structural damage	9.4	14.0	4.9	3.7
Explosion	7.6	11.0	39.6	52.6
Leak	5.8	8.4	3.0	7.1
Fire	5.3	4.8	7.9	5.3

Table 5. Percentage of tanker accidents for the leading flags

Flag	% of accidents
Liberia	28.2
Greece	15.4
United Kingdom	6.9
United States	6.7
Norway	6.1
Panama	5.1
Japan	3.6
Italy	3.4
Cyprus	2.5
France	2.5

Table 6. Classification by flag of the number of accidents (N) and amount of oil spilled (Q) considering the dwt of the petroleum fleets

Flag	N/TPL (%)	Q/TPL (%)
Cyprus	27.8	20.2
Greece	3.8	4.2
South Korea	3.7	9.5
Kuwait	2.8	0.8
Argentina	2.6	2.3
Singapore	1.5	0.5
Panama	1.4	1.2
Liberia	1.1	1.3
Italy	1.1	0.6
U.S.A.	1.0	0.6
Sweden	0.7	0.2
F.R.G.	0.6	0.4
United Kingdom	0.6	0.4
U.S.S.R.	0.5	0.4
Norway	0.5	0.3
France	0.5	0.3
Japan	0.4	0.4

Table 7. Frequency of accidents by small tankers (DWT <30,000 t) by age group, showing a critical age level

Age (in years)	N(%)/TPL(%)
0-3	0.7
4-7	0.8
8-11	0.9
12-15	2.2
16-19	6.3
20-23	8.2
24-27	4.5
>27	8.0

OCAAR: A Tool for Incident Analysis, Reporting and Processing

A. LEROY - TOTAL-CFP

M. GABORIAUD - SNEA(P)

J.C. MICHON - CEP-SYSTEMES

Summary

This paper describes a dedicated software tool to analyse incidents, to statistically process them and to edit reports. This tool, named OCAAR, is made up of two programs : one for each local unit (e.g. a plant), and one for the headquarters (to combine the information coming from each local unit). The incident analysis is based on two methods : the cause tree method and the incident chain method. They allow the incident causes to be analysed, and the succession of the disruptive factors leading to the incident to be identified.

INTRODUCTION

Over the part years, use of probabilistic assessments has steadily grown up. New methods were designed and new computer programs were prepared. Several data collections were performed, and data banks set up. However, few information is available on incidents involving operators in the petrochemical field. In order to efficiently decrease the likelihood of such incidents, ELF AQUITAINE and TOTAL-CFP initiated in 1985 a research program (Ref 1) aiming at finding methods to properly analyse incidents, and at drawing up associated computer code. The study was mainly performed by CEP-SYSTEMES over the 85-88 period, ending with the delivery of the OCAAR software. OCAAR is the accronym of OCcupational Accidents Analysis and Reporting, as its first aim was to aid the study of occupational accidents (so as to prevent them to occur again). This software is already implemented in TOTAL-CFP subsidiaries.

GENERAL DESCRIPTION

The overall structure of the OCAAR software is as shown on figure 1. This figure shows that OCAAR is made up of two computer programs :

- OCAAR/local implementation, which is the core of the OCAAR structure,
- and OCAAR/headquarters, which processes the information coming from each local unit.

As explained below, incidents are analysed at local unit level, i.e. where they occurred. Local data processing can be done, also, for local use.

There is no incident analysis performed at headquarters level, as this information **must** be provided by each local unit.

OCAAR works on "IBM-clone" (IBM-PC or IBM-PS) microcomputers, with the following specifications :

- RAM memory : 512 kbytes minimum,
- operating system : MS-DOS 2.10 or later,
- Hard disk : 10 Megabytes minimum,
- Screen : monochrome, CGA or EGA,
- Printer : "IBM-Graphics" like, or HP thinkjet.

OCAAR/LOCAL IMPLEMENTATION

The organization of OCAAR/local implementation rests on (Ref.2) :

- accident forms, each form being automatically numbered while created,
- cause tree (CT) and incident chain (IC).

Codes are extensively used to record data on accident forms. Furthermore, it is the only way to perform a statistical processing at local unit level, or at headquarters level.

Codes are defined by the headquarters for all local units, e.g. an eye injury has the same code number for OCAAR in the U.K., and in Indonesia.

The OCAAR/local implementation is adapted to the unique characteristics of each local unit. When installing OCAAR in a local unit, the sub-units making up the local unit are entered within the program and coded. A local unit is, for example, "Subsidiary in the U.K.", sub-units can be "Gas field A" and "Oil field B". This allows to perform a data processing for each sub-unit.

The software is made up of four main modules, as described below.

Entry module

The main menu allows to make a choice between five options :

- "create/modify accident form" : see figure 2,
- "modify tree or chain" : cause trees and incident chains are created only while data are input in the accident form,
- "three-month information entry" : these data (i.e. number of people and number of worked hours) are necessary to perform statistic calculations (e.g. frequency of occurrence),
- "redescription of the site" : to define a new sub-unit,
- "modification of victim identity" : self-explanatory.

More information on the use of cause trees and incident chains are given below.

Process module

This module offers four options :

- "display of accident forms",
- "ruled listings", i.e. printing of results which are to be produced according to local regulations, or to Company policy,
- "nominal analysis", which is the main qualitative analysis tool of OCAAR : any selection on accident forms or victim characteristics can be made,
- "statistical analysis" : a statistical data processing can be performed from any accident selection.

An example of ruled listing is shown in figure 3, and an example of statistical analysis in figure 4.

Tools module

It offers three options :

- "password modification", as passwords are used within OCAAR to protect confidential information and against any non-intentional error,
- "accident form cancellation",
- "files modification/check" : to modify code files, if necessary. This can be done upon approval of the headquarters, only.

Transfer module

As said above, each local unit is to send data to the headquarters. This module transfers the local unit data onto a floppy disk (5"1/4 or 3"1/2). Description of the local unit is automatically transferred, onto the floppy disk.

OCAAR/HEADQUARTERS

The aims of OCAAR/headquarters are :

- to allow to update (and to edit) code files for all local units,
- to compile data coming for all local units,
- and to perform statistical data processing, according to own headquarters needs.

This software is a small one, as the underlying philosophy of OCAAR is to allow "local unit" responsibles to identify where the problems are, and to monitor if the improvements have been useful.

CAUSE TREE AND INCIDENT CHAIN METHODS

These two methods (designed by the "french institute for research on safety", the INRS) are the methods to be used to analyse the incidents. They are an unique tool for accident analysis, and for the definition of safety recommendations.

Cause tree

Usually, a cause tree analysis is made under the graphical form of a tree, starting from the "accident" down to the "causes". Accordingly, it looks like a fault tree. This is the only similarity, as there is no logical gate in a cause tree. In order to keep the size of the software at a reasonable level, the cause tree is input using a special form as shown in figure 5. Each entry line corresponds to a particular type of accident cause and allows five entry fields. The user has to enter a code number (or several code numbers) for each entry field (helps screens are provided upon demand).

Incident chain

The purpose of the incident chain is to describe the conditions of an accident, i.e. how an unexpected event (the "incident") had led to an accident through a chain of disruptive factors. The building of the incident chain allows to take into account the sequence of the events, which is not revealed by the cause tree method. OCAAR allows to input two disruptive factors, only, as shown in figure 6.

CONCLUSION

OCAAR is an unique tool for incident analysis, reporting and processing. Its two main characteristics are :

- the use of the powerfull cause tree and incident chain methods to analyse the incidents,
- a systematic coding of all data input by the user when analysing an accident.

Accordingly, OCAAR is useful at local unit level, and at headquarters level.

By end 1988, OCAAR was already implemented in two subsidiaries of TOTAL-CFP : Indonesia and United Kingdom. The first data should then sent back to the headquarters by end 1989, allowing to check whether the incident analysis is properly made. However, this product has been well accepted, in both cases, by safety people. Additional implementations are planned to be performed in the future in ELF AQUITAINE subsidiaries and in TOTAL-CFP subsidiaries.

REFERENCES

1 - RESEARCH REPORTS ON OCAAR
TOTAL-CFP - SNEA(P)
1985-1986
(Confidential report. In french only)

2 - OCAAR/LOCAL IMPLEMENTATION
USER'S GUIDE
Release 2.0. - April 1988

3 - OCAAR/HEADQUARTERS
USER'S GUIDE
Release 2.0 - October 1988

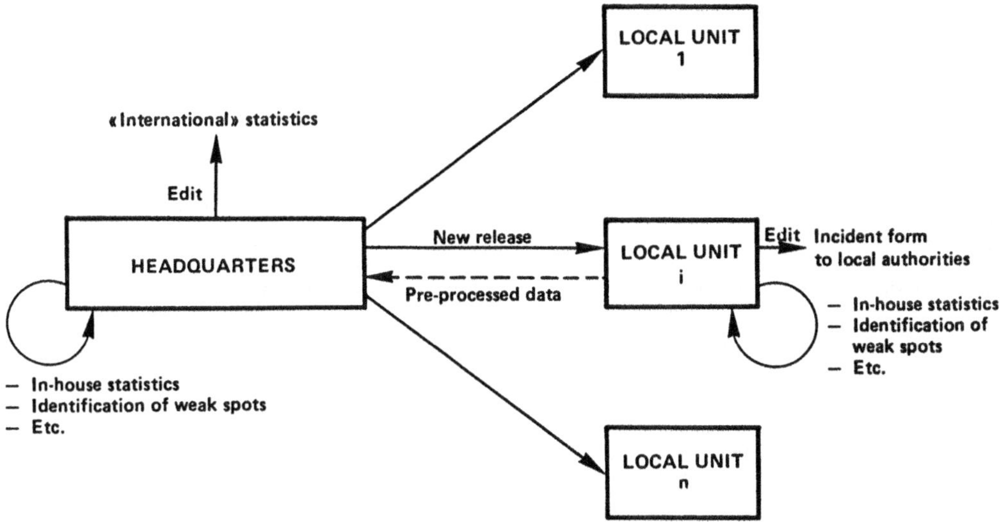

<u>OCAAR OVERALL STRUCTURE</u>

FIGURE 1

OCAAR 2.0 — ACCIDENT FORM ENTRY (Page 2) — AUSTR

Form N# 22 Identification N° |1|11|11|11|111|111|111|

Site of work On this site since

Job title Status

Hour of accident Worked hours before the accident

Activity

"Place" of the accident Running operation

Number of treatments Nature of main treatment

Injury nature Injury notification (y/n)

Existing tree/chain (y/n) Reference of tree/chain

Notes

ACCIDENT FORM

FIGURE 2

```
                          INDUSTRIAL INJURIES

1) Total of injuries for salaries....................................     0

2) Total of injuries declared to authorities/insurance companies.....     0

3) Total of injuries with days lost........ :
                                - Total ..................................     0

                        accident with stop x 1000000
4) Frequency rate    = ---------------------------------------  ....     .00
                             Total of worked hours

                      Total of days lost    x  1000
5) Severity rate    = ------------------------------------------  ...    .000
                            Total of worked days

                           ( Total of worked hours        =     55250 )

6) Total of days off                   :
                            - Total ..................................     0
                            - Injuries of  the present year..........     0
                            - Balance of the previous years..........     0

7) Number of P.D (permanent and temporary) notified during the year      0
                           Total of P.D. notified in the year .......     0

8) Number of fatal injuries    :
                            - Total ..................................     0

9) Number of injuries concerning employees from contractors              0
```

RULED LISTING

FIGURE 3

AUSTR OCAAR 06-05-88

STATISTICS FROM A SELECTION

Accident time :	tot. nbr of accid.	ac.without stop	acc.with stop	acc.with P.P.D.	deaths
fm 0 to 4 am					
fm 4 to 8 am	2 9%		2 16%		
fm 8 to 10 am	1 4%	1 10%			
fm 10 to 12 am	2 9%	1 10%	1 8%		
fm 0 to 2 pm	1 4%		1 8%		
fm 2 to 4 pm	3 13%	1 10%	2 16%		
fm 4 to 6 pm	2 9%	1 10%	1 8%		
fm 6 to 8 pm					
fm 8 to 12 pm	1 4%		1 8%		
*** undefined	10 45%	6 60%	4 33%		
SUMS :	22	10	12	0	0

To continue: <┘

STATISTICAL ANALYSIS FOR A GIVEN SELECTION

FIGURE 4

OCAAR 2.0 — CAUSE TREE N° 22 — AUSTR

HUMAN ORIGIN	* Personnal fault * Disruptive human factor
TECHNICAL ORIGIN	* Material weakness * Unreliability * Design error
PROCEDURAL ORIGIN	* Operating mode defect * Omission or misexecution
OTHER ORIGIN	* Environmental conditions * Potential hazards * Organisation constraints

Ref.

Help : F1 Next zone : ENTER End : END

<u>CAUSE TREE ENTRY FORM</u>

FIGURE 5

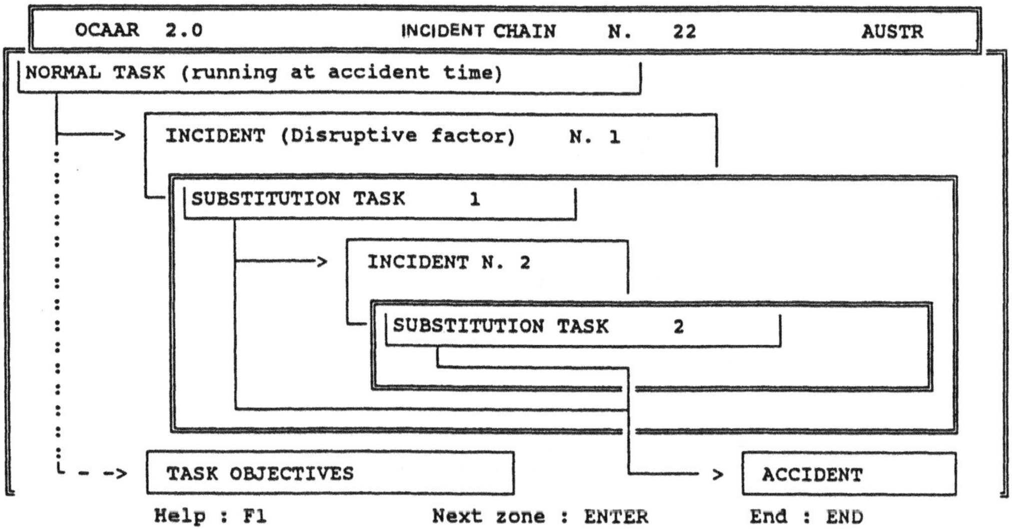

<u>INCIDENT CHAIN ENTRY FORM</u>

FIGURE 6

Common-Cause Failures – Evidence from Sellafield

T R Moss - RM Consultants Limited, Abingdon, UK.
G T Sheppard - British Nuclear Fuels plc, Warrington, UK

SUMMARY

In this study records from the Sellafield nuclear chemical plant were analysed for evidence of common-cause failure. The component-specific Beta factors estimated from these data were significantly lower than the global recommended value of 0.2 and compare favourably with estimates based on large data collected in the USA.

Component-specific Beta factors should be employed in risk assessments to predict common-cause effects. Where such data are not available a conservative global value of 0.1 appears to be appropriate for most chemical plant from the information analysed in this study.

1. INTRODUCTION

In design safety studies account needs to be taken of common-cause failures. A simple procedure using the so-called Beta factor is commonly used. The Beta factor method assumes that common-cause effects can be represented as a small proportion of the failure probability of a single channel in multi-channel redundancy systems. Normally the Beta factor is applied to the failure rate, hence it is easy to apply and avoids some of the restrictions imposed by other methods - for example, the Boundary Value model.

Early studies carried out by Edwards and Watson(1) on relatively large data indicated a range of Beta factors between 0.07 and 0.4. A recommended value of 0.2 was proposed. With such a high value for Beta it can be seen that a limit to the predicted failure probability will frequently be the common-cause contribution.

Experience within British Nuclear Fuels suggested that upper bound values of Beta of 0.1 were more likely and that attention to defences against common-cause failures could probably justify significantly lower values for Beta. A project was therefore initiated to investigate maintenance and operational records at Sellafield for evidence of common-cause failures from which estimates of Beta factor could be derived.

2. THE BETA FACTOR METHOD

Common-cause failures are defined by Edwards and Watson as follows:

> A common-cause failure is the result of an event(s) which, because of dependencies, causes a coincidence of failure states of components in two or more separate channels of a redundancy system leading to the system failing to perform its intended function.

This definition aims to exclude situations which can be predicted from existing failure statistics such as loss of electric supply. Common-cause failures thus defined generally fall within the areas shown in Fig.1, which Edwards and Watson(2), propose as the basis for evaluating defences against common-cause failures. These areas are also used by some analysts to estimate modifying factors which may be applied to the recommended mean value of Beta. It should be noted that modifying factors need to be based on judgement since by definition common-cause failures are relatively rare-events.

The Beta Factor relates the CCF rate of redundant channels to the total failure rate of one channel. Hence, if the total failure rate of one channel is λ and the independent and common-cause failure rates for that channel are λ_i and λ_{cc} then:

$$\lambda = \lambda_i + \lambda_{cc}$$

$$\beta = \frac{\lambda_{cc}}{\lambda} \qquad \text{ie } \lambda_{cc} = \beta\,\lambda \qquad (1)$$

$$\lambda_i = \lambda - \lambda_{cc} = (1-\beta)\,\lambda \qquad (2)$$

EXAMPLE

SIMPLE ACTIVE-PARALLEL REDUNDANT SYSTEM

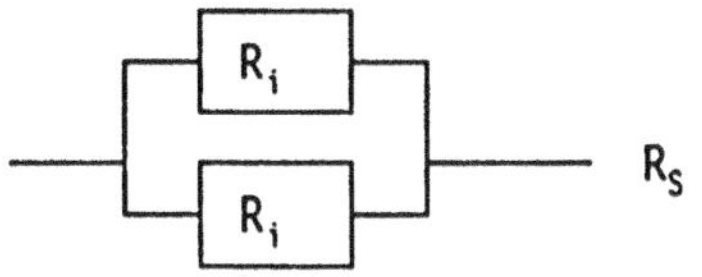

Considering only independent failures the reliability of the system:

$$R_S = 1 - [(1-R_i).(1-R_i)]$$

where R_i = Probability of success with Independent failures

For a period of time t

$$R_S(t) = [1-[(1-\exp[-\lambda_i t]).(1-\exp[-\lambda_i t])]$$

$$= 1-(1-2\exp[-\lambda_i t] + \exp[-2\lambda_i t])$$

$$= 2\exp[-\lambda_i t] - \exp[-2\lambda_i t] \qquad (3)$$

With Common Cause failures included the Reliability Block Diagram is modified as follows:

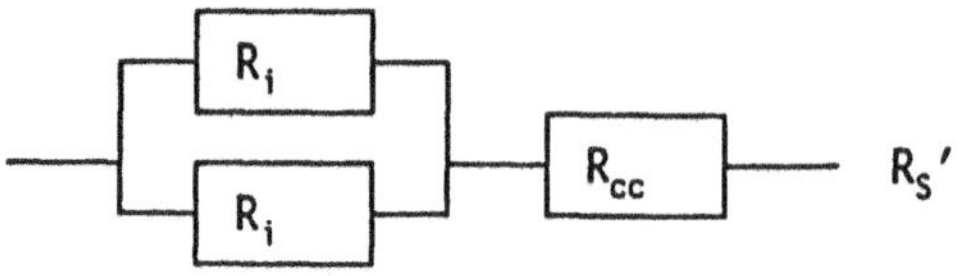

The system reliability is then

$$R_S' = R_S \,.\, R_{cc} \tag{4}$$

= (prob. of survival with independent failures) x (prob. of survival with Common Cause failures)

For a period of time t

$$R_S'(t) = R_S(t) \,.\, R_{cc}(t)$$

$$= (2\exp[-\lambda_i t] - \exp[-2\lambda_i t]).\exp[-\lambda_{cc} t]$$

$$= (2\exp[-(1-\beta)\lambda t] - \exp[-2(1-\beta)\lambda t]) \exp(-\beta \lambda t)$$

$$= 2\exp(-\lambda t) - \exp[-(2-\beta)\lambda t]$$

If ß = 1 (ie <u>ALL</u> Common Cause Failures)

$$R_S'(t) = 2\exp(-\lambda t) - \exp[-(2-1)\lambda t]$$

$$= 2\exp(-\lambda t) - \exp(-\lambda t)$$

$$= \exp(-\lambda t) \quad \text{and, since } \lambda = \lambda_{cc} \text{ in this case}$$

$$= \exp[-\lambda_{cc} t] \tag{6}$$

If $\beta = 0$ (ie NO Common Cause Failures)

$$R_S'(t) = 2\exp(-\lambda t) - \exp[-(2-0)\lambda t]$$

$$= 2\exp(-\lambda t) - \exp(-2\lambda t)$$

and, since $\lambda = \lambda_i$ in this case

$$= 2\exp(-\lambda_i t) - \exp(-2\lambda_i t) \qquad (7)$$

The analysis given above corresponds to the interpretation of Edwards and Watson(1), and is based on the original paper from the General Atomic Company in the USA(3) who are reported to have developed the method.

3. EVIDENCE FROM SELLAFIELD

Sellafield information studied included:

- Process Shift Engineers Weekly and Monthly reports
- Sellafield Safety Newsletters
- Detailed failures studies carried out by the Sellafield Plant Safety Group of electrical supplies, ventilation system equipment and data loggers.

Evidence of common-cause failure was obtained for the following systems:

Electrical distribution systems	- 23 outage events	- 2 ccf

Ventilation Fans	-	213 outage events	- 4 ccf
	=	166 running fan failures	- 2 ccf
	=	47 standby fan failures	- 2 ccf
Data Loggers	-	100 outage events	- 4 ccf
	=	66 processor failures	- 2 ccf
	=	34 drive disk failures	- 2 ccf

Estimates of Beta factors based on Equation 1, ie:

$$\lambda = \lambda_i + \lambda_{cc}$$

$$\beta = \frac{\lambda_{cc}}{\lambda} \qquad \text{ie} \quad \lambda_{cc} = \beta\lambda \tag{8}$$

are as shown below:

Sample No	Description	CCF / Total Failures	Estimated β factor
1.	Electrical distribution systems	$\frac{2}{23}$	= 0.09
2.	Ventilation fans (running failures)	$\frac{2}{166}$	= 0.01
3.	Ventilation fans (standby failures)	$\frac{2}{47}$	= 0.04
4.	Data loggers (processors)	$\frac{2}{66}$	= 0.03
5.	Data loggers (disk drives)	$\frac{2}{34}$	= 0.06

The range is thus 0.01 - 0.09 average 0.03
(cf 0.07 - 0.4 average 0.2 - Edwards and Watson)

The probability of such a large divergence from the recommended mean value of ß can be estimated by employing the χ^2 test.

χ^2 is calculated as the sum of normalised squared differences of the observed and expected frequencies, ie:

$$\chi^2 = \Sigma \frac{(O-E)^2}{E} \tag{8}$$

where O = Observed frequency
E = Expected frequency

The results are compared with published tables of the χ^2 distribution for the appropriate number of degrees of freedom.

For the 5 Sellafield samples the Expected and Observed frequencies are shown in the following table.

Sample No	1	2	3	4	5
Expected CCF	4.6	33.2	9.4	13.2	6.8
Observed CCF	2	2	2	2	2

(Expected CCF = 0.2 x total number of failures in each sample).

Thus: $$\chi^2 = \frac{2.6^2}{4.6} + \frac{31.2^2}{33.2} + \frac{7.4^2}{9.4} + \frac{11.2^2}{13.2} + \frac{4.8^2}{6.8}$$

$$= \quad 1.5 \ + 29.3 \ + 5.8 \ + 9.5 \ + 3.4$$

$$= \quad 49.5$$

The tabulated values for 5%, 1% and 0.1% probability for 4 degrees of freedom are,

Probability	χ^2
5%	9.5
1%	13.3
0.1%	18.5

and if Sample 2 (the largest value) is treated as an outlier then χ^2 = 20.2 and the appropriate tabulated value for 3 degrees of freedom are:

Probability	χ^2
5%	7.8
1%	11.3
0.1%	16.3

Since the calculated values of χ^2 are significantly greater than the tabulated values for 0.1% probability it is evident that there is less than a 1 in 1000 chance that Beta values of 0.2 are representative for Sellafield systems. In fact there is more that a 1 in 20 chance that a Beta factor of 0.1 is too high.

4. COMMON-CAUSE FAILURE MODELS

Attention has been focussed by Atwood(4), on problems of predicting common-cause failures and possible refinements to the simple Beta factor model. His approach (Binomial Failure Rate Model) assumes that a component in a system has a constant independent failure rate and a susceptibility to common-cause shocks which are assumed to hit the system at a constant rate. Vesely(5) and others have extended the BFR model to include common-cause shock which are not lethal to the component. The BFR model is equivalent to the Beta factor model in the case where all common-cause shocks are considered to be lethal.

Martin and Wright(6) have proposed a limiting Beta factor model which takes account of the levels of redundancy and diversity in the systems. A similar approach is evident in the so-called multiple Greek letter model(7) which includes an additional factor in the basic Beta factor expression to account for the level of redundancy.

There must, however, be some reservations on the validity of models employing global Beta factors. Obviously the independent failure rate of a channel will increase as the number of components increases and it may not be unreasonable to assume that the common-cause failure rate will also increase as the number of components increases. To assume that the common cause failure rate increases in direct proportion to the independent failure rate is, however, less obvious unless the parameters have been estimated from event data specifically collected and classified in a manner consistent with the assumptions built into the models. The practice of applying a global Beta factor, estimated from applications as divergent as nuclear reactor and aircraft systems, to chemical plant systems would appear to be questionable. Recent common-cause failure studies conducted for the USAEC(8) are more component-specific, as shown in Figure 2, and are generally nearer to the (component-specific) estimates derived from the small Sellafield data discussed here than the global figure proposed in reference 1.

An alternative approach to predicting common-cause failure probabilities recently proposed by Hughes(9) of CEGB appears to have a number of distinct advantages over parametric models and warrants further study. The Hughes method does not require a specific model since predictions are made directly from event data. The basis of the approach is to represent the variability of a component's failure probability by distributions which can be estimated directly from relatively-small data on multiple failures. The estimation of these distributions would require the systematic collection and analysis of site data.

If site data collection is carried out for common-cause failure prediction it is recommended that the classification system developed for EPRI, for component fault analysis(10) should be adopted. In addition to permitting prediction of common-cause failure probabilities by the Hughes method it would also allow estimation of parameters for the various Beta factor models.

5. CONCLUSIONS

1. Evidence from Sellafield indicates that Beta factors significantly lower than the global value (0.2) employed currently apply on chemical plant. Even an upper bound value of Beta = 0.1 appears to be quite pessimistic. Beta factors estimated from the small Sellafield data compare favourably with experience from the analysis of component-specific failure data in the USA.

2. The use of global Beta factors is considered questionable for chemical plant systems. Alternative approaches based on estimates from component-specific, common-cause data are likely to provide much lower predictions for the failure probability of multi-channel redundancy systems.

3. Collection and analysis of event data from sites using the failure classification system developed for EPRI is considered to be highly desirable. The information collected would provide a better basis for predicting common-cause failure probabilities and facilitate the exchange of common-cause experience with other organisations.

REFERENCES

1. Edwards, G T and Watson I A - A Study of Common-Mode Failures - SRD R146, July 1979.

2. Bourne, A J, et al - Defences Against Common-Mode Failures in Redundancy Systems. SRD R196, January 1981.

3. Fleming, K N - A Reliability model for Common-Mode Failures in Redundant Safety Systems. General Atomic Report GA 13284, December 1974.

4. Atwood, C L - Estimates for the Binomial Failure Rate Common-Cause Model : NUREG/CR-1401, EGG-EA-5112, 1980.

5. Vesely, W E - Estimating Common-Cause Failure Probability in Reliability and Risk Analysis : Marshall Olkin Specialisations, Proc. Int. Conf. on Nuclear Systems Reliability and Risk Assessment. Gattinburg, 1977.

6. Martin, B R and Wright, R I - A Practical Method of Common-Cause Failure Modelling. Proceedings Reliability '87, 14-16 April 1987.

7. Fleming, K N and Kalinowski, A M - An Extension of the Beta Factor Method to Systems with High Levels of Redundancy. Pickard, Lowe and Garrick Inc. PLG-0289, June 1983.

8. Fleming, K N, Moseleh, A, et al. - Classification and Analysis of Reactor Operating Experience Involving Dependent Failures. EPRI NP-3967, June 1985.

9. Hughes, R P - A New Approach to Common-Cause Failure. Reliability Engineering 17 (1987) 211-236.

10. Crellin, G L et al - A Study of Common-Cause Failures. Phase 2: A Comprehensive Classification System for Component Fault analysis. EPRI NP-3837, June 1985.

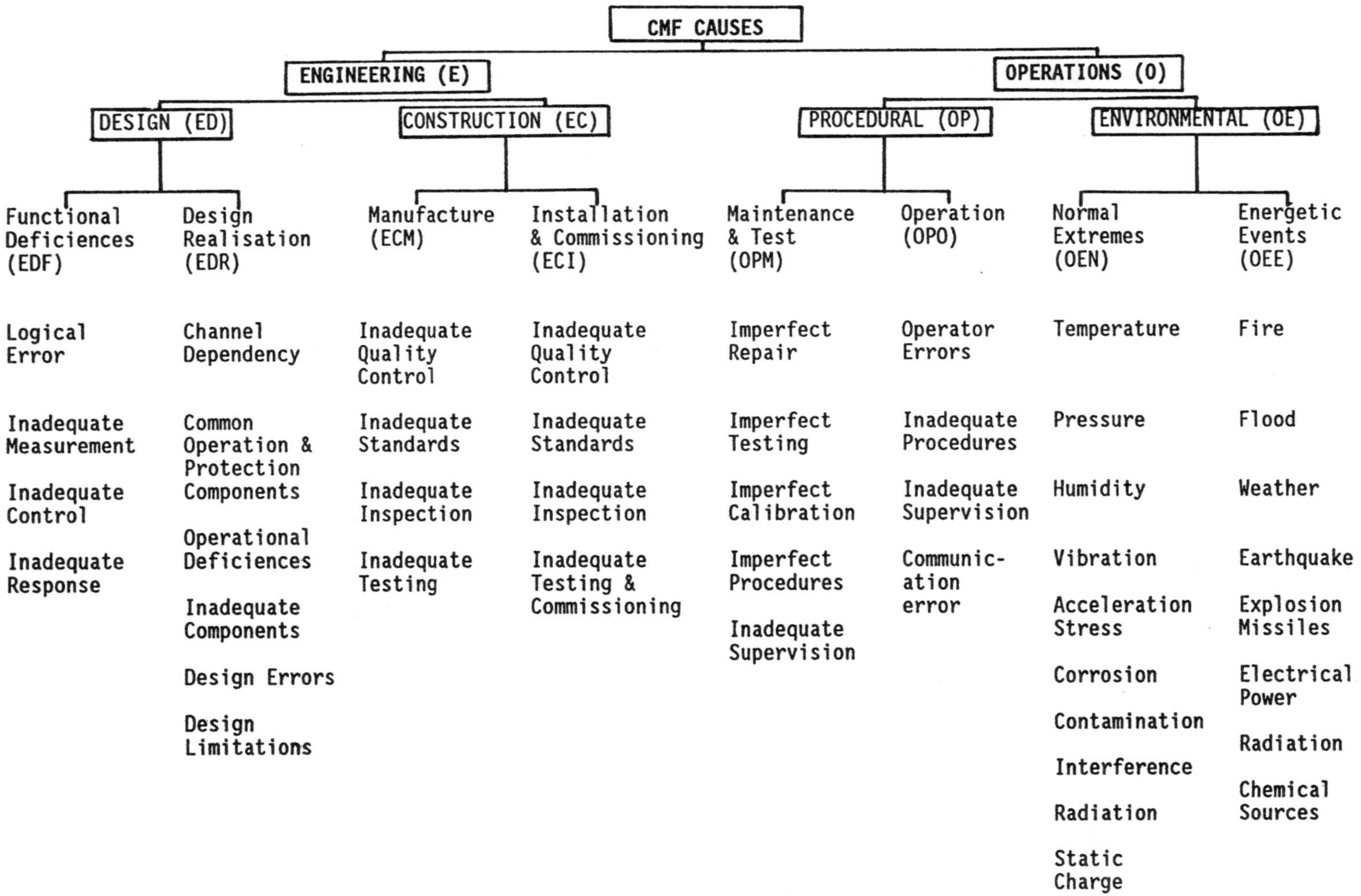

FIGURE 1: CLASSIFICATION OF COMMON CAUSE FAILURES IN REF.1.

Component	Reactor Years	Number of Events Classified	Event Distribution			Estimated Beta Factors
			Independent	Dependent	Generic Common Cause Events	
Reactor Trip Breakers	563	72	56	16	11	0.15
Diesel Generators	394	674	639	35	22	0.03
Motor-Operated Valves	394	947	842	105	42	0.04
Safety/Relief Valves						
PWR	318	54	30	24	0	-
BWR	245	172	136	36	14	0.08
Pumps						
Safety Injection	394	112	77	35	8	0.07
RHR	394	117	67	50	5	0.04
Containment Spray	394	48	32	16	2	0.04
Auxiliary Feedwater	394	255	194	61	5	0.02
Service Water	394	203	159	44	4	0.02
Total	-	2,654	2,232	422	113	0.04

*Events classified include those having one or more actual or potential component failures or functionally unavailable states.

FIGURE 2: ESTIMATES OF BETA FACTORS FOR EVENTS CLASSIFIED IN REFERENCE 8

Stochastic Reliability Analysis – Its Application to Complex Redundancy Systems

PETER DOERRE

SIEMENS AG (KWU), P.O.Box 101063, D-6050 Offenbach, FRG

Summary

Background and formalism of a new approach to reliability analysis of redundancy structures are briefly presented. An extension of the formalism from the component level to the system level is carried out and the general equation for single and multiple failure frequencies in k-redundant symmetric systems with n serial components per subsystem (redundancy) is derived.

1 Introduction

The essence of Stochastic Reliability Analysis (SRA) /1/ can be roughly described by the statement that dependent failure of redundancy structures composed of multiple components of the same type (with respect to failure behaviour) is considered as the basic phenomenon, whereas "independent failure" refers to a special limiting case, namely the perfectly homogeneous population.

The generalized Binomial distribution function is introduced as a stochastic model of dependence to describe the effects of population variability on multiple failure in redundancy structures in a unified way. It is shown that dependence is a consequence of a random process which is constituted by the necessity to select a component (type) from an inhomogeneous population.

It will be shown that this process is usually treated in the deterministic limit which disregards the possibility of different behaviours in favour of an average or generic behaviour. Such a simplified approach yields adequate results only in case of non-redundant or perfectly diverse components. To overcome inadequate results for redundant components, corrective parametric common cause failure (CCF) models are often applied. However, such models introduce extra failure processes instead of disclosing the origin of the inadequacy. The generalized Binomial distribution model creates "globally" dependent probabilities in a natural way without assuming any extra processes: there is no need for further corrections using the CCF picture.

As this approach directly applies only to component-level redundancy, it has to be extended to more complex structures. This paper is therefore organized in two parts. In the first part, formalism and background of the new approach to reliability analysis of redundancy structures is reviewed. The second part is the extension of the formalism to the general case of partial or total failure of k identical redundancies with n components each.

2 The influence of population variability on the failure behaviour of redundant components

One of the most important model assumptions is the "homogeneous population", which here means a population without variability in the failure behaviour. Especially in case of rare knowledge or data this assumption cannot be falsified. However, this is no good reason to assume that it has been verified. The usual situation is characterized by an inhomogeneous population (especially when different data bases have been merged into a "concocted" population) with unknown, or unspecified, systematic variability: such a population can be best described by a failure probability distribution with respect to **objective** influences from the physical world (in contrast to state-of-knowledge distributions /2/, which model subjective uncertainty of numerical values and apply to the case of "physical" homogeneity). "Causes" for the spread may e. g. be different "environments" /3/: components show different failure behaviours, depending on the subpopulation to which they belong, i. e. the failure behaviour is a "local" property of the special environment, not a "global" or generic property of the whole population.

The hypothesis that there could exist **more than one** value for the characteristic parameter of the failure behaviour, e. g. a failure probability per demand, is rarely taken into account adequately, although it is clear that in a design situation, even if there were a precise description of the future system environment of the component, the test/maintenance context, etc., it would be hard to predict the component's actual failure behaviour, as it could sensitively depend on "small" influences which are characteristic of the special design and environment. Even if we have good reasons to assume that any future failure behaviour is already contained in our knowledge of past operational experience, it still remains indetermined **which** of the relevant possibilities will actually be realized.

This is typical for a stochastic process, which is not described by a single "deterministic" value (which may be subjectively uncertain), but by a probability distribution of possible outcomes. Such a process can be understood as being generated by a random choice experiment, in which all possible outcomes have to be accounted for with their proper (probability) weights.

Therefore, two different types of random processes governing component failure have to be discriminated:

- in the first process, the actual failure time is random,
- for the second process, randomness refers to the special failure behaviour, when **more than one** behaviour is possible, i. e. when the population is inhomogeneous.

A physical analogy of these two random processes is the decay of atoms of a chemical element which consists of more than one radioactive isotopes. Here the decay ("failure") time represents the first process, and the different isotopes are discrete realizations of the random variable of the second process. It is clear that the decay rate of an atom (or its decay probability with respect to a given time interval) depends on the chosen isotope.

Whereas randomness in time seems to be correctly modeled by the usual probabilistic approach, the second process is modeled only in a global way "on the average" (by an expectation value, often together with its uncertainty bounds), i. e. its essential feature is neglected. Then, from a global point of view, dependencies in "the" failure process (i. e. in the remaining first process) are recognized. Their effect on redundancy structures is the partial or total **breakdown** of multiplicity, i. e. the loss of independence, or more precise: the loss of the validity of the "independent failure" approach - its results can become meaningless.

Some kinds of dependencies can be excluded from extra modeling (we will usually do that in what follows), as they can be modeled explicitly, e. g. as extra fault tree entries. This holds for:
- common or shared components (e. g. auxiliary systems),
- (excessive) external events or environmental effects,
- failure propagation to adjacent subsystems (if they are not separated by appropriate barriers).

3 Derivation of the generalized Binomial distribution model

The basic observed phenomenon to be explained and modeled is **enhanced** multiple component failure in redundancy systems in comparison to an expected behaviour, based on the "independent failure" approach. Our basic explanation is variability of the failure behaviour, present in an inhomogeneous population with different "environments" /3/. Each such (ideal) environment has zero variability. Different environments do not physically interact. An environment is therefore a subpopulation in which variability has a fixed "value": with respect to failure, this is e. g. the numerical value of the component´s failure rate or probability, i. e. the relative numbers of observed failures per time unit or per demand.

Notation. We assume a population of N components (e. g. pumps, valves), forming k-redundant "systems", which are periodically tested. In N_D simultaneous system test, $n_{r/k}$ failure events (failure of exactly r of k components) have been observed for a special failure mode (e. g. "fails to start", "fails to open"). The corresponding relative frequencies $P_{r/k}$ and $P_{\geq r/k}$ for the failure of at least r of k redundancies are given by

$$P_{r/k} = n_{r/k} / N_D \tag{1}$$

$$P_{\geq r/k} = \sum_{l=r}^{k} P_{l/k} \tag{2}$$

For a perfectly homogeneous population, the frequencies $P_{r/k}$ are calculated by the Binomial formula for "independent" failure,

$$P_{r/k} = k!/(r!\,(k-r)!)\; p^r\,(1-p)^{k-r} = b(r,k|p) \qquad (3)$$

$P_{r/k}$ is the function of only one parameter, namely the "true" failure probability p. The (conditional) probability to find (or to select) a component with a special failure behaviour in this population is always 1, as only one behaviour represents the whole population (this is what homogeneity means): any arbitrarily chosen component behaves like the "average" component. How can eq 3 be generalized to a non-homogeneous population?

A single-point approximation. If we re-define the Binomial distribution with another parameter x,

$$b(r,k|x) = \binom{k}{r}\, x^r\,(1-x)^{k-r} \qquad (4)$$

where x represents the values of a random probability variable X, $P_{r/k}$ can be rewritten as

$$P_{r/k} = b(r,k|p) = \int dx\; \delta(x-p)\; b(r,k|x) \qquad (5)$$

where $\delta(x-p)$ is a special (degenerate) probability density function (PDF), namely Dirac's delta distribution.

A simple inhomogeneous population. The simplest inhomogeneous population consists of two "locally" homogeneous subpopulations, given by the component "types" A and B (e. g. component manufacturer types or one manufacturer type in different plants). Then the relative occurrence frequency for a special failure behaviour is less than 1, as there are now two cases:

- a fraction a_1 of components (subpopulation 1) is characterized by a failure probability p_1 and hence a PDF $f_1(x) = \delta(x-p_1)$,
- a fraction $a_2 = 1-a_1$ of components (subpopulation 2) is characterized by p_2 and hence a PDF $f_2(x) = \delta(x-p_2)$.

The PDF which then describes the variability of the whole population is the linear superposition of the two PDFs,

$$f(x) = \sum_i a_i \, f_i(x) \qquad i = 1, 2 \tag{6}$$

The weights a_i can be interpreted as the probabilities of selecting a component from the i-th subpopulation by random choice from the whole population,

$$a_i = Pr(X=p_i) \tag{7}$$

The "shell" approximation. Any distribution function can be conveniently represented by a weighted sum of (sufficiently many) delta functions ("shells")

$$f(x) = \sum a_i \, \delta(x-p_i) \tag{8}$$

The "population variability parameters" (p_i, a_i) represent the features that a component might be operated by different users or in different environments, or might be a special manufacturer type, of which the failure behaviour in the special environment is not known from operational experience. In this approach these features can no longer be considered as essential limitations to predictability /4/. Usually detailed reliability data from operational experience do not refer to an "average" population, but to more or less specified subpopulations. This confirms the conclusion that the adequate representation which conserves evidence is a discrete distribution, not a continuous distribution created by fitting data to an assumed model distribution.

From eqs 5 and 8 we see that the multiple failure frequencies $P_{r/k}$ can be written as a sum of Binomial distributions,

$$P_{r/k} = \binom{k}{r} \sum_i a_i \, p_i^{\,r} \, (1-p_i)^{k-r} \tag{9}$$

which defines the generalized Binomial distribution model for multiple failure (including single failure and the intact state). We note that according to the interpretation of a_i and the normalization of the Binomial distribution, two sum rules hold:

$$\int f(x) \, dx = \sum_i a_i = 1 \tag{10}$$

$$\sum_r P_{r/k} = 1 \qquad r = 0, \ldots, k \tag{11}$$

It has been shown /1/ that a 3-shell approximation with

$$f(x) = a_0\, \delta(x-1) + a_1\, \delta(x-p_1) + (1-a_0-a_1)\, \delta(x-p_2) \qquad (12)$$

is sufficiently flexible for representing operational experience on multiple failure events. Here the occurrence frequency a_0 is the weight of the highest possible probability $p_0 = 1$ (the "zero reliability" state) and refers to "lethal shock" contributions which fail the whole system of redundant components with certainty. Alternatively, this quantity can be interpreted as the (relative) failure frequency of a shared component, if one wishes to include such a feature in the model. A lethal shock source can always be simulated by some kind of shared component.

<u>Moments</u>. The r-th moment of the PDF is given by the average (here denoted by <...>)

$$m_r = \langle x^r \rangle = \int dx\, x^r\, f(x) = \sum a_i\, p_i^{\,r} \qquad (13)$$

The expectation value for the frequency of an event in which r components failed and k-r components remained intact is given by the "Binomial moment"

$$m_r^{(k)} = P_{r/k} / \binom{k}{r} = \langle x^r\, (1-x)^{k-r} \rangle = \sum a_i\, p_i^{\,r}\, (1-p_i)^{k-r} \qquad (14)$$

The "independent failure" result is obtained as a limiting case by using a "deterministic" single-point approximation for the PDF

$$f(x) = \delta(x-\bar{p}) \quad \text{with} \quad \bar{p} = \sum a_i\, p_i \qquad (15)$$

where $\bar{p} = \langle x \rangle$ is the first moment. Multiple failure frequencies are then given by eq 3 with $p = \bar{p}$.

To conclude, we emphasize that a reliability prediction will have to account for **all possible** (discrete) environments (failure behaviours), not only for an average environment. Seeking "the true" failure probability for e. g. check valves with the requirement that it be valid for all check valves in all potential environments is not very far from gambling.

4 Extension to multi-component redundancies

We now go over to the general case of (single and) multiple failure frequencies of multi-component redundancies. Formally, the failure frequency for exactly r of k subsystems, where each subsystem consists of a series of n different components, can be written in terms of an n-component Binomial moment as

$$P_{r/k} = \binom{k}{r} m_r^{(k)}(1,\ldots,n) \quad (16)$$

with

$$m_r^{(k)}(1,\ldots,n) = \quad (17)$$
$$\langle (1-(1-x(1))\ldots(1-x(n)))^r \; ((1-x(1))\ldots(1-x(n)))^{k-r} \rangle$$

The functional form of eq 17 is determined by the Boolean structure function of an OR gate with n entries,

$$X_{OR}(1,\ldots,n) = OR(X_1,\ldots,X_n) = 1 - (1-X_1)\ldots(1-X_n) \quad (18)$$

The averaging procedure now involves an n-fold integral over the values of all components' random variables x(j), where the PDF of the j-th component is given by

$$f(x(j)) = \sum_{i(j)} a_{i(j)}(j) \; \delta(x(j)-p_{i(j)}(j)) \quad (19)$$

The summation indices are now denoted by i(j) because they may be different for different components. The double component reference by "j" will be dropped subsequently so far as this does not lead to ambiguous expressions.

There are two possible ways for the intended extension:
- decomposition of multi-component into single-component moments,
- determination of a total PDF for one redundancy with n components by defining a new suitable random variable.

First way. This way is presented because it is straightforward and illustrative. However, the result is only a definite upper bound. To proceed we make use of two definite approximations:

$$((1-x(1))\ldots(1-x(n)))^{k-r} \leq 1 \quad (20)$$

$$1 - (1-x(1))\ldots(1-x(n)) \leq x(1) + \ldots + x(n) \quad (21)$$

Then

$$m_r^{(k)}(1,\dots,n) \le \langle (x(1) + \dots + x(n))^r \rangle \tag{22}$$

A multinomial expansion gives

$$m_r^{(k)}(1,\dots,n) \le \sum_{r_1} \dots \sum_{r_n} C(r;r_1,\dots,r_n) \; \langle x(1)^{r_1} \dots x(n)^{r_n} \rangle \tag{23}$$

with the multinomial coefficient

$$C(r;r_1,\dots,r_n) = r!/(r_1! \dots r_n!) \tag{24}$$

and the summation restriction $\sum r_j = r \quad (j = 1, \dots, n)$.
The n-fold integral can be factorized

$$m_r^{(k)}(1,\dots,n) \le \sum_{r_1} \dots \sum_{r_n} C(r;r_1,\dots,r_n) \; m_{r_1}(1) \dots m_{r_n}(n) \tag{25}$$

with

$$m_{r_j}(j) = \langle x(j)^{r_j} \rangle = \sum_{i(j)} a_{i(j)}(j) \; p_{i(j)}(j)^{r_j} \tag{26}$$

where the right-hand side of the last eq is obtained in the generalized Binomial distribution model as a modification of eq 13.

Then

$$\sum_{r_1} \dots \sum_{r_n} C(r;r_1,\dots,r_n) \; \Big(\sum_{i(1)} a_{i(1)} \, p_{i(1)}^{r_1}\Big) \dots \Big(\sum_{i(n)} a_{i(n)} \, p_{i(n)}^{r_n}\Big) \tag{27}$$

$$= \sum_{i(1)} \dots \sum_{i(n)} a_{i(1)} \dots a_{i(n)} \sum_{r_1} \dots \sum_{r_n} C(r;r_1,\dots r_n) \; p_{i(1)}^{r_1} \dots p_{i(n)}^{r_n}$$

This is the multinomial expansion of the definite upper bound; therefore we obtain as a final result

$$m_r^{(k)}(1,\dots,n) \le \sum_{i(1)} \dots \sum_{i(n)} a_{i(1)} \dots a_{i(n)} \; (p_{i(1)} + \dots + p_{i(n)})^r \tag{28}$$

Second way. This efficient way is based on the introduction of a suitable random variable and a total PDF for one subsystem of n different serial components, starting with the PDFs of the n components. We begin with the two-component case (n=2).

Let X(1), X(2) be the random variables (failure probabilities) of the first and second component, respectively. Then

$$X(1) \quad \leftrightarrow \quad f(x(1)) = \sum a_{i(1)} \, \delta(x(1)-p_{i(1)})$$
$$X(2) \quad \leftrightarrow \quad f(x(2)) = \sum a_{i(2)} \, \delta(x(2)-p_{i(2)}) \qquad (29)$$

If we define a new subsystem random variable $Z(1,2) = X(1)+X(2)$, the domain of this variable is given by

$$(p_{i(1)i(2)} = p_{i(1)}+p_{i(2)} \mid \text{all given } i(1),\ i(2)) \qquad (30)$$

It contains i(1) times i(2) combinations (with double indices) with $p_{i(1)}$ for the first and $p_{i(2)}$ for the second component. Then

$$Z \quad \leftrightarrow \quad f(z) = \sum_{i(1)} \sum_{i(2)} a_{i(1)i(2)} \, \delta(z-p_{i(1)i(2)}) \qquad (31)$$

with

$$a_{i(1)i(2)} = a_{i(1)} \, a_{i(2)} \qquad (32)$$

f(z) is again a discrete PDF which fulfills the sum rule (eq 10). With regard to the structure function of an OR gate, the more precise expression for the "local" probabilities reads

$$p_{i(1)i(2)} = p_{i(1)}+p_{i(2)}-p_{i(1)}p_{i(2)} \le p_{i(1)}+p_{i(2)} \qquad (33)$$

The subsystem PDF can then be used to calculate moments as usual. A straightforward generalization to arbitrary n yields

$$m_r(1,\ldots,n) = \langle z^r \rangle = \sum_{i(1)} \ldots \sum_{i(n)} a_{i(1)\ldots i(n)} \, p_{i(1)\ldots i(n)}^{\;r} \qquad (34)$$

with

$$a_{i(1)\ldots i(n)} = a_{i(1)} \cdots a_{i(n)} \qquad (35)$$

and

$$p_{i(1)\ldots i(n)} = 1 - (1-p_{i(1)})\ldots(1-p_{i(n)}) \le p_{i(1)}+\ldots+p_{i(n)} \qquad (36)$$

If we insert the upper bound expression of eq 36 into eq 34, this eq then exactly corresponds to eq 28, which has also been derived using this approximation. However, eq 34 is already more general, and we may now transcend the limiting first approximation (eq 20) to obtain the exact expression for the multi-component Binomial moment as the desired result

$$m_r^{(k)}(1,\ldots,n) = \qquad (37)$$
$$\sum_{i(1)} \ldots \sum_{i(n)} a_{i(1)\ldots i(n)} \, p_{i(1)\ldots i(n)}^{\;r} \, (1 - p_{i(1)\ldots i(n)})^{k-r}$$

The new population variability parameters for the n-component subsystem are given by eqs 35 and 36.

5 Conclusions and recommendations

The concept of population variability and the generalized Binomial distribution model as a unified stochastic model of dependence have been introduced with the intention to replace a vague concept, namely "CCF", and its mathematical descendants, "stochastic" parametric CCF models /2/ which represent two separated artifacts: "independent" and "common cause" failure.

Our basic recommendation for a "best estimate" reliability analysis is to work in a stochastically consistent picture from the beginning, i. e. when data are evaluated, and to conserve full information about population variability throughout the analysis of redundant systems: not to "collapse" the evaluation in the beginning into simple but inadequate single-point approximations, with the forthcoming need to correct their shortcomings by again simple, (easy, credible etc.) CCF models, and so on.

"Common cause" failure is a misleading notation indicating that an extra property on the cause level is responsible for enhanced failure of redundancy structures. In our opinion the ultimate reason for the existence of the CCF concept is the initial neglect or inadequate interpretation of variability. This error in the first step creates model uncertainty which cannot be removed by accuracy in any subsequent corrective step, e. g. the use of sophisticated stochastic models of dependence or of concepts how to handle parameter uncertainty. If one does not know the relevant parameters of a problem, precise knowledge of the exact numerical values of other parameters does not help much.

Methods to improve the treatment of data uncertainty sometimes seem to solve the wrong problem. Objective population variability (inhomogeniety) caused by systematic influences cannot be represented by the subjective uncertainty of a parameter which is obtained by first "averaging out" variability. Therefore, it is clear that the state-of-knowledge correlation approach /2/ does not model population variability. On the other hand, the generalized Binomial distribution model in its present form does not describe subjective uncertainty. However, the latter concept is still necessary to determine the uncertainty bounds of the model parameters (p_i, a_i) for given operational experience.

Not being able to prove inhomogeniety (e. g. in case of rare statistics) is no proof of homogeneity. Homogeneity is an unlikely property (and hence a strong assumption) in a world in which at best the physical principle of weak causality holds (which, as is now well known, sometimes cannot prevent "deterministic chaos").

To sum up, the need to use methods which treat and conserve inhomogeneity is emphasized. This paper is dedicated to the propagation and improvement of a stochastically adequate method.

References

1. Dörre, P.: Basic Aspects of Stochastic Reliability Analysis for Redundancy Systems. Reliab. Engng. & System Safety (to be published)

2. Apostolakis, G.; Moieni, P.: The Foundation of Models of Dependence in Probabilistic Safety Assessment. Reliab. Engng. 18 (1987) 177-195

3. Hughes, R. P.: A New Approach to Common Cause Failure. Reliab. Engng. 17 (1987) 211-236

4. O'Connor, P. D. T.: Practical Reliability Engineering, 2nd edition, John Wiley & Sons 1985

(See also refs. 1 and 2 for further references.)

The Distributed Failure Probability Approach to Dependent Failure Analysis, and Its Application

R P HUGHES

Central Electricity Generating Board
Berkeley Nuclear Laboratories
Berkeley, Gloucestershire GL13 9PB
United Kingdom

Summary

The Distributed Failure Probability (DFP) approach to the problem of dependent failures in systems is presented. The basis of the approach is that the failure probability of a component is a variable. The source of this variability is the change in the "environment" of the component, where the term "environment" is used to mean not only obvious environmental factors such as temperature etc., but also such factors as the quality of maintenance and manufacture. The failure probability is distributed among these various "environments" giving rise to the Distributed Failure Probability method. Within the framework which this method represents, modelling assumptions can be made, based both on engineering judgment and on the data directly. As such, this DFP approach provides a soundly based and scrutable technique by which dependent failures can be quantitatively assessed.

1. Introduction

One of the principal design features used to achieve reliable systems is that of redundancy. If an important function is required of a component within a system, then several such components might be used in parallel in order that the function should still be performed even if one or more of those components should fail. This is sound engineering practice, which is widely followed.

In order to assess the benefits of such redundancy, it is necessary to be able to quantify the reliability of systems which rely on the working of several similar components. Simplistic techniques are based upon the assumption that the components within the system being analysed are independent of one another. This implies that the reliability of a system could be improved to an arbitrary extent by increasing the level of redundancy. This is not achieved in practice, because there are dependencies between components which lead to a level of reliability which cannot be improved upon. In effect, the reliability of the system has a cut-off value as the level of redundancy is increased. This cut-off behaviour is not a characteristic which arises out of standard fault tree methods, unless a cut-off is introduced into the tree as a separate element. It is this cut-off approach to dependent failures in systems which the CEGB has used for some years. The value of this cut-off cannot be determined by standard fault tree methods and is based upon engineering judgment. The CEGB, therefore, wishing to improve on the techniques available to it for reliability studies, identified a requirement for a method of treating dependent failures between redundant components which does allow them to be included in fault tree analysis in a consistent way, and which can properly evaluate their effects on system reliability.

Since the recognition of the significance of dependent failures for redundant systems, many approaches have been put forward with a view to evaluating their effects. The choice available is wide, and is not a benefit to the analyst, but only adds to his problems - he is faced with the question of which method to chose and why. In practice, his choice will be motivated by factors such as familiarity, data availability, ease of application etc.. Having made his choice, it will be difficult to defend over the alternatives, because the interrelationship between all the different approaches is not generally understood. Indeed, it is precisely this lack of a coherent framework in which modelling can be considered which has led to the existing diversity of models.

To meet the need for a coherent framework, the Distributed Failure Probability (DFP) approach has been developed by the CEGB. In this new approach to dependent failure analysis certain other methods can be seen as special cases, and other, more appropriate modelling assumptions can be justified directly from the data or using engineering judgment. The details of this approach have already been published {1,2}. The intention here is to outline the basic concepts, and illustrate its application.

2. Basis of the DFP Approach

In standard fault tree analysis, the basic elements are generally the components which comprise the system being analysed. The structure of the system is expressed in terms of a fault tree which is solved to produce minimal cut-sets of these components, and the reliability of the system is estimated by evaluating these cut-sets in terms of the individual failure probabilities of the components. As discussed in the introduction, this process does not account for the presence of dependent failures among the components.

The underlying reason for this, as we shall see below, is the fact that the values which we use for the failure probabilities of the components are average values. This averaging has been performed over two sources of variability, which can be illustrated as follows. Suppose we are interested in a particular type of component, a pump say. This type of pump may be manufactured by several manufacturers. One manufacturer may produce pumps which are "more reliable" than most others because of the quality of his machines, the detail of his design etc.. It is natural, therefore, to expect a variability in pump performance depending on manufacturer. Superimposed on this kind of variability will be the usual random differences in pump reliability arising out of the random variations in the manufacturing process. The difference between these two types of variation in reliability is that if several pumps are chosen from the same manufacturer, their reliabilities will be correlated as far as the first kind of variation is concerned, but uncorrelated as far as the random variations of the manufacturing process are concerned.

It is with this correlated type of variation that we are concerned here, because it has a marked effect on the reliability characteristics of systems of several similar components. As well as the example given above of the influence of a common manufacturer, the reliability of a component can be influenced by environmental factors such as temperature and humidity. It can also be affected by the quality of maintenance. We shall refer to all these correlated variabilities as due to the "environment" of the component, where "environment" means all those factors which influence the failure probability of the component as discussed above. In other words, the "environment" is identified with the root cause of failure. Some of these "environmental" factors will change with time, and in doing so will change the failure probability of the component. If this happens, we say that the "environment" is now different. Clearly, if one of the factors has no effect on the component, then any change in it is irrelevant and does not change the "environment". Thus the failure probability of a component varies with time as its "environment" varies, but within a given "environment" the failure probability is fixed.

Let us now establish our notation. We shall consider a group of components which are susceptible to dependent failures, and call it a multiple failure group (mfg). (We can think of an mfg as being a group of identical components, pumps say, in order to fix the ideas, but this is not generally necessary.) The label A will be associated with this mfg, so that A represents any single component in the group. Several components from the mfg will be represented by A with a superscript or subscript as described below. If we denote the failure probability of each component in the group A within a particular "environment" E_j by $p(A|E_j)$ and the probability that the component will be in that particular "environment" by $p(E_j)$, then the average failure probability is given by

$$p(A) = \sum_j p(A|E_j) \times p(E_j) \tag{1}$$

with the sum running over all possible different "environments". Equation (1) shows clearly how the component failure probability is distributed over various possible values associated with each of the various "environments".

If we now consider several such components, n say, together in the same "environment", then the probability that they all fail together is given by $p(A|E_j)^n$, so that averaging over all "environments" gives the total failure probability of the n components together as

$$p(A^n) = \sum_j p(A|E_j)^n \times p(E_j) \tag{2}$$

This is because, given a particular "environment" E_j in which they all exist together, the components are conditionally independent, and their failure probabilities are each $p(A|E_j)$. The components fail independently of one another (we are not considering cascade failure here for simplicity, although they could be included very straightforwardly) but the failure probabilities of each are directly correlated by the common "environment" in which they find themselves.

Generally, $p(A)^n \neq p(A^n)$, since in the first expression, the averaging over "environments" is performed before raising to the power n. The correct value for the failure of several similar components together is given by equation (2). Both equations (1) and (2) make plain the fact that the failure probability of a component is a variable, varying from "environment" to "environment", and that the proper distribution of failure probability over these "environments" is essential in order to evaluate correctly the failure probability of several components together. This key concept forms the basis of the Distributed Failure Probability (DFP) approach.

3. Dependencies Between Components

When illustrating how the DFP approach can be applied to treat dependent failures within a multiple failure group of components as in equation (2), then for simplicity, it is assumed that all of the components in the mfg exist together in the same "environment". As the "environment" changes, then all the components move together into the same new "environment". This represents the case of total dependency between the components. It is possible, however, that although two components come from the same mfg, that they may be effectively segregated, and consequently be independent of each other.

To see this more clearly, consider the example of an mfg of 4 components, labelled A. Then we can write

$$A^4 = A_4 + 4A_3A_1 + 3A_2^2 + 6A_2A_1^2 + A_1^4 \quad (5)$$

Here, A^4 represents the failure of all 4 components in the A mfg. This can occur if all 4 fail in the same "environment" which is represented by A_4, if 3 are segregated from the other 1, which is represented by $4A_3A_1$, and so on down to A_1^4 representing the complete segregation of the 4 components. This last term is equivalent to the assumption of independence between the four components. We now define the rank of a cut-set to be the minimum number of root causes necessary to fail the cut-set. This last term therefore has rank 4, and is what would be calculated by a standard fault tree technique. The first term has rank 1 since it assumes complete dependence between the components. The three middle terms have ranks 2, 2 and 3 respectively. In the simplest application of the DFP method, these intermediate rank terms are not explicitly considered when evaluating the probability of 4 failures $p(A^4)$. Strictly, when evaluating equation (5), we would have

$$p(A^4) = \lambda_1 p(A_4) + 4\lambda_2 p(A_3)p(A_1) + 3\lambda_3 p(A_2)^2 + 6\lambda_4 p(A_2)p(A_1)^2 + \lambda_5 p(A_1)^4 \quad (6)$$

where the λ_i's are the probabilities of each of the possibilities. Clearly, the λ_i's sum to one. If it is assumed that all the components always exist in the same "environment" i.e. only the rank 1 term A_4 is possible, and $\lambda_1 = 1$ with the other λ_i 's being zero.

This assumption, corresponding to the neglect of these intermediate terms is not obviously pessimistic. It can be shown, however, that it is. The details are rather involved, but the basic reason is as follows. Since the largest failure probability for the four components comes from the case when they are assumed to be totally dependent, the most pessimistic assumption is that this is always the case. In other words, since the rank 1 term dominates, any reduction in the likelihood of the rank 1 contribution in order to include higher rank contributions will outweigh these added contributions.

Symbolically, this means that instead of equation (6), we can write

$$p(A^4) \leq p(A_4) \qquad (7)$$

Generally, if the number of components in the A mfg is n and a cut-set contains r of them, where r is less than n, then we have

$$p(A^r) \leq {}^nC_r\, p(A_r) \qquad (8)$$

So, whenever we need to evaluate the probability of a term like A^r we can do so pessimistically but most simply by assuming only rank 1 contributions. Of course, we can refine this assumption and be less pessimistic by evaluating the higher rank contributions explicitly if necessary (see the example below).

4. Example

In order to illustrate the ideas presented above, we shall consider the example of a simple system which contains two pumps, the failure of both of which fails the system. The two pumps are regarded as vulnerable to dependent failures, and so are grouped together as an mfg, labelled A, with two components. The system then has a cut-set, represented by A^2 which should be evaluated using the DFP method as described above.

The first step in this process must be to establish the "environments", their probabilities $p(E_j)$, and the conditional failure probabilities $p(A|E_j)$ of the pumps in these "environments". We shall assume for this example that the relevant data on these pumps is sparse, and that we must rely on the judgment of an expert with detailed knowledge of the workings of these pumps to give us all the appropriate "environmental" information. The only 'hard' data is that there were 33 failures out of 16327 pump tests.

Generally, the knowledge held by an expert will not be in the form of probabilities and "environments". It is most likely that his views will be most readily expressed in terms of failure modes (rather than causes) and the relative proportion of failures occurring in each of these modes. When the expert has given us these pieces of information, we can then go on to extract the root causes of each failure mode, which we can then regard as being the relevant "environments" for the pumps. In order to judge the component failure probability in a particular "environment", we also need the expert's view on the proportion of failures which would be double failures if two of those components were present together.

So, we assume that the information which the expert has given us is as summarised in Table 1. Firstly, the left hand columns detail the various failure modes of the pump and the proportion of the failures which are in that mode. The next two columns give the expert's view as to the root cause of those failure modes and their respective proportions. The final column gives the probability that, in the system with two pumps, if one of the pumps is found to be failed then so also is the other. So, for example, maintenance-induced shaft failure accounts for 0.1x75% = 7.5% of the pump failures. Of these, 30% occur in pairs rather than as single failures.

Using these judgments as to fractions and root causes of failure, and the fact that 33 pump failures have been observed, we can then work out the number of these 33 failures attributable to each root cause. The results are given in Table 2. The maintenance-induced shaft failures, the third type of maintenance failure listed in Table 1, are the third set of figures appearing under "Maintenance" in Table 2. So, with a total of 33 pump failures, 33x0.075 = 2.47 failures are attributable to maintenance-induced shaft failure (non-integral failures are non-physical of course, but are quite acceptable here for working out probabilities). Of these, 2.47x0.3 = 0.74 are failures occurring in pairs (i.e. there were 0.74/2 = 0.37 events in which both the pumps failed together) and 2.47-0.74 = 1.73 are single failures. Of course, maintenance appears in several places in Table 1, and the calculation above must be repeated for all these, and the results summed to give the overall numbers of failure events due to maintenance. This is repeated for all the root causes identified, and the results are shown in Table 2. (Two different design root causes are judged to be relevant).

We are now in a position to estimate the necessary probabilities. To do this, we shall need the following formulae. If the total number of system tests is N, then the numbers of single and double failures are given by

$$N_{1/2} = 2\,N\,p(E_j)\,p(A|E_j)\,\{1 - p(A|E_j)\}$$
$$N_{2/2} = N\,p(E_j)\,p(A|E_j)^2$$

respectively. Assuming the data is gathered on two pump systems, then N = 8164 (half the total number of component tests, 16327), and these two equations yield the probabilities indicated in Table 2 for each "environment".

The final column in Table 2 uses these probabilities to give the total failure probability of two similar pumps together, and the contribution from each "environment". Clearly, the dominant contribution (more than half) comes from the effects of maintenance.

The figures in the final column can be cross-checked. For example, there were 2.74 double failures due to maintenance in 8164 trials of a two pump system. This immediately gives a failure probability (due to maintenance) for the system of 2.74/8164 = 3.36×10^{-4} as already derived. The advantage of going via the conditional failure probabilities, rather than this direct route, is that we can now work out the failure probability for a system of any redundancy, rather than just two.

We can also refine the calculations if necessary. In calculating the probability of both pumps failing, it was assumed (pessimistically as indicated by equation (8)) that the two pumps existed in the same "environments" at the same time, in particular, that they were maintained at the same time.

The dominance of the maintenance contribution suggests that we should look again at this assumption with a view to a more realistic evaluation of $p(A^2)$ than is given by $p(A_2)$. The price to be paid for any benefit arising by this is a slightly more complicated calculation.

The total figure can be reduced by imposing the requirement that the two pumps are never maintained at the same time, i.e. that the pumps are never in the "maintenance environment" together. In these circumstances, the formula for $p(A^2)$ is rather complicated, but effectively, the maintenance contribution is subtracted from the total. In fact, it can be shown that

$$p(A^2) \le p(A_2) - p(M)p(A|M)^2 + 2p(M)\{p(A) - p(M)p(A|M)\}$$

so that an upper bound on the failure probability of the two pumps is given by

$$\begin{aligned} p(A^2) &= 5.27\times10^{-4} - 3.36\times10^{-4} + 6.7\times10^{-6} \\ &= 1.98\times10^{-4} \end{aligned}$$

which is an improvement on the previous result of more than a factor of two.

5. Conclusions

The Distributed Failure Probability (DFP) method for dependent failure has been outlined, and shown to be firmly based on the root causes of component failures. The establishment of the component response to these root causes or "environments" enables the response of a system of such components to be evaluated. The practicality of the approach has been demonstrated even in the absence of all the relevant data, since then the approach provides the natural framework for expert judgment. The flexibility of the approach has also been demonstrated, in that simple bounding calculations of system reliabilities can be refined with more realistic assumptions, as required. The significant benefits of the DFP method as viewed by the CEGB can be judged by the fact that it will form the basis of the dependent failures assessments for the safety case being prepared for its new PWR at Sizewell.

References

1. Hughes, R. P., 1986, CEGB report TPRD/B/0813/R86

2. Hughes, R. P., Reliability Engineering 17 (1987) 211

AKNOWLEDGEMENTS

This paper is published with the permission of the Central Electricity Generating Board.

The author is grateful to Miss A. J. Addis for providing the example in Section 4.

Failure Mode	Proportion of Total Failures	Failure Cause	Proportion of Mode Failures	Probability of Second Failure
Poor Alignment	2%	Maintenance	99%	1%
		Q A	1%	95%
Air Binding	1%	Design	100%	5%
Int. Structural Failure	1%	Corrosion	80%	80%
		Design	20%	95%
Leakage past Seals	10%	Maintenance	100%	1%
Shaft Failure	10%	Maintenance	75%	30%
		Corrosion	25%	90%
Loss of motor Cooling	1%	Maintenance	50%	20%
		Design	50%	90%
Mis-set Torque Switch	15%	Maintenance	70%	30%
		Vibration	30%	30%
Circuit Breaker Failure	50%	Maintenance	55%	40%
		Others	45%	20%
Random	10%	Random	-	-

Mean failure rate:- $2.0\text{x}10^{-3}$ per demand
Operating experience:- 16327 demands
Number of failures:- 33

Table 1 - Data and Judgments on Pump Failures

Environment	No of failure events in Env			P(A)	P(A|E)	Contribution to CCF value for 2 pumps ($p(A_2)$)
	Total	Single	Double			
Maintenance	0.65 + 3.30 + 2.47 + 0.17 + 3.47 + 9.08 = 19.14	0.64 + 3.26 + 1.73 + 0.13 + 2.42 + 5.45 = 13.65	0.005 + 0.017 + 0.370 + 0.017 + 0.520 + 1.815 = 2.74	$3.99x10^{-3}$	0.29	$3.99x10^{-3}$ x $0.29^2 = 3.36x10^{-4}$
Q A	0.007	$3.3x10^{-4}$	0.003	$4.20x10^{-7}$	0.95	$4.20x10^{-7}$ x $0.95^2 = 3.79x10^{-7}$
Design i	0.33	0.31	0.008	$3.90x10^{-4}$	0.05	$3.90x10^{-4}$ x $0.05^2 = 9.85x10^{-7}$
Design ii	0.06 + 0.17 = 0.23	0.003 + 0.017 = 0.02	0.031 + 0.074 = 0.105	$1.55x10^{-5}$	0.91	$1.55x10^{-5}$ x $0.91^2 = 1.28x10^{-5}$
Corrosion	0.26 + 0.83 = 1.09	0.05 + 0.08 = 0.13	0.105 + 0.375 = 0.477	$7.54x10^{-5}$	0.88	$7.54x10^{-5}$ x $0.88^2 = 5.84x10^{-5}$
Vibration	1.49	1.04	0.223	$3.03x10^{-4}$	0.30	$3.03x10^{-4}$ x $0.30^2 = 2.73x10^{-5}$
Others	7.43	5.94	0.743	$2.28x10^{-3}$	0.20	$2.28x10^{-3}$ x $0.20^2 = 9.12x10^{-5}$
'Random'	3.30	3.30	0	1.0	$2.0x10^{-4}$	1.0 x $(2.0x10^{-4})^2 = 4.0x10^{-8}$
						total = $5.27x10^{-4}$

Table 2 - Component and System Failure Probability Information

DEFEND – A Dependent Failures Database

A M GAMES

UKAEA
National Centre of Systems Reliability (NCSR)
Culcheth, Warrington, UK

Summary

The traditional sources of operational reliability data do not record system and component unavailabilities in a way which is amenable to the support of dependent failures modelling. Progress has been made in developing techniques for analysing the data so as to detect and express dependencies, but different data sources require different techniques and express differing views of component and system availability.

Given a large amount of inhomogeneous data from event reports and component failure data, NCSR undertook the development of a new dependent failures database which would contain all of the available data in an upgraded and rationalised form. This database, DEFEND, was designed so that it could potentially support a range of dependent failures models.

Introduction

Difficulties in dependent failures modelling are not entirely due to a lack of accurate models. They can arise instead from supporting data not being readily available in sufficient quantity and in the required format.

The large number of parametric and partial parameter models commonly used, together with the more recent concepts of a trigger-coupling mechanism (TCM) model (1) and a distributed failure probability (DFP) method (2), require data in different forms. There is, however, a large body of underlying failure information which is a common requirement for all models. This arises because most of the models are closely related.

It was decided that the only reasonable approach to the data problem was

i) to ensure that the maximum amount of relevant information would be extracted from the source data and would be recorded in a logical and systematic way;

and ii) to store the data in a database, so that different types of dependent failures model could be supported by means of a flexible data retrieval system.

Data Sources

In the nuclear industry, two sources of operational reliability data can be identified

a) Component failure event data from maintenance records.

b) Event reports such as Licensee Event Reports (LERs) and Nuclear Power Experience Reports (NPERs).

Neither type of data is collected specifically for quantifying dependent failures. For this reason, it has been necessary in recent years to devise methods of data analysis to identify dependent failure events in the source data. These methods depend upon the nature of the source data.

Component failure event data consists entirely of individual component failure descriptions. Ostensibly, all of the events are independent. In order to detect dependencies between these failures a systematic method of searching and sorting data has to be applied so that failures with specific common features and attributes can be drawn together. Such a method has been proposed by Games (3)(4).

The detection of dependent failures in descriptive event reports requires

"A method for logically dissecting multiple component unavailability scenarios into individual component unavailabilities" (5)

The Electric Power Research Institute (EPRI) in the US sponsored Los Alamos Technical Associates (LATA) to develop a classification scheme for dependent failures.

The scheme is based on cause-effect logic, that is each component unavailability is expressed in terms of a cause and an ultimate component state:

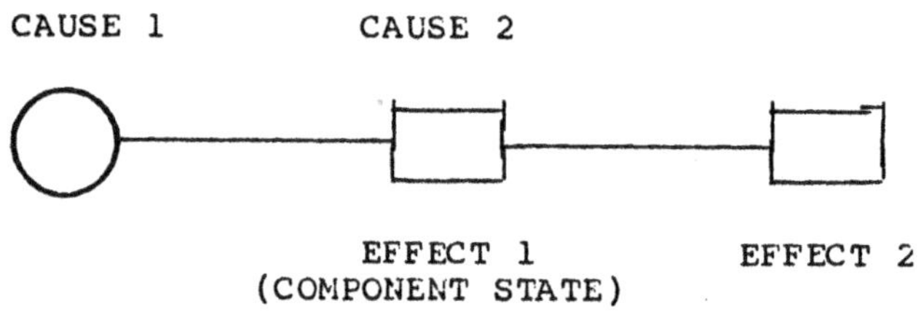

The cause of unavailability may be another component unavailability as shown above. It is also possible for a single cause to have multiple effects:

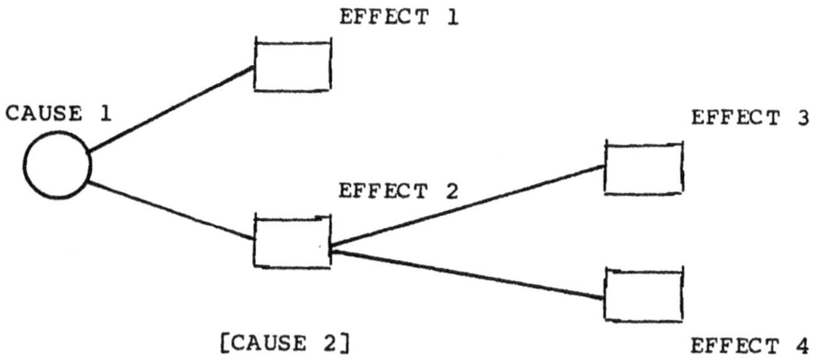

Typically, root cause of failure might be defined as poor design, poor manufacture, maintenance error, inadequate operating procedures and so on. Component states (effects)

might be failed, functionally unavailable, potentially failed, potentially functionally unavailable etc.

Unavailability is defined as the inability of a component to perform its intended function (either due to damage or lack of proper input or lack of support function). Failure and functional unavailability are two mutually exclusive and exhaustive subsets of component unavailability.

The value of this structured approach is that each event scenario can be dissected and expressed in a consistent form. There still remain problems of interpretation regarding what the analyst perceives as the ultimate root cause or the ultimate effects and so on, as well as the residual difficulties imposed by lack of certain types of information in, for example, the Licencee Event Reports. In a comparison of the results of the analysis of a set of events by four different analysts it was shown that differences were due almost entirely to interpretation, some of which could be overcome by adding some specific rules and guidelines to the classification scheme (6). It has also been suggested that inclusion of post- event actions, system structure and more detailed cause description in event reporting would facilitate data analysis.

It has been found, however, that this method is a very valuable analysis tool. The scheme (of which there have been two or three variants) has been applied to LER and NPER data. Pickard Lowe and Garrick (PLG) applied their version of the LATA scheme to NPERs with a view to supporting parametric dependent failure models (7). Another US company, SENCON, also used the LATA method, in its original form, to classify a large number of LERs.

The cause-effect logic method can also be used to express dependent events which are detected in component failure data.

NCSR, through various collaborative projects, therefore, found itself in possession of data in a variety of formats and from event and component failure data. The structure, quality, accuracy, scrutability and presentation of the data was very variable. A single method of recording the data was clearly required so that this data could be rationalised and a computer database was devised to achieve this.

Database Development

Fundamental Data Structure

It was decided that the LATA cause-effect logic method was the best available method for describing all types of unavailability scenario. This method was therefore used as the basis for the new database.

The variants of the LATA method were analysed and compared by NCSR (6). It was decided that the PLG method of describing events on a component-by-component basis was the clearest and this technique has been expanded to form the new database.

Several cause-effect logic chains may be present in each LER/NPER. That is, there may be several independent and/or dependent events. It was decided that all independent and dependent events from the NPERs and LERs would be recorded in the new database.

Each failure report in a component event data bank constitutes a cause-effect logic event in itself. Little would be achieved, however in simply transferring all of these independent events into the new database. Instead, the dependent events, found by data sorting and searching can be expressed in the database, without reference to an individual source event report but with cross-referencing of component and failure numbers in the source data.

Each event or cause-effect logic chain will contain one or more component unavailabilities. The fundamental structure of the information is, therefore, as shown in Figure 1.

Certain descriptors can be applied to an event as a whole, such as where it occurred, when it occurred, what effect it had on the system and so on. Other descriptors are specific to each component unavailability such as the root cause of unavailability, mode of failure, component type and so on.

The information to be recorded about each event was grouped under six general headings. The first five relate to the event as a whole and the sixth group relates to component unavailability description.

1 Source Identification and Referencing

A group of attributes was specified which would identify the sources of the information and which would allow events to be precisely identified both in the source data and the new database.

2 Event Location

In order to use data for certain dependent failures models, it would be necessary to be able to screen events for their applicability to the problem in hand. Part of this would include identifying the specific plant and system affected to see whether there were similarities between the system being analysed and the system(s) from which the data arise. This could help in generating less generic and more specific parameters.

3 Essential System Design Information

It was noted that some essential system design features need to be known in order to support certain parametric models. In particular, fields were provided for recording

- the number of channels in the system,
- the voting logic, and
- the minimum number of channels required for operability (under operating regulations)

where this information is available. Unfortunately, these details are frequently not recorded in event reports.

It was recognised in the earliest stages of this work that this design information was scant and that it would be desirable to have a system design information database connected to the event database. The two databases would be linked through plant and system codes in 2 above. Full descriptions of the plants and systems from which the event data arose would be stored. The development of this database would be considered as a separate project and an initial feasibility study demonstrated that considerable resources would be required.

The linking of databases, existing and projected, is illustrated in Figure 2.

4 QA Checks

The analysis of event reports, although carried out under a classification scheme, requires interpretation by the data analysts. For checking consistency, therefore, it was thought necessary to record who the analyst was and note that his assumptions were. His interpretation would also be encapsulated in a very short textual summary of the event.

5 Event Description

Dependent failures studies are especially concerned with the effect of component failures on systems and plants.

Attributes were identified which describe these effects for each event.

a) The state of the system and plant before the event may be noted together with the consequences of the event in terms of changes of plant and system state. The status of the system at the time the event occurred may be a relevant screening device since events which occur during commissioning or pre-op testing may not be considered as relevant to the system being analysed.

b) If the system has been rendered unavailable then the reason for that unavailability should be noted. Provision was, therefore, made for recording the coupling mechanisms which have propagated the trigger event to affecting multiple channels. These coupling mechanisms are essential in the support of NCSR's TCM Model.

c) Another aspect of dependent failures work is concerned with defensive tactics against triggers and coupling mechanisms. From event reports it is only possible to deduce what **prospective** defensive tactics would have been effective in preventing the event. (Details of **applied** defensive tactics would form part of the system design information).

The partial beta factor model requires a knowledge of the defensive tactics.

Fields were provided for storing information on defences.

d) Related closely with defensive tactics are specific corrective actions and the means of discovery of failures and fields were provided to allow this information to be recorded when available.

e) The complete event may then be classified as independent or dependent (cascade or shared cause) and as an actual or potential failure event.

6 Component Description

Having specified attributes which are relevant to the whole event as viewed by the plant or system, the components which were affected must be examined in turn. Each component in the event needs to be uniquely referenced so that, in theory, the cause-effect logic diagram can be reconstructed.

For each component in the event, its type and function, its final state, and the cause and mode of failure are recorded. (It is at this level that the particular component failure can be cross-referenced with the source data when that source is a component failure event databank).

Component type and function may be used in event screening and the root cause information is used specifically in models.

It is believed that the attributes which were identified and included represent the very maximum amount of information which can be extracted from event reports. Indeed, it has been found that it is difficult to fill in 100% of every form. It is suggested, therefore, that any shortcomings of the database in providing support for models will tend to lie with the limitations of the source data (ie the event reports).

The structure developed and the contents identified for the database recognised the need for a careful balance between what was desirable and what is achievable given the state of the source data. The coding schemes for attributes such as causes, modes of failure, defences were selected from currently available schemes. In this case too a balance had to be achieved between the ability to fully describe the event and the utility of any greater detail in the ultimate use of the data. Too many codes can result in difficulty of choice for the analyst when interpretation of an event is required, whilst too few codes will result in a loss of information in transcoding.

Requirements for Data Retrieval

The new dependent failures database contains qualitative descriptions of occurrences. The qualitative information may be extracted in a structured way or quantitative information in terms of frequencies, or relative frequencies, of occurrence of certain attributes may be derived. Different dependent failures models will require different types of information from the database.

The majority of current dependent failures models can be usefully classified into

a) Parametric models requiring identification of dependent events and subsequent screening

and b) Partial parameter models which require once-and-for-all calibration

Parametric models require

1) Identification of dependencies with branched, shared root cause logic:

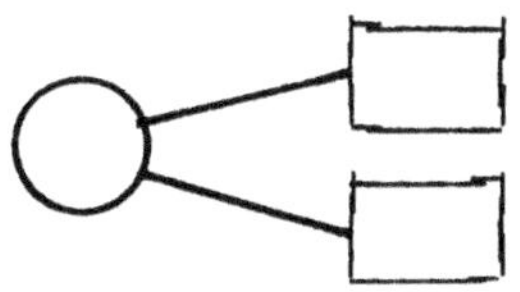

for each component type

2) Division of events into actual and potential dependencies.

3) Enumeration of the independent events.

4) Screening of the dependent events by removal of events such as those occurring prior to commercial

operation or during shutdown conditions, or events which would be modelled explicitly in systems analysis or events involving failures or potential failures that do not have a significant impact in analyses for PRA applications.

5) Determination of impact vectors.

Many queries may be made of the database in relation to parametric models. The nature of the specific system being assessed will determine which parameters are required and for which systems or components.

The partial parameter models, on the other hand, use the complete set of available data **once only.** Partial parameter models are simply calibrated.

The same set of permissible ranges of partial β values which were originally estimated are still being used. It is conceivable that the new database could be used to support or alter these values. The new coupling mechanism model, on the other hand, has yet to be calibrated properly.

The extended partial beta factor models require calibration using two attributes - causes and defences.

A great deal of analysis was envisaged, once the database was complete, to establish links and relationships between

defences
causes
and coupling mechanisms

It was also thought that there would be a need for great flexibility in terms of selecting subsets of events and histogramming relative frequencies. (3-D histograms may also be required so that more than one variable can be examined).

Some basic requirements for the satisfactory retrieval of data from the database were therefore specified based on the above observations:

1) It must be possible to SELECT subsets of events based on any of the fields which have been specified. This selection process may be carried out at event and component failure description levels and it should be possible to move up and down from one level to another in subsequent selections.

2) It should be possible to HISTOGRAM the contents of every field for which a list of identifiable codes exists. Histogramming may need to be carried out at any point in a sequence of subset selections.

It should be possible to send data to a graphics package for histogram plots.

3) At any stage in a selection of events it should be possible to generate an event summary based on

a) standard forms
b) fields selected by the analyst.

4) A browse facility is required.
The layout and content of screens for data entry, browsing and retrieval should be user friendly. That is, they should be clear and should not be filled only with coded information. The contents of each screen should also be logical from an engineering viewpoint.

If it is not possible to de-code information for browsing it must be possible for the analyst to query what a code is if he does not understand it. This involves the use of look-up lists. These would serve a dual purpose in that they can assist in data entry thus ensuring good data quality.

5) The ability to print out information is essential

i to record results
ii to keep track of queries

It should be possible to print screen dumps and files.

Other general requirements were specified as follows:

6) The database package to be employed should operate on a desktop microcomputer and it should be IMB compatible.

7) In the event that the database will be distributed to interested parties, it should be possible to transfer programmes and data to diskettes.

8) The database should be designed for future growth in terms of the number of records which can be held. At present, in the region of 2000 events are to be loaded but it is expected that this will be increased in future years.

9) The database data entry and retrieval should be menu driven and robust.

The database package used to achieve the above requirements was DATAEASE. The database is relational and data entry is greatly assisted by look-up lists for attribute coding schemes. Data retrieval can be carried out using the DATAEASE Quick Reporting facility for routine enquiries and statistics. It is possible to carry out more detailed enquiries by writing specific programs.

Conclusions

The DEpendent Failures ENgineering Database (DEFEND) is a step forward in the storage of dependent failure events in a systematic and relevant format. It incorporates the best and most essential features of current procedures for the analysis of source data.

The major difficulties experienced by NCSR have been in

a) Determining an appropriate level of detail in terms of event attributes and related coding schemes.

b) Ensuring consistency in event analysis and coding given that much personal interpretation and judgement must be employed in applying the cause-effect logic method in particular when event analysis is being carried out without detailed plant and operational knowledge.

The former issue has been dealt with by applying considerable experience, gained in event analysis and review of coding schemes to achieve a compromise design. The latter issue is being dealt with by employing an independent engineer (who was not associated with event analysis) who is applying a formal quality assurance procedure to the encoded data.

Acknowledgement

The assistance of Mr B Stevens, Mr S Saunders and Miss S Mennell is gratefully acknowledged.

References

1 Ballard, G M; Dependent Failure Analysis in PSA, to be presented at IAEA International Conference on Nuclear Power Performance and Safety, Vienna, Austria, 28 September - 2 October 1987.

2 Hughes, R P; A New Approach to Common Cause Failure, CEGB Berkley Nuclear Laboratories, TPRD/B/0813/R86, May 1986.

3 Games, A M; Some Aspects of Common Cause Failure Analysis in Engineering Systems, PhD Thesis, University of Liverpool, October 1986.

4 Games, A M et al; Exploitation of a Component Event Data Base per Common Cause Failure Analysis, International ANS/CNS Topical Meeting on Probabilistic Safety Methods and Applications, San Francisco, February 1985.

5 Los Alamos Technical Associates; A Study of Common Cause Failures; Phase 2 A Comprehensive Classification System for Component Fault Analysis, EPRI NP-3837, Project 2169-1, Interim Report, June 1985.

6 Humphreys, P et al; Progress Towards a Better Understanding of Dependent Failures by Data Collection, Classification and Improved Modelling Techniques, Reliability '87 Conference, Birmingham, April 1987.

7 Pickard, Lowe and Garrick; Classification and Analysis of Reactor Operating Experience Involving Dependent Failures, EPRI NP-3967, Project 2169-4, Interim Report, June 1985.

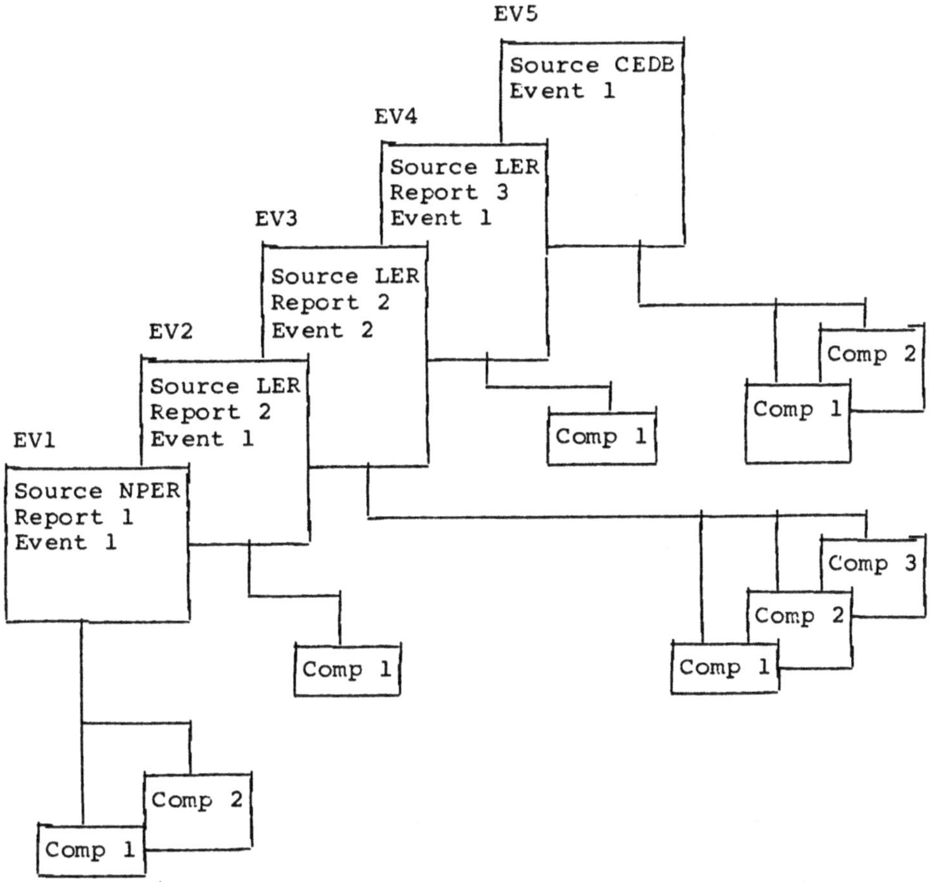

Figure 1. DEFEND Database Structure

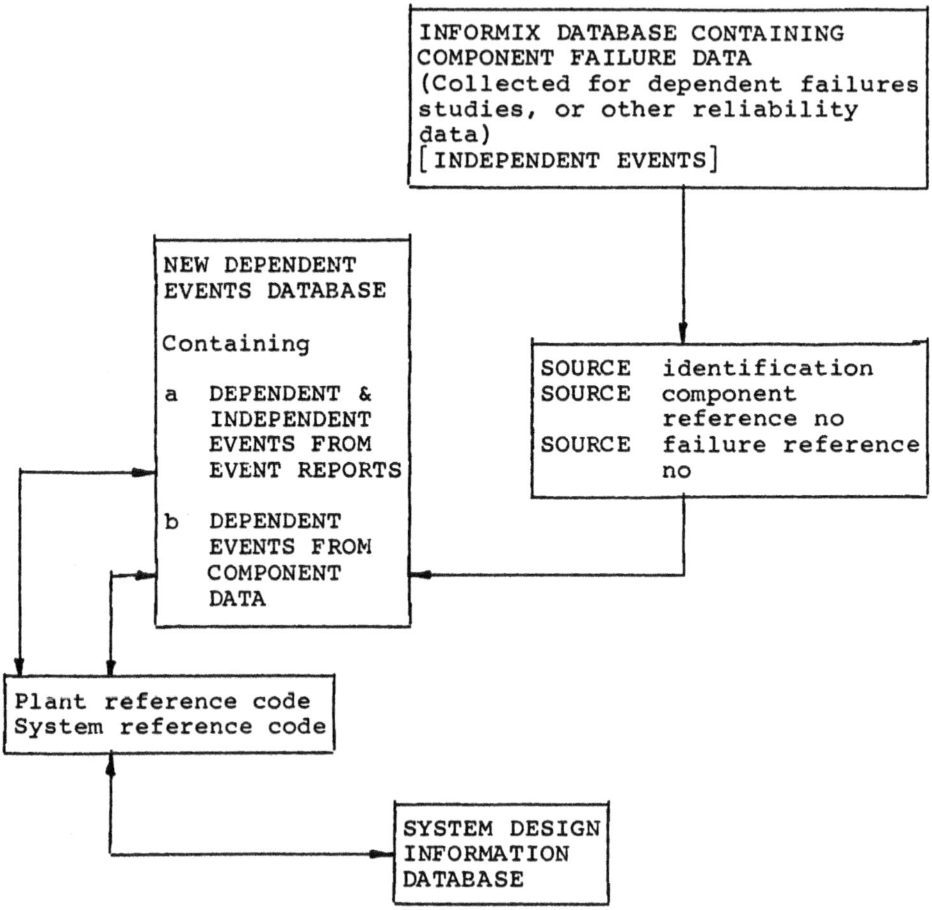

Figure 2 Linking of Databases

An Analytic Method to Estimate Uncertainty in a Risk Function

Henrik Kortner
Norsk Hydro, Research Centre
N-3901 Porsgrunn
Norway

Abstract

This paper presents a method for estimating the uncertainty in a probabilistic risk analysis. Based on the estimation of probability intervals of some basic parameters of a risk function, the method estimates a probability interval for the risk function. An example of risk analysis of releases of flammable materials is shown.

1. INTRODUCTION

As risk assessments are often used to make decisions of great importance to our society, the uncertainties in risk assessments are of great importance. This paper presents a method that has been used to estimate uncertainties in risk analyses of flammable releases. The method is partly based on expert judgements and partly on an analytic mathematical method.

2. SOURCES OF UNCERTAINTY IN RISK ASSESSMENTS

The different sources of uncertainty in risk analyses may be divided into three main groups:

- Parameter uncertainties
- Modelling uncertainties
- Completeness uncertainties

Parameter uncertainties refer to the uncertainties in the constants which are input to the estimation of frequencies and consequences of an accident.

Modelling uncertainties refer to the applicability and the quality of the models which are used.

Completeness uncertainties refer to the fact that normally there will be accident scenarios which are not identified and consequently not considered in a risk analysis.

The aim of the present paper is to develop a method that can handle parameter uncertainties and modelling uncertainties. Completeness uncertainties are more difficult, if not impossible, to quantify. Some interesting ideas to solve this problem was presented by Kaplan and Garrick (1981).

3. MATHEMATICAL FRAMEWORK

This chapter reviews some important properties of the lognormal distribution and develops formulae that will be used in the case study in chapter 5. The mathematics may be well known to statisticians, but are included here for the sake of completeness.

3.1 The lognormal distribution

A stochastic variable, X, is said to have a lognormal distribution if lnX is normally distributed.

The constants υ and τ^2 are defined as the expectation and variance of lnX.

$$\upsilon = E(\ln X) \tag{1}$$

$$\tau^2 = \mathrm{var}(\ln X) \tag{2}$$

The probability density function of the lognormally distributed X may then be expressed as

$$f(x) = \frac{1}{\sqrt{2\pi}} \frac{1}{\tau} \frac{1}{x} \exp\left[-\frac{1}{2\tau^2} (\ln x - \upsilon)^2\right] \text{for } x>0 \tag{3}$$

The expectation and variance of X are

$$\mu = E(X) = \exp\left(\upsilon + \frac{1}{2}\tau^2\right) \tag{4}$$

$$\sigma^2 = \text{var}(X) = \exp(2\upsilon)\left[\exp(2\tau^2)-\exp(\tau^2)\right] \tag{5}$$

The lower and upper bounds of a 100 α % probability interval can be calculated from the equations:

$$\Pr\left\{x \le \lambda_1\right\} = \frac{\alpha}{2} \tag{6}$$

$$\Pr\left\{x > \lambda_u\right\} = \frac{\alpha}{2} \tag{7}$$

Using the normal distribution we get

$$\lambda_1 = \exp(\upsilon - k\tau) \tag{8}$$

$$\lambda_u = \exp(\upsilon + k\tau) \tag{9}$$

where k can be found from a table of the cumulative normal distribution according to the formula

$$k = \Phi^{-1}(1 - \frac{\alpha}{2})$$

According to this a 90 % probability interval is given by

$$\lambda_1 = \exp(\upsilon - 1.645\tau) \tag{10}$$

$$\lambda_u = \exp(\upsilon + 1.645\tau) \tag{11}$$

It is often convenient to express the lognormal distribution by the two parameters, median, M, and error factor, f. The relations between these and υ and τ are

$$M = \exp(\upsilon) \tag{12}$$

$$f = \exp(k\tau) \tag{13}$$

Some useful relations which can be derived from 8,9,12 and 13 are:

$$\lambda_1 = \frac{M}{f} \tag{14}$$

$$\lambda_u = Mf \tag{15}$$

$$\upsilon = \frac{1}{2}(\ln\lambda_u + \ln\lambda_l) \tag{16}$$

$$\tau = \frac{1}{2k}(\ln\lambda_u - \ln\lambda_l) \tag{17}$$

3.2 Product of lognormal distributed variables

Let $X_1, X_2, \ldots X_n$ be n independent lognormally distributed variables. We consider the function

$$Y = X_1^{a_1} X_2^{a_2} \ldots X_n^{a_n} \tag{18}$$

where a_i are constants.

From (18) it follows that

$$\ln Y = \sum_{i=1}^{n} a_i \ln X_i \tag{19}$$

As $\ln X_i$ $(i=1,2,..n)$ are normally distributed independent variables, lnY is normally distributed and

$$E(\ln Y) = \sum_{i=1}^{n} a_i\, E(\ln X_i) \tag{20}$$

$$\mathrm{var}(\ln Y) = \sum_{i=1}^{n} a_i^2\, \mathrm{var}(\ln X_i) \tag{21}$$

We now assume that we know two parameters of the distribution functions of the X_i's $(i=1,2,\ldots n)$, for instance the upper and lower bounds of some probability interval, λ_{iu} and λ_{il}. If in fact we know two other parameters of the distribution, λ_{iu} and λ_{il} are easily calculated by the formulae in 3.1.

Using the formulae in 3.1 we get

$$E(\ln X_i) = \upsilon_i = \frac{1}{2}(\ln\lambda_{iu} + \ln\lambda_{il}) \tag{22}$$

$$\mathrm{var}(\ln X_i) = \tau_i^2 = \left[\frac{\ln\lambda_{iu} - \ln\lambda_{il}}{2k}\right]^2 \tag{23}$$

and (20) and (21) may be written as

$$E(\ln Y) = \sum_{i=1}^{n} a_i \upsilon_i = \upsilon \tag{24}$$

$$\mathrm{var}(\ln Y) = \sum_{i=1}^{n} a_i^2 \tau_i^2 = \tau^2 \tag{25}$$

As lnY is normally distributed, Y is lognormally distributed, and the parameters of its distribution may be calculated by the formulae in 3.1.

Expectation and variance are given by

$$E(Y) = \exp(\upsilon + \frac{1}{2}\tau^2) \tag{26}$$

$$\mathrm{var}(Y) = \exp(2\upsilon)\left[\exp(2\tau^2) - \exp(\tau^2)\right] \tag{27}$$

The median and error factor are given by

$$M = \exp(\upsilon) \tag{28}$$

$$f = \exp(k\tau) \tag{29}$$

and the probability interval is defined by the bounds

$$Y_l = \exp(\upsilon - k\tau) \tag{30}$$

$$Y_u = \exp(\upsilon + k\tau) \tag{31}$$

As before, k is a constant and equal to 1.645 for a 90 % interval. Another interesting result can be derived as follows:

$$f = \exp\left\{k\ \tau\right\} = \exp\left\{\sqrt{\sum_{i=1}^{n} a_i^2 (\ln f_i)^2}\right\} \tag{32}$$

If all the a_i's are equal to 1, and all the f_i's are equal, we get the simple formula

$$f = f_i^{\sqrt{n}} \tag{33}$$

4. QUANTIFICATION OF UNCERTAINTY IN RISK ASSESSMENTS

The subject of risk has received a lot of attention in the past years and several kinds of risk have been discussed in the literature, in industry and in public debate.

It is not the intention of this paper to present a general quantitative definition of risk, but rather an example, related to accidental releases of flammable materials. This is worked out in the following chapter.

As for the quantification of uncertainty in a risk analysis the following method based on the mathematics in chapter 3 is proposed:

1. We assume that the risk we estimate can be expressed as a function of the form

$$R(x_1 \ldots x_n) = x_1^{a_1} x_2^{a_2} \ldots x_n^{a_n} \tag{34}$$

   where x_i $(i=1,2,\ldots n)$ are some variable parameters and a_i $(i=1,2,\ldots n)$ are constants.

2. We assume that the uncertainty in the parameter x_i $(i=1,2\ldots n)$ can be described by saying that x_i is a random variable, which is lognormally distributed. Furthermore we are able to define a certain, for instance 90 %, probability interval for x_i.

3. We assume that all x_i's are statistically independent.

If these three basic assumptions are fulfilled the formulae developed in chapter 3 are valid. If we denote the lower and upper bounds of the probability interval of x_i by λ_{il} and λ_{iu}, the lower and upper bounds of the risk function are

$$R_l = \exp(\upsilon - k\tau) \quad (35)$$

$$R_u = \exp(\upsilon - k\tau) \quad (36)$$

Where

$$\upsilon = \sum_{i=1}^{n} \frac{a_i}{2} (\ln\lambda_{iu} + \ln\lambda_{il}) \quad (37)$$

$$\tau^2 = \sum_{i=1}^{n} \left[\frac{a_i (\ln\lambda_{in} - \ln\lambda_{il})}{2k}\right]^2 \quad (38)$$

$$k = \Phi^{-1}(1 - \frac{\alpha}{2}) \quad (39)$$

If we are looking for a 90 % interval, $k = 1.645$.

Using the formulae 26, 27, 28 and 29 we are also able to estimate the expectation, variance, median and error factor of the risk function.

5. UNCERTAINTIES IN RISK ANALYSIS, AN EXAMPLE

This case study is based on a risk analysis, that we performed at one of Norsk Hydro's chemical plants.

This plant is relatively simple to analyse, as the only hazardous material is vinylchloride. Furthermore we are only concerned with the risk from accidental releases that may be ignited. The carcinogenic or other health effects are not considered.

5.1 Quantification of risk

The question we intend to answer is: What is the probability that a person standing at a certain location, A, will be engulfed in flames due to the ignition of a gas cloud from an accidental release at our plant?

This probability is a kind of individual risk. However, it is not related to a particular person, but to a certain location, A. It might be extended by another factor which is the probability of a person being present at A.

The probability we want to estimate may be quantified as

$$p = p_1 \; p_2 \; p_3 \qquad (40)$$

where

p_1 = probability of a release with a release rate large enough to cause a flammable mixture at A.

p_2 = probability of a wind direction that will cause a flammable mixture at A given a large enough release.

p_3 = probability of ignition when there is a flammable mixture at A (i.e., there must not be ignition before the cloud reaches A).

If we regard p_1, p_2 and p_3 as lognormally distributed random variables and are able to estimate a 90 % probability interval for each of them, we may use the methodology described in chapter 4. This is outlined in some detail below.

5.2 Probability of a large enough release

The estimation of this probability is quite complex and consists of the following two steps.

1. The plant is regarded as a point source with a failure rate equal to the sum of the failure rates of all the equipment within the plant. As the release rate may differ within the plant due to dimensions of pipelines, pressure, etc. the failure rate is expressed as a function of the release rate.

2. The hazard distance for a given release rate is calculated in accordance with some dispersion model.

These two calculations are combined to give the probability of a release rate large enough to create a flammable mixture at A.

Some of the important uncertainties in the estimation are:

- Uncertainties in failure rate data.
- Uncertainties in the dispersion model.
- Uncertainties about where the actual release is ,due to our approximation of the plant as a point source.

Expert judgement or more sophisticated methods may be used to estimate a 90 % interal for this probability.

5.3 Probability of wind direction towards A

This probability may be estimated from local weather statistics. An important source of uncertainty here is the width of the drifting cloud which is dependent on the wind speed. Another source is the location of the release, as described in 5.2.

5.4 Probability of ignition

Ignition probabilities are estimated by identifying specific ignition sources, like cars or hot surfaces, and adding them to a general background ignition probability depending on whether we are looking at plant area, housing area, industrial area or any other area type. Again uncertainties may be estimated by expert judgement to give a 90 % interval.

5.5 Calculation of total probability interval

Having estimated 90 % probability intervals for each of p_1, p_2 and p_3, a 90 % interval of p is easily calculated in accordance with the formulaes in chapter 4. For instance, for one particular point, A, the following probability intervals are estimated:

p_1 : 0,0001 - 0,01
p_2 : 0,05 - 0,3
p_3 : 0,05 - 0,5

The probability interval of p is then

$1 \cdot 10^{-6} - 3 \cdot 10^{-4}$

We remark that the factor between the upper and lower bounds is

300 which is quite small compared to the product of the factors for p_1, p_2 and p_3, which is 6000.

6. SOME COMMENTS ON THE ASSUMPTIONS OF THE METHOD

6.1 The lognormal distribution

One of our assumptions in the model is that the basic probabilities of (34) are lognormally distributed. Reactor Safety Study (1975) gives some arguments in favour of lognormally distributed failure rates. As for ignition probabilities for instance, these are obviously not lognormally distributed as they are never larger than one. Apostolakies and Kaplan (1981) concluded that in estimating interval bounds the use of the lognormal distribution is satisfactory, whereas estimation of expectation may be significantly biased by the impossibly high probabilities.

6.2 Independence

We have assumed independence between the basic probabilities. This is not quite correct. For instance the hazard distance calculated by the dispersion model is dependent on the wind speed. Estimation of the wind direction probability is also dependent on the wind speed as the width of the cloud decreases with increasing wind speed. This dependence is, however, believed to be weak.

7. CONCLUSIONS

We have proposed a method for uncertainty estimations in risk analyses. The method is based on expert judgement and some quite simple mathematics. The method is easy to use and not very laborious. More research is needed to improve the method in order to make more precise estimations of the probability intervals of the basic probabilities in the risk equation (34).

8. REFERENCES

S. Kaplan and B.J. Garrick: On the Quantitative Definition of Risk, Risk Analysis, volume 1, no. 1, 1981.

U.S. Nuclear Regulatory Commission: Reactor Safety Study: An Assessment of Accident Risks in U.S. Commercial Nuclear Power Plants, WASH-1400, 1975.

G. Apostolakis and S. Kaplan: Pitfalls in Risk Calculations, Reliability Engineering, 2, 1981.

Modeling of Fuzzy Measurements in Reliability Estimation

Reinhard VIERTL

Institut für Statistik und Wahrscheinlichkeitstheorie
Technische Universität Wien

Abstract

There are two main sources of uncertainty in reliability data. The first kind of uncertainty is stochastic variation analyzed by stochastic models. The second kind of uncertainty is fuzziness of observed life times. This kind of uncertainty is neglected in standard statistical inference. Therefore statistical inference procedures have to be adapted to take care of fuzzy life time data. Models to analyze fuzzy life time data are given in classical frequentist and Bayesian context.

1. INTRODUCTION

Collected reliability data are usually not precise numbers or vectors. In most of all cases the observed life times are fuzzy. This kind of uncertainty is different from life time variation in samples from a stochastic quantity and can be modelled by so called fuzzy numbers. Therefore we distinguish two types of uncertainty:

a) randomness

b) imprecision .

The first kind of uncertainty is described by stochastic models. In reliability analysis the life time is modelled by a stochastic quantity X with corresponding probability distribution P. In the parametric case $X \sim P_\theta$, $\theta \in \Theta$, the set Θ of all possible parameters is called parameter space. In symbols

$$X \sim \bar{F}(.|\theta),\ \theta \in \Theta \tag{1.1}$$

where $\bar{F}(.|\theta)$ denotes the reliability function of the life time X, i.e.

$$\bar{F}(t|\theta) = Pr(X > t) \quad \text{for} \quad t \geq 0. \tag{1.2}$$

The second kind of uncertainty, imprecision, is usually neglected in statistical inference. But real observed life times (data) are usually not precise. A possibility to model this kind of uncertainty is given by so called fuzzy numbers x^*

$$x^* = \varphi(x) \tag{1.3}$$

where $\varphi(.)$ is a measurable function defined for $x \geq 0$ obeying $0 \leq \varphi(x) \leq 1$ for all $x \geq 0$. These fuzzy numbers are describing the uncertainty of the observations. They are generalizations of indicator functions $I_A(x)$. For precise observed life times

$$x^* = t \in \mathbb{R}$$

we have a precise number and the corresponding specialized function is the one-point-indicator

$$\varphi(x) = I_{\{t\}}(x) \tag{1.4}$$

which is a special case of (1.3).

In figure 1 examples of fuzzy numbers are depicted.

Figure 1: Fuzzy numbers

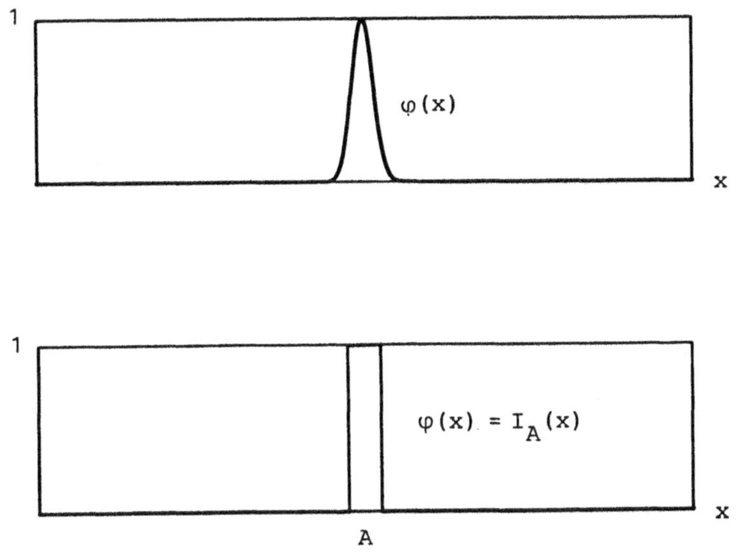

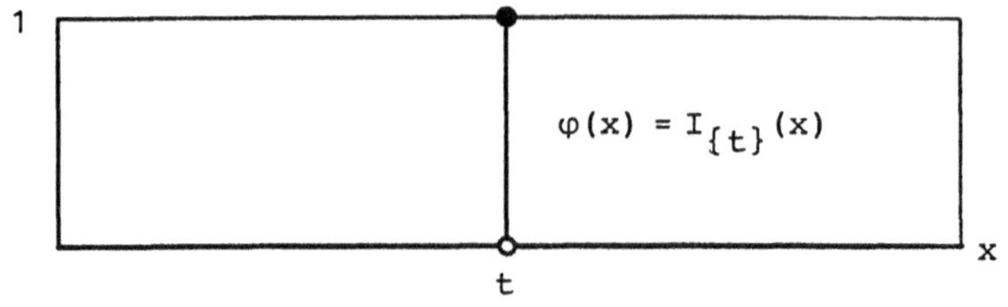

2. CLASSICAL STATISTICAL INFERENCE FOR FUZZY DATA

In classical statistical inference for samples $X_1,\dots,X_n$ from a stochastic quantity X, describing the life time of a component, with stochastic model $X \sim P_\theta$ the following procedures are used to estimate the life time distribution.

Point estimators
Confidence regions
Nonparametric estimators

2.1 Point Estimators

Let $T(X_1 \dots, X_n)$ be a suitable estimation function for the parameter θ in a life time model which yields the point estimator $\hat{\theta} = T(x_1,\dots,x_n)$ for precise life time data $x_1,\dots,x_n$.

For fuzzy data $(x_1^*,\dots,x_n^*) = (\varphi_1(x),\dots,\varphi_n(x))$ this estimation function T can be used to obtain a corresponding fuzzy estimator $\theta^* = \psi(\theta)$ in the following way.

The combined fuzzy sample point $\underline{x}^*$ in the sample space $M_X^n = [0,\infty)^n$ is defined by

$$\varphi(\underline{x}) = \prod_{i=1}^{n} \varphi_i(x_i) \qquad \text{for} \quad \underline{x} \in M_X^n \tag{2.1}$$

and the fuzzy subset $\psi(\theta)$ of the parameter space Θ denoting the fuzzy estimator θ^* of the (true) parameter is given by variation of $\underline{x}$ according to $\varphi(\underline{x})$ with corresponding grades of membership.

$$\psi(\theta) = \sup\{\varphi(\underline{x}) : T(\underline{x}) = \theta\}$$

By this procedure a fuzzy estimator for the parameter is obtained which takes care of the fuzziness of observed life times.

2.2 Confidence Regions

Let $\kappa(X_1,\ldots,X_n)$ be a confidence function for θ with confidence level $1-\alpha$. Then for each precise data set $\underline{x} = (x_1,\ldots,x_n) \in M_X^n$ we obtain a non-fuzzy confidence region $\kappa(x_1,\ldots,x_n) \subseteq \Theta$. For fuzzy data $(x_1^*,\ldots,x_n^*) = (\varphi_1(x),\ldots,\varphi_n(x))$ for all $\underline{x}$ the non-fuzzy confidence region $\kappa(x_1,\ldots,x_n)$ is calculated. From these different confidence regions a fuzzy confidence region Θ^* for θ is constructed using the combined fuzzy sample $\varphi(\underline{x})$ from equation (2.1). The membership function $\xi(\theta)$ of the fuzzy confidence region Θ^* is given by

$$\xi(\theta) = \sup\{\varphi(\underline{x}) : \theta \in \kappa(\underline{x})\} .$$

2.3 Nonparametric Estimation of the Reliability Function

Similar to the empirical reliability function $\bar{F}_n(.)$ for precise samples $X_1,\ldots,X_n$, given by

$$\bar{F}_n(t) = \frac{1}{n} \sum_{i=1}^{n} I_{(t,\infty)}(X_i) \quad \text{for } t \geq 0,$$

a nonparametric estimation for the reliability function using fuzzy samples can be constructed.

For fuzzy sample $\varphi_1(x),\ldots,\varphi_n(x)$ with integrable membership functions $\varphi_i(x)$, $x \geq 0$, the estimation $\bar{F}_n^*(.)$ of the reliability function (1.2) can be defined by

$$\bar{F}_n^*(t) = \frac{1}{n} \sum_{i=1}^{n} \frac{\int_t^\infty \varphi_i(x)\,dx}{\int_0^\infty \varphi_i(x)\,dx} \quad \text{for } t \geq 0 .$$

3. BAYESIAN INFERENCE FOR FUZZY DATA

In Bayesian Statistics the most important problem is the calculation of the a-posteriori distribution $\pi(\theta|D)$ of the parameter θ using sample information from the data D.

For precise data the a-posteriori distribution is calculated from the a-priori distribution $\pi(\theta)$ of the parameter and the likelihood function $l(\theta;D)$ by Bayes' theorem

$$\pi(\theta|D) \propto \pi(\theta).l(\theta;D). \qquad (3.1)$$

The analogue to confidence regions in classical statistics are so called highest a-posteriori density regions (HPD-regions).

3.1 A-posteriori Distribution for Fuzzy Data

For fuzzy data $D^* = (\varphi_1(x),\dots,\varphi_n(x))$ Bayes' theorem can be extended (see [5]) and a fuzzy a-posteriori distribution $\pi^*(\theta|D^*)$ is obtained in the following way. For independent observations $\varphi_i(x)$, $i = 1,\dots,n$ the combined fuzzy sample point $\underline{x}^* = \varphi(x)$ for $\underline{x} = (x_1,\dots,x_n)$ is given by

$$\underline{x}^* = \varphi(\underline{x}) = \prod_{i=1}^{n} \varphi_i(x_i) \ . \qquad (3.2)$$

For fixed $\theta \in \Theta$ by variation of $\underline{x}$ according to $\varphi(x)$ different values of $\pi(\theta|D)$ are obtained from (3.1). Therefore these values form a fuzzy number $\psi_\theta(.)$ obtained by

$$\psi_\theta(y) = \sup\{\varphi(\underline{x}) : y = \pi(\theta|\underline{x})\} \ . \qquad (3.3)$$

The fuzziness in the a-posteriori distribution is described by the family of fuzzy numbers $\psi_\theta(.)$ defining the fuzzy a-posteriori distribution

$$\pi^*(\theta|D^*) := (\psi_\theta(.);\theta \in \Theta). \qquad (3.4)$$

3.2 HPD-Regions for Fuzzy Data

The construction of an analogue to HPD-regions for fuzzy data is possible but a little more complicated. For details see [5] .

3.3 Fuzzy Predictive Reliability Function

Let $\overline{F}(t) = Pr\{X > t\}$ be the reliability function of the life time X of a unit. Then for precise data D the predictive reliability function

$$\bar{F}(t|D) = \int_{t}^{\infty}\int_{\Theta} f(x|\theta)\,\pi(\theta|D)\,d\theta \qquad (3.5)$$

is obtained in the usual way.
For fuzzy data $D^* = (\varphi_1(x),\ldots,\varphi_n(x))$ the combined fuzzy sample point $\underline{x}^* = \varphi(\underline{x})$ from equation (3.2) is used.
To fixed $\underline{x}$ the predictive reliability function $\bar{F}(t|\underline{x})$ is obtained by equation (3.5). By variation of $\underline{x}$ according to $\varphi(\underline{x})$ with corresponding grade of fuzziness a fuzzy number $\zeta_t(.)$ is obtained by

$$\zeta_t(y) = \sup\{\varphi(\underline{x})\,|\,\bar{F}(t|\underline{x}) = y\} \; .$$

The family of fuzzy numbers $\zeta_t(.)$ defines the fuzzy predictive reliability function

$$\bar{F}^*(t|D^*) := (\zeta_t(.);t \geq 0)$$

based on observed fuzzy life time data $D^* = (\varphi_1(x),\ldots,\varphi_n(x))$.

REFERENCES

1. J.O. Berger: Statistical Decision Theory and Bayesian Analysis, Springer-Verlag, New York, 1985

2. G.J. Klir and T.A. Folger: Fuzzy Sets, Uncertainty, and Information, Prentice Hall, Englewood Cliffs, N.J., 1988

3. R. Viertl: Is it necessary to develop a Fuzzy Bayesian Inference, in: Probability and Bayesian Statistics, Plenum, New York, 1987

4. R. Viertl: Statistical Methods in Accelerated Life Testing, Vandenhoeck & Ruprecht, Göttingen, 1988

5. R. Viertl and H. Hule: On Bayes' Theorem for Fuzzy Data, Res. Report RIS-1988-6, Inst. f. Statistik u. Wahrscheinlichkeitsth., Techn. Univ. Wien, 1988

Evaluation of the Probability Distribution of Evacuation Time from a Light Railway Transport Station

B. DORE, L. LAMBARDI, L. LAZZARI and V. SICILIANO*

Ansaldo S.p.A, Via G. D'Annunzio 113 - 16121 Genova - Italy

* Ansaldo Trasporti, Corso Perrone 25 - 16161 Genova - Italy

SUMMARY

In order to support the Naples Light Railway Transport design, Ansaldo performed a study aimed at evaluating the evacuation time t_e from the stations. This time depends on the parameters that determine station lay-out, accident propagation and people motion which are affected from considerable uncertainties. In this paper the methodology followed to evaluate the t_e probability distribution is presented. The analysis was performed through the following steps: uncertainty sources identification and estimation, sampling of the input variables of the deterministic evacuation model, numerical simulation of the evacuation process with the EPDES code, response surface synthesis using stepwise regression, uncertainty propagation by means of Latin Hypercube Sampling. Finally the numerical results were presented and the major findings of the study were discussed.

1. INTRODUCTION

In the last years many incidents occurred involving underground stations and some of them, like the fire at the London Uhderground King's Cross Road Station in 1987, caused injuries and fatalities. In order to prevent the occurrence of such accidents safety analysis is usually performed during the design phase of the project to identify possible system weak points and to propose system modifications that allow to improve design reliability and safety. The above mentioned analysis can be performed through the following two steps: identification of the relevant accident scenarios and quantitative evaluation of their consequences on people and on the system. For the accident consequences estimation three models have to be set up: an accident propagation model (aimed at evaluating the physical parameters to be used to assess the damaging effects of the accident), a people movement model (aimed at evaluating people time-dependent positions during the accident) and a vulnerability model for both people and the system.

In this paper an application of the above quoted approach to the Piazza della Vittoria station of the Light Railway Transport (LRT) line in Naples (the design of which is ongoing) is presented. The review of past accident reports pointed out that fires are by far the most relevant accidents occurring in underground stations. Due to the fact that only the preliminary design was available (with consequent lack of data about fire loads, fire detection and extinguishing systems, etc.) a simplified fire model was used. It was conservatively supposed that the fire could happen in every part of the station and it istantaneously causes the interruption of the exit routes passing by the zone affected by the fire.

An analysis of the evacuation from the station following the above mentioned accident scenario was performed in order to gain insight about the safety characteristics of current station lay-out. Evacuation analysis was aimed at evaluating the maximum evacuation time t_e which is the time all people initially in the station take to reach a safe place. It is to be noted that t_e depends on a number of parameters related to station lay-out, accident scenario (like fire position), system operating conditions (number of people in the station, train frequency, etc.) and human behaviour during the emergency. The values of the above mentioned parameters are affected from uncertainties due to 3 sources: inherent randomness of people behaviour and fire position, state of knowledge uncertainties about the model describing people motion and uncertainties in the input parameters of such model due to lack of data about people movement capability and the preliminary stage of the design. So t_e was regarded as a random variable and its probability distribution (PD) was evaluated by means of the propagation of the above mentioned uncertainties through the people movement model. It is to be noted that such PD could be used together with the PD for the time t_d a fire takes to harm people to evaluate the success probability of an evacuation given a fire in the station.

In the following the station is described and the deterministic people movement model used is summarized (section 2). Then the uncertainty propagation methodology followed is outlined (section 3 and 4), with emphasis on the various steps: input parameters identification, estimation of input parameter ranges, input parameters sampling and simulation of the evacuation process with the deterministic model, response surface construction, estimation of the input parameter probability density functions (PDFs), uncertainty propagation through the response surface. Finally the numerical results are presented (section 5) and the major findings of the study are discussed (section 6).

2. DESCRIPTION OF THE STATION AND OF THE EVACUATION MODEL

The station can be divided into three main parts (see fig. 1): the two platforms, the exit/entrance corridors and the compartments for personnel only. The station has four entrances, two for each platform, and a subway linking the two platforms. The corridors have a minimum width of 1.80 m. The minimum width of the platforms is 3.0 m and their rail-side length is 76.5 m.

In order to simulate people movement following a fire in the station the EPDES code (Emergency Plans DESign), developed by Ansaldo (see 8.) was used. The code has to be provided with a topological model of the system to be analysed in which the lay-out is divided into several zones linked together with routes. People's motion is simulated by calculating at each time step for each group of individuals the velocity as a function of people density, taking into account several parameters that affect the route practicability (like the presence of obstacles, stairs, etc.), people movement capability (like age and physical fitness) and the presence of narrow passages (like doors) on the escape routes. EPDES treats in a coherent way the interferences between different traffic modes (in this case pedestrians and trains). The code is also capable to search alternative routes to safe zones, allowing reversings and route changes, depending on the situation as known by single persons.

In order to set up an input deck available to EPDES the station lay-out (see fig. 1) was divided into 46 zones (zones n. 45 and 46 represent the places where the trains stand while zone 1 represents the safe place) linked together with 26 routes. Ten fictious gates were introduced in order to simulate the narrow passages of the exit routes as well as train doors.

3. IDENTIFICATION OF THE RELEVANT INPUT VARIABLES RANGES AND DENSITY FUNCTIONS

In this section the approach followed to identify the ranges and the PDFs of the parameters that significantly affect the evacuation time is presented. A preliminary analysis of station design and of the models implemented in the EPDES code showed that such parameters can be divided into two groups:

- model parameters (depending directly on the motion model implemented in the EPDES code), like the coefficients of the correlation between people velocity and people density, gate exit rate, etc.;
- input parameters (supplied by the analyst according to the system to be examined) like the user inlet flow, train frequency, fire position, etc.

For some of the input parameters selected in the preliminary analysis the influence on t_e was estimated by means of pilot EPDES runs. This led to exclude from further analysis the coefficients of the correlation velocity-density, and the stairway routability factor because the corresponding evacuation time variations were found negligible (i.e. less than 2%) for variations in the input parameters in the [-20%,+20%] range.

For the screened parameters (see table 1) the ranges were estimated on the grounds of data available in the literature, station and line design characteristics, and engineering judgement. It is to be noted that the number of people on the platforms or in the train at the moment of the accident varies during the day. The station occupancy state can be expressed in terms of the following flows: inlet flow in the station from one entrance (F_e), flow of people getting out of a train (F_d) on a platform, maximum flow of people getting on a train (F_{sm}) from a platform. F_d and F_{sm} were calculated as functions of train frequency (f), train utilization factor (u) (ratio between the number of people on the train when it stops and its maximum capacity) and station utilization factor (s) (ratio between the number of people getting off one train and the number of people on it) by means of the following relations:

$$F_d = f*C*u*s \quad \text{<1>}$$

$$F_{sm} = f*C*[1-u*(1-s)] \quad \text{<2>}$$

Due to the lack of information about the degree of symmetry of train and station utilization the same values of u and s were assumed for the two directions and equal flows Fe were used for the four entrances. It is to be noted that the number of people in the station also varies, about the daily trend, during a train period as people get on or get off from a train so a normalized initial time (i.e. the fraction of the train period at which the fire begins), uniformly distributed in the range [0,1] was introduced.

In the following some considerations is reported on the physical meaning attached to the range of variation of the input variables and their PDF. First the gate exit rate PDF takes into account a faulty knowledge of people dynamic behaviour near narrow passages and doors. For the various exits the gate exit rates were supposed to be equal with average value equal to 56 persons/(m*min) (see 2.-3.). In the case of people velocity the PDF accounts for different user movement capabilities due to age, physical fitness etc. The value of 1.3 m/s was taken as people mean velocity according to almost all existing references (see 1.-2.). Due to the fact that in this project phase information about the fire load distribution and the maintenance procedures was not available a uniform PDF over the user accessible parts of the station was assumed for the fire position. The distribution for the time t of the accident was used to account for the fact that a fire could happen during the various station operation phases (rush hours, normal operation, etc.). According to the subdivision of daily operation in intervals Dt_i, with train frequency f_i, the probability that a fire would happen during the ith interval was evaluated with the following formula:

$$P_i = Dt_i*f_i/(\sum_i Dt_i*f_i) \quad \text{<3>}$$

It must be pointed out that also the train utilization factor u and the entrance flow F_e were modelled as time dependent functions according to the laws reported on table 2.

Finally in order to model people behaviour in occupying the station platform an 'over-crowding repulsion factor' r was introduced in EPDES. So as people move in the station EPDES makes a comparison between the crowding of the jth zone A(j) and that one of the j+1 zone A(j+1). If the ratio A(j+1)/A(j) is smaller than r people occupying the jth zone move to the j+1 zone. For r a uniform distribution in the range [0,1] was assumed.

4. RESPONSE SURFACE SYNTHESIS AND UNCERTAINTY PROPAGATION

The response surface method was used to develop a reduced polynomial model that approximates the deterministic model implemented in EPDES when the input variables X_i (i=1,2,...,10) (see table 1) vary in the ranges determined in the previous section. Use was made of such response surface both to gain insight about the trends of evacuation time as related to the X_i's and to propagate the uncertainties by means of the Latin Hypercube Sampling (LHS) technique (see 6.-7.).

As a first step in the response surface synthesis the X_i's were sampled in the ranges listed in table 1 using LHS to develop a set of N (in this case N=28) 'experimental' input points W^j in the 10-dimension input space D. It is to be noted that the sampling of the X_i's was made according to uniform distributions in order to adequately account for the scenarios associated to points in D corresponding to the distribution tails. Then a set of N evacuation times t_{ei} was evaluated by means of N EPDES simulations performed using the input points W^j's. Each simulation was made through the following two steps:

- evaluation of people distribution in the station at the accident start as a function of time, gate exit rate, people velocity, station utilization factor, over-crowding repulsion factor and normalized initial time;
- quantification of the evacuation time from the station following a fire in a given position.

The EPDES responses were used to set up, by means of the stepwise regression technique (see 6.-7.), the response surface. It was found that two different functions had to be used to estimate the evacuation times corresponding to the following station configurations: A) the fire is occurring while the train is in the station; B) the fire is occurring while the train is not in the station. Due to the fact that a train stands in the station for t_o = 30 s the station configuration varies within each train period according to the value of the product of the train period times the normalized initial time. So the response surface was expressed by means of the following relationships (the values of the coefficients a_i's and b_i's are listed in table 3):

$$t_e = \begin{cases} a_0+a_1*X_{11}+a_2/X_2+a_3*X_9*(1-X_4*X_7*X_8/t_o) & \text{for conf. A} \\ b_0+b_1*X_1+b_2/X_2+b_3*X_8+b_4*X_{10}+b_5*X_7 + & \text{for conf. B} \\ \quad b_6*(X_{11}*X_1/X_2)+b_7*X_8*X_{10}+b_8*X_1*X_8 & \end{cases} \quad <4>$$

Finally the LHS technique was again used to propagate the uncertainties in the input variables (see table 1) to the evacuation time through the response surface. Account was taken for the constraints on the variation of the time dependent variables like train frequency, train utilization factor and entrance flow.

5. ANALYSIS OF THE RESULTS

In this section the numerical results of the study are presented. As mentioned before two different behaviours were found whether the fire happen in station configuration A or B. In the first case t_e strongly depends on the distance X_{11} from the ignition point to the nearest practicable exit. As 'correction' factors were introduced in the model people velocity and the parameter

$$Z = X_9*(1 - X_4*X_7*X_8/t_o)$$

that is proportional to the number of people in the train when the fire starts. The adjusted coefficient of determination R^2 of the regression equation <4.1> is equal to 0.968. In order to develop a good response surface for the cases corresponding to fires in station configuration B it was found necessary to insert in the reduced model 8 predictors. The first 5 were gate exit rate, people velocity, train period, entrance flow, and normalized initial time. The last 3 (see <4.2>) were proportional, respectively, to the time necessary for one person to reach a safe place from the fire position, the number of people that enter the station in a period, and to the number of people that can get out of the station in a period. R^2 for equation <4.2> is equal to 0.888.

The PDF and PD for the evacuation time are contained in figures 2 and 3 while the 10,50 and 90% percentiles are listed in table 4. The PDFs and the PDs for t_e conditional to fires in configuration A and B were also evaluated and the percentiles are listed in table 4. From the above quoted results it turned out that a fire when the train is in the station (this event has a conditional probability equal to 0.257) leads to an evacuation slower than in the other configuration (see table 4 and the two peaks in fig. 2).

The analysis showed that station lay-out allows to keep t_e finite (i.e. no people are trapped in the station only as a consequence of some escape route interruption due to fire) for every combination of station occupancy state and fire position. It was found that the most critical fire position (see the tail distribution in fig. 2) is located in zone 16 because in such case people located in zone 43 have to escape through the emergency sidewalk along the railway until they reach the other station.

5. CONCLUSIONS AND REMARKS

In this paper the methodology used to evaluate the probability distribution for the evacuation time from an underground station was presented together with numerical results. The above mentioned application pointed out the peculiarities of the methodology and the capabilities of the EPDES code. First of all the use of the response surface method to develop a reduced model allows to locally preserve the following features of the deterministic model implemented in EPDES:

- the capability to cope with complex lay-outs;
- the possibility of evaluating the time dependent position for individuals or groups;
- the capability to account for the architectural characteristics of the system (obstacles, narrow passages, stairs, doors, etc.);

- the possibility to take into account accident scenario related information (like escape routes practicability, etc.).

The outlined methodology also permits to use one of the typical Monte Carlo techniques (Latin Hypercube Sampling) and to incorporate in the t_e PD new information about the input variable PDFs, as it will become available, because the uncertainty propagation is made on the response surface.

Due to its modular form the proposed approach also allows to extract, during the various phases of the analysis, a lot of useful information about the sensitivities of t$ to the variations in the input variables caused by changes in the operating environment (like the time dependent trend of user flow) or in the station and line design (like the train frequency). From the above quoted considerations it follows that the described methodology can give suggestions to the designer in order to identify possible system bugs and further can be used to quantitatively evaluate the improvements of design modifications on system safety.

Finally it must be noted that further experimental and theoretical efforts should be addressed in the future to develop robust fire propagation models (taking into account radiation effects and the diffusion of the toxic combustion products) in order to be able to evaluate the success probability of the evacuation from an underground station given a fire.

REFERENCES

1. Fruin J.J.: Pedestrian planning and design. Metropolitan Association of Urban Designers and Environment Planners Inc., New York, 1971.

2. Melinek, S.J.; Booth, S.: An analysis of evacuation times and the movement of crowds in buildings . Building Research Establishment, Fire Research Station, Borehamwood, 1975.

3. Second report of the operational research team on the capacity of footways . London Transport Board Research Report No. 95, London, 1958.

4. D.M. 11 Gennaio 1988. Norme di prevenzione degli incendi nelle metropolitane .

5. Dati di Progetto per la Metropolitana Leggera di Napoli . Ansaldo Trasporti, July 1988.

6. Iman, R.L.; Helton, J.C.; Campbell, J.E.: An Approach to Sensitivity Analysis of Computer Models: Part I . Journal of Quality Technology. Vol. 13, No. 3, July 1981.

7. Iman, R.L.; Helton, J.C.; Campbell, J.E.: An Approach to Sensitivity Analysis of Computer Models: Part II. Journal of Quality Technology. Vol. 13, No. 3, July 1981.

8. Donati, A.; Lambardi, L.; Siciliano, V.; Silvestri, E.: Effective Emergency Planning Design by means of Risk Analysis Models. Proc. Int. Conf. on Emergency Planning for Industrial Hazards, Varese, November 4-6th, 1987.

TABLE 1 INPUT VARIABLE RANGES AND PROBABILITY DENSITY FUNCTIONS

N	VARIABLE NAME	UNITS	MIN. VALUE	MAX. VALUE	REF.	PDF	PARAMETERS a	b
1	Gate Exit Rate	$(m*min)^{-1}$	40.0	80.0	2-3	Beta	2.	3.
2	Group Basic Velocity	m/s	0.5	2.0	1-2	Beta	3.	2.63
3	Fire Position	-	2	27	*	Uniform	-	-
4	Station Utilization Factor	-	0.05	0.15	*	Uniform	-	-
5	Time	h	5.5	23.5	5	Histogram	-	-
6	Over-Crowding Repulsion Factor	-	0.	1.0	*	Uniform	-	-
7	Normalized Initial Time	-	0.	1.0	*	Uniform	-	-
8	Train Period	s	90.0	360.0	5.	**	-	-
9	Train Utilization Factor	-	0.3	0.9	*	**	-	-
10	Entrance flow	h^{-1}	700.0	1250.0	5.	**	-	-

LEGENDA REF. = References; PDF = Probability Density Function.
NOTE * This range was estimated using engineering judgement.
NOTE ** Train frequency and utilization factor,and entrance flow are time depende

TABLE 2 TRAIN PERIOD, ENTRANCE FLOW AND TRAIN UTILIZATION FACTORS FOR THE VARIOUS TIME INTERVALS

NUMBER	INTERVAL LOWER LIMIT (h)	UPPER LIMIT	TRAIN PERIOD (s)	ENTRANCE FLOW LOWER LIMIT (person/h)	UPPER LIMIT	TRAIN UTILIZATION FACTOR UPPER LIMIT	LOWER LIMIT
1	5.5	6.5	360	700	700	.30	.42
2	6.5	7.5	180	700	750	.42	.54
3	7.5	9.0	90	1000	1250	.78	.90
4	9.0	12.0	120	750	875	.54	.66
5	12.0	15.5	108	875	1000	.66	.78
6	15.5	17.0	120	750	875	.54	.66
7	17.0	18.5	90	1000	1250	.78	.90
8	18.5	21.0	180	700	750	.42	.54
9	21.0	23.5	360	700	700	.30	.42

TABLE 3 RESPONSE SURFACE COEFFICIENTS

COEFFICIENT INDEX i	COEFFICIENT VALUE a_i	b_i
0	-50.044	39.510
1	87.378	3.123
2	0.737	34.790
3	608.922	-2.515
4	-	-2168.530
5	-	-50.188
6	-	0.018
7	-	33.550
8	-	-0.045

TABLE 4 PERCENTILES OF THE EVACUATION TIME PROBABILITY DISTRIBUTION

PROBABILITY %	PERCENTILES (s) CONF. A	CONF. B	GLOBAL
10.	353.	42.	50.
50.	475.	100.	122.
90.	586.	350.	532.

NOTE CONF.A and CONF.B correspond to the evacuation time probability distribution associated to a fire given, respectively, station configuration A and B. GLOBAL correspond to the final evacuation time probability distribution that account for the probabilities of the above quoted configurations.

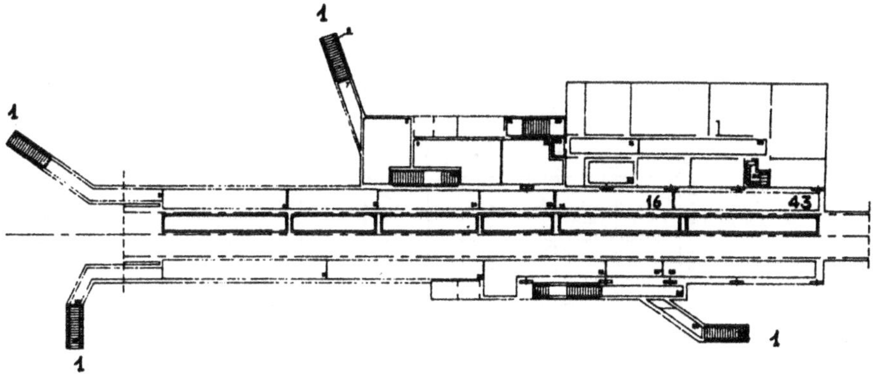

FIGURE 1   LAY-OUT OF THE LIGHT RAILWAY TRANSPORT STATION

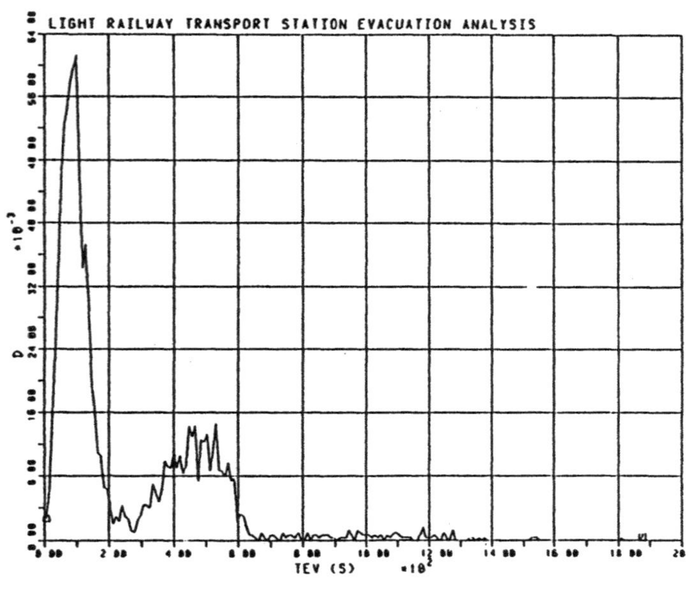

FIGURE 2 EVACUATION TIME PROBABILITY DENSITY FUNCTION

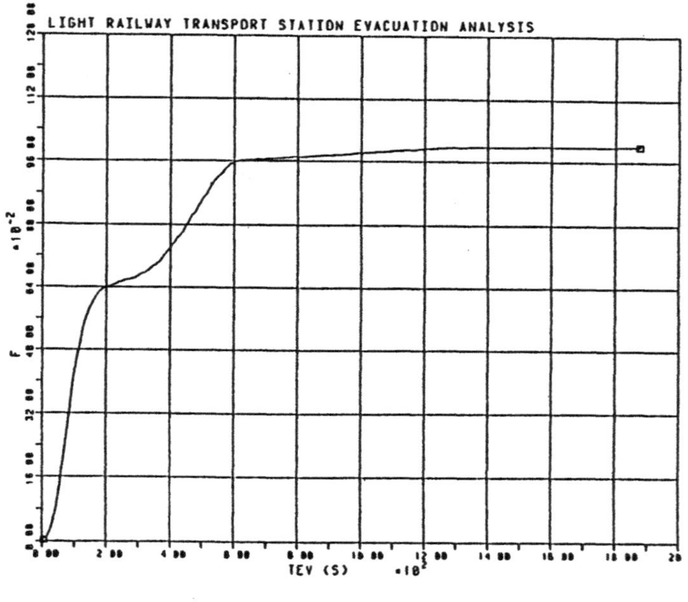

FIGURE 3 EVACUATION TIME PROBABILITY DISTRIBUTION

Computer Aided Reliability and Risk Assessment

R. Leicht and H.J. Wingender

NUKEM GmbH

P.O. Box 1313

D-8755 Alzenau, FRG

Abstract

Activities in the fields of reliability and risk analyses have led to the development of particular software tools which now are combined in the PC-based integrated CARARA system. The options available in this system cover a wide range of reliability-oriented tasks, like organizing raw failure data in the component/event data bank FDB, performing statistical analysis of those data with the program FDA, managing the resulting parameters in the reliability data bank RDB, and performing fault tree analysis with the fault tree code FTL or evaluating the risk of toxic or radioactive material release with the STAR code.

1 Introduction

The requirements of reliability and risk analyses have initiated the in-house development of a variety of computer codes which are tailor-made for the specific needs of the analyst. However, much information and valuable data were recorded on paper only in the past, and are sometimes difficult to retrieve. This was the situation when the need for an integrated software system comprising all these codes and data was recognized.

It was obvious that such a system should provide

- interactive or dialogue use,
- easy and fast access to desired information,
- low hard and software costs, and
- ability for an optimal design of the user-surface.

All these requirements are fully met by the modern Personal Computer (AT), which was therefore chosen as hardware basis of the system. Today, the integrated system called CARARA (Computer Aided Reliability And Risk Assessment) is realized to a great extent, and is continuously being further developed. The structure of this system and the modules forming it are described in the following.

2 Structure of the CARARA system

The structure of the CARARA system is graphically represented in fig. 1. The system consists of interlinked data banks and programs. The basis of all reliability work is the failure behaviour of components as recorded from either field or test experience. These data are organized and managed in the component/failure event data bank FDB. Homogeneous component subsets are identified, and the life data file is prepared for the statistical analysis with the program FDA.

The results of this analysis - or reliability parameters from external sources - are stored in the reliability data bank RDB and are available for use in fault tree or risk analyses.

The reliability of technical systems which are composed of a variety of components is evaluated with the fault tree code FTL. Input data required are information on the structure of the system to be analysed and reliability data of the basic events of the tree.

For evaluating the risk of accidental release of toxic or radioactive material into the environment with the STAR code, physical/mathematical models describing the transport of this materials must in addition be provided.

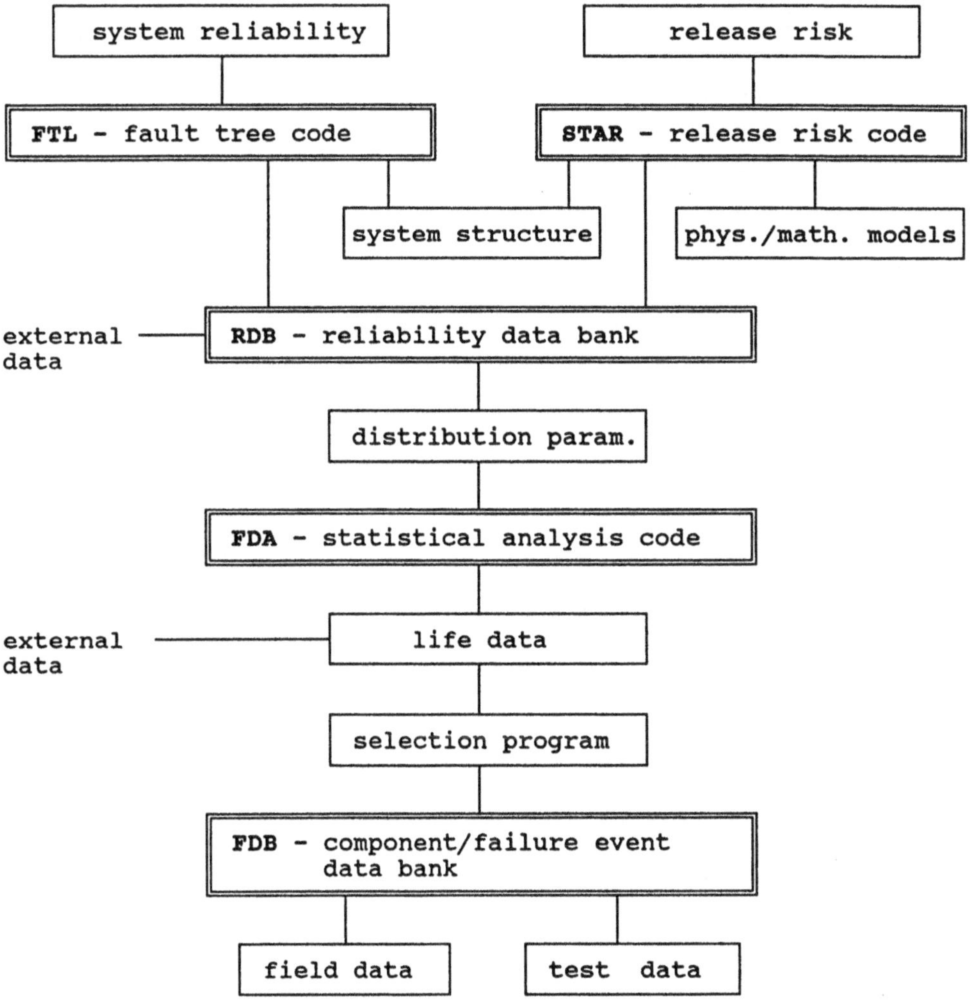

Figure 1: Schematical representation of the CARARA structure

3 The CARARA Modules

The five main modules of CARARA are described in the following. Although designed to operate within the integrated system, each of these modules is able to work independently, and is easily adaptable to specific requirements not offered by the system.

3.1 FDB - Data Bank for Component/Failure Event Data

The failure data bank consists of two principal data files:

- the component data base, and
- the failure data base.

Component specific information, such as its specification, design and operating conditions, and life history (i.e. a chronological list of operating and non-operating times) is stored in the component data base. In general, two failure modes are to be considered:

- failure during operation, and
- failure on demand.

In order to allow the evaluation of the reliability parameters, the observation period, the total operating time and the number of demands or operating cycles is recorded for every component. Since the components are assumed to be repairable, more than one failure (and different failure modes) can occur during the observation period.

Failure related information is stored in the failure data base. This information includes the chronological order of observed failures for each component, the particular operating times and numbers of demand, the failure cause, mode and degree, the failed parts, the resulting unavailability and repair time, the consequences of the failure, etc.. The component and the failure data base are linked via the component identifier.

Appropriate selection programs are used for identifying homogeneous component subsets - a task which cannot be performed without engineering judgement and experience. Finally, life data are provided for the statistical analysis with FDA.

3.2 FDA - Program for Statistical Analysis of Life Data

The statistical analysis is based on component life data such as:

- times to first failure, times between failures, and
- times without failure until the end of the observation period or replacement of the intact component (censoring times).

The aim is to describe these data with simple lifetime distributions such as

- exponential distributions and
- 2- or 3-parameter Weibull distributions.

The best fit parameters of these distributions are numerically evaluated according to the maximum likelihood method or by a least squares fit, in some cases also by a combination of both. The goodness of the fit is checked according to the Kolmogorow-Smirnow and Chi-square tests.

The results can be presented graphically as
- cumulative probabilities (see example plot in fig. 2),
- histograms used for the least squares fit,
- distribution density functions, and
- hazard rates.

In all these cases, the fit results are shown in comparison with the corresponding quantities derived from the observed data. Automatic scaling is provided as well as the option of linear or logarithmic axis and as exponential or Weibull paper plot.

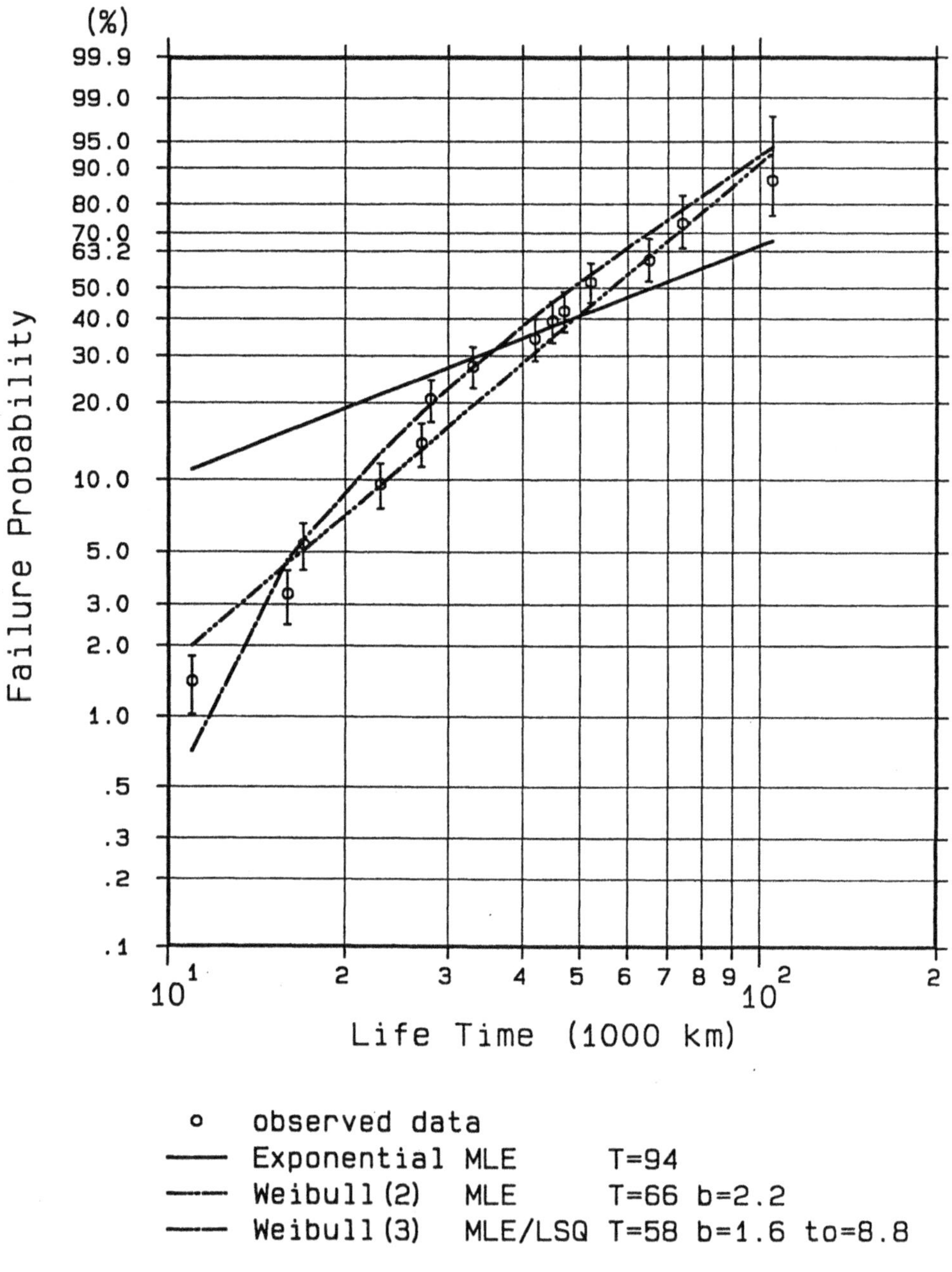

Figure 2: Cumulative probability example plot (FDA)

3.3 RDB - Data Bank for Reliability Data

The reliability data bank RDB provides component reliability data, such as failure rates, failure probabilities on demand, and repair times, for use in reliability or risk assessments. These data are either the result of a statistical analysis of field or test data with FDA, or are taken from other sources.

As far as available, the uncertainties related to these data are also stored, i.e. mean and variance (in the case of normal distributions) or median and error factor (in the case of log-normal distributions).

In addition to the reliability data, a characterization is given of the component, of the underlying operating conditions, and of various aspects concerning the failure mode.

Since published data from various reliability data sources in general do not meet the same information standards, it has been found appropriate to create a separate data base with a particular structure for some of these sources.

Access to the desired information on reliability parameters for a certain component is provided in a retrieval by menue-guided iterative selection and increasing degree of specification.

3.4 FTL - Program for Fault Tree Analysis

The fault tree code FTL is used to evaluate the reliability of technical systems which consist of a variety of repairable components. The following options are offered:

Fault tree data management

All fault tree-related data are taken, stored, and managed in a special data base. These data include:

- the system structure, i.e. the gates and gate inputs,
- the component type of the basic events, and
- describing texts for the gates and events.

Fault tree plot

Graphical output of the fault tree is provided in various sizes and in international representation or according to DIN. An interface to a VAX-CAD station allows plotting very large fault trees on a CAD plotter.

Qualitative fault tree evaluation

The minimum cut sets of arbitrarily selectable gates of the fault tree are evaluated, thus giving qualitative information on the importance of certain basic events or on critical failure combinations. No reliability data are required for the basic events, and the applied cut-off criterion is the number of cut set elements.

Quantitative fault tree evaluation

Based on the reliability data of the basic events, the following quantities are calculated for the selected top event:

- annual failure frequency,
- unavailability,
- mean system down time,
- probability of exceeding tolerable down times.

The computing time required for the minimum cut set evaluation is essentially reduced by applying effective cut-off procedures. Those cut sets which contribute less than a specified amount to the top event failure frequency or unavailability are rejected in the early stage of evaluation.

The following 3 component models are available:

- repairable components which fail during operation, the failure being immediately indicated or recognized by the operators, described by the failure rate and repair time,

- periodically inspected stand-by components, with failures detected only by inspection, described by failure rate and inspection interval, and

- stand-by components or components with intermittent operation, failing on demand, described by a failure probability on demand and by a repair time.

3.5 STAR - Program for Evaluating the Release Risk

The risk of accidental release of toxic or radioactive substances from technical installations into the environment is generally caused either by the failure of technical components, or by human failure. In the scenario treated by STAR, the toxic agent is contained in a system of barriers, which consist of technical components and thus can fail. In the case of failure of a barrier, a flow of toxic material from the inventory (the source) into the environment is either created, or an already existing flow is increased.

Source and barriers form a system, which most of the time is in its intact state. Sometimes, however, transitions into failed system states occur, i.e. into states with at least one barrier failed. There is a released flow of toxic material associated with every system state, and the quantity released then depends on the duration of this state. Physical/mathematical models must be used to calculate the material flow and the effects of the barriers in the various states. As a simplification, the time dependent flow is appproximated by quasistationary phases of constant flow. This is the deterministic part of the risk evaluation.

The probabilistic part evaluates the frequency of a certain failed system state and the distribution of its duration. This evaluation is based on fault trees for the failures of the particular barriers, similar to the procedure used in FTL.

The release risk of a certain failed system state is the amount of toxic material released in this state, multiplied by the annual frequency of its occurrence. The release risk of the total system is given by the sum of the contributions of all failed system states.

The results of the risk evaluation with STAR can be provided graphically in form of complementary cumulative distribution functions (CCDF's), or as frequency-consequence diagrams, as shown in figs. 3 and 4.

4 Conclusion

The modules of the CARARA system cover a wide range of reliability tasks. The integration and improvement of formerly independent codes and data bases has eased reliability engineering tasks, which for instance require the design and realization of a reliability test programme. Although the system is quite a support for the engineer, it has its limitations. For instance the treatment of time dependent failure rates and the consideration of uncertainties according to the Monte Carlo method has not been included yet.

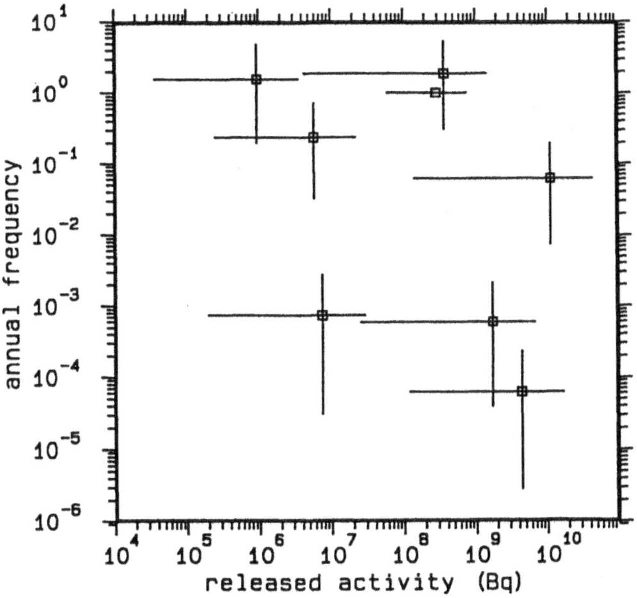

Figure 3: Example of a STAR frequency-consequence diagram of a radioactive waste tank system, each point representing a system state, and error bars indicating uncertainties

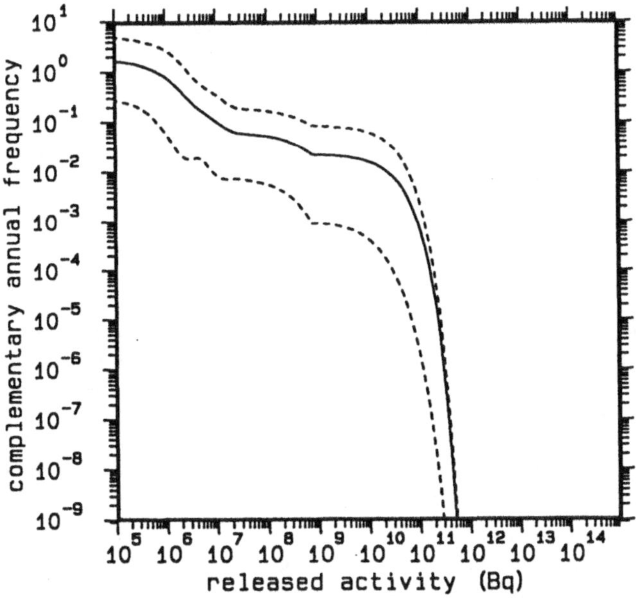

Figure 4: Example of a complementary cumulative distribution function (CCDF) and its uncertainty band, provided by STAR for a radioactive waste tank system

Hazard Identification: A Proposal for a New Development

S. Contini
Commission of the European Communities
Joint Research Centre, Ispra Establishment
21020 Ispra, Varese, Italy

N. Labath
Consultant, Via Cee 18, Ispra, Varese, Italy

Summary

Risk assessment is a complex task performed to quantify and minimise the hazards associated with potentially dangerous installations. The hazard identification represents the first step and can be performed by means of different methodologies. Some of these methodologies allow the collection of information on the possible plant behaviours in a structured way. However, this information cannot be easily used for further investigations (i.e. accident quantification).
The paper suggests the application of a simple technique for hazard identification and fault tree construction, that can easily be implemented in a computer program.

1. HAZARD IDENTIFICATION AND QUANTIFICATION METHODOLOGIES

Risk assessment is a complex task performed to quantify and minimise the hazard associated with potentially dangerous installations. Broadly speaking, the risk assessment procedure can be subdivided into the following phases:
- hazard identification (what can go wrong)
- frequency estimation (accident occurrence probability)
- consequence evaluation (damage to people, environment, property)
- risk quantification

The hazard identification represents a very important phase, since only knowing the possible hazards one can proceed with the evaluation of the efficacy of the existing preventive and protective measures. It has to be stressed that those hazards that are not identified in this step cannot be found in the subsequent phases. In practice, due to the complexity of the hazard identification problem, the limitations of the available methodologies and the not perfect knowledge of the plant, it is difficult to find out all existing hazard.

The hazard identification can be performed by means of different methodologies /1/ such as:
- Process / System Checklist
- Safety Review
- Preliminary Hazard Analysis
- Hazard and Operability Analysis
- Failure Mode, Effect and Criticality Analysis
- What - if Analysis

Other complementary techniques can be applied to get a more complete list of hazardous situations. For instance, the AEA (Action Error Analysis) /2/ is useful to investigate about the effects of human errors in operation and maintenance tasks, especially for batch plants.

A very useful tool in hazard identification is the accident data base. For plants for which a significant experience on past accidents or near misses is available, the analyst can check whether the accidents recorded have been identified in the study.

The second part in risk analysis is the hazard quantification. It can be subdivided into two main steps:
- determination of the major hazard frequencies
- determination of the potential damage to people, environment and property

The first step requires the construction of a model for each major hazard; the fault tree technique is generally applied to this purpose. This step rather time consuming, and therefore it is desirable to be able to easily use the information already collected during the hazard identification phase.

Tab I contains the main features of the methodologies for hazard identification.
Being both the completeness and systematicity two very important features from the hazard identification viewpoint, the structures of both FMEA and HAZOP may represent a good starting point for developing a new systematic procedure which allows an easy quantification step.
The FMEA allows the identification of the causes and consequences of predefined component failure modes. This methodology, widely applied in the nuclear industry, is suitable in the early design stages, and the information collected can be periodically updated to describe design modifications during the project evolution. Even if this method presents a high degree of systematicity, it does not allow the straightforward use of information collected for the frequency estimation.
The aim of the HAZOP is the identification of safety and operability problems through the analysis of causes and consequences of predefined abnormal process variable deviations.
The methodology is generally applied when the plant is in a well defined stage. For this reason the study is performed by a multidisciplinary team who, through a series of brainstormings, examines the plant, section by section, to find out any type of safety and operability problems. However, it is not the purpose of the study to perform a detailed description of causes and consequences of the process variable deviations, and therefore it is difficult to construct any fault tree from the information collected. In practice, whenever the frequency quantification is needed, a separate fault tree analysis has to be performed.

Very often, however, there are situations in which it is not feasible to perform a complete HAZOP analysis as in the classical way because of the difficulty of keeping the multidisciplinary team for the time needed to perform the study. Think for instance of the case of existing plants for which the safety report is required. In these situations a suitable methodology should allow a systematic generation of all sequences leading to the hazardous situations, and in such a way so as to be able to use the same information for easily constructing the fault trees. Therefore, the examination of the actions to be taken to improve the plant safety has necessarily to rely on the results of the quantitative analysis (i.e identification of the most critical components and/or on human errors).

The characteristic features of a technique for hazard identification and quantification can be summarized as follows:

- systematicity;
- degree of completeness of the results;
- ease of application;
- systematic construction of fault trees;
- suitability of being implemented in a computer program to enhance the completness and to reduce the cost.

An attempt in this direction is described in /3/, where the operability analysis is followed by the generation of fault trees.

The key point of the procedure is the preliminary subdivision of the plant into disjoint subsystems whose boundaries are identified by a set of nodes on the input / output flow lines. These nodes also represent the points in which the different subsystems interact with each other. A subsystem can be identified by grouping together components performing the same process function, belonging to the same protective system, etc. Experience generally results in optimum decomposition schemes.

The propagation of the perturbation is then performed by studying the causes and consequences of process variable deviations at the input and output nodes. When necessary, internal nodes can be defined to better describe how the deviation can propagate within the subsystem. During the study, the attention of the analyst is mainly focussed on the components belonging to the subsystem, even though the overall view of the plant must be kept in mind to avoid misleading interpretations.

A special notation is adopted to describe the causes and consequences of the process variable deviations as well as the failure of the protective devices.

Once the study has been completed it is possible to select the hazardous situations of interest and to generate the fault trees for quantification. The experience gained with the application of this procedure has shown that the time needed to perform the hazard identification is greatly compensated by the little effort required to construct the fault trees.

In the next section the main concepts of a simple, systematic and easily automated technique for hazard identification are presented. Details of the technique can be found in /4/.

2. BASIC CONCEPTS OF THE PROPOSED TECHNIQUE

The basic concepts can be shown by looking at the typical structure of a fault tree (see Fig. 1a)
The top-event, describing the hazardous situation of interest, is related to its basic causes by means of AND - OR operators.
Let a path be defined as a sequence of intermediate events starting from a basic event towards the top-event.
The generic intermediate events in the path can be either of AND or OR type, according to the operator that links it to its direct causes.
An example of path, for the primary event C, is shown by heavy lines. Branches represented by dotted lines indicate conditions that must be satisfied for the top-event to occur once the event C is occurred.
It can easily be realised that any fault tree can be decomposed into paths of this type. Therefore, it follows that starting from failure modes of components (initiating events) and developing, for each of them, the corresponding paths (see Fig. 1 b) it is possible to identify the hazardous situations (top-events) and to collect the necessary information for constructing the fault trees in a systematic way.

Selection of the initiating event

An initiating event can be the failure mode of a relevant component. A component is relevant if at least one of its failure modes directly generates a perturbation in the plant in terms of deviation of one or more process variables.
For example, in a flow control loop the component that, in case of failure, directly produces a deviation of the controlled process variable is the valve. Indeed, the failure of the other elements have the effect of moving the valve to the wrong state: in other words the direct effect is on the valve and not on the process variable. The same considerations hold for a protective circuit : only the final actuators have to be considered as relevant components.
This criterion allows to limit the analysis to the smallest number of components without affecting the completness of the study. Furthermore, as the information collected is specifically related with the process, the effort needed to identify the hazards is the minimum required.
For the relevant components the study of the perturbations in the plant must be performed for each failure mode. For this purpose a data base of failure modes of generic components, to be used as a check list, can significantly improve the completeness of the analysis.
Apart from component failures, perturbations can also be initiated by external events and deviations of the process variables on the input flow lines of the plant under study.
However, for the sake of simplicity and without loss of generality, in the following discussion these initiating events will not be explicitely considered.

Description of the perturbations

The failure mode of a relevant component generates a deviation on one or more process variables in some particular points (nodes) of the plant. Each of these points is described by means of a triplet < Deviation, Variable, Node >. These deviations are then properly propagated up to a hazardous situation is reached. This procedure is obviously applied to all failure modes of all relevant components. The main rules to be applied to initiate and to propagate the perturbations on the P&ID are listed below.

1). A triplet has to be associated with the component output.
2). The other points of the P&ID in which the triplets have to be described are those where at least another condition must be satisfied for the deviation to propagate. (Looking at the fault tree structure these correspond to the AND intermediate events). Note that the analyst, however, need not to verify the real existance of the necessary condition(s): this will be checked at the end of the analysis in a way that can easily be automated.
3). When the propagation of a deviation results in a triplet previously defined it is no longer necessary to proceed (except in some particular cases).
It follows that the effort needed to propagate the deviations decreases as the number of initiating events already examined increases.
4). A path terminates when the consequence of the deviation being studied is either a hazard or an existing triplet.

Fault tree construction procedure

Once all paths have been developed the fault tree construction procedure can be initiated for all the interesting hazards.
Briefly speaking, for a selected top-event all paths containing it are ORed and those ending with the necessary conditions are searched for among the remaining paths.
During this process some rules are to be applied to prevent the generation of both logical loops and redundant subtrees.
Actually, these fault trees contain only the failure modes of relevant components (as well as external events and plant input variable deviations), and therefore they have to be completed by developing all their causes. This step of the fault tree generation procedure can be automated by using the existing components failure models /5/.

3. CONCLUSIONS

In risk analysis the hazard identification represents a fundamental step. Different techniques were developed for performing this task, but they do not allow the rapid identification of hazardous situations and fault tree construction for quantification.
A proposal for applying a technique to deal with this specific problem has been described. Compared with existing techniques, the main advantages are a reduction of the effort needed to identify the significant hazards and the use of the information

collected for constructing the fault trees for the quantification stage.
The proposed approach can easily be automated to obtain an interactive computer aided tool to be integrated with a fault tree analysis code.

4. REFERENCES

/1/ Lees F., " Loss Prevention in the Process Industries ", Volume 1, 1980

/2/ Suokas J., Accident Analysis and Prevention, 20, 67, 1988

/3/ Messina S., " FMEA e Analisi di Operabilita' " Ispra Course AAS, 1985

/4/ Contini S. and Labath N., " A Simple Technique for Hazard Identification and Fault Tree Construction " CEC-JRC internal report in preparation

/5/ Poucet A., "CAFTS : Conputer aided fault tree analysis" ANS / ENS Int. Top. Meet. on Probab. Safety Methods and Applications, San Francisco, 1985

Table 1. Main Hazard Identification Methodologies

METHODOLOGY	PURPOSE	TYPES OF RESULTS	APPLICABILITY	COMPLETENESS	ADVANTAGES	LIMITATIONS
Process / System Check list	Identify common hazards Ensure compliance with standard procedures	Common hazards Deviation from procedures	Design Construction Sturt-up Operation Shut-down	Depends on the completness of the check list	Ease of application Applicable to the different project stages	Not detailed results
Safety review	Verify that design operation and maintenance satisfy design goals	Deviations from any kind of procedures	Operation (periodically performed depending on risk level)	Limited to major hazardous situat. Depends on staff experience.	Ease of application	Existing plants only. Experienced people require.
Preliminary hazard analysis	Early identification of hazards	Guidance for subsequent design phases	Early design phase	Limited to hazard associated with substances and main equipments	Ease of application	
Hazars and operability analysis (HAZOP)	Identification of hazards and operability problems	Reccomendations of plant modifications to solve identified probl.	Nearly firm design New and exsisting facilities	High	Very detailed results	Considerable manpower and time required. Not all causes investigat.
Failure mode and effects analysis (FMEA)	Identification of effects of single component failure modes	Reccomendations of design modifications to solve identified probl.	Design(all phases) Construction Operation	High	Very detailed results	Only single components failure are considered. Considerable manpower and time
What - if analysis	Identification of accident sequences hazards and consequences	List of accident scenarious and consequences	Process develop. stages Pre start-up stage Existing plants	Depends on staff experience	Powerful procedure for experienced exgineers	Considerable manpower and time required

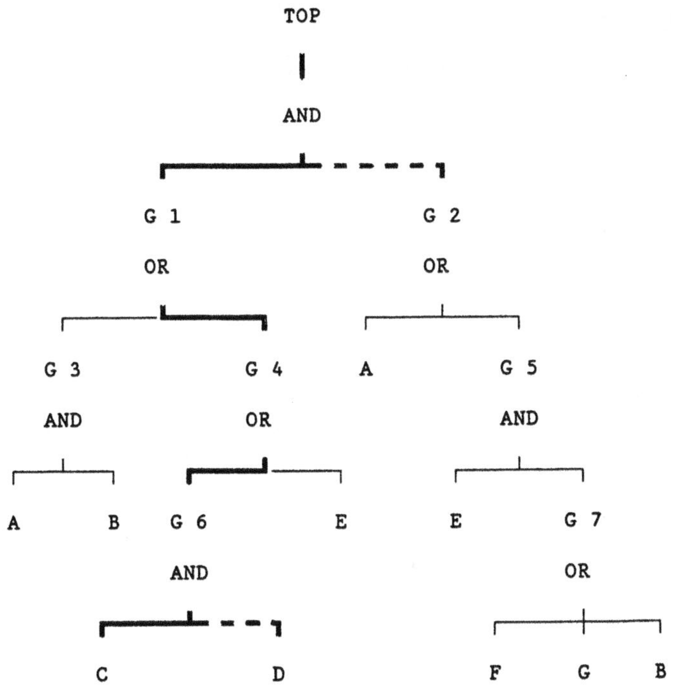

Fig. 1.a) Example of a fault tree.

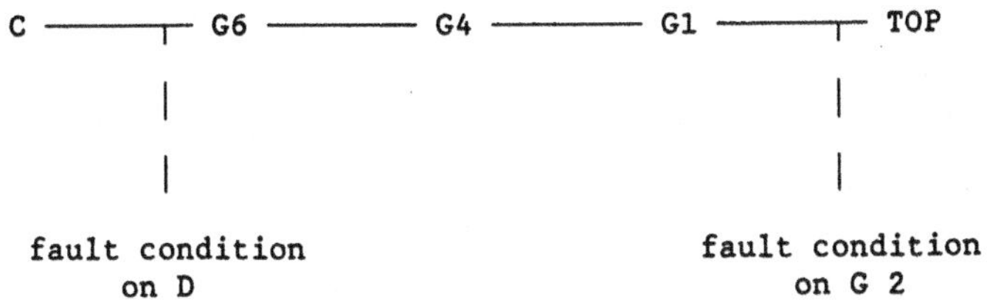

Fig. 1.b) Path from C to the TOP event.

A Prototype Expert System to Perform Safety Evaluation of Alternative Architectures for Offshore Platforms

Lancia A. (Tecsa Srl - Levate - Italy)

Scataglini L. (Agip Spa - Milano - Italy)

Bortolazzo L. (Agip Spa - Milano - Italy)

Romano A. (Tecsa Srl - Levate - Italy)

Summary: Basic safety evaluations should be carried out before all the key design decisions are taken. This doument describes the architecture of an expert system prototype in order to perform safety analysis of alternative architectures for offshore gas platforms.

Keywords: Expert Systems, Probabilistic Risk Assessment, Safety, Offshore Platforms, Layout Safety Evaluation

Introduction

In the standard design procedure for offshore installations, safety is usually not sistematically considered during the early stage of the work.
In some cases, e.g. when following NPD regulations [1], some guidelines are followed but often safety principles are only taken into account at variable extent according to the attitudes and the background of the engineers preparing the platform layout.
Layout evaluation guidelines are anyway restricted to very basic statements such as the need for safe escape routes etc.
Systematic safety evaluation is therefore carried out only at the last stage of designing sothat any suggestion for safety improvement which involves a substantial change in the platform architecture, is not considered.
This traditional way of working does not necessarily limit the platform safety, but the need to meet a certain maximum risk level, when working at a late design stage can lead to a substantial cost increase which could be avoided with a first optimation of platform basics.
Although if these concepts have been expressed plenty of times by safety people [2,3] and they are also related to the majority of the industrial installations characterized by

potential high risks, they are not currently put in practice for many reasons, including the will to avoid a further time -taking analysis procedure in the design iter.

These considerations brought Agip to consider the possibility of using a software system devoted to perform a fast preliminar safety analysis of alternative platform architectures, in order to select the best basic layout without a substantial slowdown of the job process.
The idea of such a system is not new and some efforts have been made by others, mainly using pure statistical approaches or building traditional "software tools" [4] to make the analyst's job easier.
Agip was seeking something more i.e. a system being able to substitute as much as possible the analyst and expanding the analysis toward the limits of the system description at the layout stage and of the project cost.
The feasibility of the system was stated by Tecsa which was then commissioned to develop a prototype.

Problem constraints and system requirements

The main constraint for the development of a system to perform the analytical evaluation of a platform layout is a rough knowlegde about the design and the subsequent need to assume standard or statistical evaluations for many parameters affecting the probability and the expected magnitude for the possible accidents.
The system is infact bound to perform the analysis only when the following information is available:

- general layout of the platform including main structures, decks floors, stairs, escape routes, lifeboats, helydeck, living quarters, flare stack, etc.
- location and main characteristics of the whoule production, process and service equipments (separators, pumps, compressors, wellheads, risers, generators, etc.)
- simplified P&I diagrams
- location of the main flowlines and valves
- position and classes of the fire-resistant and blast-proof walls
- types and locations for detection and fire protection systems.

This rough knowledge affects the possible accuracy for many parameters including failure rates, availability, vulnerability to fires and explosions, etc.
Such introduced uncertainties, the precision limits inherent in the scope of work and

the development cost, all suggest together that no major effort should be made to bring the system performance very close to the theoretical maximum accuracy compatible with the design details.
We can summarize some of the main requirements for the system in the following items:

- accounting for a series of standard safety regulations and estabilished design principles
- performing automatic classification of areas
- constructing automatically the event trees and calculating the risk levels associated to accidents
- allowing human intervention to modify part of system default data and procedures
- producing a well readable output pointing out the key points principally contributing to the most relevant accident risk levels
- allowing short input times for platform description
- facilitating the modification of an already described design
- possessing a user friendly interface.

General Description of the system

The system is being built within an Expert System Shell and is made out of two main modules; the first module is devoted to interact with the user to set up the platform description and performing the basic design checks while the second one performs the PRA with minimum interaction with the human analyst.

Among the many conceptual and technical problems to be solved for designing the system, we had to find a representation of the platform being suitable for the analysis process and convenient for the processing structures.
We decided to divide the platform in areas to be defined as "compartments", where there is no internal separation for accidents such as fires and explosions.
So, a platform area or block does not coincide with plant units or systems and is purely safety oriented.
Plant units and flowlines are therefore positioned within or across the above defined platform areas.
Accidents chains are defined, generated and quantified in terms of initiating events, vulnerable items, consequences models and propagation barriers.

Functional description of the system

The first module begins its dialogue with the user by asking a series of general questions about the platform typology such as manned or unmanned, single platform or cluster, production only or simultaneous production and workover, geographical location, regulations to be considered, etc.
The answers to these questions are used to trigger different sets of rules and subprocedures by the subsequent steps.
The system then goes to the phase of interactive definition of the platform design: this is presently made by a graphic software interface within the expert system and it will be replaced by a transcoding procedure getting the design data from a CAD file.
The description includes the plans of the decks, the separation elements and all the equipments; all the characteristics of the items can be taken from default definitions and/or edited.
For system components which are not defined in quantity and position at layout stage but give a major contribution to accident probabilities (e.g. flanges and instrumentation piping), mean values are taken on the base of statistics on existing platforms.
As far as detection and fireprotection systems are concerned, they are defined in terms of reliability and performance standard requirements.
The simplified P&ID's are input at a third stage after the definition of the relevant hardware. Items included in P&ID's are labelled with information concerning their vulnerability and dependance on electrical or compressed air supply, sothat fault trees can be automatically generated.
Area definition is made by the user and validated by the system applicating sets of rules deriving from the relevant regulations or the internal safety policy (e.g. incompatibility within an area, ignition sources, etc.).
Areas are automatically classified (e.g. according to CEI 886 norms) and the user is asked for the request of possible exceptions in the classification process.
During the execution of the tasks so far described, the system performs series of checks according to coherence of description and accordance with safety and design rules.
After the description stage, the system performs all the other programmed diagnostics of the layout before giving the greenlight for the execution of the PRA module.

The second module builds all the relevant event trees by expanding the initiating events associated to the equipments. The expansion of a tree branch is truncated by the existence of an effective chain breaker such as a firewall. This is accomplished by an action shading algorithm that identifies for each event the items exposed or sheltered. Fault trees are built from P&ID's to calculate the probabilities of certain initiating events

and the consequences of certain damages from fires or explosions on the plants behaviour.
Probabilities of occurences are taken from an internal database unless the user already overrode this by having supplied alternative values. For the calculation of the probabilities of chain transfer the system uses both statistical values and calculation procedures implemented in FORTRAN and automatically called by the expert system.
The procedure ends with the preparation of a printed report in which the system is highlighting the points which determine high risk levels; the analyst is so guided in the subsequent steps consisting in modifying the design and reexecuting the system to quantify the associated benefits.

Software Implementation

For the implementation of our system we have chosen the shell "ART (Automated Reasoning Tool by Inference Corp.); our choice was motivated by several reasons including our positive experiences in the application of this package to the construction of other expert systems related to safety analysis problems; ART revealed to be particularly appealing for features like:

- completness of development facilities
- easy integration with external programs
- availability of alternative built-in inference procedures
- transferability to different hardwares and operating systems
- simplicity of use
- easy building of user friendly program interfaces
- high speed and partial self optimation.

At the moment we are running our prototype system on a DEC Microvax GPX-II under VM operating system.
External programs are written in FORTRAN 77 ANSI.

State of development

As of today (November 88) we have completed some parts of the first module and a series of general submodules.

In particular the following software parts have been implemented and tested:

- preliminary input of platform basic characteristics
- interactive graphic definition/editing of platform layout
- event shading algorithm
- automatic event tree expander.

We currently foresee to complete the first module within May 89 and the whole system by November '89.
Some documents will be published after the project completion.

Conclusion

We have given a short description of an expert system prototype we are developing for performing comparative safety analysis of alternative platform architectures.
The system seems to be a cost effective solution to make preliminary safety evaluations at layout stage incorporating both conventional evaluation rules and simplified PRA.
The system is now written for analysing gas offshore platforms but it could be the precursor of a family of systems addressing oil platform and on-shore installations safety; moreover it could evolve in a much more complex and accurate system to perform second stage analysis based on detailed design documents, produced in a computerized form and automatically acquired by the expert system.

Aknowledgments

We would like to thank the colleagues who contributed to the feasibility study and to those who are currently cooperating with us in the development; among others Aldo Maurino, Daniel Grazioli, Giorgio Gentile, Carlo Fiorentini and Gianfranco Fassera.
We express our gratitude to Carlo Iacone of Agip for his kind permission to publish this paper.

Reference:

[1] Norwegian Petroleum Directorate: "Guidelines for Safety Evaluation of Platform Conceptual Design", 1.9.81

[2] Andersen R.S., Engen G., Vinnem J.E., Schmidt Pedersen K., Emblem K., Amdahl J. : "Risk Analysis in Offshore Development Projects", Sintef Report STF18 A83503

[3] Vinnem J.E.: "Quantitative Risk Analysis in the Design of Offshore Installation", Reliability Engineering 6, pp 1-12 (1983)

[4] Haugen S., Vinnem J.E.: "Safety Assessment of Offshore Installations Using a Computer Model", in "Reliability Data Collection and Use in Risk and Availability Assessment"; Proceedings of the 5th EuReDatA Conference, Edited by H.J.Wingender, Springer Verlag, 1986.

Report on the On-Going EUREDATA Benchmark Exercise on Data Analysis

A. BESI, A.G. COLOMBO

CEC-JRC-Ispra (Va) - Italy

1. Introduction

In April 1987 the JRC was charged by the Assembly of the EuReDatA members with the organization and the coordination of a Benchmark Exercise (BE) on data analysis. The main aim of the BE is a comparison of the methods used by the various organizations to estimate reliability parameters and functions from field data. The reference data set was to be constituted by raw data taken from the Component Event Data Bank (CEDB), managed by the JRC. The CEDB is a centralized bank, which collects data describing the operational behaviour of components of nuclear power plants operating in various European Countries.

In July 1987, the terms of reference of the BE (see Appendix 1) and a description of the main features of the chosen data set were sent by the JRC to the EuReDatA members as an invitation to participate.

The Organizations who agreed to participate are: INTERATOM, NUKEM (FRG), VTT (SF), UKAEA, University of Bradford (UK), SINTEF (N), ENEA VEL Bologna (I), JRC (ATI Division), VATTENFALL (S), a composite working group ENEA TIB AQ Roma (I) - Ed F (F) - JRC Ispra (SER Division). The coordination of the BE was assigned to the authors of this paper that were also charged to draw up this report.

2. First phase of the BE

2.1 Terms of reference of the BE and data set used for the analyses

2.1.1 The terms of reference for the first phase of the BE, which were suggested by the JRC in July 1987, are given in the appendix. They specify the aim of the BE, the methods proposed for the analyses and the main features of the reference data set.

2.1.2 In January 1988 a first set of raw data was distributed to the participants; an up-dated version of the data set, in a matrix format, more suitable to be treated by PC, was mailed at the end of June 1988. This reference data set concerns about 450 pumps pertaining to 16 European power plant-units (10 PWR, 2 BWR, 4 conventional), monitored for an average time period of 5 years, during which about 1200 events occurred. For each pump, identified by its engineering and operating characteristics, a description of the occurred events (complete failures, or performance degradations, or simple operation anomalies, and related repairs) is given. The operating rules of the CEDB define some precise restrictions in the information that can be made available to non-data supplier organizations; this to meet the basic requirement established by data suppliers that the plant has to remain anonymous. As a consequence, the participants received a data set where the plant name is masked (plant 1, 2, 3, ... , 16), the plant power level is not indicated, the free (i.e. uncoded) failure descriptions have been partially censored.

2.2 Short description of the approaches adopted in the intermediate reports

2.2.1 On the occasion of the EuReData Assembly meeting in Stockholm on September 1988, a meeting of the Organizations participating in the BE was held.

The aims of the meeting were discussing the problems which had arisen in the interpretation of the raw data, comparing the preliminary analysis results and possibly re-orientating the activities of the second phase of the BE.

As to the data interpretation, the participants in the meeting remarked on some incompleteness of information or inconsistency in the set of attributes characterizing the components and the events (inconsistency sometimes due to obvious errors). Some problems in data interpretation arose as a consequence of the restrictions in the information made available to the partecipants in the BE (see 2.1.2).

Concerning the comparison of the preliminary analysis results, it is to be noted that the approaches adopted, and the subsets of data analyzed by the participants have been so varied that a comparison of the results is not possible. For the definition of the activities of the second phase of the BE, it was decided that the JRC would identify some objectives (such as the estimate of reliability figures) and define a smaller revised data set with the aim of obtaining comparable results from the analyses of the various working groups.

2.2.2 Written reports have been produced and presented at the Shockholm meeting by JRC-Ispra (ATI Division), INTERATOM, NUKEM, VTT and SINTEF. A written report, in a draft form, has been produced by ENEA VEL Bologna; as they did not attend the meeting, some comments on the contents of their report were made by the coordinators. As regards the contribution of the composite group ENEA TIB AQ Roma - Ed F - JRC Ispra (SER Division), only an oral description of the main lines of the approach that they intend to follow has been given.

JRC-Ispra (ATI Division) suggests a unified model to estimate constant failure rates, constant repair rates and failure probabilities on demand (assumed to be independent on the number of demands). They suggest to assume a constant failure rate at component level and a binominal - beta model.
The distribution of the failure rate of a group of pumps (pumps homogeneous from the engineering point of view) is assumed to be beta, as in the case of a single pump. The shape of the distribution is defined by two percentiles (50th and 95th) which are computed by a weighted arithmetic averaging of the corresponding percentile of each pump distribution. In addition they tackle the problem of outliers detection.

INTERATOM is the Organization that has particularly enphasized the problem of data interpretation (see 2.2.1), focusing their attention on pumps performing two specific functions, the primary cooling pumps and the main feed-water pumps. They remark on the fact (already known to CEDB staff) that automatic data processing cannot be used to derive failure rates for specific failure modes, due to both the limited number of events for specific failure modes available in the bank (very limited for some failure modes, such as sudden loss of function) and the necessity to interpret some events. They generally assume a constant failure rate and, for the main feed pumps, consider the Weibull model to estimate the distribution of operating times to the first external leakage, on the basis of a sample of 18 events from 8 plants.

In the NUKEM Report, pumps have been grouped according to the six following criteria: all pumps, pump nominal power range (lower or higher than 100 kW), pump mode of operation (continuous or non-continuous), pump type (centrifugal or non-centrifugal), plant mean-time-to-failure (less than 5000 h, between 5000 and 10,000 h, higher than 10,000 h), medium handled (water or other), mean-time-to-failure of the engineering system the pump belong to. An analysis of the times to failure, repair times and restoration times (i.e. unavailability times due to repair) has been performed for each group assuming a Weibull model.

In the VTT Report, a descriptive analysis of the data related to pumps pertaining to systems "condensate and feed-water" and "auxiliary feed-water" is firstly illustrated. With reference to these two sets of pumps, they have estimated times to failure, repair times and restoration times (i.e. unavailability times due to repair). They have referred to various models: exponential, two-mixture exponential, conditional exponential, Weibull, log-normal and gamma. Both parametric and non-parametric statistical tests have been applied to check the adequacy of the various models. A renewal model has also been used with reference to the complete failures which occurred in plant 3. By a Bayesian approach assuming a gamma prior, they have estimated failure rates for a subset of continuously operating pumps. An original approach to study the availability performance of pumps is presented, by its practical application to the pumps of the condensate and feed-water system.

SINTEF has analyzed the complete set of pumps. They have performed a non-parametric analysis, mainly based on the Kaplan-Meier method. Furthermore they have applied a proportional hazard model, considering as covariates some engineering and operating characteristics, such as the pump nominal power, nominal flow, developed head, and the plant (to which the pump belongs). As a result of their analyses, they give tables showing the relevant covariates for a given failure mode, for the pumps pertaining to a set of plants.

ENEA VEL has firstly performed a correspondance analysis to identify homogeneous subsets of pumps. They have identified 8 groups. For each group, by applying both a classical and a Bayesian method they compute a point estimate and an interval estimate of the failure rate. In the Bayesian approach they assume a "so-called" non-informative prior.
They plan to develop and apply models based on the interpretation of the basic physical process of the degradation leading to a failure. The models refer to the operating conditions of the component and the physical laws which govern its behaviour.

The representative of the composite group criticized the use of failure rate estimates at component level. He said that he is developing a Bayesian approach to assess reliability functions of a system on the basis of a prior information at system level and raw data at the component level, without estimating any parameter at the component level.

3. Second phase of the BE; identification of common objectives for analyses to be performed on a smaller data set

As a conclusion of the meeting held in Stockholm on September 30th, 1988, it has been agreed that the second phase of the BE will consist of analyses focused on a smaller reference data set and having some common well-identified objectives, all this with the aim of getting comparable results from the various participants.

3.1 Characteristics of the new reference data set

Following the discussions held during the meeting, the new data set will refer to the centrifugal pumps of the condensate and feed-water system, handling water, located along the secondary cycle line from the condenser to the steam generators. These pumps number one hundred and their related failures to more than 400. This data set will be revised by the JRC before its use for new analyses; some checks (such as the assessment of the codes defining the function and the operating mode of the component) will be performed and some obvious coding errors will be eliminated.

To aid with data interpretation, the participants will receive a conceptual functional flow sheet for each plant, where the pumps in question are represented.

3.2 Reference objectives for the analysis

As common objectives for the analyses, the JRC suggest:

- the assessment of
  - pump failure and survival figures with reference both to a sudden loss of function and to a degraded performance;
  - repair and restoration figures;
  - unavailability figures;
- investigating their possible dependency on
  - plant status at the failure detection (where relevant);
  - function performed by the pump (extraction or feed);
  - engineering characteristics of the pump;
  - operating characteristics of the pump;
  - plant to which the pump pertains;
  - plant grouping (PWR's, BWR's, conventional units, and possibly other groupings suggested by the analyses);
- and identifying, when possible, time dependency of the assessed figures.

Suggested time interval for survival and unavailability analysis: ½ year.
Suggested time unit for repair and restoration figures: one hour.

4. Conclusions

- The intermediate reports produced by the participants and presented in the Stockholm meeting describe analyses of great interest, based on a large spectrum of methodologies adopted.

As a consequence, we consider the results of this first phase of the BE to be very positive.

- The diversity of the approaches adopted is the primary cause of the impossibility to compare the results obtained by the participants.

  The fairly large data set used has been a second cause of the spread of the results. In fact some analyses refer to different subsets of raw data. However the comparison of the results was not intended to be an objective of the first phase of the BE.

- The definition of a smaller revised reference data set, combined with the identification of common objectives for the analyses to be performed during the second phase of the BE should facilitate the attainment of comparable results. Comparing results should favour a better understanding of the various methods used by the participants.

Appendix

Terms of reference of the EuReDatA Benchmark Exercise (BE) on Component Reliability Data Analysis (July 1987).

1. Aim of the BE

Comparison of the methods used by the various Organizations to estimate reliability parameters from raw data.

The reliability parameters of interest are:

- failure and repair rates;
- failure probabilities on demand;
- repair time distributions;

for the relevant component failure modes. The reference data set will be taken from the component Event Data Bank.

2. Methods proposed

a. Exploratory data analysis (e.g. principal component analysis, correspondence analysis, cluster analysis, covariate analysis).

b. Statistical inference

   b.1 test of hypothesis on failure and repair distributions (e.g. exponential, Weibull, lognormal, gamma);

   b.2 classical and Bayesian estimation (parametric and non parametric).

Note 1: Data available refers to components monitored for a number of years. Observations end either with a failure causing a replacement or with a suspension. A component may have zero or one or more failures during its observation time.

Note 2: The above mentioned exploratory data analysis methods may be used for a preliminary screening of the data and/or to perform special multivariate analyses.

3. Main features of the reference data set

3.1 According to the CEDB structure, the raw data of each component contains the following information items:

a) engineering and operating characteristics;

b) operational history: i.e. annual operating times and/or annual number of demands, failure-event reports.

3.2 The reference data set concerns pumps - see Appendix "Reference Data Sets for EuReDatA BE". Each Organization participating in the BE will receive raw data taken from the CEDB (see example, item 3 of the appendix) and the information necessary for a correct understanding of the CEDB coding. This raw data can be supplied by the JRC either on a magnetic support or in a printed form.

Expert Opinions as Data Source: Methods and Experiences

JACQUES F.J. VAN STEEN
Department of Industrial Safety
Netherlands Organization for Applied Scientific Research TNO
Apeldoorn, Netherlands

ROGER M. COOKE
Department of Mathematics and Informatics
Delft University of Technology
Delft, Netherlands

Abstract

Formalized procedures for using expert judgment are expected to contribute substantially to improving the quality of reliability data such as the failure rates of mechanical components. Key requirements for these procedures are that they provide solutions to such questions as: How do we encode, evaluate and combine expert opinions?
Three different categories of models for eliciting and combining expert opinions have been selected for detailed investigation: weighted averaging, Bayesian models and psychological scaling models. Within each category, one model has been selected/developed that is hoped to be an improvement over the models suggested so far within that category. These models have been made operational and have been tested and evaluated in case studies on real-world problems in industry.
The paper discusses the models developed and presents some results of the case studies.

INTRODUCTION

Expert opinion has always played a role in various parts of risk and reliability analysis. Important areas of expert opinion use include the assessment of human error probabilities and of failure rates of mechanical components. However, a number of specific problems are associated with the use of expert opinion [1, 2]:

- The percentage spread of opinion in estimating the probabilities of rare events generally appears to be extremely large. Thus, probability distributions instead of point estimates are required in order to reflect the uncertainty surrounding the quantities in question.
- Probability assessments, in particular in the range of small probabilities, are characterized by various natural and per-

sistent "biases".

Consequently, the need exists for formalized procedures for using expert opinion in risk and reliability analysis in order to improve the quality of the data. Key requirements for such procedures are that they provide solutions to the following questions:

- How do we encode and evaluate expert opinion?
- How, if at all, do we combine the opinions obtained from several experts?

On the basis of an extensive literature study, which is reported in [3] and briefly summarized in [2], three different categories of models for eliciting and combining expert opinions have been selected for detailed investigation. These categories are: weighted averaging, Bayesian models and psychological scaling models. Within each category, one model has been selected/developed that is hoped to be an improvement over the models suggested so far within that category. These models, the classical model (weighted averaging), the Mendel-Sheridan Bayesian model and the NEL model (paired comparisons), have been made operational and have been tested and evaluated in case studies on real-world problems, in particular in the chemical process and space industry.
This paper is mainly devoted to a discussion of the models developed. In conclusion, some results of the case studies are presented.

THE MODELS

This section discusses the classical, Bayesian and paired comparison models. The classical and Bayesian models proceed from the following simple fact. When an expert quantifies his uncertainty for a continuous random variable x by giving his cumulative distribution function F, he effectively gives us a transformation such that the transformed variable is uniformly distributed on the unit interval. The transformation is simply F(x), i.e. the quantile function of x. When he quantifies his uncertainty for a set of variables, the transformed variables are all identically distributed, and upon observing realizations for these variables, we can treat the expert as a statistical hypothesis and apply the vast resources of classical or Bayesian statistical analysis. The paired comparison models are rather different in character, and lend themselves to situations in which quantitative assessments from experts are difficult to obtain.

In formulating models for using expert opinion in science, it was found necessary to formulate five general principles to guide the selection of models. These principles are designed to guarantee that the use of expert opinion advances the fundamental aim of science: rational consensus.

Reproducibility:

> It must be possible for scientific peers to review and if necessary reproduce all calculations.

This entails that the calculational models must be fully specified and the ingredient data must be made available.

Accountability:

> The source of expert subjective probabilities must be traceable for the decision maker.

Empirical control:

> Expert probability assessments must in principle be susceptible to empirical control.

The requirement of empirical control through observations is universally recognized as a cornerstone of the scientific methodology. In contemporary methodology this requirement is given the following expression: scientific statements and scientific theories should be falsifiable in principle. It is recognized that theories can never be conclusively verified, but at least it should be possible in principle to discover a reproducible conflict with observations, if the theory is in fact false. The necessary experiments need not be feasible, but they must be physically possible.
In the same spirit, a methodology for using expert opinion should incorporate some form of empirical control, at least in principle. In other words, it must be possible in principle to evaluate expert probabilistic opinion on the basis of possible observations. Moreover, this evaluation should be reflected in the degree to which an expert's opinion influences the end results.

Neutrality:

> The method for eliciting/evaluating/combining expert opinion should encourage experts to state their true opinions.

Various systems of self-weights or peer weights allow the experts to rate their own or each other's expertise. Such systems raise obvious problems with respect to neutrality, as "influence on the decision maker" is obviously a form of reward for the experts. Problems of confidentiality arising when experts' judgments may be perceived as damaging for the commercial interests of their employer must also be addressed under the principle of neutrality.

Fairness:

All experts must be treated equally, prior to processing the results of observations.

A further discussion of these principles, and their relation to existing practice may be found in [4].

The Classical model

The classical model involves eliciting probabilistic expert assessments for discrete or continuous items, and combining these assessments via weighted averaging. Letting P_{DM} denote the decision maker's distribution for an uncertain item, and p_e, $e = 1,...E$, the distributions of experts $1,...E$ for the same item, then P_{DM} is a weighted combination of $p_1,...p_E$ if:

$$P_{DM} = \Sigma_{e=1,...E}\, w_e p_e; \quad \Sigma\, w_e = 1; \quad w_e \geq 0. \tag{1}$$

The classical model provides a mechanism for determining the weights w_e in the above formula. The weights introduced below reward good calibration and low entropy, and satisfy constraints deriving from the theory of proper scoring rules.

Notation and definitions

By a "probability distribution" we shall mean either a probability mass function (in the case of discrete variables) or a probability density function (in the case of continuous variables). A probability distribution is not to be confused with a cumulative distribution function.

Let $p = p_1,...p_n$ be a probability distribution. The entropy or negative information $H(p)$ of p is defined as

$$H(p) := -\, \Sigma\, p_i \ln(p_i) \tag{2}$$

where "ln" denotes the natural logarithm. $H(p)$ assumes its minimal value, 0, if $p_i = 1$ for some i, and assumes its maximal value $\ln(n)$ if $p_i = 1/n$, $i = 1,...n$. $H(p)$ is commonly regarded as an index of the lack of information in the distribution p; high values of H indicate low informativeness.

Let $q = q_1,\dots q_m$ be another distribution, and let pq denote the product distribution; pq is the joint distribution of two independent variables having marginal distributions p and q. It is easily verified that:

$$H(pq) = H(p) + H(q). \tag{3}$$

(3) generalizes for products of finitely many independent variables.

Let $s = s_1,\dots s_n$ be a probability distribution, and assume $p_i > 0$, $i = 1,\dots n$; then the relative information of s with respect to p (also called discriminal information, cross entropy or directed divergence) $I(s,p)$ is defined as:

$$I(s,p) := \Sigma\, s_i \ln(s_i/p_i). \tag{4}$$

$I(s,p)$ is always non-negative and $I(s,p) = 0$ if and only if $s = p$. $I(s,p)$ is commonly taken as an index of the information learned if one initially believes that p is correct and subsequently learns that s is correct. Note that

$$H(p) = \ln(n) - I(p,u) \tag{5}$$

where u is the uniform distribution over $1,\dots n$; i.e. $u_i = 1/n$, $i = 1,\dots n$.

Let s denote a sample distribution generated by M independent samples from the distribution p. Let Chi^2_B denote the cumulative distribution function of a Chi-square variable with B degrees of freedom. Then

$$Prob\{2MI(s,p) \le x\} \to Chi^2_{n-1}(x) \text{ as } M \to \infty. \tag{6}$$

(6) says that the statistic $2MI(s,p)$ is asymptotically Chi-square distributed as the number M of independent samples from p goes to infinity. The sum of independent Chi-square variables is again a Chi-square variable and the number of degrees of freedom of the sum is the sum of the numbers of degrees of freedom.

The classical model can be applied in a wide variety of situations. We discuss below only the case where all experts assess the same quantities, where the quantities are practically continuous over some bounded "intrinsic ranges" (not necessarily the same for each quantity), and where the experts express their assessments by stating fixed fractiles or quantiles of

their subjective distributions. It is assumed that the true values can be observed after the assessments are given. Alternatively the true values may be known to the analyst beforehand, but not to the experts. In the latter case we speak of "seed variables" (to seed the model). Calibration and entropy indices will be used to define weights, which in turn will be used to combine expert assessments for future variables. To recall, if x is a continuously distributed quantity with cumulative distribution function F, then x_r is the r-th fractile of x (or of F) if $F(x_r) = r$.
The following notation will be used:

$x_1,\dots x_M$: practically continuous variables

$f_1,\dots f_R$, $0 < f_1 < \dots f_R < 1$: fractiles solicited from distributions

G_{me}, $m = 1,\dots M$, $e = 1,\dots E$: cumulative minimal information distribution for x_m satisfying the constraint that the $f_1,\dots f_R$-th fractiles agree with expert e's assessment for variable m (strictly speaking, this definition is meaningful only if the range of the variables is bounded; for situations in which this is not the case, see below). If x_{mr} is the f_r-th fractile of expert e for x_m, then $G_{me}(x_{mr}) = f_r$.

b_r, $r = 1,\dots R+1$: probability bin associated with $(f_{r-1}, f_r]$
(for $r = 1$, take $f_{r-1} = 0$, for $r = R+1$, take $f_r = 1$)

p_r, $r = 1,\dots R+1$: theoretical probability associated with b_r ($p_1 = f_1$, $p_2 = f_3 - f_2$, etc.)

$p = \{p_1,\dots p_{R+1}\}$

Chi^2_R = cumulative Chi^2 distribution function with R degrees of freedom.

When the value x_m is observed, then for each expert exactly one probability bin is "hit", namely the unique bin b_r such that $G_{me}(x_m) \in b_r$. In this way, observations of x_m, $m = 1,\dots M$, generate a sample distribution over the probability bins for each expert. For the notation in (7)-(9) we restrict attention to a single expert e.

$$C(e) = 1 - Chi^2_R(2MI(s,p)) \tag{7}$$

where $s = (s_0,\ldots s_R)$ = sample distribution over $b_0,\ldots b_R$.

$$H(e) = \sum_{m=1}^{M} \frac{H(G_{me})}{MH(U_m)} \tag{8}$$

where $H(G_{me})$ = entropy of density of G_{me} and U_m is the uniform distribution over the intrinsic range of x_m, $m = 1,\ldots M$.

$$w'_e = 1_\alpha(C(e)) \times C(e)/H(e). \tag{9}$$

$$w_e = w'_e/W \tag{10}$$

where $W = \Sigma_{e=1,\ldots E}\, w'_e$ and assuming $W > 0$.

(10) expresses the basic weight in the classical model. This weight has the following important feature. For $M \geq 10$, the terms $C(e)$ will typically vary over two or more orders of magnitude. The terms $H(e)$, on the other hand, seldom vary by more than a factor 2. This means that the calibration term $C(e)$ dominates in (10), and the entropy term $H(e)$ serves to modulate the calibration term, i.e. it becomes important in differentiating the weights of experts who are more or less equally calibrated. This is the type of behavior which the weights should have, as we do not want poor calibration to be compensated by very low entropy scores of "overconfident" assessors. In [5] it is proved that, under suitable conditions, w'_e is an asymptotically strictly proper scoring rule. Briefly, this means that in the long run an expert attains his maximal expected (unnormalized) score by stating his true opinion.
Further, note that (10) contains the free parameter α. It will be shown below how α can be optimally chosen.

In [4] many variations and refinements of the basic weight (10) are discussed. The most important of these will be briefly indicated.
Other measures of information may be more appropriate than $H(e)^{-1}$ in some contexts. In particular, this holds when the experts have assessed different variables in the past, and are now called upon to assess the same variables. In this case their weights must be based on their assessments of different variables.
When the ranges of the variables are not bounded, bounds must be chosen by the analyst. The choice of bounds affects the measure of information, though not dramatically. The choice of bounds will not greatly affect the decision maker's distribu-

tion P_{DM} between the f_1 and f_R fractiles. In general, because of the dominance of the term C(e) in (10), or any of its variations, the classical model is robust with respect to choices of the analyst.
One variation of (10) deserves mention here. Instead of using the average H(e), it is possible to compute a separate weight for each variable using the term $H(G_{me})/H(U_m)$. This has the effect of allowing the experts to "downweight" or "upweight" themselves on the individual variables, according to the entropy computed for the distribution G_{me} in question.

Optimization; virtual weights

This paragraph shows how the significance level α can be chosen so as to optimize the decision maker's distribution. For each choice of α a distribution P_α of the decision maker is defined. This distribution can also be scored with respect to calibration and entropy. Hence, a "virtual weight" for the decision maker is defined as a function of α. This is the weight which a "virtual expert" would receive when he gives the decision maker's distribution, and is scored and normalized along with the actual experts. The decision maker's distribution is optimized by choosing α such that his unnormalized virtual weight is maximal. In practice, it will be prudent to forego a small gain in unnormalized virtual weight in order to assign non-zero weights to more experts.

Suppose we add the virtual expert to the set of experts and calculate the decision maker's probability anew. Let P be the decision maker's probability based on the original experts and P' the new probability. A simple calculation shows that P=P'.

The use of virtual weight optimization to determine α is a significant feature of the classical model. It replaces the negative grammar of hypothesis testing with the positive grammar of optimization. Moreover, it gives us a way of assessing model performance. Barring exceptional situations in which the calibration terms for different experts coincide, the decision maker's virtual weight can never be less than that of the "heaviest" expert. Experience with the model to date indicates that the decision maker's virtual weight is generally higher than the weight of the "heaviest" expert.

In many optimization contexts, a suboptimal solution may prove more robust with respect to variations of "initial conditions" than the optimal solution. The same may apply here; hence a suboptimal value of α may be chosen at the discretion of the analyst, particularly in cases when the number M of variables whose true values are observed is small.

The Bayesian model

Bayesian models require that the user supply prior probability distributions, and process expert assessments by updating these distributions via Bayes' theorem. The model of Mendel and Sheridan [6] uses Bayes' theorem to recalibrate and combine expert assessments of continuous variables. The only other model which putatively accomplishes the same and which was proposed by Morris [7] has serious computational and philosophical drawbacks (see e.g. [8] and [9]). The Mendel-Sheridan model compares favorably with other Bayesian models in the following respects:

(1) Default egalitarian prior distributions are given by the model itself.

(2) The expert assessments are not restricted to a given class of distributions, but rather the experts give quantiles of their distributions for continuous variables (which makes this model compatible with the classical model).

(3) Bayes' theorem automatically accounts for correlations in the expert assessments - correlation coefficients need not be assessed by the decision maker.

(4) The computational algorithm has been streamlined for ease of implementation, and software is available.

(5) The model has demonstrated its worth in experiment.

Because of the features mentioned above, the Mendel-Sheridan model fully conforms to the principles formulated at the begining of this section. There are features of this model which restrict its practical application at this moment. In particular, the algorithm for combining experts' opinions is unyieldy, and rests on assumptions which are less convincing than the single expert case. We confine the present discussion to the single expert case and refer to [4] for derivations.

The notation of the classical model will be maintained, however, we restrict attention to a single expert, and drop the parameter e. We assume that the expert has given fractile assessments for continuous variables $x_0, x_1, \ldots x_M$, that the true values for $x_1, \ldots x_M$ have been observed and that x_0 is the current variable for which the decision maker's distribution must be determined. For $m = 0, \ldots M$, we define the multinomial variable z_m, and the counting variable y_j, $j = 1,\ldots R+1$:

$z_m = r$ if the realization x_m hits the r-th probability bin

$y_j = |\{z_m = j \mid m=1,\ldots M\}|$

y_j counts the number of times that j has been observed for the variables preceeding the current decision variable.

$y = y_1,\ldots y_{R+1}$

P = the decision maker's probability.

The Mendel and Sheridan model assumes that the variables $z_0,\ldots z_M$ are exchangeable with respect to P, and hence that y contains all information necessary to update the decision maker's distribution for z_0, given the realizations $x_1,\ldots x_M$.

Let $\pi = \langle\pi_1,\ldots\pi_{R+1}\rangle$, $\pi_j \geq 0$, $\Sigma\pi_j = 1$. De Finetti's representation theorem entails that:

$$P(z_0 = j \mid y) = \int \pi_j \prod_{k=1}^{R+1} \pi_k^{y_k} \, dF(\pi) \qquad (11)$$

for some suitable prior distribution f = dF over the possible values of π.

Mendel and Sheridan choose F to be a Dirichlet distribution with parameters r_j, $j = 1,\ldots R+1$, $T = \Sigma r_j$. The parameters of the Dirichlet distribution may be interpreted in terms of "equivalent observations". T corresponds to the total number of "equivalent observations", and r_j to the number of "equivalent observations" of the events "$z_m = j$", $m = 1,\ldots M$. In Bayesian multinomial inference, the Dirichlet is a conjugate prior for the multinomial likelihood function. By choosing the parameters of the Dirichlet, the user expresses his prior confidence in the outcome j by saying, in effect, "I am as confident in j as I would be if I had no previous information and then observed r_j occurrences of j out of T trials". The left hand side of (11) is then also a Dirichlet, and

$$P(z_0 = j \mid y) = (y_j + r_j)/(M + T). \qquad (12)$$

Setting M = 0, we derive the prior probability $P(z_0=j) = r_j/T$.

Mendel and Sheridan adduce arguments for setting $r_j = 1$, $j = 1, \ldots R+1$. This yields:

$$P(z_0 = j \mid y) = (y_j + 1)/(M + [P(z_0 = j)]^{-1}). \qquad (13)$$

(13) represents the decision maker's updated probability of seeing the realization for x_0 fall between the (j-1)-th and j-th fractiles for x_0 given by the expert. As M gets large, this approaches the relative frequency y_j/M with which this was observed for variables $x_1, \ldots x_M$.

It remains to determine $P(x_0 \mid y, G_0)$. If the range of x_0 is bounded, a minimal information solution subject to the constraints expressed by (13) is chosen. The probability mass for $z_0 = j$ in (13) is uniformly distributed over the corresponding interval $(x_{0,j-1}, x_{0,j}]$, where $G_0(x_{0,j-1}) = f_{j-1}$ and $G_0(x_{0,j}) = f_j$.

If the range of x_0 is not bounded, the density in the "tails" cannot be determined by a minimal information argument. In this case the analyst must choose to cut the range of x_0 off at some arbitrarily chosen minimal and maximal values. The decision maker's updated distribution for x_0 is very sensitive to this choice. For example, suppose that the lowest fractile solicited is the 5% fractile, and suppose that the decision maker's updated probability of seeing the realization fall beneath the expert's 5% fractile for x_0 is 10%. Following the above recipe, the decision maker's updated 5% fractile is located at one half the distance between the updated 10% fractile and the lower cut-off point. Hence, the updated 5% fractile depends linearly on the lower cut-off point.

Paired comparison models

Psychological scaling models, in particular the method of paired comparisons, are discussed in [10] and considered suitable for the estimation of human error probabilities. In the method of paired comparisons, objects are presented in pairs to one or more experts. For each pair, the experts are asked to indicate their preference for one of the objects concerning the attribute under investigation. In the present context, the attribute may be "probability of occurrence", but the attribute could in principle be any subjective quality of the objects. This is done for a number of objects, and usually with a relatively large number of experts. The strength of the method lies in its redundancy: each object is evaluated several times in comparisons with other objects. The method picks up the underlying trend in such comparisons and is robust with respect to distractions and fluctuations in judgment.

The responses of the experts are processed in order to obtain values for the objects considered. This processing is performed in two steps: (1) using modeling assumptions, scale values for the attribute in question are derived, and (2) using reference values for one or two objects, depending on the model, the scale values are transformed into absolute values.
Various models are available for this processing, and the most important of these are discussed below: the Thurstone model which was taken from the literature and the NEL model which was developed at the Delft University of Technology and reported by Stobbelaar [11]. Some of the models allow for the investigation of goodness-of-fit. Confidence intervals can be determined analytically and/or by simulation. If these confidence intervals are interpreted as Bayesian confidence intervals, i.e. as fractiles of a subjective distribution, and if appropriate reference values are available, then the output can be regarded as probabilistic assessments of the attributes in question.
Firstly, two general issues are introduced which apply to all models: within-expert consistency (experts may not always be consistent in their preferences) and between-expert agreement (to what extent is there agreement between the experts consulted?). A more detailed summary of the general issues and the models can be found in [4], whereas references to the literature are given where appropriate.

The general notation is as follows:
- t objects A(i), i = 1,...t,
- n experts,
- scale values V(i), i = 1,...t for objects A(i).
- a(ij): the number of experts who prefer A(i) to A(j).

Within-expert consistency

Inconsistency of an expert is measured by the number of circular triads occurring in the responses. A circular triad exists if A(i) is preferred to A(j), A(j) to A(k), and A(k) to A(i). The number of circular triads C(e) occurring in the responses of expert e is calculated by

$$C(e) = (t/24)(t^2-1) - \tfrac{1}{2} \Sigma_i \, [a(i,e) - \tfrac{1}{2}(t-1)]^2 \qquad (14)$$

where $a(i,e)$ is the score of $A(i)$ with expert e: the number of times expert e preferred $A(i)$ to the other objects [12]. Statistical tests involving $C(e)$ are given by Kendall [13]. These tests allow for investigation of the question whether the preferences have been given at random.

Between-expert agreement: (1) coefficient of agreement u

Kendall [13] defines the coefficient of agreement u as follows:

$$u = \frac{2 \sum_{i \neq j} \binom{a(ij)}{2}}{\binom{n}{2}\binom{t}{2}} - 1 \qquad (15)$$

where the Σ-term in the numerator denotes the sum of the number of agreements between pairs of experts and summation is performed over $t(t-1)$ terms.
In case of complete agreement u equals 1, whereas u gets smaller if the number of agreements between pairs of experts diminishes.
Statistical tests involving u are given by Kendall [13]. These tests allow for investigation of the question whether the preferences have been given at random.

(2) coefficient of concordance W

The sum of ranks $R(i)$ is calculated by $\Sigma_e\ R(i,e)$ where $R(i,e)$ is the rank of $A(i)$ obtained from the responses of expert e; the value of $R(i,e)$ ranges from 1 to t. According to Siegel [14],

$$W = \frac{S}{(1/12)n^2(t^3-t)} \qquad (16)$$

where $S = \Sigma_i\ [R(i) - t^{-1}\Sigma_j\ R(j)]^2$ is the sum of squares of the observed deviations from the mean of $R(i)$.
In case of complete agreement W equals 1, whereas W gets smaller when agreement diminishes.
Statistical tests involving W are given by Siegel [14]. These tests allow for investigation of the question whether the preferences have been given at random.

Thurstone model: (1) principle

The Thurstone model is built upon the assumption that the value V(i) associated with object A(i) is described by the normal distribution. Torgerson [15] distinguishes three ways of using this model with respect to the number of experts and calls these ways "classes". In class II, several experts are consulted each of whom makes every possible paired comparison just once.
In the Thurstone model (class II), the judgments of the various experts with respect to V(i) are assumed to be normally distributed with expectation $\mu(i)$ and variance $\sigma^2(i)$, where $\mu(i)$ is considered to be the estimator for V(i). This assumption implies, that the judgments with respect to V(ij) = V(i) - V(j) are also normally distributed, with expectation

$$\mu(ij) = \mu(i) - \mu(j)$$

and variance

$$\sigma^2(ij) = \sigma^2(i) + \sigma^2(j) - 2\rho(ij)\sigma(i)\sigma(j)$$

where $\rho(ij)$ is the correlation coefficient between V(i) and V(j).

P[V(ij) > 0] equals the proportion of the number of times A(i) is preferred to A(j). Since only the standard normal distribution is available in tabular form, a transformation is required. Let X(ij) have a standard normal distribution. Standardization leads to the so-called Law of Comparative Judgment [16]:

$$\mu(i) - \mu(j) = x(ij)[\sigma^2(i) + \sigma^2(j) - 2\rho(ij)\sigma(i)\sigma(j)]^{1/2} \tag{17}$$

where x(ij) is such that $\Phi(x(ij))$, the value of the standard normal cumulative distribution function Φ at x(ij), equals the proportion of the number of times A(i) is preferred to A(j).

There is no solution to the Law of Comparative Judgment in its general form. In one of the ways of simplification considered by Torgerson [15] it is assumed that the variances $\sigma^2(ij)$ are constant. As a consequence, (17) is simplified to

$$\mu(i) - \mu(j) = cx(ij). \tag{18}$$

(2) processing of responses

From the responses of the experts the matrix P' is derived, where $p'(ij)$ equals the proportion of the number of times A(i) is preferred to A(j): $p'(ij) = a(ij)/n$. (Real values are indicated by letters without accent, whereas the corresponding estimators are indicated by letters with an accent.)
From the matrix P' and a table of the standard normal distribution the matrix X' is derived, where $x'(ij)$ is such that $p'(ij) = \Phi(x'(ij))$. The estimators for the scale values are then calculated by

$$\mu'(i) = (1/t)\, \Sigma_{j=1,\dots t}\, x'(ij) \qquad (19)$$

Two ways exist for calculating confidence intervals for the scale values: an analytical one and one which makes use of simulation.

The analytical 90% confidence interval for $\mu'(i)$ is calculated by

$$[\mu'(i) - (10 \text{ Var } \mu'(i))^{1/2} ,\ \mu'(i) + (10 \text{ Var } \mu'(i))^{1/2}] \qquad (20)$$

Var $\mu'(i)$ is defined by

$$(1/t^2) \sum_{j=1}^{t} \sum_{y=0}^{t} [\Phi^{-1}(y/n)]^2\, P[a(ij)=y|p(ij)] - [E\, \mu'(i)] \qquad (21)$$

where $P[b(ij)=y|p(ij)] = \binom{n}{y} [p(ij)]^y\, [1-p(ij)]^{(n-y)}$

> $p(ij) = b(ij)/N$, where $b(ij)$ is the number of experts out of N preferring A(i) to A(j); N is assumed to be the total number of experts with respect to the objects in question, whereas only n experts are consulted, $n < N$.

$$E\, \mu'(i) = (1/t) \sum_{j=1}^{t} \sum_{y=0}^{t} \Phi^{-1}(y/n)\, P[a(ij)=y|p(ij)]$$

Since $p(ij)$ is unknown, $p'(ij)$ is used as estimator for $p(ij)$. A problem arises when all experts agree at a certain paired comparison, as this leads to $\Phi^{-1}(0) = -\infty$ and $\Phi^{-1}(1) = +\infty$.

A second way for calculating confidence intervals for the scale values is via simulation. The scale values obtained can be used to generate for each expert from the distribution $N(\mu'(i),1)$ a new value ("opinion") for each object. These "opinions" can then be used to calculate new estimates for the $\mu(i)$. By repeating this operation a large number of times, say 1000 times, confidence intervals for the scale values $\mu'(i)$ can be derived from these 1000 estimates.

The observed proportions $p'(ij)$ are used to calculate estimates for the unknown parameters. This procedure can also be applied the other way around: given the estimates, values $p''(ij)$ can be calculated: $p''(ij) = \Phi(x''(ij))$, where $x''(ij) = \mu'(i) - \mu'(j)$.
Goodness-of-fit tests involving $p'(ij)$ and $p''(ij)$ are given by Torgerson [15] and Mosteller [17].

Transformation involves calculation of absolute values by applying the following relationship between the initial scale and the absolute scale:

$$f(i) = bV(i) + c \tag{22}$$

where $f(i)$ is the absolute value of interest, and b and c are the scaling constants. In order to be able to determine the scaling constants, the absolute values of two objects have to be known. Once the scaling constants are determined, both the absolute values of the other objects and the corresponding confidence intervals can be calculated by applying the relationship assumed.
Several researchers advocate the use of a logarithmic relationship (see e.g. [10]):

$$\log f(i) = bV(i) + c \tag{23}$$

which basically implies a modification of the assumption underlying the Thurstone model: the logarithms of the values (instead of the values themselves) associated with the objects are described by the normal distribution.

NEL model: (1) principle

The NEL model (NEL is short for Negative Exponential Lifetime) is especially suited to determining the failure rates of mechanical components. In the NEL model the lifetime $L(i)$ of $A(i)$ is assumed to be characterized by a negative exponential distribution with expectation $1/\tau(i)$:

$$f_{L(i)}(t) = \tau(i)\exp(-\tau(i)t) \qquad (24)$$

where f is the density function and $\tau(i)$ the failure rate. This assumption implies that the probability that A(i) fails before A(j), P[L(i) < L(j)], equals $\tau(i)/[\tau(i) + \tau(j)]$.

The probability r(ij) that any expert prefers A(i) to A(j) is assumed to equal the probability that A(i) fails before A(j). Thus, r(ij) is represented by

$$r(ij) = \frac{\tau(i)}{\tau(i) + \tau(j)} . \qquad (25)$$

It appears that r(ij) in the NEL model is identical to r(ij) in the Bradley-Terry model [18], where the probability r(ij) that A(i) is preferred to A(j) is defined as follows:

$$r(ij) = \frac{V(i)}{V(i) + V(j)} . \qquad (26)$$

This implies that the solution technique for determining V(i) in the Bradley-Terry model can be applied for determining $\tau(i)$ in the NEL model.
It is assumed in the Bradley-Terry model that $\Sigma\, V(i) = 1$. The values V(i) are the unknown parameters. An estimator for r(ij) is a(ij)/n.

(2) processing of responses

The probability that a(ij) experts prefer A(i) to A(j) is represented by

$$\binom{n}{a(ij)} \left[\frac{V(i)}{V(i)+V(j)}\right]^{a(ij)} \left[\frac{V(i)}{V(i)+V(j)}\right]^{n-a(ij)} . \qquad (27)$$

Applying the method of maximum likelihood leads to a set of equations. From these equations, together with $\Sigma\, V'(i) = 1$, the maximum likelihood estimators V'(i) for the scale values V(i) can be obtained by applying an iteration process as described by David [12].
The condition that the scale values add up to 1 is convenient for two reasons: it leads to absolute values instead of ratios and it guarantees that the numbers remain easy to handle

throughout the (computerized) calculation process. It should be noted, however, that the choice of the number 1 is an arbitrary one; consequently, the need for transformation still exists in order to obtain "real" absolute values.

In the NEL model confidence intervals for the scale values can be calculated by simulation, in a way similar to that described for the Thurstone model. In this case, new values ("opinions") are generated from the negative exponential distribution.
In the Bradley-Terry model confidence intervals for the scale values can be calculated by generating new a(ij)-values for each pair of objects with the binomial distribution.

A goodness-of-fit test is given by Bradley [18].

Transformation in the Bradley-Terry and NEL models requires that the absolute value of just one of the objects is known, since the following relationship is applied:

$$f(i) = \frac{V(i)}{V(j)} f(j). \qquad (28)$$

Similarly, the confidence intervals of the absolute values can be calculated.

SOME EXPERIENCES

The above models have been made operational and have been applied in case studies on a number of real-world problems in industry. These applications involved the following areas:
- risk analysis: failure rates of basic events in the fault tree of an advanced technological system in the space industry;
- reliability engineering and maintenance management: failure rates of two different components of gas pressure regulators, relative contributions of failure causes of flanged connections in the chemical process industry, reliability and maintainability characteristics of gaskets in a chemical process plant, contamination causes and frequencies in a pilot plant scale fermentation process;
- project management: subjective assessments of the probabilities that projects will be realized.

Some results of two applications are described very briefly below.

Application 1

The first application to be discussed concerned the failure rates of basic events in the fault tree of an advanced technological system in the space industry (see also [19]. It was decided to apply the classical model, as it best fit the requirements of the problem. There were 4 experts, 35 variables of interest, and 13 seed variables, whose frequencies of occurrence were known to the analyst, but not to the experts.

Table 1 shows the results of the classical model: calibration scores, entropy scores and weights of the four experts and of the decision maker. Optimization of the decision maker's virtual weight is reached at the significance level of 5%. At this level, two of the four experts receive zero weight. Reducing the significance level to 0.1% enables all experts to receive positive weight, but degrades the decision maker's calibration and entropy scores. Even more degradation results from assigning the experts equal weights.

Table 1. Results of the classical model.

expert	calibration score	mean seed entropy score	normalized weight	
			α = 5%	α = 0.1%
1	0.025	0.527	0	0.113
2	0.001	0.625	0	0.004
3	0.1	0.556	0.484	0.428
4	0.1	0.521	0.516	0.456
			virtual weight	
DM (5%)	0.7	0.615	0.754	
DM (0.1%)	0.4	0.654		0.593
DM (equal weights)	0.2	0.714	0.4	

The results of the fault tree quantification are shown in table 2. The median and 90% confidence bounds of the frequency distribution of the top event are calculated for the optimized decision maker. Also shown are the median and 90% confidence bounds resulting from the individual experts' assessments. The median assessment of the decision maker agrees with that of expert 3, but the decision maker's confidence bounds are narrower by an order of magnitude than those of expert 3. If the experts had input their distributions into the fault tree individually, the resulting median assessments would span 2 orders of magnitude and the confidence bounds would span 5 orders of magnitude.

Table 2. Results of fault tree quantification using the decision maker's and the experts' assessments.

expert	top event		
	5%	median	95%
DM	9.10^{-6}	1.10^{-4}	1.10^{-2}
1	5.10^{-6}	8.10^{-5}	3.10^{-3}
2	1.10^{-7}	2.10^{-6}	3.10^{-5}
3	9.10^{-7}	1.10^{-4}	1.10^{-2}
4	4.10^{-5}	2.10^{-4}	1.10^{-3}

Since the Bayesian model is compatible with the classical model, it is possible to feed the expert data from this application into the Bayesian model and evaluate the model's performance.
The Bayesian analysis was performed for each of the four experts individually. The Bayesian updated experts' assessments can then be fed back into the classical model. The results of this exercise are shown in table 3, which also contains the calibration, entropy and weights of the four original experts and of the optimized decision maker. For each variable, the 5%, 50% and 95% quantiles are extracted from the Bayesian updated distribution, and the calibration and entropy scores are determined corresponding to these assessments. (The entropy scores differ slightly from those in table 1, because the additional experts influence slightly the intrinsic ranges for each variable.) In each case, the Bayesian updating improves calibration at the expense of entropy. It appears that the updating has the effect of driving the outer quantiles further apart, making the updated expert less overconfident but also less informative.

Table 3. Results of feeding the Bayesian updated experts' assessments into the classical model and comparing the calibration, entropy and weights with the four original experts and with the optimized decision maker (DM).

expert	calibration	mean entropy	norm'd weight
1	0.025	0.521	0.012
1 upd.	0.4	0.687	0.144
2	0.001	0.645	0.0004
2 upd.	0.1	0.825	0.03
3	0.1	0.548	0.045
3 upd.	0.4	0.730	0.135
4	0.1	0.520	0.048
4 upd.	0.925	0.768	0.298
DM	0.7	0.6	0.288

Application 2

The second application to be discussed concerned irregularities with flanged connections at DSM, Geleen, Netherlands. In particular, the problem was to determine the most important causes of flange failure.
Different categories of failure causes can be distinguished. In consultation with DSM, it was decided to investigate procedural and environmental failure causes, since these categories are the most interesting with respect to the flange problem. Firstly, 6 environmental failure causes were distinguished, to which "wrong mounting" was added as representing the procedural subgroup (causes 1 to 7). Secondly, the failure cause "wrong mounting" was split up into 10 more specific procedural failure causes (causes 7.1 to 7.10).
It was decided to apply the method of paired comparisons to obtain both a ranking and relative contributions of the failure causes. At each comparison of two failure causes, the following question was asked:

> Which of these causes has led more often to a failure of flanged connections at the plant?

The case study was split up into two rounds. Round 1 involved comparisons of the causes 1 to 7 (21 pairs), whereas round 2 concerned the causes 7.1 to 7.10 (45 pairs).
DSM selected 14 experts. They were operators, mechanics, maintenance engineers, mechanical engineers and their chiefs. The experts were divided into a maintenance group and a production group. The education of 10 experts was comparable to a bachelor of science level. The remaining 4 experts were

operators or mechanics. Three subgroups of experts were distinguished: experts with high education level in the maintenance group (M, 5 persons), experts with high education level in the production group (P, 5 persons) and the other experts in the maintenance and production groups (R, 4 persons).

It appeared that the experts were quite consistent in their answers. The hypothesis that the answers have been given at random was not rejected for one expert in round 1, which points to qualitatively good experts and/or distinct differences between the failure causes.
Scale values, indicating the relative contributions of the various failure causes, were calculated for the group of all experts and for the subgroups by applying the NEL/Bradley-Terry model. The principal results are shown in table 4; numbers in parentheses indicate the rankings of the most important failure causes. The main conclusion from table 4 is that in both rounds a number of corresponding opinions appear to exist despite differences in opinion between subgroups of experts.

Table 4. Results of failure causes analysis.

failure cause	all experts	subgroup		
		M	P	R
round 1				
1	0.19 (2)	0.15 (3)	0.15 (2)	0.12 (3)
2	0.08	0.05	0.04	0.12 (3)
3	0.04	0.05	0.01	0.06
4	0.03	0.02	0.01	0.01
5	0.31 (1)	0.42 (1)	0.10 (3)	0.44 (1)
6	0.05	0.03	0.01	0.10
7	0.31 (1)	0.28 (2)	0.68 (1)	0.15 (2)
round 2				
7.1	0.07	0.05	0.01	0.11 (2)
7.2	0.07	0.13 (4)	0.02	0.02
7.3	0.02	0.004	0.002	0.04
7.4	0.29 (1)	0.24 (1)	0.13 (2)	0.40 (1)
7.5	0.02	0.01	0.003	0.05
7.6	0.17 (3)	0.19 (3)	0.11 (3)	0.10 (3)
7.7	0.21 (2)	0.21 (2)	0.69 (1)	0.06
7.8	0.05	0.09	0.003	0.06
7.9	0.02	0.004	0.002	0.05
7.1	0.10 (4)	0.07	0.04	0.10 (3)

CONCLUSION

This paper presented a discussion of three models which seem promising for incorporating expert opinion in risk and reliability analysis in a formalized way. Some results of two case studies were described. More results of these and other applications and a discussion of the practical implications are presented elsewhere [20, 21].

ACKNOWLEDGMENT

This paper is based on a research project which was performed for the Ministry of Housing, Physical Planning and the Environment of the Dutch Government. The authors are grateful to this Ministry for permission to publish this paper. The authors are also grateful to DSM Limburg B.V. for permission to discuss the case study which was performed at one of DSM's plants and to Mark Stobbelaar who provided important contributions to this case study. The idea of choosing the significance level so as to optimize the decision maker's virtual weight is due to Simon French.

REFERENCES

1. Van Steen, J.F.J., Expert opinion use for probability assessment in safety studies: Main topics and elements of an application-oriented research program. European Journal of Operational Research 32 (1987) 225-230.

2. Van Steen, J.F.J., Expert opinion in probabilistic safety assessment. In: Libberton, G.P. (ed.), 10th Advances in Reliability Technology Symposium, pp. 13-26. London, New York: Elsevier Applied Science (1988).

3. Van Steen, J.F.J. and Oortman Gerlings, P.D., Expert opinion in safety studies, Vol. 2: Literature study. TNO/Delft University of Technology, Apeldoorn/Delft, Netherlands (1988).

4. Cooke, R.M., Van Steen, J.F.J. and Mendel, M.B., Expert opinion in safety studies, Vol. 3: Model description report. Delft University of Technology/TNO, Delft/Apeldoorn, Netherlands (1988).

5. Cooke, R.M., A theory of weights for combining expert opinion. Report 87-25, Faculty of Mathematics and Informatics, Delft University of Technology, Delft, Netherlands (1987).

6. Mendel, M.B. and Sheridan, T.B., Optimal estimation using human experts. Department of Mechanical Engineering, MIT, Cambridge, Massachusetts, USA (1987).

7. Morris, P.A., Combining expert judgments: A Bayesian approach. Management Science 23 (1977) 679-693.

8. Winkler, R.L., Expert resolution (with discussion by Lindley, Schervish, Clemen, French and Morris). Management Science 32 (1986) 298-328.

9. Kempthorne, P.J. and Mendel, M.B., Adjusting experts' probabilistic forecasts: The use and implications of exchangeability assumptions in the calibration problem. Department of Mechanical Engineering, MIT, Cambridge, Massachusetts, USA (1987).

10. Seaver, D.A. and Stillwell, W.G., Procedures for using expert judgment to estimate human error probabilities in nuclear power plant operations. NUREG/CR-2743, U.S. Nuclear Regulatory Commission, Washington, DC, USA (1983).

11. Stobbelaar, M.F., Paired comparisons. Report 87-137, Division of Technology for Society TNO, Apeldoorn, Netherlands (1987) (in Dutch).

12. David, H.A., The Method of Paired Comparisons. London: Griffin (1963).

13. Kendall, M.G., Rank Correlation Methods. London: Griffin (1962).

14. Siegel, S., Nonparametric Statistics for the Behavioral Sciences. New York, Toronto, London: McGraw-Hill (1956).

15. Torgerson, W.S., Theory and Methods of Scaling. New York: Wiley (1958).

16. Thurstone, L.L., A law of comparative judgment. Psychological Review 34 (1927) 273-286.

17. Mosteller, F., Remarks on the method of paired comparisons. Psychometrika 16 (1951) 3-9, 203-218.

18. Bradley, R.A., Some statistical methods in taste testing and quality evaluation. Biometrics 9 (1953) 22-38.

19. Cooke, R.M., Uncertainty in risk assessment: A probabilist's manifesto. To appear in Reliability Engineering and System Safety (1988).

20. Goossens, L.H.J., Cooke, R.M. and Van Steen, J.F.J., Protocols for expert opinion use in risk analysis. 6th International Symposium on Loss Prevention and Safety Promotion in the Process Industries, Oslo, Norway (June 19-22, 1989).

21. Goossens, L.H.J., Cooke, R.M. and Van Steen, J.F.J., Expert opinion in safety studies, Vol. 1: Final report. Delft University of Technology/TNO, Delft/Apeldoorn, Netherlands (1988).

The Analysis of Electronic Component Reliability Data

J. M. MARSHALL, J. A. HAYES, D. S. CAMPBELL

Component Technology Group,
Department of Electronic & Electrical Engineering,
University of Technology,
Loughborough,
Leicestershire LE11 3TU, U.K.

and

A. BENDELL

Trent Polytechnic,
Mathematics, Statistics & Operational Research,
Nottingham NG1 4BU, U.K.

Abstract

The paper describes the development of analysis procedures for data within the Electronic Component Reliability Database at Loughborough University of Technology. The database includes data acquisition from Plessey, GEC, STC and two Danish Companies.

Earlier papers have described the content and administration of the database and the path to establishing the database. The current paper concentrates instead on the analysis of the data within the base. The structure of the base has facilitated an exploratory approach to analysis and enabled the identification and calibration of relevant environmental, mounting, screening and other influences upon failure behaviour to be made from the data itself rather than from prior assumptions.

1. INTRODUCTION

The establishment of an electronic component field failure database centred at Loughborough University in the United Kingdom has been described in two papers [1,2]. At present several data sources both in the U.K. and Denmark are actively involved in this field failure study. A detailed reporting format has been developed using statistical and engineering expertise, and is used to assess the dependability of the data collected and processed by the companies. Because of the number of individual companies involved it is possible to follow a wide range of equipments, both military and commercial, in several operational environments. Historical data, at present up to seven years old, is also processed if shown to be dependable by meeting the requirements of the reporting format.

The database has been designed around the detailed reporting format and, in practice, this means that component behaviour can be evaluated not only with respect to time but with regard to other important parameters such as environment, screening level, encapsulation, mounting etc. Data has been coming into the database from the sources for approximately two years and the database is expanding both in terms of the volume of data and the number of sources actively involved.

2. CONVENTIONAL METHODS OF ANALYSIS

Electronic component field failure and population data is essential for systems reliability prediction. The accuracy of such predictions depends on the failure data that is used. Blanks [3] has stated that no matter how excellent the mathematical modelling, how precise the analysis and how detailed the resulting formulae, ultimately the prediction can be no more accurate than the numerical data substituted in them. The majority of reliability studies on electronic components unfortunately, but understandably, have been carried out under accelerated conditions since data collection is expensive and time consuming. Failure rates are than extrapolated from such tests to what might be expected under field conditions. These extrapolations are often subject to doubt if only because multiple failure mechanisms are usually involved.

The most widely used failure rate models which incorporate π factors to take account of effects such as environment, screening etc. are those published in MIL-HDBK-217, [4]. It is not uncommon for reliability predictions based on the Parts Stress Analysis method to be a contractual obligation for system suppliers. Much criticism has been voiced with regard to predictions based on MIL-HDBK-217, particularly with regard to the wildly pessimistic predictions for some semiconductor devices when compared with observed field performance data. This is not entirely unexpected as, in general, the models are derived from test bed, accelerated life and limited fragmented field failure data. In addition the concept of a constant failure rate has not been substantiated, in practice, for all electronic components and indeed is not accepted by many [5,6]. However, reliability models based on a constant failure rate assumption are attractive from a systems engineers stand point as they are in principle simple and easily interpreted.

Electronic component field failure data is available, notably that presented in the British Telecom 'Handbook of Reliability Data for Components Used in Telecommunication Systems' [7]. The failure data presented is limited, however, in that it is derived from inland telephone exchanges only, operating in benign environments. In order to give failure rate data on a wide range of components it has been necessary to obtain data in some cases from 'one or more alternative sources with minimal supporting evidence'. Although the data is of use to the suppliers of telecommunication equipment, it is of limited use to manufacturers of equipments operating in environments other than benign as again the models presented are of an empirical nature.

The need to establish an electronic component field failure database to allow more detailed analysis and to study the behaviour of components under different environments to determine the factors affecting their field failure behaviour is the basis of the Loughborough database.

3. DATA IMPLICATION OF IMPROVEMENTS TO CONVENTIONAL ANALYSIS METHODS

Historically, most reliability databases have reflected the type of data structure and implicit assumptions underlying the conventional methods of analysis [8]. This is a "chicken-and-egg" situation, since it can be claimed that the analysis methods in conventional use have evolved to make use of data available. On the other hand, it has frequently also been claimed that there is not much point in collecting extensive detailed failure data since conventional analysis methodologies are unable to make use of its richness of structure.

To break this closed loop of conservatism, it was necessary at the start of the project to identify the data requirements that would be necessary to extend beyond conventional analysis. The intention was to realistically identify the nature and extent of data that would be necessary to allow analysis without the restrictive assumptions of constant hazard and pre-determined π-factors, and to allow empirical comparison to be made between sets of data that were previously inaccessible.

The process of defining the data requirement was a long and arduous one, and is discussed in some detail elsewhere [2]. The important points are that it was undertaken by a Working Group including both data-providers and would-be users, and that it was fundamentally iterative in nature with proposals being challenged for their adequacy and consistency in data provision, as well as the realistic achievement of data capture. To this end, whilst some data requirements were soon identified as essential to the intentions of the project, other data features were regarded as desirable only, and relegated to "non-essential" status.

The essential features of the data was that instead of capturing data on a failure-count or failure-rate basis (i.e. number of failures/total of exposure times), data was required in a format that provided individual component time to failure information. This needed to be directly cross-referenced to detailed component and environmental specification for both failures and non-failures. For the various forms of exploratory analysis foreseen, the data structure needed to contain information on component encapsulation, mounting technology, screened level and circuit reference position as well as a detailed categorisation of component type. At the equipment level, environment was essential as well as detailed up and running date, observation end date and failure time information on equipments and boards.

The definitions of all such terms evolved out of the divisions of the Working Group and were (and are) well documented. Such definitions needed to change over time as new data sources became available and conflicts appeared.

The problem with all reliability data collection is that unlike other forms of statistical data, reliability field data is the product of at least three distinct processes. These both contribute to and mask its information content.

1. Firstly, there is the failure process itself that we wish to find out about. Unfortunately we must remember that in practice the failure information available to be obtained is a direct product of the deployment employed and monitored.

2. Secondly, there is the logistic support system, since failure information is typically only obtained as a by-product of repair and replacement activities. In a similar way, population at risk information is frequently only available, if at all, from deployment data.

3. Finally, there is the data collection system to provide the data for analysis. Typically, this must make use of partial collection already in existence for diverse or unclear purposes.

In the current context, the data requirements were realistically identified given the restrictions imposed by equipment structures and deployments, logistic support systems, and already existing partial data collection systems that varied considerably. In a sense, the essential data requirements represented the minimum that would satisfy the requirements of analysis desired. It was a minimal set of what was regarded as standard features of field reliability data.

Nevertheless, problems have still arisen in the operation of the data system, due to the complexities of reliability field data. In places these could be resolved by further investigation of the data in the database, in other cases they had to be resolved by sometimes likely and sometimes arbitrary assumptions. For example, can one be sure that screened level of replacement components is the same as that of the components they replace or whether a component in exactly the same position in nominally identical items of equipment have the same screened level? The latter question could be partially resolved by searching for mixed screening levels at equipment and then board level. Similarly, mounting technology of replacement components had to be assumed identical to the components they replace. The most major assumption, of course, is to work with component 'failure' data that effectively arises from multiple component replacements upon board repair, often when the components are still in working condition. Investigative approaches are needed when this is suspected to minimize the error.

4. ANALYSIS METHODS

The database design facilitates an exploratory approach to the analysis of the data.

When a component fails in a system, the information stored in the database details how and when it failed, together with descriptive details of the device itself. An abbreviated version of the database design can be found in [2]. The database contains information, not only on the failure characteristics of a component, but also on systems out in the field. This data consists of information about the environment the system is working in, the percentage of time the system is operational and the full component listing broken down by circuit board part number. One of the aims of the study is to detect some of the possible factors which may affect the reliability of electronic components, for example, screening, environment, encapsulation etc.

As discussed in earlier sections, in general, the industry standard for the reliability of electronic components is indicated by the failure rate or inversely as the mean time to failures (MTTF), and is given by:

$$\lambda = \text{total number of failures/total component hours}$$

or

$$MTTF = 1/\lambda$$

These reliability indicators assume that a constant failure rate exists, excluding the phenomena of "infant mortality" and "wear out".

Figure 1 shows a flowchart of the two stages of analysis to be discussed below.

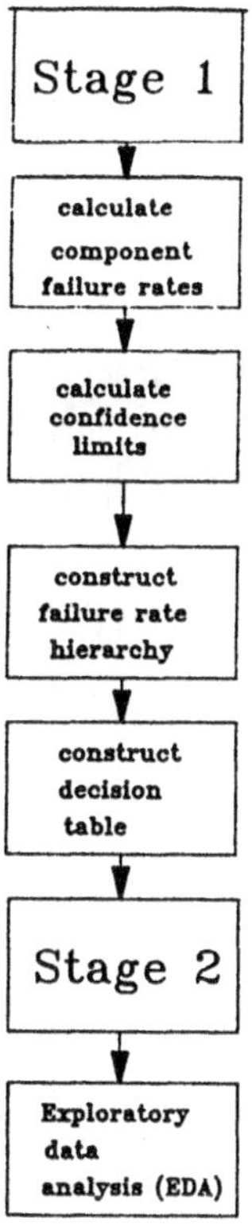

Fig. 1. Flowchart of the stages of analysis

4.1 The First Stage of Analysis - Failure Rates

Preliminary calculations carried out on the data are the failure rates per component type. This gives the first impressions of the reliability of each component type in the field, and from these an hierarchy of problem components can be obtained as shown in Figure 2. The component categories shown in Figure 2 have previously been identified [1]. X^2 confidence limits for these failure rates have been calculated in an attempt to qualify this hierarchy and to indicate those components which may require additional date [2,9].

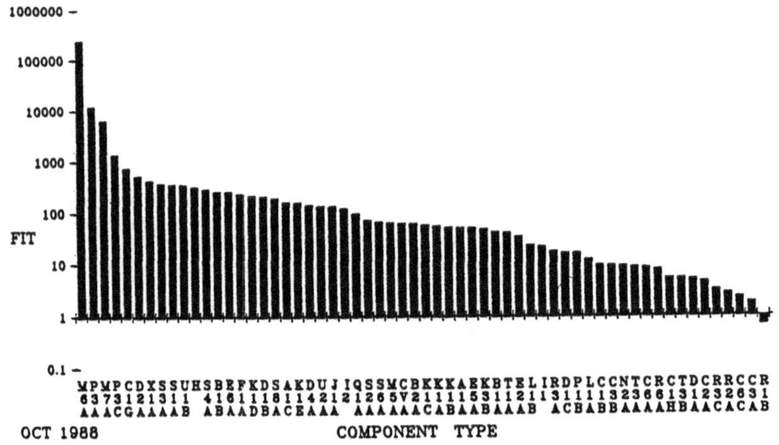

Fig. 2. Bar chart of failure rates for each component type

A table has been established to give a more explicit view of the data (Table 1). This table shows a breakdown of the failure rates into total number of failures and total component hours. For example, a component type which is at the top of the hierarchy implies that this component type is the most unreliable. However, looking at how this failure rate is calculated may give a better impression. Thus a component with small component hours and small number of failures may appear at the top of the hierarchy but, in reality the data is not sufficient to make any statement about the reliability of this component type. This hierarchy can, therefore, be misleading in some instances. Similarly, a component with a large number of failures and small component hours may indicate a need to investigate for design problems or replacement policy.

Table 1. Table for identifying the component type order for analysis (Failure rate in FITS)

	High Fits	Low Fits
HF, HCH	2	4
HF, LCH	1	3
LF, HCH	5	7
LF, LCH	6	8

HF – High Failures **HCH – High Component Hours**
LF – Low Failures **LCH – Low Component Hours**

The model in Table 1 indicates the order in which component types can be analysed. Deciding what represents a high 'Fit' (i.e. the number of failures in 10^9 component hours), high number of failures etc. is purely subjective and based upon the actual data in the database. Each box contains those component types which have, for example, high number of failures, high component hours and high Fits etc. When this table is complete the first stage of investigation can begin.

This first stage of investigation is concerned with checking for design problems, establishing the repair policy of multiple failures, together with any source related problems, before any further exploratory data analysis (EDA) takes place. For example:

1. Check the number of sources providing the information on the component type under investigation;

2. Breakdown number of failures and total component hours by source;

3. Check for design problems and multiple failures.

Highlighting possible design problems is important not only to the company manufacturing the equipment, but in order to understand the failure pattern of the data. Similarly, if there are multiple 'failures' of the same component type it is important to know whether these are in fact real failures or are a consequence of the replacement policy carried out by the company. The flowchart in Figure 3 shows the order of this stage of investigation. Basically, if any problems relating to a particular source are found then the company involved is contacted and asked for further information relating to the component type under investigation.

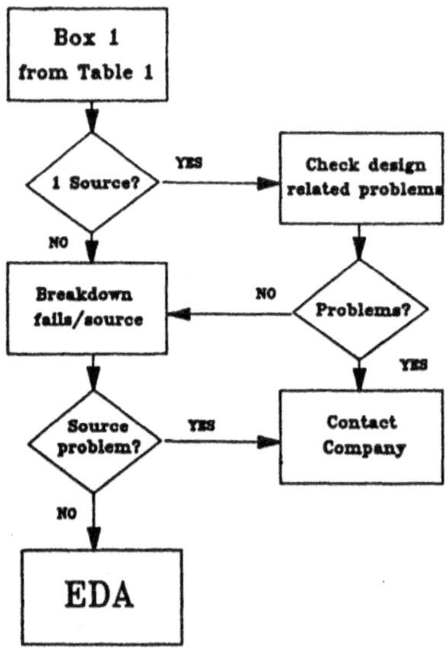

Fig. 3. Flowchart of order of analysis

4.2 Stage 2 - Exploratory Data Analysis (EDA)

EDA involves using simple graphical and tabular methods to highlight any patterns or irregularities in the data. It increases not only the understanding of the nature of the data and failure phenomena, but also indicates the type of analysis which is appropriate. When patterns or trends are identified they can be built into models for analysis, so that results will be more meaningful.

Standard techniques for analysing reliability data assume that the time between events are independently and identically distributed (iid). Exploratory techniques such as checking for trend and serial correlation in the data will reveal where these assumptions are upheld.

Past papers have discussed [10] the need for EDA and the techniques employed. However, they are mostly concerned with repairable systems. This data is based on non-repairable electronic components coming from numerous different repairable systems. This introduces a more complex approach to EDA. In order to pool this component data from different sources, it is important to explore the data for different time patterns between sources, environments, screening etc. Checking for design and multiple replacements have to be previously carried out, thus eliminating some of the dependencies within the database.

The time metric in the database is operating time to failure, not calendar time. This is the time a piece of equipment has been switched on. The time between events is only relevant for a given item of equipment and is meaningless for pooled data. Thus a different approach to exploring this data is required.

There are 4 stages of exploratory analysis currently employed i.e.:-

1. Failure rates broken down by source, environment etc.;

2. Operating time to failure against failure number plots for pooled data, for different equipment types (sources) and different environments;

3. Hazard plotting;

4. Proportional Hazards Modelling (PHM);

Item 1 above is basically calculating component failure rates for each source etc. and comparing them. If, for example, out of 4 sources, 1 source is consistently different from the other 3 over all component types, then investigation of that source is required. As explained earlier, failure rate calculations have their limitations; they do not give any indication of when the failures occurred and only give an average value. However, they do give a reasonable first impression of the data.

After exploring any of the inconsistencies of source data, item 2 is carried out. This involves simply plotting the operating time to failure of a given component type against failure number. Again, breaking down by source or environment, gives an indication of the failure pattern associated with this component type.

By only considering failures, as in the previous items, information on the population at risk is lost. However, hazard plotting of the component types takes into consideration the censoring times and gives a more optimistic representation of the data. Similarly, plotting time against cumulative hazard for different sources or environments gives an indication of any differing failure distributions between these factors. Hazard plotting also highlights whether the assumption of constant failure rate is in fact valid or not [11]. Examples of these figures are discussed in section 5.

The fourth item is a further development of this exploratory approach. Proportional Hazards Modelling (PHM), as discussed in many texts [12,13,14], incorporates explanatory variables such as source, environments, screening etc. into the model and consequently highlights whether they are in fact significant contributory factors of the failure distribution of the component. Hazard plotting alone merely takes account of the censorings when calculating the hazard contribution, but PHM actually uses the censorings in the calculation of the likelihood, thus improving our understanding of the failure pattern. Furthermore, it assumes no distributional form and empirically estimates the hazard distribution, taking into account the significant covariates established by the model. By including PHM as an exploratory tool, different explanatory variables can be incorporated into the model, consequently giving more information. Thus for example, having found environment and source are significant factors contributing to the hazard of a component type, then the model can be extended to include further variables such as screening level of the component, batch number, circuit reference position etc. Many different explanatory variables can be incorporated into the models thus approaching a more accurate representation of the failure of an electronic component.

4.3 Problems Encountered in the Analysis

The database to date contains information on 18 million components and 98 billion component hours, representing a vast amount of data. However, there are in fact only 4,500 failures of which about 50% are due to "no fault found". Thus the amount of censorings is vast compared to the number of failures. This causes several problems when analysing the data, two of which are:

1. Disaggregation - sparseness of the failure data does not allow the analysis of component types to be broken down by source, environment, screening etc. It appears that, apart from a few component types, information across sources is sparse.

2. Vast number of censorings causes numeric problems in PHM software [15].

These problem can be overcome by:-

1. Encouraging the companies to provide data from other sources thus building up the amount of data in the database. However, this problem will only be overcome in time.

2. Developing the software to cope with the large censorings and make use of the natural groupings of population data which appear in the database. It is believed that this problem has now partly been solved.

5. PRELIMINARY RESULTS

As described in section 4, the first stage of analysis is to calculate the failure rates for each component type and produce a barchart displaying the failure rate hierarchy. Figure 2 has shown such a failure rate hierarchy. The failure rate scale is log, since the difference in failure rates from the highest to the lowest is large. From this figure it is apparent that the first 4 worst component types are M6A (microwave cavity), P3A (wirewound variable resistor), M7A (microwave switch) and P3C (metal film variable resistor).

However, further information categorised as in Table 1 is shown in Table 2. This table shows that although there are very high failure rates for P3C and M7A, they have a small number of failures. Thus further analysis can be concentrated on the other two components at the top of the hierarchy. The other component types, namely P3A and M6A, have a high number of failures but a small number of component hours. This highlights a need for investigation into the number of sources providing the data and consequently checking for design errors. In this case there is only one source providing the data and they have been contacted and asked to investigate this problem.

Table 2. Illustrative example of table for identifying the component type order for analysis

Failures	Component Hours	Fit > 500	Fit < 500
>= 10	>= 10e7		I1 D1A T1A R2A B2A I2 A1A H
>= 10	< 10e7	D2A M6A P3A	K1D K1E
< 10	>= 10e7		C3B C6A C3B C2A L1A C6C CVA K3B R6A L1B
< 10	< 10e7	P3C M7A C1G	A1C D1B B1B S4A C1B D4A S1A S8A

Further design problems have been identified by using this iterative approach. However, they will not be discussed in this paper.

The next stage of analysis involves exploratory data analysis. The component under investigation in the following examples are bipolar transistors (T1A). The reason for choosing this component is historical. However, there is information from 4 sources on this component.

Figure 4 shows aggregated operating time to failure against failure number for T1A. This plot indicates that as time progresses the time between events increases. By breaking down the data by environment, i.e. GB - ground benign and GM - ground mobile, we can see more clearly which environmental condition is contributing to this increasing trend.

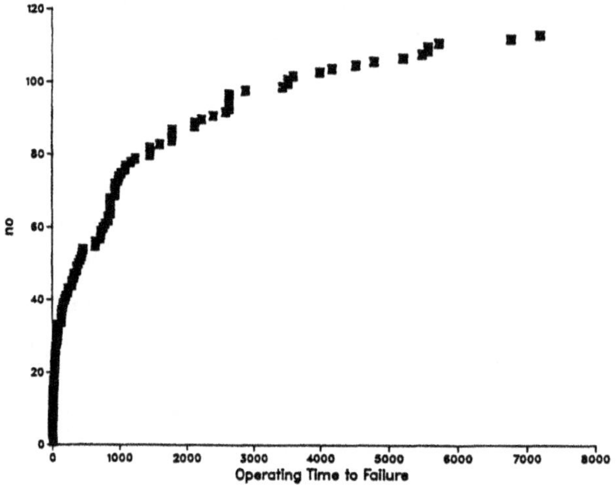

Fig. 4. Failure number against operating time to failure for T1A

Figure 5 shows that bipolar transistors under GM conditions fail with a different pattern from those under GB conditions. In the GB environment the time between failures increases earlier than under GM conditions. In this case, the more severe environment of ground mobile is showing an increasing trend in times between failures which may not be surprising.

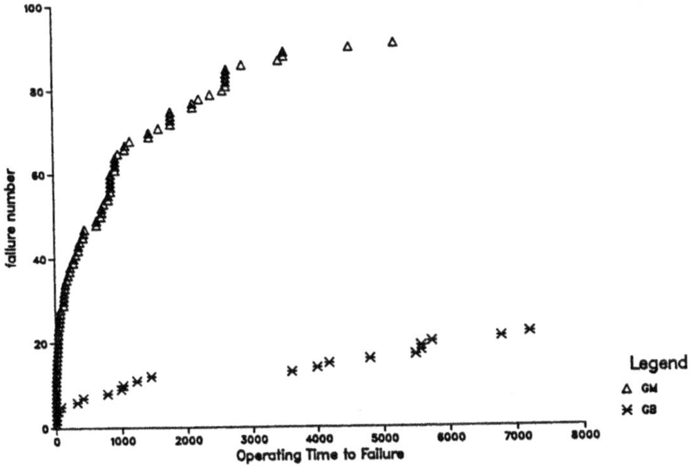

Fig. 5. Failure number against operating time to failure for T1A for two different environments

Although we have highlighted the presence of a trend in this particular data set, cumulative hazard plots may give us some further indication of the nature of this trend. Figure 6 shows time against cumulative hazard for T1A from the aggregate data. This plot exhibits a decreasing hazard rate corresponding to infant mortality.

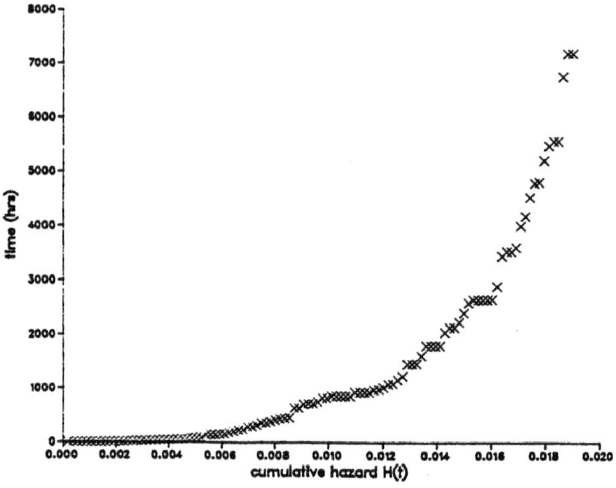

Fig. 6. Cumulative hazard plot on linear scale for T1A

This data is then plotted on Weibull hazard paper (log/log scale) in Figure 7 to estimate some of the parameters. On Weibull hazard paper the shape parameter β is equal to the reciprocal of the gradient of the fitted line [9]. However, it is important to note that, at this stage of the analysis, we are not particularly interested in the exact values of the parameters but in the shape and trend of the distribution.

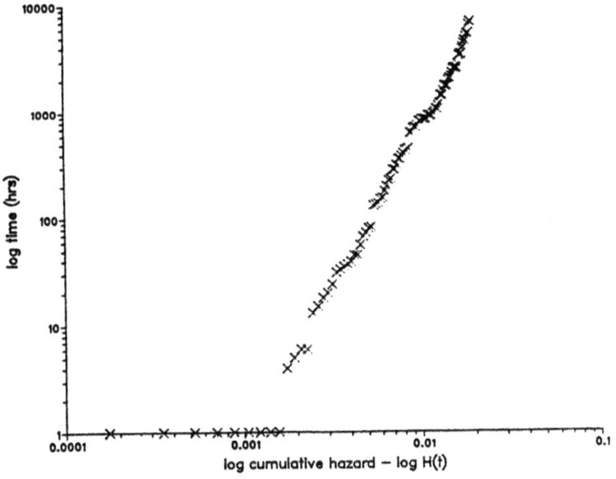

Fig. 7. Cumulative hazard plot on Weibull hazard paper for T1A

In Figure 7, apart from those points at 0 hours, the data appears to fit a Weibull distribution with a β value of approximately 0.625, indicating a decreasing hazard rate. Comparing the hazard plots of the two differing environments may give further information about the contributory factors of the failure distribution of bipolar transistors.

Figure 8 shows time against cumulative hazard for GB and GM environmental conditions for T1A.

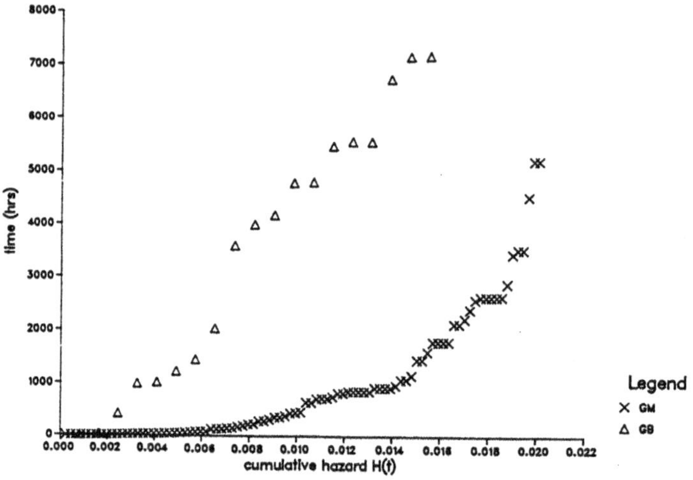

Fig. 8. Cumulative hazard plot on linear scale for T1A for two different environments

The curvature of the GM plot implies a decreasing failure rate which is similar to that in Figure 6. However, the curvature of GB is not so obvious. Plotting the data on Weibull hazard paper shows the difference in the failure distributions of bipolar transistors in GM and GB conditions.

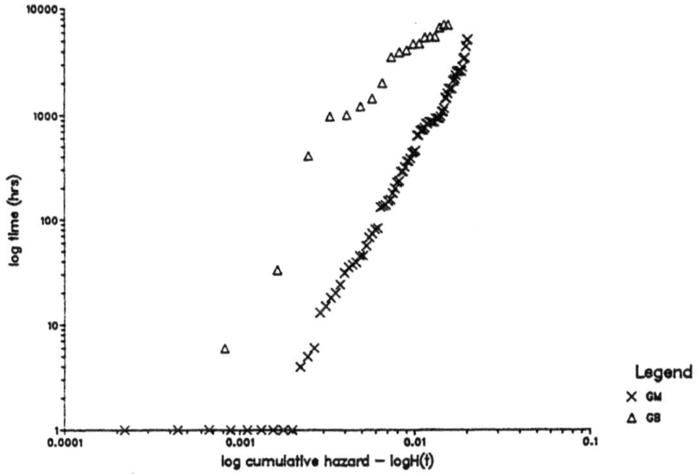

Fig. 9. Cumulative hazard plot on Weibull hazard paper for T1A for two different environments

The β value for GM is approximately 0.5, ignoring the 0 hour failures, thus indicating a decreasing hazard rate. However, it may be dubious to fit one line to the GB plot.

However, one must be careful in coming to firm conclusions on the basis of this limited analysis because:-

1. There is less data on GB than on GM in the database for this component type;

2. There may be significant external factors which affect the failure distribution of this component type for both GM and GB environments;

3. There may be different failure modes mixed in the data.

4. There are different component screening levels in the data.

5. Since GB is a less severe environmental condition than GB, we would expect differing failure distributions;

6. The GB plot in Figure 8 may in fact approximate a straight line which would imply a constant hazard rate;

7. This analysis takes no account of different technologies; thus further breakdown of the part numbers may explain the different trends.

From the exploratory analysis above it is obvious that further exploration is required, in order to model the failure distribution of this component type. PHM may help to highlight further explanatory variables and thus bring us closer to a fuller representation of the failure distribution of bipolar transistors.

6. CONCLUSION

This paper has developed a methodology for the analysis of field failure data of electronic components. It has shown that a field failure hierarchy can be established, and the data used to obtain an analysis of the effect of various aspects such as environment etc. on component failure. A detailed example has been given with regard to discrete bipolar transistors. This analysis shows that bipolar transistors have a complex failure pattern and do not, in general, fulfil the constant hazard rate assumption. These graphs also show that environment affects the failure distribution of these devices.

ACKNOWLEDGEMENTS

The work is being carried out with the support of the Procurement Executive, Ministry of Defence, and we are grateful to them for permission to publish this paper.

REFERENCES

1. Campbell, D.S.; Hayes, J.A.; Hetherington, D.R.: The organisation of the field failure of electronic components. Quality & Reliability Engineering International. 3 (1987) 251-258.

2. Marshall, J.; Hayes, J.A.; Campbell, D.S.; Bendell, A.: An electronic component reliability database. 10th ARTS, Bradford, (1988) 40-53.

3. Blanks, H.S.: The generation and use of component failure rate data. Quality Assurance. 3 (1977) 85-95.

4. MIL-Handbook-217E. Reliability prediction of electronic equipment. October 1986.

5. Williams, V.: The working life and reliability of electronic components. Proc. of Eurocon '82, North Holland. (1982) 1110.

6. Wong, K.L.: The bathtub does not hold water any more. Quality & Reliability Engineering International. 4 (1988) 279-282.

7. British Telecom: Handbook of reliability data for electronic components used in telecommunication systems. Issue 4.

8. Bendell, A.; Cannon, G.: Reliability database. Elsevier Applied Science 1989.

9. Nelson, W.: Applied life data analysis. Wiley & Sons, 1982.

10. Bendell, A.; Walls, L.A.: Exploring reliability data. Quality & Reliability Engineering International. 1 (1985) 37-51.

11. King, J.R.: Probability charts for decision making. New York: Industrial Press 1971.

12. Cox, D.R.: Regression models and life-tables (with discussion). J. R. Statist. Soc. 30 (1972) 248-275.

13. Lawless, J.F.: Statistical models and methods for lifetime data. Wiley & Sons, 1982.

14. Kalbfleish, J.D.; Prentice, R.L.: The statistical analysis of failure time data. Chichester: Wiley & Sons 1980.

15. Wightman, D.W.: The application of proportional hazards modelling to reliability problems. Ph.D. Thesis, Trent Polytechnic, Nottingham, 1987.

An Intelligent End-User Interface for the Collection and Processing of Accident Data

A.R. HALE[1], J. KARCZEWSKI[1], F. KOORNNEEF[1], E. OTTO[1] & L. BURDORF[2]

Safety Science Group
Delft University of Technology, The Netherlands

SYNOPSIS

Incident data collection is hindered by the need to use bulky standard forms with many questions designed to cover a wide range of varying types of incident. The majority of such questions are irrelevant for any one incident.

Interactive software offers the opportunity to collect all the data relevant to any one incident while presenting the person completing the report only with relevant questions. A program will be described with a tree structure of questions which can be programmed by an experienced user at one level of abstraction, for use by naive personnel who are prompted by a series of questions based upon the structure of their previous answers. The program can therefore be adapted to the recording of incidents with different specific characteristics in a wide range of situations.

Experience with the use of the program in practical applications will be reported.

INTRODUCTION

The collection and analysis of accident and incident data for use in research, in decision making in companies, and as the raw material for risk assessment has always been a problem area. The data normally collected for national statistics is limited in scope and depth ([1]), and subject to severe restrictions in quality and compatibility across countries ([2]). Only a few variables are collected, mostly those which concern the injury, the injured person, the immediate cause of injury and some data on location. The underlying causes of the accidents are rarely touched upon ([3]). Company statistics suffer from many of the same limitations, with the result that those who collect them are often at a loss to know how they can fruitfully use the data they have collected ([4]).

---

1. Safety Science Group, Delft University of Technology, Netherlands.

2. Institute of Occupational Health, Erasmus University of Rotterdam, Netherlands.

At the other end of the scale highly detailed information about single incidents is collected in official enquiries into serious accidents and disasters ([5]), in the protocols collected by some research projects ([6]), and by safety specialists or reporting schemes in some organisations ([7]). This data is seldom fully analysed partly because it is so rich and diverse and has been collected as unstructured narrative. Analysis is therefore often limited to a few incidents which are studied in very considerable depth ([8]).

A structured method of capturing information which taps some of the richness of the detailed incident reports and so goes further than the shallow analysis of routine statistics offers the promise of much more fruitful support for decisions about accident prevention and risk assessment. Some attempts have been made to collect this sort of data with the use of structured accident reporting forms ([9]), but a major hindrance to the success of such schemes has been the problem of designing suitable reporting forms which are manageable. Because the factors leading to individual accidents are very varied, the majority of questions on a comprehensive and detailed form are not applicable to any given accident. The result is that the person filling in the form gets lost in its complexities, or becomes frustrated, bored and de-motivated by the problems of finding the boxes which do have to be filled in for any given accident.

AN INTERACTIVE COMPUTER PROGRAM

A research team from the Universities of Delft and Rotterdam decided to tackle this problem of accident registration by developing a flexible, interactive computer program for accident data collection. The concept behind the program was that the user should be taken through a tree of questions which would cover the same topics for each accident, but which would present differently worded questions depending upon the information collected in answer to previous questions. In other words the program would steer the user down different branches of the total tree depending upon the type of accident being dealt with. An additional requirement formulated for the program was that it should be possible to modify the tree structure easily, to add or subtract branches, so that priorities or shortcomings revealed in initial analysis of data could be reflected in more or less detailed questioning about particular aspects of the accident determinants.

To develop such a program it was necessary to carry out the following steps:

1. Choose a suitable programming language and structure to allow for maximum flexibility in programming and modifying the question tree.
2. Choose an appropriate accident model to act as the basis for the question tree.
3. Couple the program to a flexible database management system for the analysis of the complex data collected. The chosen DBMS (EVA) has been described elsewhere ([10]) and will not be dealt with further in this paper.
4. Carry out trials with the program in real life situations to test the practicability and user-friendliness of the system.

DESCRIPTION OF IDA

The design of the software was subject to a number of objectives:

1) The interface with the person filling in the accident information should be as user-friendly as possible so that it could be used by non-specialists with the minimum of training. No knowledge or experience of either safety or computing was assumed to be present by the end-user.
2) The software itself should choose the question path through the question tree dependent upon the answers to earlier questions.
3) The structure of the question tree should be easily modifiable by a trained safety expert, to allow regular adaptation of the questions asked on the basis of changing priorities or uses.
4) Updating of information about accidents should be simple, given that all relevant information may often not be collectable at one time by one end-user.

The resulting software (IDA - Intelligent DAtabase interface or Interactive Data on Accidents) is written in Turbo Pascal and is made up of 9 modules. The end-user comes into contact only with the modules for input and correction of data, from which he can gain access to the information on help screens and for print-out options. The end-user is locked out of all other modules to prevent any inexpert or unintended tampering. A trained safety expert (subsequently called the 'system-user') is equipped with the password to interact with the set-up modules to alter the question tree, codes, and form of print out, and to carry out statistical analysis.

To meet these objectives we have chosen a pre-programmed conversation between the system and the end-user who has to input the data. The first version of IDA provided a pre-programmed tool-box with related procedures. The structure of the conversation could be programmed within certain constraints. We are now working on a new version which makes it possible for system-users to define on-line and interactively their own structure for the database (records and fields), and later if necessary to redefine the schemas for the conversation between the end-user and the system, and their own output documents to be printed by the system.

Structure and possibilities.

System IDA consists of the following main modules :

- initialisation and set up of the system,
- design and update of the question-trees,
- input of the accident data,
- documentation and correction,
- manual and help,
- statistics,
- output,
- options,
- user groups and privileges.

1. Initialisation and set up

This module serves to define the structure of the database and the things necessary for the end-user conversation and input. The system-user can define not only the structure of the database (records and fields) but also the questions related to each of the fields, the tables with values related to fields and the coding of those values and finally the help screens related to particular questions. Every record consists of several fields. For each field the user has to set up :

- type of the field (coded, boolean, real, string,...),

- related table with codes and values (if necessary) (e.g. types of injuries, workplaces, activities),
- related question (e.g "What was the injury received?" "Where was the victim working?" "What was he/she doing at the time?")
- related help (made by system user for end-user).

All the names of the records, of the fields, questions, coded values and defined links between fields and other elements (help, question, table with codes) can be changed later if necessary. This information forms the basis for the other functions and modules of the system.

2. Design and update of the question-trees

This module allows the system-user to define his own conversation schemas, based on the accident model chosen, and the depth of investigation considered necessary. Using a specially devised simple menu the user is able to define the course of the conversation between system IDA and the end-user who has to complete the accident report. The system-user is able to define a question tree with all the desired decision points. These decision points determine the path which is followed through the tree for any given accident. The path choice can be based on:

- information already gathered (e.g. type-of-hazard - noise),
- an answer to an additional question (e.g. are there more injured persons?),
- a test of a proper predicate (e.g. injured-person-age < 21).

The system-user is also able to program such actions as:

- open new record,
- verify information in already filled record,
- stop the investigation, etc.

Every question tree is given its own unique name, which is used by the end-user to call up trees other than the default-tree for the input of the accident data.

3. Input of the accident data

With this module the end-user types in the accident data. The question form, the codes and the help screens or prompts are controlled by the chosen question tree (from module 2). The user can call up help screens at any point in the conversation (using function key 10) if he is not sure what a particular term or the whole question means, if he has doubts which of the coded answers is appropriate for his case, etc.). The task of the user is limited to answering the questions appearing on the screen. If the system-user so chooses the end-user can be forced to answer a question before going on (necessary for questions which determine the choice of the subsequent path through the tree, or to prevent the end-user skipping over important information); if the end- user does not know the answer to such a question immediately he can exit the program, saving the information already entered, and return to the same point at a later date. If the question does not determine the path through the system, the end-user can be allowed to skip a question where the answer is not immediately available and to return to fill it in later, using module 4.

4. Documentation and correction

This module can be used in two ways:

- to give a print out of the database contents. The system is able to print out all or only selected records in a standard form as hard copy.
- to verify and if need be update the data. This allows for corrections which do not affect the path through the question tree, e.g. correction of personal data (name, place of birth), alteration of textual data, inputting of missing values, etc. Errors, or new information which would alter the decision path through the question tree cannot be corrected in this way; the accident must be deleted and typed in again.

5. Manual and help

There are two ways to learn and to get help about the system. The first is to use the interactive interface with the user instructions, which can be read or printed out as a manual. The second is to use the interactive help screens (function key 10). This gives the user access to on-screen information about the current module, its purpose, possibilities and use. These two options are standard and intended for the system-user. The help screens for the end-user are created in module 1 and called up by function key 10 during the input of accident data.

6. Statistics

This module provides a variety of analyses of the collected data. Our module is limited to problems relevant to the statistical analysis of accident data. If a more sophisticated analysis is necessary the contents of the database can be exported as an ASCII file, to be used by such standard statistical packages as SPSS, SAS, etc.) or by more specialised database packages such as EVA (10).

7. Output

This module permits the system-user to define standard documents, such as registration forms, annual reports, statistical reports, etc., which are required on a regular basis by the organisation either for internal use or for reporting to other bodies, such as inspectorates or insurance companies. The system-user defines the document format and the data which should be extracted from the database to fill it.

8. Options

This module enables the system-user to customize the system and to set up necessary default values for:

- colours of the different menus, pop-up windows, etc.
- names of the records, fields, parameters and different system texts i.e. names of the commands, control posts in the system where authority must be checked, etc.
- user-related parameters, i.e. year, number of the accident, default question tree, etc.
- directories in which the database, layout of the documents, helps, question trees, log-file (history of the input), etc. reside. Using different combinations of the directories the system-user can easily produce different question trees, or versions of the same tree in different languages.

9. User groups and privileges

In system IDA there can be several different user groups, who need to be given access to different parts of the data or to different modules, either from considerations of privacy or to protect the data or the structure of the question trees against unauthorized tampering. The system-user can define groups of users and the privileges of each group (which functions of the system are available to them). For example:

- system-user - full access to all the modules, to the database structure, question-trees, documents, etc.
- accident reporter (end-user) - access only to "Input of the accident data".
- chief of safety department - access to documentation and statistical modules, etc.

Users get their own user-i.d and password. Log-on procedures check the user-i.d and password and establish the privileges of the user. The password is coded using a one-way algorithm ([11]), which can authenticate the user without itself being found in the system. If forgotten it must be re-declared.

Application of the system.

System IDA can easily be adapted to any accident registration system. Its flexibility guarantees full compatibility with already existing documents, input procedures and accident forms, which can easily be programmed as question trees.

CHOICE OF AN ACCIDENT MODEL

System IDA has been developed to provide the maximum flexibility and user-friendliness for the end-user, coupled with the maximum freedom for the system-user to program whatever question tree structure he wishes. We have carried out a number of projects in industry using the system to program more far-reaching registration and investigation of accidents and incidents than has been possible with paper forms. The starting point of the research was a dissatisfaction with the quality of data collected about accidents on current forms and with its suitability for decisions about deeper research needs or prevention priorities and actions. To overcome this we chose to use an accident model combining the energy-barrier concept found in MORT ([12]) with the concept of accidents as deviations from normal, or desired, circumstances ([13]). Such a model encourages collection of data about the complete accident process from the initial deviation to the damage process and about the barriers and controls which failed or were absent. The latter are susceptible to far more effective and systematic intervention than the events immediately preceding the injury or damage, which have been traditionally recorded on accident forms.

The general structure of the question tree used in the practical projects is given in table 1.
The tree starts with some general questions about the physical circumstances at the time of the accident, about the victim, the activity and the injury and treatment. The questions then work backwards from the material cause of injury through the events which led up to it, and finally to the initial deviations from
normal or desired procedures, activities and circumstances. The technical and procedural controls and barriers which failed are identified on the way. This backwards questioning order is chosen for two reasons:

- It provides an unequivocal starting point for the questioning and avoids the considerable uncertainty and individual differences which arise if a victim is asked 'to begin at the beginning'. If that is asked, some victims will begin only seconds before the injury, others will go back hours and sometimes days.
- The second essential reason lies in the need to structure a tree which asks different questions about different types of accidents. The logic of the choice of questions is much easier to construct based upon the nature of the proximal cause of the injury and the way in which the victim came into contact with the damaging energy. This question must therefore come early in the tree. For example questions about the place from which someone fell and how they came to fall are relevant to falls from heights, while questions about how toxic materials came to escape are relevant to gassing accidents (but not vice versa).

The disparate question paths about the different types of accident come together again at the end of tree when the failed barriers and controls have been identified. Those questions establish in general terms what management factors led to the failures. These factors are potentially very similar for all types of technical failure (12). The structure of the question tree we have used also allows the management-oriented questions to be answered by a different person from the one who answers the simpler questions about what

Table 1: QUESTION TREE STRUCTURE

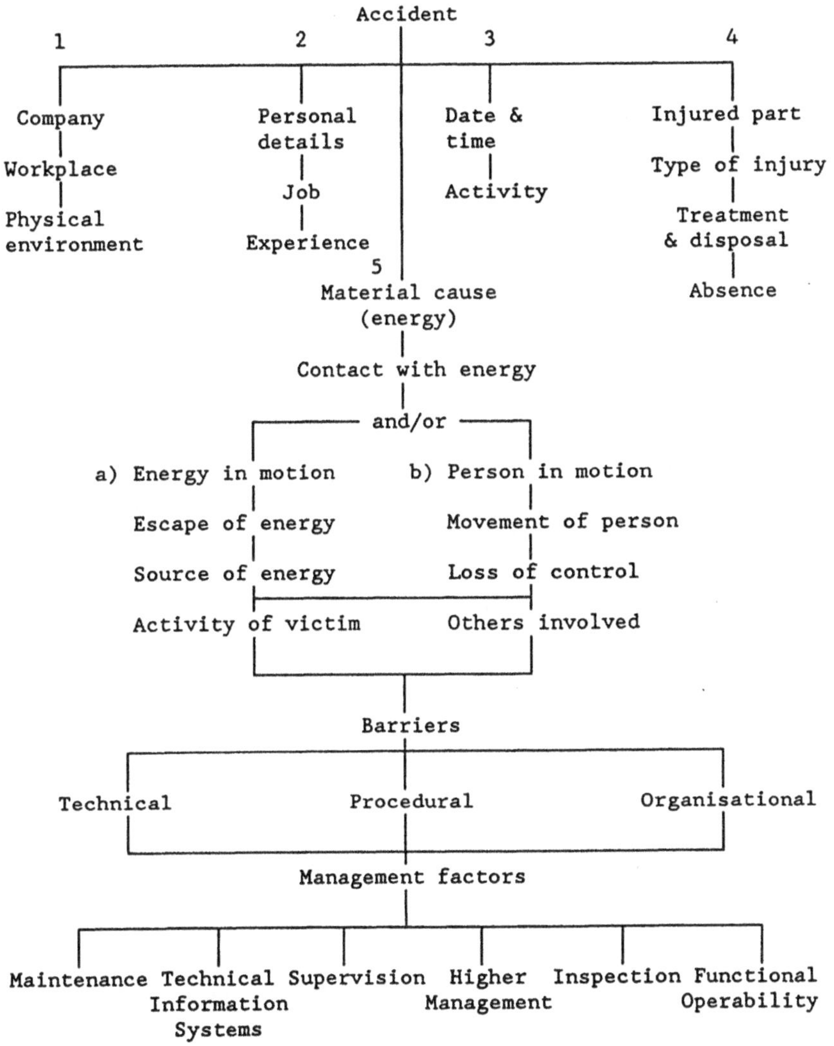

happened. In one project the split was between ambulance personnel and safety staff; in another between the accident victim plus his supervisor and a more senior departmental manager. The victims and personnel in the lower echelons of the organisation often do not know about the background factors to the accident and have little chance to find them out. Our experience is also that they are inclined to give biassed answers to questions about responsibility for particular procedures or their own use of them ([14]).

PRACTICAL APPLICATIONS

The program was developed with the help of a project partly financed by the European Community in the harbour of Rotterdam, in collaboration with Regional Occupational Health Service ([15]). The program has been used subsequently in the chemical industry.

The projects follow a common pattern:

1. Definition of the objectives of the organisation in the registration and analysis of its accident and incident data, including the breadth of incidents to be recorded (injury, damage, near miss, etc.), the depth of investigation of different types of incident, etc.
2. Definition of the organisational structure within which the reporting and analysis will take place; central or distributed reporting? Which information will be input by which end-user (the victim, first aider, supervisor, manager, safety staff)? What analyses at what levels?
3. Programming of the question tree and the coding framework (e.g. work places, job titles, machinery types). This has been done upto now by the research team, but modifications will be done by the company personnel.
4. Testing of the tree with existing records of past accidents and modification.
5. Introduction of the system to collect data on current and future accidents.

None of the schemes has been running long enough yet to evaluate the usefulness of the data collected, but a number of issues, both fundamental and practical have been encountered:

- there is a tension between two possible objectives of accident reporting, namely the need to be able to reconstruct a given accident in as much detail as possible at a later date and the need to analyse patterns of accident determinants. The former calls for highly detailed recording in many sub-categories, the latter for far greater aggregation of data to obtain significant differences between categories. This problem is in principle soluble at the analysis stage by data aggregation.

- each organisation has the need for some special branches to its question tree, or for the elaboration of branches representing specific problems for the organisation. A common core of branches can however be built up, and gives promise that the program can be offered in future with a number of default trees.

- the exercise of developing and installing the program uncovers shortcomings in the existing recording and analysis systems of the organisations, particularly in the availability and accuracy of some information. The program cannot itself solve most of these problems, but can be a catalyst for change.

- the energy-barrier concept appears to be easily explainable to, and usable by end-users with no specialist knowledge of safety. The questions related to underlying organisational and management factors require a greater level of training and the ability to stand back and take a more objective and comprehensive view of the incident.

- the nature of the recording system generates the need for different analysis techniques. Unlike recording systems with fixed questions, each accident type contains answers which are specific to the type; most data categories therefore have a large number of cases for which the question was not relevant. Analysis must therefore be in terms not of data categories, but of paths taken through the total question tree. Such an analysis is particularly appropriate for the investigation of accident scenarios ([16]).

FUTURE DEVELOPMENTS

We have plans to expand the technical possibilities of the system to make its use still greater:

- new field types (document, multi-value fields, etc.),

- incorporation of MORT-like analysis (12) of accidents with the possibility of going through the question tree both forwards and backwards, and of re-entering the analysis at different places.
- an easy but powerful query language for deep investigation of specific data.

The data generated will present a challenge to the development of better analysis techniques to get the best out of the richer information generated.

Finally, the structure of IDA is such that its use is not confined to the collection of accident data. It can be used to program any interactive conversation with an end-user in which the questioning structure can be reduced to a tree form with decision points determining the path through the system. It is therefore appropriate to many applications where standard but complex questionnaires have to be filled in by varied groups of people.

References

1. Smit H.A. Epidemiology of occupational accidents. Part 2. The adequacy of existing statistics for epidemiological research into occupational accidents. (Epidemiologie van bedrijfsongevallen. Deel 2. De bruikbaarheid van bestaande statistieken voor epidemiologisch bedrijfsongevallen-onderzoek). TNO-NIPG. Leiden. 1984.

2. Glendon A.I., Hale A.R., Booth R.T., Carroll C.H., & Clarke R.C. Occupational accidents and diseases: a review of data sources: consolidated report. European Foundation for the Improvement of Living and Working Conditions, Dublin, 1986.

3. The Swedish accident reporting scheme is a welcome exception to this rule (see Carlsson J. How to use the Swedish Occupational Injuries Information System, Proc. Int. Seminar on Occupational Accident Research, Saltsjobaden, Sweden, Elsevier, 1983) and has served as a partial model for the development of the program described in this paper.

4. Van Vliet L. Preventie staat of valt bij betrouwbare informatie. (Prevention stands or falls with the reliability of information) Maandblad voor Arbeidsomstandigheden 62 (1986) 478-482.

5. See for example published reports of civil aviation disasters under the rules of the International Civil Aviation Organisation, the reports of the Dutch Shipping Board (Raad voor de Scheepvart) on ship collisions and the reports of railway accidents produced by the British Railway Inspectorate.

6. See for example Hale A.R. Accidents during high voltage electrical switching. Report to Electricity Council, National Institute of Industrial Psychology, London, 1970, or Powell P.I., Hale M., Martin J. & Simon M. 2000 Accidents, National Institute of Industrial Psychology, London 1970.

7. For example the Licensee Event Reports of the Nuclear Regulatory Commission in the USA (see e.g Rasmussen J. What can you learn from human error reports, in Duncan K.D., Gruneberg M.M. & Wallis DJ. (Eds) Changes in working life. Wiley 1980, for analyses of this sort of data), or the reports of accidents in the UK Youth Training Scheme (Glendon A.I. & Hale A.R. A study of 1700 accidents on the Youth Opportunities Programme, Manpower Services Commission, Sheffield, 1985)

8. See for example Turner B. Man-made disasters. Wykeham, London 1978, or Reason J.T. The Chernobyl errors. Bulletin of the British Psychological Society. 40 (1987) 201-206.

9. Manning has been a pioneer in this field with his reporting schemes in Ford UK (see e.g. Manning D.P. & Shannon H.S. Injuries to the lumbosacral region in a gearbox factory. J. Society of Occupational Medicine. 29 (1979) 144-148.

10. Karczewski J.T. Navigational high-level interface for an expert database. Proceedings of 5th EUREDATA conference: Reliability Data Collection and Use in Risk and Availability, Heidelberg, April 1986. Springer-Verlag. 691-698.

    Karczewski J.T. Experimental Data Base Management System EVA. User manual version 1.3. Safety Science Group. Delft University of Technology, January 1988.

11. Denning D.E.R. Cryptography and Data Security. Addison-Wesley Publishing Company, 1982.

12. Johnson W.G. MORT: the management oversight and risk tree. J. Safety Research. 7 (1) (1975) 5-15.

13. Kjellén U. The deviation concept in occupational accident control theory and method. Occupational Accident Group, Royal Technological University, Stockholm, 1983.

    Hale A.R. & Glendon A.I. Individual behaviour in the control of danger. Elsevier. 1987.

14. See also Adams N.L. & Hartwell N.M. Accident reporting systems: a basic problem area in industrial society. J. Occupational Psychology. 50 (1977) 285-298.

15. Hale A.R., Otto E., Vroege D. & Burdorf A. Accidents in Ports: report of a pilot study to set up an accident registration system in the Port of Rotterdam. Safety Science Group. Delft University of Technology. October 1987.

16. Heimplaetzer P., Goossens L.H.J., Musson Y. & Clement R. Accident scenarios for domestic stairs accidents: characteristics of households, stairs & dwellings as risk factors, in Safety in the Built Environment. J.D. Sime (ed). London. Spon. 1988. for a presentation and use of the scenario concept.

Physics Models for Reliability Data Processing

R. RIGHINI

ENEA (Comitato Nazionale per la Ricerca e per lo Sviluppo dell'Energia Nucleare e delle Energie Alternative)
VEL Department - Bologna

G. ZAPPELLINI

NIER (Nuovi Interventi Energetici)
Bologna

Summary

Normally, in examining and processing failure reliability data, people apply the flat part of the "bathtube" curve; experience has verified that for many simple electric and electronic components such a distribution occurs: they are characterized by a long period with few and random failures. But experience has also verified that many types of components have not a constant hazard rate; f. e., normally mechanical components have a wearout characteristic and a continuously increasing hazard rate during all the life. Besides, if only two or three items are tested for a short time, little in the way of accurate quantitative conclusions can be drawn applying the classical statistics laws and assuming a bathtube distribution; bayesian methods may be taken into consideration, but these methods may be applied only when significative data on a set of similar items (components) is available; on the contrary we must remark that the preferred method to compute the reliability of an item is to consider the basic physics of the failure: it is, to predict the component failure rate in terms of a statistical variation of the physics parameters characterizing its operating conditions. On this subject, we must say that if one agrees to characterize a component by a hazard rate function and to evaluate the hazard parameters by testing, the result is a macroscopic model. Such a model is not concerned with what happens inside the black box containing the part but only with the statistics of the part performance. To delve inside the black box, one must postulate various hypotheses about the microscopic behaviour of the device and to prove validity; in this type of reliability model, the causes of the component failure are considered. We have called "physics models" this type of methods to process reliability data; they mainly consist of the following three basic steps:

1- Analysis of the physics laws characterizing the component(part) normal operation.

2- Singling out of the parameters of the above laws affected by stocastic variations in order to choose and to define the failure distributions to be applied.

3- Evaluation, on the basis of the material data and of the component characteristics, of the limit variation range of the parameters mentioned in point 2 in order to exactly evaluate the foreseen failure rate distribution of the component.

ENEA has carried out a data collection on typical components of the French fast reactors Rapsodie and Phenix; these data have been registered on an informatics system. In consideration of the experimental type of these components, of the small number of followed items and of occurred failures, the classical and bayesian methods do not allow the evaluation of too significative reliability data. At this aim ENEA-VEL has set-up some physics models on the ground of the physics laws governing the component operation and of the failure causes. A brief description of these methods, the results obtained by their application on some components and a comparison of these results with the ones obtained applying the classical and bayesian methods will be the subject of the present paper.

1. General remarks.

In reliability studies, the availability of processed data (i. e. failure rate, repair time, etc.), to be applied in system analysis, has a fundamental importance.

Component failure data are generally obtained from two sources: the failure times of various items in a population placed on a life test or repair reports listing operating hours of replaced parts in equipments already in field use.With reference to failure times, experience has shown that a very good way to present these data is to compute and to plot the failure density function; if we assume that our data describe a set of N items placed in operation at t=0, that at time t the number of survivors is n(t), the "failure density function" may be defined as:

$$f(t) = \frac{[\,n(t_i) - n(t_i+\Delta t_i)\,]\,/\,N}{\Delta t_i} \qquad \text{per } t_i < t \leq t_i + \Delta t_i$$

We may also define an other parameter (the "data hazard rate") as:

$$z(t) = \frac{[\,n(t_i) - n(t_i+\Delta t_i)\,]\,/\,n(t_i)}{\Delta t_i} \qquad \text{per } t_i < t \leq t_i + \Delta t_i$$

After plotting the failure number without considering the cause of the failure and examining the results, people began to recognize, for many component types, the following main modes of failure: early, in the lifetime of the component, there are a large number of failures due to initial weakness or defects: poor insulation, weak parts, bad assembly, etc.; these early failures are also called "initial failures" or "infant mortality". During the middle period of equipment operation, fewer failures take place, but it is difficult to determine their cause. In general they seem to occur when the environmental stresses exceed the design strengths of the part or component. It is difficult to predict the environmental stress amplitudes or the part strengths as deterministic functions of time; thus the middle life failures are often called "random failures"; it is characterized by an approximately constant z(t) and a companion f(t) which is exponential. As the item reaches old age, things begin to deteriorate and many failures occur. This failure region is quite naturally called the "wear region". In this region, also called "rising failure-rate region", the z(t) function increases whereas f(t) has a humped appearance.
Typical f(t) and z(t) curves illustrating these three modes of behaviour are shown in the following figure 1 and is often called the "bathtub curve".

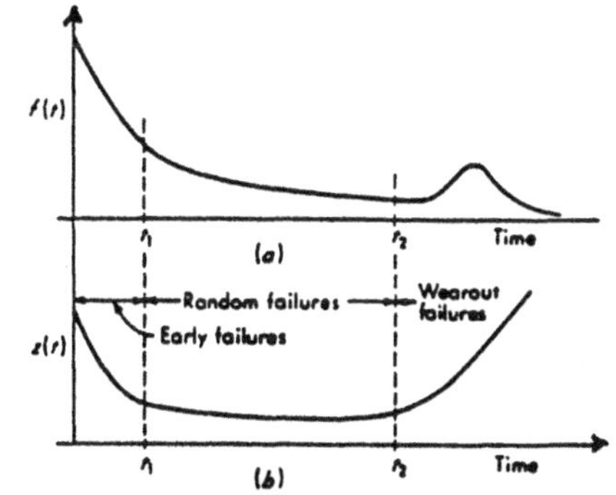

Fig. 1 General form of failure curves. a)Failure density b)Failure rate

Experience has verified that for many electric and electronic components such a distribution occurs: they are characterized by a long life period with few and random failures with time constant failure rate.
But experience has also verified that many types of components have not a constant hazard rate; f.e., normally mechanical components have a characteristic wear and a continuously increasing hazard rate during all the life.
Beside, a word of common sense is in order before delving in mathematics. If only two or three items are tested for a short time, little in the way of accurate quantitative conclusions can be drawn applying the classical statistical laws and assuming a bathtub distribution; bayesian methods (applying Bayes theorem) may be applied only when significative processed data on a set of similar components is available.
Now we must remark that the preferred method to compute the reliability of a part (or component) and to predict its hazard parameters is to consider the basic physics of failure. For example, given a capacitor which is constructed in a known manner with a dielectric material

operating under known conditions of temperature, pressure, humidity and voltage, one should be able to predict the component failure rate in terms of the above parameters and a "basic" capacitor failure rate. On this subject, we must say that if one agrees to characterize a component by a hazard rate function and to evaluate the hazard parameters by testing, the result is a "macroscopic model". Such a model is not concerned with what happens inside the black box containing the part but only with the statistics of the part performance. To delve inside the black box, one must postulate various hypotheses about the microscopic behaviour of the device and to prove validity; in this type of reliability model, the causes of the component failure are considered. We have called "physics models" this type of methods to process reliability data and their description is the subject of this report.
These methods resulted very important to process the data collected by ENEA on typical components of the French fast reactors Rapsodie and Phenix, in consideration of the small number of followed items and of the occurred failures; these data have been registered on an informatics system (the ENEA-CEA data bank). In the following a short description of this system and of the collected data type may be found; a demonstration of the unusuless of classical and bayesian methods will be reported. A short description of the set-up "physics methods" and of their profitable application in the processing of the available data will follow.

2. ENEA-CEA data bank : available data.

ENEA has set-up an informatics system (1,2,3) to register data collected on Fast Breeder Reactor components; the data bank may be considered an arrangement of the Euratom-CEDB (4) which has been adapted to the particular characteristics of the fast reactors.
The following data have been registered on each component taken in consideration:
-Identification codes.
-Engineering data.
-Operating and maintenance data.
-Failure data.
-Annual operating data.
The informatics system has been provided with an original and very flexible computer method to entry and select the data (1).
The failure data have been processed in order to obtain reliability data to be used in reliability system analysis; in consideration of the low number of operating components (for the majority of the followed component types), it resulted that a classical approach cannot be significatively applied in reliability data processing. The examples of the following figure 2 reporting the failure density relative to the data collected on some typical components such as the centrifugal pumps, may contribute to demonstrate how a classical exponential distribution cannot be applied: a clear trend of the failure density cannot be singled out and the number of failure data is too low in order to apply classical methods.
It resulted necessary to develop particular methods to process the data; these methods, called "physics methods", need detailed data on the cause of the failure and on the influence of the operation (maintenance, operating cycles, etc.) and of the material properties on the failure. It follows that particular importance has been dedicated to the collection and the research of such a type of data.
The following main causes (5) of the component failures have been singled out:
-Fatigue
-Wear
-Erosion
-Corrosion
-Operator error
-Operator error during the maintenance
-Design
-Electronic cause
-Electric cause
Excluded the cases of electric and electronic components where the classical approach results practicable (5) in consideration of the quantity and of the type of the available data, in the other cases the above mentioned particular data processing methods had to be developed in order to calculate the failure probability of the component part on the basis of the probability that the

physical parameters characterizing the normal operation of the interested part exceed the normal fluctuation range causing the failure. These methods, called physics methods, have been set-up referring to the above listed physical phenomena which have caused the failure as it results from the collected data; they will be described in detail in the following paragraphs. This description will be preceded by some short notices on the classical and bayesian methods set-up by ENEA and directly connected with the data bank.

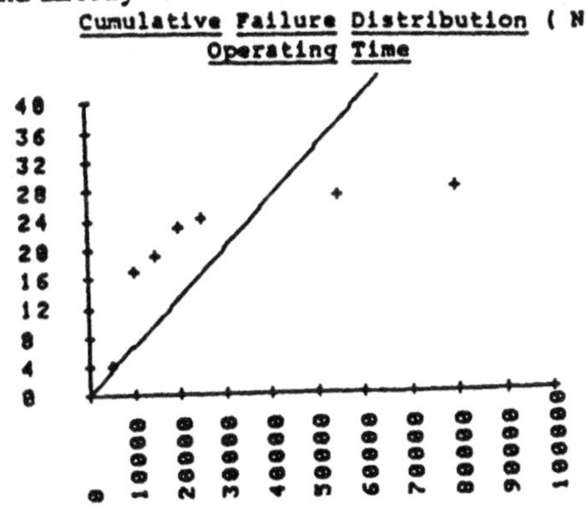

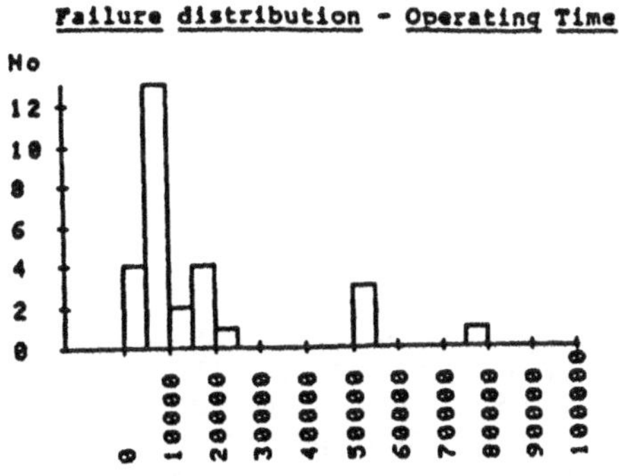

Fig. 2 Rapsodie reactor: centrifugal pumps; exponential test. a) Cumulative failure distribution. b) Failure distribution.

3. Classical and bayesian methods set-up by ENEA.

As said above, if at least 20 to 100 items have been tested, "classical statistical techniques" will help to extract the most information from such a situation. These "classical methods" are "macroscopic models" which are not concerned with what happens inside the black box constituting the part but only with the statistics of the past performance: they calculate the failure distribution function on the basis of the curve which better approximate the time distribution of the happened events. At this aim ENEA has set-up trend tests in order to find the failure distribution curve which better approximate the available failure data (6,7); after the approximating distribution has been found (exponential, Weibull, binomial,etc.), it is extremely simple (5,6,7) to calculate the reliability parameters such as the failure rate etc.. It is also clear that the so found approximating curve allows to understand if the failures are totally random or some particular factors (such as wear or infant mortality) influence the failure occurrence.
But, as said in the previous paragraph, if we refer to the data available in the ENEA-CEA data bank, only in few cases the approximating failure distribution was found; in the majority, or because few data were available (not enough to apply classical methods) or because no available curve was approximated, the exponential distribution was applied assuming the failure occurrence as totally random even if it was not realistic. In the case of few available data, bayesian methods (6,8,9)were applied. They are based on the concept that statistical inference doesn't consist in the research of a unique frequency limit for a certain event but that it must be the progressive adjustment of the cumulated information on the basis of the original prevision.The basic instrument of this progressive adjustment is the Bayes' theorem; if the parameter of interest is the failure rate (number of failures per unit time), the Bayes' theorem may be expressed by the following equation:

$$f(\lambda/E) = \frac{f(\lambda)\,L(E/\lambda)}{\int_0^\infty f(\lambda)\,L(E/\lambda)\,d\lambda}$$

where:

$f(\lambda)$ is the prior distribution, the probability density function of λ based on the generic information but not including any specific evidence E.

$f(\lambda / E)$ is the posterior distribution, the probability density function of λ, conditional on the specific evidence E.

$L(E/\lambda)$ is the likelihood function, the probability distribution of the specific evidence E for a given value of λ.

The application of these methods in order to obtain significative results, clearly requires the knowledge of $f(\lambda)$: it means that it is necessary to have significative information on components similar (in terms of engineering characteristics, operating and environmental conditions)to the one taken in consideration; it is also clear that for many sodium components, it is not possible to find such a type of information in the literature.

4. Physics failure models.

4.1. Generals.

These methods attempt to predict the component failure rate in terms of a statistical variation of the physics parameters characterizing its operating conditions. As said above, the classical methods are "macroscopic models"; they are not concerned with what happens inside the black box constituting the part but only with the statistics of the past performance. To delve inside the black box, one must postulate various hypotheses about the microscopic behaviour of the device and to prove validity: this involves the development of a physics failure model. In this type of reliability model, the causes of the component failure are considered. This type of approach considers the failure as the loss of a precise function to be accomplished by the component (endurance, seal, energy transmission, etc.): it is normally the result of a progressive damage resulting from the accomplishment of its functions and from the influence of the environment and operating conditions. In consideration of the different functions to be accomplished by the component and of the factors which influence its wear up to the failure (loss of the function), the failure of the component may be treated as the result of different basic physics failures each independent of the others and having its own particular failure probability P_i to be considered as a fraction ε_i of the total failure probability of the component as a series system (7) where the independence hypothesis allows to write

$$P_{tot} = \Sigma (P_i \, \varepsilon_i)$$

The independence may be assured grouping the common failure causes in the physics failure which is more affected by this common cause. In the case the failure probabilities have different distributions, the total failure probability of the component may be introduced by a fault tree where the single failure causes are considered as primary events.
It follows that we may consider independently the single physics failure and identify, on the basis of the existing uncertainties in the parameters which characterize the phenomenon, the failure distribution to be applied and the relative failure rate.
As said above, referring to the available data in the ENEA-CEA data bank, for some components, specially for the electric or electronic ones, it is possible to apply a classical approach in the data processing; in any case for such a type of components it is really difficult to set-up a physics model in consideration of the totally random variation of the environment parameters affecting the failure.
It follows that in the following, when we speak of physics models, we shall exclusively deal with mechanical components.
ASME codes (10) identify the following main physics failure modes for a mechanical component: corrosion, erosion, lubrication, elastic-plastic deformation, surface fatigue, radiation damage, spalling corrosion, wear, buckling. In order to allow the analysis of these phenomena by synthetical and general physics models to be represented by univocal probability distributions, the above listed failure modes have been reduced in the following ones:
-Fatigue (also corrosion and temperature related).
-Corrosion (chemical,electric-chemical,etc.)
-Erosion
-Stress (temperature picks).

Referring to the chapter 2 of this paper, we may see that the above listed failure models include the main accident causes affecting the failure of the components considered in the ENEA-CEA data bank (2,5), excluded the failures caused by error of the operator (during the operation or during the maintenance): a suitable physics model has been set-up in order to consider also this cause of failure.
Consequently, each mechanical component was split-up in its main parts: for each part, the most significative failure causes were singled out and it was associated to one of the above listed failure models; for each-one of these failure models a computer program was developed on the basis of the physics laws governing the phenomenon and of the information on the materials, the maintenance, the failure causes and consequences, the supported stress cycle, etc. available in the data bank. The development of each failure model provides the three following main steps:

1- Analysis of the physics laws characterizing the component (part) normal operation.
2- Singling out of the parameters of the above laws affected by stocastic variations in order to choose and define the failure probabilistic distributions to be applied.
3- Evaluation, on the basis of the material data and component characteristics, of the limit variation range of the parameters mentioned in point 2 in order to exactly evaluate the foreseen component failure rate distribution.

The failure models allow to calculate the foreseen failure parameters (failure rate, failure distribution,etc.) of the component; if some data on failures effectively occurred on the component type are available, they have been used to better approximate the theoretically calculated failure rate of the component by applying the bayesian methods where the failure distribution calculated by the physics model is assumed as prior distribution; the times of the happened failures constitute the information cumulated by the experience which may be used to define the posterior distribution.

4.2. Stress-strength model.

This very sample model may be applied to calculate the failure probability of a component (or part) which must support a constant mechanical load. Normally, in structural mechanics, the solidity of a construction is assured by a parameter called "security factor" (S) and defined by:

$$S = \frac{\text{strength}}{\text{applied stress}} = \frac{\rho}{C}$$

where as strength, the material elasticity limit (11) multiplied by a security factor is normally used.
In the reality this security factor is not a constant (middle value), but, (cause the material heterogeneity, the inaccuracy in the dimensions and in the measures, the surface state, etc.), it is not a deterministic value: it can be represented by a statistic distribution.
In the same way, also the applied stresses may fluctuate during the mission time.

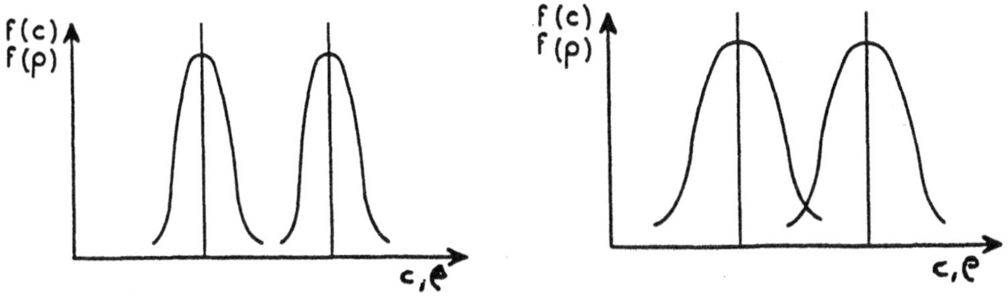

Fig. 3 Stress and strain model. a) $C/\rho < 1$; b) $C/\rho > 1$.

It follows that the reality is represented by the figure 3: stresses and strengths are represented by two distribution functions whose mean values are, respectively, μ_c and μ_p It is therefore possible to calculate the reliability of the structure; this computation consists in the probability that the structure strengths exceed, during the considered mission time, all the effectively applied stresses. It is:

$$R = \text{reliability} = \text{probability that } C < \rho$$
$$\text{or } R = \text{probability that } C/\rho < 1.$$

This condition is clearly specified in fig. 3a; it is not satisfied in fig. 3b. In this last case we must treat the success and failure probabilities as functions of the random variable "y" and then integrate over all values "y>x". A more direct procedure is to define a new variable z=y-x such that, if z>0., the part fails. If f(y) and g(x) are normal, the computation is very easy: the density function for z, $\Phi(z)$, is normal with E(z) = E(x)-E(y) and var.z = var.x+var.y. The probabilities of success and failure are:

$$P_f = \int_{-\infty}^{0} \Phi(z)\, dz$$

$$P_s = \int_{0}^{+\infty} \Phi(z)\, dz$$

The set-up computation method also considers the case in which f(y) and g(x) are not normal.

4.3. Fatigue.

The mechanical phenomenon consisting in the application of repeated cycles on a component part is called fatigue; the cyclic repetition of the strains generates plastic micro-strains which may deteriorate in micro-cracks: their propagation may lead to a breakage of the interested part also in the case of stresses much lower than the elastic strength limit.
The applied cyclic stresses may be of the following main three types:
-Synusoidal type (middle value = 0.).
-Repeated type (middle value Sm=S/2.).
-Undulatory type (middle value Sm=(Smax.+Smin.)/2.)
Referring to each type, the stress is represented by:

Sa=1./2. (Pm+Pb+F) range Erif./E

where:
Pm = membrane stress
Pb = bending stress
F = pick stress
E = coefficient of elasticity of the material
Erif = coefficient of elasticity in the actual loading conditions.
The considered stresses are simple stresses (traction, flexion, etc.); in the case of compound stresses, the equivalent stress is evaluated on the basis of the Van Mises hypothesis assuming that the breakage happens when the volumetric value exceeds the critical value:

$$Seq. = \sqrt{\frac{\Sigma_{ij} (\sigma_i - \sigma_j)^2}{2}}$$

The overlap of the different strains by different cycles on the structure causes a damage on it; if we apply on a structure n_i cycles whose stress intensity is Ci and if the number of cycles necessary to break the structure is Ni, the increase of the damage in the structure is given by:

$$n_i / N_i$$

and we have the breakage of the structure when:

$$\Sigma_i \, n_i / N_i = 1$$

Successive studies have shown that this equation is not exact because it doesn't totally consider all the phenomena which effectively arise in the structure. Miner, who gave the name to this model, has then corrected this expression as it follows:

$$\Sigma_i \, n_i / N_i = \gamma$$

Where "γ" is a variable quantity having a normal distribution. The experimentally obtained values of "γ" are included between 0.18 and 23.

On the basis of this equation, many laws have been developed (Henry, Corten et Delon, etc.).

In the "physics model" set-up by ENEA, the fatigue was treated referring to the "Wohler diagram": if we submit a material sample to periodical stress cycles having an established maximum amplitude and a constant frequency, it fails when a certain delimited number of cycles is exceeded; considering equal samples submitted to equal cycles characterized by different maximum stresses, we obtain a diagram usually called "Wohler diagram" represented in the following figure 4.

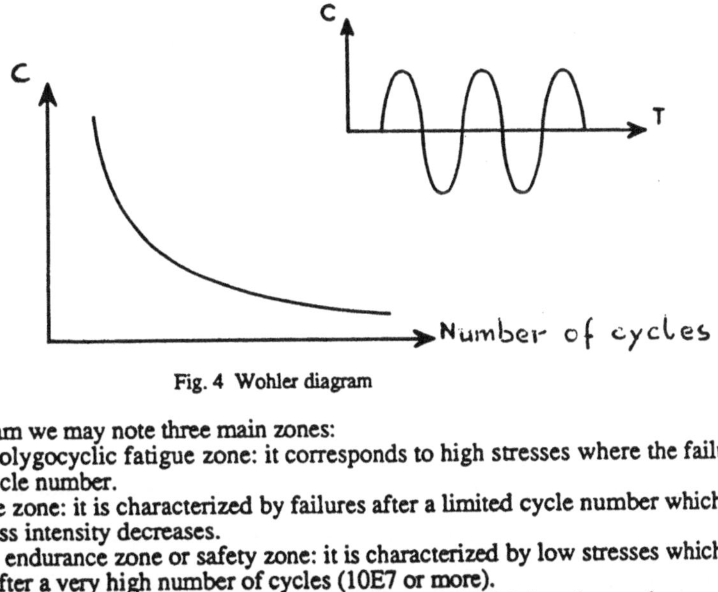

Fig. 4 Wohler diagram

On this diagram we may note three main zones:

a - Plastic or olygocyclic fatigue zone: it corresponds to high stresses where the failure occurs after a low cycle number.

b - Endurance zone: it is characterized by failures after a limited cycle number which increases when the stress intensity decreases.

c - Unlimited endurance zone or safety zone: it is characterized by low stresses which cause the failure only after a very high number of cycles (10E7 or more).

Many studies have been performed on this phenomenon (12,13,14) and very deep experimental researches (Morrow, Manson, Hirshberg) on many samples of different materials allowed to obtain the following general equation which may be applied to each material whose normal stress characteristics are known:

$$\Delta \varepsilon / 2. = 0.76 \, (D)^{0.6} \, (2N)^{-06} + 1.9 \, (Rm / E) \, (2N)^{-0.12}$$

where:

$\Delta \varepsilon$ = material strain

D = material ductility

N = applied cycle number
Rm / E = ratio between strength and elasticity module

This equation may be only applied for low temperatures; it has been modified for an application at high temperature in the following one:

$$\Delta \varepsilon = 0.25\,(D)^{0.26}\,(N)^{-0.6} + 2.66\,(Rm / E)\,(N)^{-0.12}$$

Two important factors cause the introduction of statistic and probabilistic concepts in fatigue phenomena:
- the fluctuating nature of the applied stresses;
- the dispersion of the fatigue strength values of the component even if they were obtained in controlled laboratory tests.

The following fluctuation causes may be singled out:
- External causes:
  - -environment fluctuations;
  - -test machine errors;
  - -operating factors.
- Material causes:
  - -thermal treatment;
  - -cooling temperatures;
  - -material strain-hardening.

In consideration of these fluctuation causes, it follows that in the Wohler diagram, the life length of the component cannot be represented by a unique point on the diagram, but by the distribution of N cycles. Then the diagram becomes a family of curves as represented in the following figure 5.

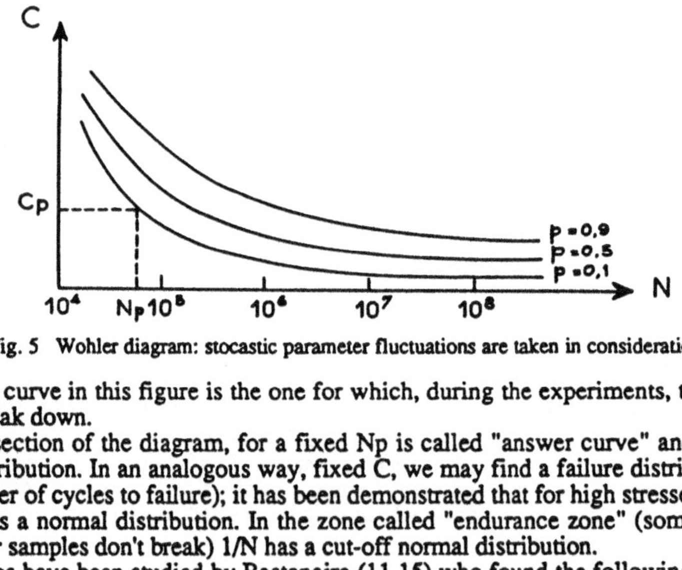

Fig. 5 Wohler diagram: stocastic parameter fluctuations are taken in consideration.

The middle curve in this figure is the one for which, during the experiments, the 50 % of the samples break down.
A vertical section of the diagram, for a fixed Np is called "answer curve" and usually has a normal distribution. In an analogous way, fixed C, we may find a failure distribution in terms of N (number of cycles to failure); it has been demonstrated that for high stresses, Log N (for a fixed C) has a normal distribution. In the zone called "endurance zone" (some samples may break, other samples don't break) 1/N has a cut-off normal distribution.
These curves have been studied by Bastenaire (11,15) who found the following expression by which it is possible to draw the curves on a (C,N) diagram as it is represented in the previous figure 5.

$$N_1 = \frac{A}{C - \rho_e} \exp\left[-B(C - \rho_e)\right]$$

where:
N_1 = breakage cycle number

C = cycle stress intensity

ρ_e = endurance limit

In order to solve the problem of the statistical prevision of the life length of a mechanical component, from the data bank we may obtain the constant alternating stress intensity (Cm) and the number of supported cycles. For many components the supported cycles are the ones generated by the reactor operation (start-up, operation and shut-down); other types of cycles may be determined (f.e., for the hoses of the rod mechanisms the supported cycles are the ones generated by the vertical movement of the mechanism). In any case, fixed the type of the cycle, the supported stress type and, consequentely, the Wohler diagram may be defined and the Bastenaire curves may be drawn. In this diagrams, since we know the alternating stress intensity, we may draw the parallel to the axis of the abscissa and so obtain the distribution of the failure cycle number , f(N), which is a lognormal function. If we remember that the cycle number corresponds to a lifetime, the component failure distribution, f(t), may be obtained. Afterwards, if we have some data on occurred failures, we may apply the bayesian methods in order to obtain a more exact failure distribution: as "prior", the distribution calculated applying the physics model is assumed; the information to be inserted in the likelihood function is given by the occurred failures.

4.4. Erosion.

This phenomenon refers to the erosion of mechanical fragments by friction between moving facing surfaces; it may assume the two following aspects:

-"abrasion" consisting in a quick removal of the produced debris;

-"pitting", characteristic of not together fitting surfaces (for the presence of lubrication) and consisting in the detaching of small particles causing deformations in the facing surfaces.

In order to calculate the probability that a component fails because the erosion phenomenon arises, we must consider the limit conditions at which this phenomenon arises; at this aim the equations of the dynamic lubrication must be analysed in order to obtain the parameters and equations necessary to define the load limit conditions.

In this logic the "middle friction" between two facing and lubricated surfaces is expressed by the Newton-Petroff equation obtained on the basis of the Reynolds lubrication theory. Two cases are generally considered: the trust pin and the rotating pin. In both cases the main parameters of the laws which govern the component operation are:

Pm=mean pressure between the surfaces

f =kinetic friction factor

The physics laws governing the operating behaviour of the trust pin are (16):

$$Pm = 6\,\mu\,\frac{V\,a}{h_0^2}\,\psi\,\xi$$

$$f = \lambda\sqrt{\frac{\mu V}{Pm\ a}\ \frac{1}{\xi}}$$

where:

ψ depends from the surface shape

ξ depends from the lubricating type

For the rotating pin the above parameters (Pm and f) are expressed by (17):

$$f = \frac{K_0}{\sqrt{K_1}}\ \frac{\delta}{V}\ \sqrt{C_1\ C_0}$$

$$Pm \leq \frac{C_l}{W_r}$$

The ISO codes don't directly use the above parameters (Pm,f) but the integrated loads(C,Pm,υ) which are related to those parameters by elementary relations (18).
The principal probabilistic factors which affect the preservation of a kinetic lubrication and therefore may cause the eventual beginning of the erosion may be resumed in what follows:
- the "fluctuations" in kinetic lubrication conditions specially with reference to the variable steady state conditions (loading, winding thickness, loss of alignment);
- alteration of physical and classical characteristics of the lubricant for causes which cannot be totally controlled by periodical tests and maintenance.
Therefore the following two main probabilistic aspects of the phenomenon may be singled out:
- variation of the parameters which may influence the phenomenon;
- the cumulative aspect of the phenomenon must be considered as the progressive succession of erosion phenomena caused by the exceeding of the limit conditions in the normal kinetic operation.
The probabilistic aspects and parameters which have been defined above result enough to determine the failure distribution model to be applied in the probabilistic analysis of the erosion phenomenon; even if the variety of the involved parameters and of the mechanical characteristics, a unique basis model may be assumed consisting in the application of a two parameter Weibull distribution.

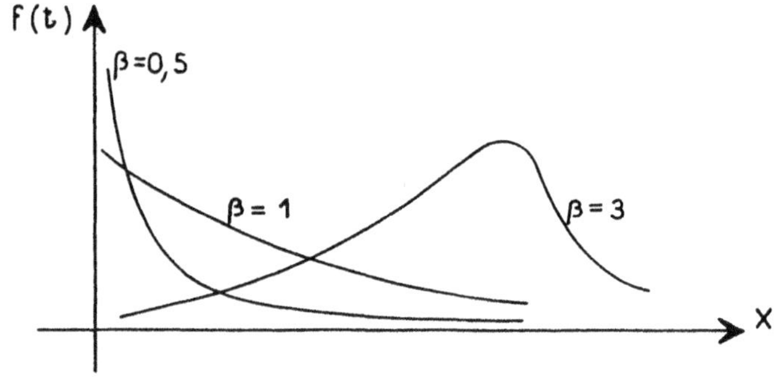

Fig. 6 Weibull distribution.

It follows that the following distribution may be assumed:

$$f(t) = \frac{\beta}{\mu} \left(\frac{t}{\mu} \right)^{\beta - 1}$$

$$F(t) = 1 - e^{-\left(\frac{t}{\mu}\right)^{\beta}}$$

$$R(t) = e^{-\left(\frac{t}{\mu}\right)^{\beta}}$$

where:
β =shape parameter
μ = scale parameter

ß is commonly obtained on the basis of an arrangement of the experience failure time data (deriving from the component manuals); it normally has a constant value.
μ has the following expression:

$$\mu = a \cdot \left(\frac{C}{P}\right)^{x}$$

This expression may be obtained so from mechanical reliability (11) as from the ISO codes (18).
For example in the case of rotating bearings, this expression becomes:

$$\mu = a_{23}\ (L_{10})_h = a_{23}\ \frac{10E6}{60\,n}\left(\frac{C}{P}\right)^{p}$$

where:

a_{23} is a technical coefficient referring to the lubricant type; introducing a factor K which is directly connected to the oil viscosity, the a_{23} coefficient may be expressed as it follows (19):
$a_{23} = 1.33\,K - 0.33$ for $0.4 < K < 1$
$a_{23} = 0.375\,K + 0.623$ for $1 < K < 5$

n = round number per minute;

C = dynamic load coefficient (i.e. the dynamic load at which the 90% of bearing breakages takes place);

P = real dynamic load;

p = coefficient related to the bearing (or rotating part) type.

For each bearing type the tabulated values of C/P may be obtained from the ISO codes.
It is clear that, if the bearing and lubricant type are fixed, the factor μ depends from the cycle number.
For the usual bearings, the experimental value of C/P is equal to 1.5 as it may be obtained from the ISO codes.
Also in this case the value of the failure rate calculated by theory may be adjusted on the basis of the occurred failures applying the bayesian methods.

4.5. Corrosion.

The corrosion physics model refers to the progressive and spontaneous technological decay (with the consequent loss of its main properties) of a mechanical material caused by the environment where it operates. The following main types of corrosion may be normally singled out:
-"humid corrosion" caused by condensed water (f.e. iron in saturated steam);
-"dry corrosion" caused by non-condensed water or other liquids or gases (f.e. iron in overheated steam, in gas, in liquid metal, etc.);
-"intermediate corrosion" caused by a non-watery chemical solvent.
The theory of corrosion phenomena has shown that they are affected by many chemical and physical factors referring to the metallic material, the environment, etc.. In general, the corrosion phenomenon is expressed and measured by the following parameters:
a) corrosion intensity, expressed in terms of load variation on time and surface unite (gr/m2.y);
b)corrosion speed expressed by the penetration of the corrosive process on the time. It is normally measured in mm/year or inch/year. In laboratory tests, the corrosion speed is normally related to the erosion intensity by the following expression:

$$v\ (\text{mm/year}) = \frac{8760\ 10\ \Delta P}{A\, d\, t}\ (\text{mm/year})$$

where:
ΔP = load variation (gr.)
t = exposition time (h)
d = metal density (gr/cm3)
A = metal surface (cm2)

Generally the corrosion speed depends from the physical and electrochemical phenomena which intervene in the corrosion process.

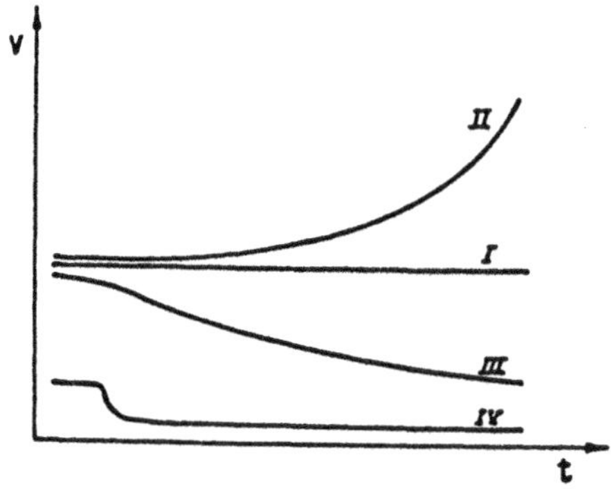

Fig. 7 Typical trends of the corrosion velocity in function of the time.

The following main trends of development of the phenomenon result from the experience (see fig. 7):
1 - Constant speed processes (curve 1): they consist in the thinning of the surface without accelerating or decelerating phenomena; it follows that the life length may be provided on the basis of simple kinetic laws. A typical example of this case is given by the iron in hydrochloric acid.
2 - Self-accelerating processes (curve 2): they refer to the intervention of factors accelerating the existing corrosion phenomenon. A typical example is given by the cathodic action of impurities which cumulate on the metal surface during its dissolution.
3 - Self-decelerating processes (curve 3): they refer to the intervention of factors which slacken the metal dissolution. These factors are produced by the dissolution itself.
4 - Self- humbling phenomena (curve 4): it refers to the intervention of passive phenomena such that the corrosion stops after a certain time.
The humid corrosion is normally the responsible for the component failure; if the corrosion intensity is measured by an electrical current intensity (see the case 2 above), the measure instrument must be calibrated on the basis of the measured " v" and "i" by the following relation (see figures 8a, 8b):

$$i_{corr} = \frac{\Delta i_{appl}}{\Delta E} \; \frac{b_c \, b_A}{2.3\,(b_c + b_A)}$$

From the current intensity, we may obtain the corrosion intensity (speed) as follows:

$$\Delta p = \frac{i_{corr}}{Z\,Q_e} \; \frac{A}{N_0} \; 3600 \;(\text{gr/h})$$

$$v = \frac{8760 \; 10 \; \Delta P}{S \; d} \;(\text{mm/sec})$$

where:
Δi= applied current
b_c = cathodic slope
b_a = anodic slope
Z = valency

S = etched surface
Q_e = electron charge (coulomb)
d = metal density
N_0= Avogadro number
The corrosion depends from the material type, the nature of the liquid causing the corrosion, the report between the anodic and the cathodic surfaces and from the temperature.

Fig. 8a

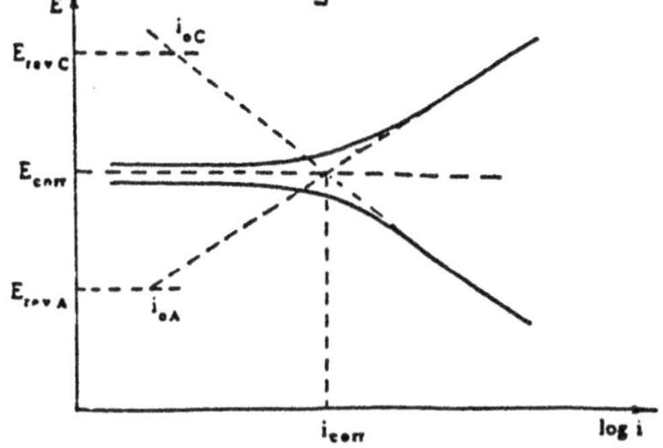

Fig. 8b

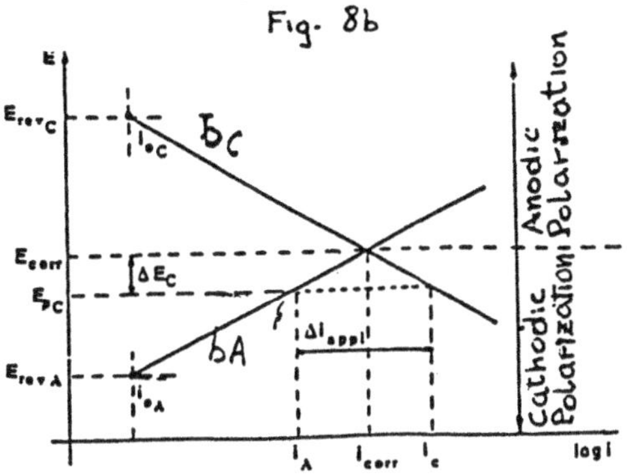

Fig. 8a Extrapolation of the Tafel straight lines at the Ecorr electric potential for the determination of icorr.
Fig. 8b Establishment of an electric potential Epc < E for the circulation of an outer cathdic current.

Many aleatory factors affecting the advance velocity (intensity) of the process, may be found in the corrosion phenomenon. Among them, the following two main aleatory aspects are typical of the corrosion and may be exactly differentiated:
-a) a first aspect, referring to a process of uniform and diffused corrosion in its totality, is related to the variability of the general electric-chemical conditions as represented in the previous figures 8;
-b) a second aleatory aspect refers to localized processes connected to the leak of homogeneity of the phenomenon. In the following figure 9, the reported trends show the aleatory (probabilistic) distribution of the penetration in deepness on a metal surface expressed as a function of time; this penetration is placed on a middle intensity value of corrosion developed with a self-accelerating law.
An analogous aleatory presence may be described for the "pitting" corrosion as distribution around the middle value of deepness in the single craters. Such aleatory values are particularly important for failures such as a crack followed by a leakage, very common in our data bank.

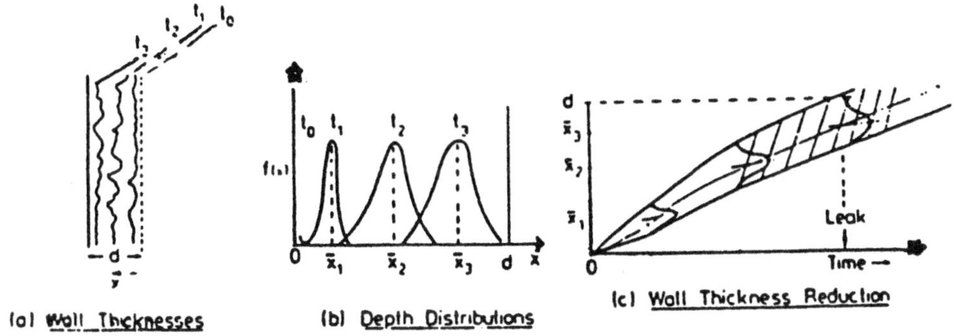

Fig. 9 Corrosion: leak models.

Corrosion is a cumulative process; it follows that the probabilistic distribution characterizing this type of process is the Weibull distribution: the two types of the above defined aleatory parameters characterize two different types of Weibull distribution.

This distribution is characterized by the two parameters ß and μ ; ß is directly related to the time corrosion behaviour:

ß is > 1 (for example = 1.5) when the corrosion process is linear or accelerated;

ß is <1 (for example = 0.5) in the case of a self-saturating process.

μ has been evaluated so for large and uniformly corroded surfaces (Sc/Sa=1) as in processes with localized corrosion (Sc/Sa>1). The general expression normally assumed for such a parameter is:

$$\mu = \frac{s}{v}$$

where:

s = thickness of the metal wall;

$$v = \text{median rate of the corrosion process} = \frac{8760\ 10\ 360\ \text{Icorr.}\ A}{6\ 0.013\ 10E23\ 1.6\ 10E\text{-}19\ z\ s\ d}\ \text{(mm / year)}$$

where:

A = atomic weight

z = element valence

d = metal density

Icorr. = current intensity

The data on the current intensity of the above relation may be obtained from the experimental tables (f. e. see the following ones for some values of Sc/Sa and of the temperature).

table a)

pH	Ea (V)	Ioo(mA)
< 3	0.8	25
= 7	0.4	1.5
= 8-9	0.1	0.02

table b)

pH	Ea (V)	Ioo(mA)
< 3	~ 1	35
= 7	0.8	8
= 9-8	~ 0.2	0.1

Table 1 Corrosion: experimental current intensity: a) Sc/Sa = 3-5, t = 25°C; b) Sc/Sa = 1, t = 120°C

In fact they report the current intensity for different fluid acidity values at different temperatures (20,21).

With reference to the localized corrosion, for values of Sc/Sa included between 1. and 10. the data are usually expressed in terms of mg/month (21).

These values may be found in experimental tables similar to the ones reported above; they refer to the stein-iron weldings at various temperatures. On the basis of these tabulated values, the parameter may be obtained.

The processing of eventual experimental data may be carried out applying the bayesian methods where the theoretical distribution calculated by the "physics models" is assumed as prior function.

4.6. Human error.

The failures of the components may be caused by errors of the operator during the component operation (not requested intervention, etc.) or during the maintenance; these last ones are the most common cases of failure caused by the operator: the operator assembles the component in a wrong way, puts on the component not suitable parts or doesn't use suitable lubricants, etc.

and consequentely causes the failure of the component after its start-up. Normally the errors of the operator during the maintenance may be classified in the following main groups:

a) omission errors:
   - a procedure is not applied (rule based error);
   - during a maintenance without procedure a requested operation is not performed (skilled based error);

b) errors during the application of a procedure;

c) errors during a manual operation without procedure rules (f. e. the placing of a bearing in its seat).

Data on the human error probability in terms of "skill or rule based operating errors" have been found in THERP as developed in the Human Reliability Analysis Handbook by M. Swain (22). THERP defines an "operator failure probability" called "nominal EHP value" expressed by a lognormal distribution, having a fixed median value and represented in the following figure.Naturally THERP gives different EHP values for each of the above listed human errors. These "probabilities of error" have been introduced by ENEA in the Bayes' theorem expression as "generic information" (prior function); as likelihood, the binomial distribution is assumed: it is "tested" on the basis of the experimental failure data available in the data bank. By this application of the bayesian methods it is then possible to obtain the new lognormal distribution which allows to calculate the failure probability that the component fails in consequence of the operator error.

5. An application example.

In order to give an example, the above described methods have been applied to the data collected and registered in the ENEA-CEA data bank on the centrifugal pumps of Rapsodie reactor. These data are resumed in the following table 3.

PART NAME	CAUSE	FAILURE RATE (exp) lower	medium	upper	PHYSICS MOD. lower	medium	upper
SEALING	WEAR	5.85 E-6	1.28 E-5	2.78 E-5	1.26 E-5	2.27 E-4	3.61 E-3
LUBRIC. DEVICES	WEAR	2.18 E-6	7.69 E-6	1.99 E-5	1.21 E-5	2.12 E-5	4.25 E-5
BEARING	WEAR	2.18 E-6	7.69 E-6	1.99 E-5	5.55 E-6	1.22 E-5	2.88 E-5
H.E. MODE	**PART NAME**	**FAILURE RATE (exp) lower**	**medium**	**upper**	**PHYSICS MOD. lower**	**medium**	**upper**
COMM. ERROR SKILL BASED	BEAR. SEAL.	8.28 E-6	1.75 E-5	3.29 E-5	4.37 E-7	7.48 E-7	8.85 E-7
COMM. ERROR RULE BASED	VALVE INSTR.	2.84 E-7	7.58 E-6	1.94 E-5	6.92 E-7	3.77 E-6	6.92 E-6
OMISS. ERROR RULE BASED	VALVE INSTR.	2.84 E-7	7.58 E-6	1.94 E-5	2.87 E-7	1.83 E-6	2.26 E-6
TOTAL FAILURE RATE		4.24 E-5	6.16 E-5	8.66 E-5	4.19 E-5	6.87 E-5	8.12 E-5

Table 2 Results obtained by "physics methods" on the centrifugal pumps of Rapsodie reactor

As it results from the figure 1 of this report, the trend of the collected data is not exponential and no other failure distribution is approximated; we must also say that the available data are not so many to allow the application of the classical methods. It follows that the physics models may be considered the most profitable method to obtain significative data. As it results from the table 3, the failed parts (bearings, sealings, valves, lubricating devices, instrumentation) and the failure causes (wear, operator error during the maintenance, design) may be exactly singled out; the following failure models have been therefore easily applied to the singled-out parts: wear and error of the operator during the maintenance. The obtained results are resumed in the previous table 2; in the same table, we have also reported the values obtained applying the classical methods (since no other distribution was approximated, an exponential distribution has been assumed even if it is not too realistic): these values may contribute to demonstrate the importance of the physics models in order to obtain significative data. At the same aim, in the same table a "total stocastic value" is reported: this value has been obtained on the basis of all the data collected on the centrifugal pumps unwrapping the "infant failures" and the failures caused by design errors; also in this case an exponential distribution has been assumed, even if it is not realistic, because no other distribution was approximated; the same value has been calculated applying the bayesian methods (in lack of significative data on particular components as the sodium centrifugal pumps, a "non-informal" distribution has been assumed). With reference to the operator error, the obtained value is a failure rate obtained multiplying the failure probability obtained from THERP (see the previous paragraph 4.6) by the maintenance intervention frequency; all the failures caused by errors of the operator are considered together independently of the part:: they have been separated in function of the type of error as resumed in the list of paragraph 4.6. of this report.

part	cause	conseq. on part	conseq. on comp./system	conseq. on other system	interv.	repair time	oper. time	part oper. time	fail. mode
sealing	wear	fracture	degraded	no effect	replace	20:00	16625	16625	incip.
"	"	"	"	"	"	20:00	7875	7875	sudden
"	"	"	"	"	"	20:00	11075	11075	incip.
"	"	"	"	"	no act.	20:00	7660	7660	incip.
"	"	"	"	"	repair	20:00	8150	8150	incip.
"	oper. error	"	"	"	"	20:00	9040	9040	incip.
"	"	"	"	"	replace	20:00	18180	18180	incip.
"	mainten.	"	no effect	contamin.	replace	20:00	7750	7750	sudden
lubricat. devices	wear	degraded	degraded	no effect	repair		78267	78267	sudden
"	"	"	"	unavail.	"		7880	7880	incip.
"	"	"	"	no effect	"		8100	8100	
"	mainten.	"	"	"	"		5900	5900	incip.
valves	"	blockage	loss of flow	"	replace		120	120	sudden
bearings	wear	seizure	"	"	replace	40:00	13225	13225	sudden
"	"	degraded	"	"	"	20:00	18180	10170	
"	"	"	"	"	"	20:00	20075	20075	
bearings	mainten.	seizure	degraded	unavail.	repair	100:00	10	10	sudden
"	"	"	"	no effect	replace	40:00	51040	32860	sudden
"	"	"	"	"	"	40:00	8005	8005	sudden
"	"	"	"	"	"	60:00	8015	5	sudden
"	"	"	"	"	"	40:00	51500	31425	sudden
"	"	degraded	"	"	"	20:00	170	170	
instrum.	electr.	failure	no effect	"	"		15750	15750	
"	"	"	"	"	no act.		7225	7225	incip.
"	mainten.	"	"	"	"		8225	8225	incip.
"	manufact.	"	"	"	replace		51700	51700	sudden
hydrostat. bearing	externe	blockage	loss of flow	unavail.	replace comp.		190	190	sudden
//	unknown		leakage	contamin.	no act.		6940		incip.

Table 3 Data collected on the sodium centrifugal pumps of Rapsodie reactor

6. Conclusions.

In this report a preliminary description of some "physics methods" is reported: they result very important to process data when few rough data are available or few components have been followed; when the data are enough to apply the classical methods , the physics methods may be really useful to verify the significance of the obtained results. It is clear that many other "physics methods" should be developed; in particular, with reference to the mechanical components other efforts must be applied: some "physics models" must be totally developed (f. e. the one referring to "temperature pick stresses") and some other models treated in this report must be analysed with more deepness (see fatigue, corrosion, etc.) in accordance with the particular applications in safety and design. In any case, further information on the subject may be found on the report (7) of bibliography.

BIBLIOGRAPHY

1 - ENEA -RT /VEL/87/3-R. Righini, G. Sola - "How-to-use the ENEA data bank for the classification and reliability processing of fast reactor component event data. Data entry and inquiry procedures."

2 - ENEA;FT.WLC.00005 - R. Righini, G. Sola, L. Briati - "How-to-use the ENEA data bank for the classification and reliability processing of fast reactor component event data. Structure and codes."

3 - Pubblicazione ENEA - "The ENEA-CEA data bank: processing of reliability data on FBR components - Proceedings of the Seminar organized by ENEA in collaboration with CEA and NIER on: ENEA-CEA data bank, Methods for system analysis, Updating of processed reliability data on Rapsodie and Phenix components" - Bologna, March 22-23 1988.

- Pubblicazione ENEA - Atti del Seminario su "Raccolta ed elaborazione dati di affidabilità di componenti di reattori veloci su un sistema informativo" - Bologna, 17-10-1984

4 - J.R.C. - EURATOM Ispra
"Component Event Data Bank: handbook"
T.N. 1.05 C1.84.66

5 - ENEA-FT.WLC.00006 - R. Righini, V. Guerrini, G. Zappellini - "Updating Journal of the processed reliability data on Phenix and Rapsodie reactor components."

6 - ENEA-RT/VEL/85/8 - G. Cambi, R. Righini, P. G. Sola - "Il programma ABC: un codice di calcolo per la stima di parametri statistici secondo l'analisi bayesiana e classica."

7 - ENEA - RT report to be issued - R. Righini, G. Zappellini - "New methods to process reliability data."

8 - PRA - Procedures Guides" - NUREG/CR - 2300 (1982).

9 - Commission of the European Communities - Nuclear Science and Technology - A. G. Colombo, R. J. Jaarsma - "BAESNUM. A conversational computer program for the bayesian extimation of a parameter by a numerical method." - EUR 7900 EN.

10 - ASME BOILER AND PRESSURE VESSEL CODE - SECTION III - "Rules for the construction of Nuclear Power Plant Components." - ANSSI/ASME-BPV-III-1-NB.

11 - J. G. Ligeron - "La fiabilitè en Mecanique" - Desforges Publisher - Paris.

12 - L. F. Coffin - "A study of the effects of cyclic thermal stresses on a ductile metal" - Trans. ASME 76, 923.

13 - S. S. Nelson - "Thermal stress in design. Part cyclic life of ductile materials" - Mechanical Design 32 (14) - 7.

14 - S.S. Nelson - "Fatigue: a complex subject, some simple approximations" - Exp. Mach. 5 (7) July 75.

15 - F. Bastenaire - "Etude de la rupture par fatigue par les methodes statistques. Bulletin de l'Institut International de Statistique" 1961.
-F. Bastenaire - "Aspects probabiliste et statisique sur la rupture par fatigue." - Revue française de la Mecanique N° 37, 1971

16 -"Mechanical Engineerig Design" Mc Graw Hill Book Co. - N. Y., 72 J.E. Shigley.

17 - "Interaction among the various phenomena included in the design of dynamic one rotary machinery" Ad 671 780, April 68.

18 - ISO Rule Number 1000.

19 - RIV - SKF Catalogo 3200 I - 1982

20 - a) "Centro Ricerche FIAT" International Report N. C-304
b) "Corrosion Data Survey" - Metal Section 5ª Edizione
G. A. Nelson, NACE, Houston, Texas.

21 - a) ENI "Enciclopedia del Petrolio" - Milano 1978
b) "Corrosion Guide" Elsevier, London 1978
E. Rabald.

22 - A. D. Swain, H. E. Guttmann - "Handbook of human reliability with emphasis on Nuclear Power Plant application" - NUREG/CR - 1278 (August 1983)

Human Reliability Assessment and Probabilistic Risk Assessment

D.E. Embrey and D.A. Lucas

Human Reliability Associates Ltd.,
1, School House, Higher Lane,
Dalton, Wigan, Lancs, WN8 7RP, UK

Abstract

Human reliability assessment (HRA) is used within Probabilistic Risk Assessment (PRA) to identify the human errors (both omission and commission) which have a significant effect on the overall safety of the system and to quantify the probability of their occurrence. There exist a variey of HRA techniques and the selection of an appropriate one is often difficult. This paper reviews a number of available HRA techniques and discusses their strengths and weaknesses. The techniques reviewed include: decompositional methods, time-reliability curves and systematic expert judgement techniques.

Introduction

Human reliability techniques were first developed in the context of military and aerospace systems in the late fifties, where human error was found to be a cause of a large proportion of system failures. Early approaches to human reliability analysed the human in the same way as a hardware component, and the primary emphasis was on quantification of human error probabilities for insertion in fault trees as part of larger system reliability assessments. More recently, approaches derived from cognitive psychology have analysed the human as a goal directed processor of information, and have emphasised the importance of his or her role in higher level functions such as problem solving and decision making. There has also been an increased interest in the organisational and the managerial variables that influence errors and which have been implicated in such major disasters as Chernobyl and Challenger (see, for example, Reason (ref.13) and Woods (ref. 18)).

The majority of the currently available human reliability assessment (HRA) techniques have been developed in the context of nuclear power generation. Increasingly, analysts in other high-risk socio-technical systems are searching for methods of estimating and improving human reliability. However, it is not always clear whether every HRA technique may be applied in another industrial context. The aim of this paper is to provide an overview of four of the main approaches to HRA to allow risk analysts to compare and contrast such techniques.

An Overview of Quantification Techniques.

Four major approaches to human reliability quantification are described in this section. These are: decomposition methods, time based modelling of human error, and two approaches utilising quantified expert judgement: direct methods, and methods based on Decision Analysis. This paper contains extracts from a more comprehensive state-of-the-art review of quantitative techniques by Embrey (ref. 7). Each approach is outlined very briefly and the main advantages and disadvantages of each are summarised.

1. Decomposition Approaches.

These techniques are analogous to hardware techniques. The task to be performed is broken down into a series of elements, and a failure model (analogous to an event tree or fault tree) is drawn up which defines the various failures which could occur and how they might be recovered. Human error probabilities from a data bank are modified to reflect the operational conditions and are then assigned to each of the task elements in the failure model. They may be further modified to reflect dependencies between task elements. These probabilities are then combined using the standard mathematical rules to give an overall probability for the task as a whole.

The most extensively developed decomposition technique is THERP (Technique for Human Error Rate Prediction) which has been applied mainly in the nuclear power industry (Swain and Guttman, ref. 16). It has a comprehensive database of human error probabilities. Mostly these are relevant to the power and process industries only.

Summary of advantages of THERP

- Comparisons - Design and Risk Assessments: THERP can be used for design, risk and reliability assessments at all stages of design. The level of detail of the THERP analysis can be tailored according to the depth of the overall assessment being performed e.g. early design conceptual assessments would only require a crude THERP analysis. Thus, the resulting HEPs are of a similar quality to the other data used in the comparisons.

- Integration into PRAs: THERP can be easily integrated into Probabilistic Risk Assessments. Its form and approach make it compatible with fault tree methodologies.

- THERP provides a structured, logical, well-documented record of the factors and errors considered in the human reliability assessment. This allows the results to be reviewed easily and assumptions used to be examined. This can be particularly important in sensitive areas (for safety or design changes).

- THERP has a high perceived validity to reliability assessors as it is of a similar form to hardware reliability assessment approaches.

- THERP can be used either with the handbook data or any other data that is considered to be relevant, e.g. simulator data; known plant-specific data; or data produced from other human assessment methods for a particular task.

Summary of disadvantages of THERP

- A very detailed THERP analysis can require a large amount of work to generate Human Error Probability (HEP) values. In practice by tailoring the detail for the THERP analysis to the appropriate level for each application, the amount of work can be reduced.

- THERP is poor for evaluating errors concerned with diagnosis or high level decisions (i.e. those requiring significant conscious thought). Although the most recent version of the THERP handbook provides procedures for assigning HEPs to diagnosis errors on the basis of the time available to the operator, this process is regarded as dubious by many human reliability specialists.

- THERP is a relatively crude human reliability evaluation tool, best used in support of engineering assessments. It is not an "ergonomic tool" in that it is nt sensitive to many possible Performance Shaping Factors (PSFs) nor does it attempt to identify the underlying psychological causes of a particular error.

- Several problems were encountered during a peer review study of THERP where several analysts evaluated the same scenarios. There was a lack of uniformity in the modelling of the scenarios and in the use of the data base. This led in some cases to very large discrepancies in different analyst's estimates for the same task. THERP does not therefore provide a very explicit modelling framework.

2. Time Dependent Modelling of Human Reliability

The major technique for quantifying human reliability using time dependent modelling is the HCR (Human Cognitive Response) technique (Hannaman et al, refs. 8, 9, 10). In this technique a relationship between the time available for an operator to respond to an event and the probability that the response will be made within that time is established empirically by means of simulator experiments. Different relationships are derived for skill, rule and knowledge based behaviour. To use the method, tasks are first classified into one of these categories and an appropriate time/non-response probability curve selected. The time available for the response is then evaluated and the ratio of the time available to the median response time is then used to read off a non-response probability for the selected curve.

Although this method is being extensively developed in the nuclear industry it is not recommended for other industrial applications, since it is closely linked to nuclear power simulators, and is insufficiently flexible to address the range of situations that would need to be evaluated in other industrial contexts.

Summary of advantages of the HCR correlation

- The HCR correlation is a quick and relatively easy method to use for the PRA analyst who may have a large number of assessments to make.
- The approach takes account of the time-dependent nature of many operator actions modelled in PRA.
- It attempts to cover knowledge-based behaviour as well as skill and rule-based.
- According to its originators, the final HCR curves will be based on documented simulator data. This will constitute a verifiable database. The context of the data collection will be both known and appropriate to the assessment situation.
- The method uses the same input data (time available from the onset of the problem) as is used by hardware reliability assessment techniques.

Summary of disadvantages of the HCR correlation.

- A serious deficiency of the HCR correlation is that it addresses only a limited aspect of operator performance in emergency responses i.e. the probability that he or she will not process information relating to diagnosis and action formulation within the available time. However, in virtually all of the major incidents that have occurred, (particularly in the nuclear industry where the HCR correlation is targetted), the critical operator errors were either misdiagnoses, choosing non-viable courses of action, or violations of laid down rules or procedures. The quantification provided by the HCR model is therefore of limited value unless it can be combined with other techniques and sources of data.

- The non-response probability predicted from the HCR correlation curves could arise from slow processing of acquired information or from failing to detect the onset of the situation. These are very different processes and it seems unlikely that they would be described by the same time available/non-response probability curves. The originators of the model provide no information with regard to the proportion of the experiments used to generate the HCR curves in which failure to detect the event occurred compared with failure to process the information in time. Thus, it is not possible to establish if the curves would apply to situations where detection failures or processing difficulties would predominate.

- The rules for assigning tasks to skill, rule or knowledge based categories are not comprehensive. Studies of mental processing during actual incidents (e.g. Woods, ref. 17) show that operating teams are constantly changing their levels and modes of processing as an event develops. If an event is assigned to the wrong category, this can have serious consequences, particularly at longer time intervals.

- The HCR correlation has been shown to be extremely sensitive to estimates of available time. Thus, the median response time must be estimated extremely accurately. This estimate requires the availability of a simulator for the system being evaluated, and an accurate estimate of when cognitive processing has been completed. This is not necessarily when the required action commences because the operators may begin actions which are common for a variety of different situations before they have made an unambiguous diagnosis. This makes the evaluation of median response times difficult and would adversely affect the accuracy of non-response probability estimates.

- The resources required to collect simulator median response times for all categories of an event in a PRA are considerable, especially as simulator time is very expensive and evaluations would be required for all new systems being assessed.

- A limited number of PSFs are modelled in the present version of the HCR correlation, and certain critical parameters e.g. quality of procedures, team variables, are not currently addressed. The method of simply modifying the median response time as a result of the PSFs seems to be inadequate, since certain PSFs, e.g. degree of training, will directly influence the categorisation of a situation as skill, rule or knowledge based. This will in turn affect the quantification as discussed above. The method used to incorporate PSFs into the response times has not yet been fully explicated or validated.

- It is likely that simulator response times and responses to real incidents could be very different, and that this difference could not necessarily be taken into account simply by modifying PSFs such as stress or training. It is not clear how the relationship between the response times for simulated and real events can be established.

- The HCR method does not handle situations where slips or errors occur, which are subsequently recovered.

- The method does not provide any explicit outputs for specifying <u>how</u> human reliability should be improved or optimised in order to meet specific performance goals. It provides no inputs into the process of design for error reduction.

- The model is relatively insensitive to changes in the PSFs compared with changes in the normalised time parameter.

- The range of median response times and system time windows over which the model was developed has not been stated. It is therefore difficult to assess the generality of the results.

- In significant events, more than one diagnostic activity may be taking place simultaneously, since the operating crew may be concerned with evaluating a wide range of possible hypotheses after the initial indications of a problem. In practice, the response to an event is likely to consist of a complex interaction between the three paths modelled by the HCR correlation (successful/faulty/no cognitive processing). For example, the team may make an initial misdiagnosis which is subsequently recovered, but which may slow down the response to the event. It seems unlikely that global non-response curves of the sort adopted in HCR can capture the highly interactive nature of problem solving in specific situations.

3. Expert Judgement Approaches: Direct Methods

The most important direct methods for estimating human error probabilities are Direct Numerical Estimation (DNE) and Paired Comparisons.

DNE techniques involve the direct estimation of probabilities by expert judges. Various techniques are available for minimising the biases inherent in direct estimation of probabilities. Usually, group elicitation methods are employed to enable measures of inter-judge agreement to be calculated (for an example see Comer et al, ref. 2).

Paired comparisons involve judges making pairwise comparisons of the form: "task A is more likely to fail than task B". If these comparisons are made for all possible combinations of tasks, a likelihood of success scale can be developed for which each task has a scale value. If two tasks with known error probabilities are included in the evaluation, and a logarithmic transformation between scale values and error probabilities can be assumed, then the scale can be calibrated and a human error probability derived for each task (see Blanchard et al, ref. 1).

The main advantages of direct expert judgement techniques are that they have been well researched and are generally simple to perform. However, they often require considerable resources in terms of time and availability of experts. They are also difficult to audit, since they do not allow the rationale for the elicited probabilities to be evaluated. (Further discussion of expert judgement techniques is given in Seaver and Stillwell, ref. 15.)

4. Expert Judgement Approaches: Decision Analysis Methods

The main quantification techniques derived from Decision Analysis approaches are the Success Likelihood Index Method (SLIM) (see Embrey et al, refs. 3, 4, 5, 6) and the Influence Diagram Approach (IDA) (see Phillips et al, refs. 11, 12, 13). Both of these techniques work on the principle of experts developing a model which connects the error probability in a particular situation with the factors that influence that probability. For example, the probability that the operator would make an error in a maintenance task might be a function of the time available to do the task, the quality of training and procedures, the number of distractions, etc.

In the case of SLIM, a model is developed in which the factors influencing the set of tasks are first defined, and the tasks are then rated on these factors. Weights which define the importance of each factor are then derived, and the ratings and weights then combined to give a Success Likelihood Index (SLI) for each task. If at least two tasks with known human error probabilities are included in the set being evaluated, this allows the SLI scale to be calibrated, such that the SLI for each task can be converted to a probability of error.

Advantages of the SLIM approach

- The assumptions underlying SLIM are explicit, and the relationship between the numerical evaluations and the resulting HEPs can be readily audited.

- The available comprehensive interactive software (SLIM-SAM) reduces the need for a facilitator, and enables the methodology to be easily used in-house after a relatively small amount of training.

- One of the important assumptions of the underlying SLIM model is that the ratings assigned to each performance influencing factor (PIF) are independent. The SLIM-SAM software checks the degree of shared variance between the ratings generated by the judges, and provides feedback if the ratings on two PIFs appear to be correlated.

- The technique used within SLIM-SAM to place the tasks on a likelihood of success scale conforms with the theoretical model.

- Up to ten tasks can be evaluated within a single SLIM session, which greatly reduces the time required to carry out assessments.

- Comprehensive 'what if' analyses are possible to determine the most effective error reduction strategies.

- The data requirements for SLIM are relatively modest, since only two tasks with quantified empirical data are required for each category of tasks being evaluated.

Disadvantages of SLIM

- In order to use SLIM it is necessary to have judges with detailed knowledge of the tasks being evaluated. Although the technique can be used manually it is more effective when used in conjunction with the SLIM-SAM software.

- The calibration equation log HEP = a SLI + b has not been unequivocally validated.

For the Influence Diagram Approach (IDA), a model is set up which defines the factors which influence error probability for a task (or a broad category of tasks) and the interrelationships between these factors.

Assessment of probabilities are then made at each level of the Influence Diagram, and these probabilities are combined together by the model to give the overall probability of success or failure for the task or category of tasks under consideration. The Influence Diagram is particularly useful for evaluating the efforts of socio-technical variables such as organisational structures and management policies.

Advantages of IDA

- Independence between PIFs is not assumed, and in fact dependence is explicitly modelled.
- It can consider large task 'chunks' (i.e. any level of task decomposition can be assessed).
- It does not require calibration.
- Organisational and management variables influencing human reliability can be modelled.
- IDA does not assume that all influences impinging on success or failure are identified within the model. Only influences having a significant effect on error probability are considered.
- Sensitivity analysis is possible.

Disadvantages of IDA

- It is costly in terms of resources. IDA requires access to a variety of experts for fairly extensive periods of time.
- The elicitation of unbiased probabilities in the technique requires more research.

With both SLIM and IDA, it is possible to perform sophisticated 'what if' analyses, where the effects of error probability of changing variables

such as level of training, time available, etc., can be investigated. These facilities are very valuable for cost benefit analysis. Another advantage of these techniques is their auditability. The underlying rationale
behind an assessment can easily be evaluated, which is not the case with the direct methods of judgement based probability evaluation.

Conclusions

The objective of this paper has been to survey the range of human reliability quantification techniques from the point of view of their usefulness in a range of industrial applications. The conclusion which emerges from this survey is that the techniques based on the use of the quantified expert judgement are likely to be the most widely applicable. The reasons for this conclusion are as follows:

The decomposition techniques such as THERP are ruled out primarily on the basis of their data requirements. It is unlikely that data of the required quality will be available for the whole range of skill, rule, knowledge based, communication and other errors that will need to be assessed. In addition, the decomposition techniques do not handle diagnostic situations effectively and cannot take into account organisational variables. They also require a considerable analytical effort in order to be applied effectively.

With regard to time dependent methods such as HCR, the data requirements are again prohibitive. In addition the methods are very restrictive in their application, in that a separate model has to be developed for each system to which the techniques are applied. There are also a considerable number of theoretical problems associated with these techniques, not least being the fact that they only provide a partial error probability, i.e. the probability of non-response. Other error modes are not considered by this approach.

The two remaining approaches, both of which employ quantified expert judgement, are the Direct Methods and those based on Decision Analysis techniques. The Direct Methods have the great attraction of simplicity, and are supported by a considerable body of research. However, Decision Analysis based approaches are unique in that they generate an underlying model which relates the assessed probabilities to the underlying factors that determine these probabilities. From the point of view of practical assessments, this has the considerable advantage that it allows design recommendations to be generated and sensitivity analyses to be conducted, where the effects of changing parameters such as the time available to the operator can be investigated. This allows the results of risk assessment to be fed back to designers in the form of error reduction recommendations. In addition, cost-benefit analyses can be carried out, to evaluate the cost of achieving a particular reduction in error rates.

Another major advantage of the Decision Analysis approaches is that they are readily auditable. The reasons why a particular assessment has been generated can easily be established. This is not the case with the direct methods, where the output of the assessment process is simply a numerical probability estimate. A final argument for the Decision Analysis methods is that of resources. Although the resources required to set up the initial SLIM or Influence Diagram models may be significant, once this work has to be done, the models can be used by a single trained analyst without the need for the original expert groups to be convened. This should allow considerable economies of effort if large numbers of assessments have to be made, or the technique needs to be available on demand in an analyst's office.

List of abbreviations

DNE	Direct Numerical Estimation
HCR	Human Cognitive Response (correlation)
HEP	Human Error Probability
HRA	Human Reliability Assessment
IDA	Influence Diagram Approach
PIF	Performance Influencing Factor
PRA	Probabilistic Risk Assessment
PSF	Performance Shaping Factor
SLIM	Success Likelihood Index Method
THERP	Technique for Human Error Rate Prediction

References.

1. Blanchard, R.E., Mitchell, M.B., and Smith, R.L. (1966) Likelihood of Accomplishment Scale for a Sample of Man-Machine Activities. Dunlap and Associates, Santa Monica, California, USA.

2. Comer, M.K., Seaver, D.S., Stillwell, W.G., and Gaddy, C.D. (1984) Generating Human Reliability Estimates using Expert Judgements. NUREG/CR- 3688, SAND-84-7115.

3. Embrey, D.E., (1983a). The Use of Performance Shaping Factors and Quantified Expert Judgement in the Evaluation of Human Reliability: An Initial Appraisal. NUREG/CR-2986, Brookhaven National Laboratory, Upton, New York. USA.

4. Embrey, D.E. (1983b). The Quantification of Human Reliability Using Expert Judgement: Current Findings and Future Developments. Paper presented at the Institute of Chemical Engineers Symposium "Human Reliability", in the Process Control Centre, Manchester, England.

5. Embrey, D.E., Humphreys, P., Rosa, E.A., Kirwan, B., Rea, K., (1984a). SLIM-MAUD: An Approach to Assessing Human Error Probabilities Using Structured Expert Judgement. Volume I. Brookhaven National Laboratory, Upton, New York 11973, USA.

6. Embrey, D.E., Humphreys, P., Rosa, E.A., Kirwan, B., Rea, K., (1984b). SLIM-MAUD: An Approach to Assessing Human Error Probabilities using Structured Expert Judgement, Volume II. Brookhaven National Laboratory, Upton, New York, 11973, USA.

7. Embrey, D.E. (1988) State of the Art Review of Quantitative Techniques. Report prepared for the European Space Agency.

8. Hannaman, G.W., Spurgin, A.J. and Lukic, Y.D. (1984) Human Cognitive Reliability Model for PRA Analysis, Draft Report NUS-4531, EPRI Project RP2170-3, Electric Power and Research Institute, Palo Alto, California.

9. Hannaman, G.W., and Spurgin, A.J. (1984) Systematic Human Action Reliability Procedure (SHARP). EPRI NP-3583, Electric Power Research Institute, Palo Alto, California.

10. Hannaman, G.W. and Worledge, D.H. (1987) Some Developments in Human Reliability Analysis Approaches and Tools. International Post-Smirt 9 Seminar on Accident Sequence Modelling: Human Actions, System Response, Intelligent Decision Support, Munich, West Germany, August 24-25.

11. Phillips, L.D., Humphreys, P., Embrey, D.E., (1983). A Socio-Technical Approach to Assessing Human Reliability. Technical Report 83-4. Decision Analysis Unit, London School of Economics and Political Science, Houghton Street, London, England.

12. Phillips, Humphreys and Embrey (1985a), "Appendix D: A Socio-Technical Approach to Assessing Human Reliability (STAHR)" in Selby, D., Pressurized Thermal Shock Evaluation of the Calvert Cliffs Unit 1 Nuclear Power Plant, Research Report on DOE Contract 105840R21400, Oak Ridge National Laboratory, Oak Ridge, TN.

13. Phillips, L.D. and Embrey, D.E. (1985b), "Appendix E: Quantification of Operator Actions by STAHR," in Selby, D., Pressurized Thermal Shock Evaluation of the Calvert Cliffs Unit 1 Nuclear Power Plant, Research Report on DOE Contract 105840R21400, Oak Ridge National Laboratory, Oak Ridge, TN.

14. Reason, J. (1987) The Chernobyl Errors. British Psychological Society Bulletin June.

15. Seaver, D.A. and Stillwell, W.G. (1983) Procedures for Using Expert Judgement to Estimate Human Error Probabilities in Nuclear Power Plant Operations. UREG/CR-2743. Sandia National Laboratories, Albuquerque, New Mexico 87185, USA.

16. Swain A.D. and Guttman, H.E. (1983). A Handbook of Human Reliability Analysis with Emphasis on Nuclear Power Plant Applications. NUREG/CR-1298, Washington D.C. 20555.

17. Woods, D.D. (1982) Operator decision making behaviour during the steam generator tube rupture at the Ginna nuclear power station. Brown, W. and Wyrick, R. (eds.) Institute of Nuclear Power Operations, Atlanta, Georgia, USA.

18. Woods, D.D. (1987) Technology Alone is not enough: reducing the potential for disaster in risky technologies. in: Proceedings of an international conference: Human Reliability in Nuclear Power. 22nd/23rd October 1987. IBC Technical Services Ltd., IBC House, Canada Rd., Byflect Surrey KT14 7JL England.

Human Reliability Data Collection for Qualitative Modelling and Quantitative Assessment

D.A. Lucas and D.E. Embrey

Human Reliability Associates Ltd.,
1, School House, Higher Lane,
Dalton, Wigan, UK. WN8 7RP.

Abstract

Effective human reliability assessment requires both qualitative modelling of possible errors and their causes, and quantitative assessment of their likelihood. This paper considers the available sources for both qualitative and quantitative data collection. A classification for different types of data is proposed. Currently used methods of gathering data using operational experience and simulators are discussed in relation to these data types.

Introduction

The traditional approach to human reliability analysis has emphasised the quantification of error probabilities for proceduralised operator actions in the fault tree analyses in Probablistic Safety Assessment (PSA). Discussions of the need for human reliability data have therefore concentrated on the issue of how numerical data on human error probabilities (HEPs) can be obtained from various sources. Although this aspect of human reliability data is important, the human reliability concept also covers qualitative modelling. The data needs for both areas need to be considered. This paper addresses both qualitative and quantitative modelling.

Various areas for which human error data are required can be defined as follows:

a. Data for use in the design of new systems to ensure that human reliability in areas such as operations, maintenance and testing is optimised.

b. Data for use in devising error reduction strategies for existing systems.

c. Data for qualitatively modelling the types of error expected to occur in emergency and other situations as part of PSA.

d. Quantitative data for performing cost benefit analyses in areas (a) and (b).

e. Quantitative data in the form of absolute human error probabilities for use primarily in PSA, but also applicable to (a) to (d), if available.

Human error data can be broadly divided into qualitative and quantitative groups. Qualitative data can be subdivided into two further categories. The first of these are data that are to be used either at the design stage, or retrospectively, for error minimisation purposes. The second category is data for error modelling within a PSA structure. This involves the analyst postulating various ways in which the operator could fail. These 'error modes' are included in the event or fault tree structure for subsequent quantification.

Quantitative data can also be broken down into two categories. The first category is taken to include all numerical data relevant to human reliability assessment <u>excluding</u> data on absolute error probabilities. This category is typified by data such as "the probability of incorrect operation of a control is doubled on subsequent attempts if the operator has failed on the first attempt". Alternatively, "detection of a faint, infrequently occurring signal will decrease to 25% of its initial level after about half an hour of watchkeeping". The other class of quantitative data is the absolute estimates of human error probabilities encountered in PSA, e.g. "the probability of the operator failing to operate valve 27B is 3.6×10^{-4}". In subsequent discussions we shall refer to the above categories of data as QUAL1, QUAL2 and QUANT1, QUANT2 respectively. The definitions of these types of data are summarised in table 1 below:

Data Type	Definition
QUAL1	Qualitative data to be used for design or error reduction purposes.
QUAL2	Qualitative data for error modelling in PSA.
QUANT1	Quantitative human error data excluding absolute probability estimates.
QUANT2	Absolute estimates of human error probabilities.

Table 1 Definitions of data categories.

In the remainder of this paper, each of these data categories will be considered from the point of view of data sources, the feasibility and cost of obtaining data, and possible application areas.

Qualitative error reduction data (type QUAL1)

Nature of the data

Type QUAL1 data consists of three main subcategories. The first of these is specific human factors information derived from experimentation under laboratory or, (more rarely) field conditions. Such experimental data tends to be specific to a particular combination of variables where only one (or, infrequently, two) factors are changed as part of the experimental design. For example, the increase in response time may be tabulated when an individual has to respond to an increasing number of alternative switches which need to be operated when an alarm occurs.

The second subcategory is data derived from the systematic analysis of operational experience. For example, in a particular installation, it may be observed that test engineers frequently lose their place when carrying out long written procedures. An obvious conclusion from this finding is that the written procedures should be broken down into smaller units.

The third subcategory of qualitative data are general error reduction theories, principles or models which can be applied to a wide range of systems and situations. Examples of principles of this type are those associated with Rasmussen's Skill, Rule, Knowledge (SRK) model (ref. 8) or the GEMS model proposed by Reason (ref. 9). It is often assumed that these general error reduction principles are generalisations from the specific laboratory studies and operational experience analyses of the first two categories. Although this is true up to a point, the general models are often used to provide a structure within which specific experiments are conducted to generate data. Similarly, particular models of human error can be used to structure the collection of operational data.

Application areas

All three subcategories of data can be applied during the design of new systems. Usually relevant experimental data is condensed into tables in data handbooks. However, such data handbooks are not in themselves sufficient to achieve designs which minimise human error. A systematic design methodology is necessary within which the data can be applied. An example of such a methodology is SHERPA (Systematic Human Error Reduction and Prediction Approach, Embrey, ref.3).

The reduction of human error in operating establishments and systems also utilises data from all three of the subcategories discussed up to this point. However, the existence of a systematic error data collection system is the most effective means of developing a comprehensive error reduction programme. Such a system allows the nature of human reliability problems to be identified at source and the effectiveness of any error reduction measures to be monitored.

Data sources

a. Data handbooks

There are considerable amounts of qualitative data available which could potentially be applied to design and error reduction in a number of industrial contexts. However, these data are often in specialised research reports and journals which may be inaccessible to the general user. For such users, the best sources of data are some of the many handbooks that have become available. For example, in the space industry, comprehensive design data sources such as the NASA human factors handbook are already extensively used. A text which is specifically orientated towards error reduction is the Guide to Reducing Human Error in Process Operations (Ball et. al, ref. 1) which was developed by the data subgroup of the Human Factors in Reliability Group sponsored by the United Kingdom Atomic Energy Authority. Another document which provides qualitative guidelines for error reduction (although it is primarily orientated towards quantification) is Swain and Guttmann (ref. 10).

b. Analysis of operational experience

The most cost effective method of error control is the existence of a comprehensive human performance data collection system. Such a system should allow the collection of data on human performance deficiencies such that the underlying causes of errors can be identified and appropriate error reduction measures prescribed. In order to achieve these objectives at least five requirements are necessary:

- A climate of opinion at a contractors plant and in an operational system that views error as a normal part of human behaviour and hence (by means of a non-punitive reporting policy) encourages operators to report not only errors that give rise to undesirable consequences but also 'near misses'.

- A systematic and structured data collection approach that assigns to specific individuals the responsibility to investigate incidents and collect error data.

- A classification scheme for error data that is based on a human error model (or models) such that the causes and contributing factors can be identified.

- Management acceptance (at all levels) of the importance of human reliability as a major determinant of plant safety and profitability. Also required is a willingness to act on the results of an error data collection programme by supporting (financially and otherwise) the implementation of error reduction strategies which arise as recommendations from these data.

- Regular feedback to all levels of employee giving details of the safety measures implemented as a result of the system.

These requirements are, of course, very stringent, and it is therefore not surprising that no error data collection currently exists which fulfills all these criteria. The nuclear power industry (particularly in the USA) and the aerospace industry have attempted to establish systematic approaches to the reporting of operational experience. One of the most widely established reporting schemes is the Licensee Event Report (LER) system. This is a mandatory reporting scheme operated by the Nuclear Regulatory Commission (NRC) as part of its licensing requirements for nuclear power stations in the U.S. A comprehensive report has to be submitted to the NRC in the event of any abnormal occurrence that violates the technical specification of the plant. The LER system is therefore not primarily intended as a means of reporting human errors. Nevertheless, because many abnormal occurrences originate from or implicate human performance deficiencies, LERs contain much material that is of interest to the human reliability analyst. In the aviation industry, a system of confidential reporting exists which allows anonymous feedback of problems (including "near miss" reporting). Unfortunately, this system does not provide a framework which allows the causes of incidents to be determined.

The only scheme which is orientated solely to the collection of human performance data is also operated by the U.S. nuclear industry. This is the Human Performance Evaluation System (HPES) administered by the Institute for Nuclear Power Operations (INPO). A particular advantage of HPES is that it is run by the industry for the industry, and thus there is less of a tendency to "sanitize" the data that is collected. In addition, the system involves voluntary reporting of actual problems and potential problems. The HPES programme fulfills several of the criteria discussed above. One or more individuals at each plant are designated as plant co-ordinators and are mandated specifically to investigate human performance problems. Plant coordinators receive extensive training in incident analysis techniques at INPO, and regular feedback sessions are conducted where plant coordinators discuss problems in operating the scheme. The plant coordinator attempts to investigate the causes of human errors and near misses at the plant and recommends appropriate corrective measures. Data which are sufficiently generic to be of general interest are circulated on an industry wide basis, and are added to the central INPO data base.

The HPES programme is at a relatively early stage of development and does not yet fulfil all of the criteria discussed earlier. For example, the error classification system requires further elaboration. Nevertheless, the HPES programme has achieved widespread acceptance, as nearly half of all the U.S. power utilities now subscribe to the scheme, and this growth has occurred over a period of less than a year. The major reason for this success is that the scheme is perceived to provide real benefits at individual plants in reducing the incidence of errors and other human performance problems. The widespread acceptance of HPES demonstrates that error data collection schemes are both possible and useful, provided sufficient committment and financial resources are made available. It is also encouraging that the nuclear industry in Canada and in France are trying to implement the HPES programme.

c. Practical aspects of analysing error data from operational experience

The collection and reporting of incidents involving human errors is not an activity which most people take to naturally. The use of a cognitive model of human error could provide considerable help in this key area. In particular, the use of such a model could facilitate two of the most important uses of such a data collection scheme. Firstly, it could assist in the derivation of accurate information on the causes of operator error. Secondly, it could help in the generation of effective error reduction strategies from the information in the database. These error reduction strategies might typically consist of, for example, changes in the design of equipment, a revised training programme, or the redesign of procedures. The information collected needs to allow the specification of these and other methods of reducing those human performance problems which have potentially serious consequences.

Reviews (e.g. Lucas, ref. 6, Embrey et al, ref. 4) have revealed a number of commonly occurring problems with human performance data collection schemes.

- Reports tend to be very variable in quality. Many reports are vague and incomplete. The level of information reported is very variable. One analysis by Lucas and Embrey (ref. 5) showed that at least 20% of incident reports involving human error are essentially unusable from the point of view of analysing the nature of the error or near miss. Those reports which are both clear and detailed tend to be the rare exceptions.

- The underlying causes of human errors are not derived. Another very serious problem which has been found in reviews of current reporting schemes is that the underlying causes of human errors and near misses are not adequately assessed. The writers of reports concentrate on describing what went wrong, often at the level of the behaviour of the system. The reason why the human performance problem occurred is very rarely established. We are therefore left with a clear description of what occurred, when and to whom, without the important analysis of the cause of the human error.

- Analysts experience difficulty in assessing root causes of human errors. A third and related problem is that analysts find it difficult to assess the root cause of a human performance problem. This is partly due to the variable nature of the information collected and also to the lack of job aids which would help analysts in assigning one or more possible underlying causes to a particular error type. Most analysts have not received extensive training in psychology and without explicit guidance on establishing the cause(s) and contributing factors of a human performance problem their assessments will be unreliable.

The use of a cognitive model of error would considerably assist the analyst in all of these three aspects. However, there is an urgent need to provide an appropriate "interface" between the theoretical models of human error causation and the pragmatic concerns of accident investigators in industry. A number of possible interface devices are feasible ranging from paper-based classifications of error causes, through the provision of decision aids such as flow charts, to the use of knowledge based systems for data gathering and analysis. These ideas are discussed in more detail in Lucas (ref. 6).

d. Simulator data

In the aerospace, nuclear, and certain areas of offshore activities (e.g. transport of crude oil in bulk carriers) sophisticated high fidelity simulators have been developed, mainly to train and certify pilots and operators. In theory these systems should constitute excellent sources of both qualitative and quantitative human error data. The major difficulty arises because of the high cost and therefore heavy usage of most of the simulators that exist in the nuclear and aerospace industries. They often tend to be operated round the clock for use in training applications. This severely limits their availability for human factors work. As a result, most attempts to use such simulators to obtain human error data have had to be combined with training exercises. Examples of data collection efforts of this type are those described in Beare et. al (ref. 2). Such studies have often involved the instrumentation of the simulator to enable comprehensive computer based data collection of the operators actions. However, much useful information can be gathered without this expense, particularly if scenarios are pre-analysed to identify aspects of a task that the operators are likely to find difficult.

Some studies have also been carried out specifically to collect qualitative data concerning the cognitive aspects (i.e. decision making, problem solving functions) of operator performance. An example of such a study using a plant simulator is described in Woods (ref. 11). Encouraging results have also been obtained by Norros and Sammatti (ref. 7).

It is by no means certain that large scale replica simulators are essential for providing data relevant for human error reduction. Useful results have been produced by simulators based on microcomputers and even paper and pencil simulations.

Qualitative data for error modelling in PSA (type QUAL2)

The concept of predictive error modelling in PSA is basically concerned with identifying the nature of the errors that are likely to be committed by an operator in the situation of interest. For example, will the operator choose the wrong switch, will he misdiagnose the pattern of indicators, etc. In order to make these predictions, it is helpful to utilise one of the available models of human error causation to prompt the analyst. As before, there is a problem of translating such models so that they may be used effectively by risk assessors.

Quantitative data from all of the sources considered in the previous category (QUAL1) can obviously also be applied to assist the analyst in identifying error modes. One notable, but not widely available, data base in this context is the Confucious data base being constructed by Electricite de France.

Quantitative data excluding absolute probability estimates (type QUANT1)

The data sources discussed in detail for data type QUAL1 can be used to provide quantitative data regarding the relative magnitudes of human errors when comparing two situations with differing conditions. The quantitative impact on human error rates of different factors such as quality of procedures, time available to perform operations, etc. can also be derived from these sources. An alternative approach is to use techniques which systematically elicit expert judgements regarding the relative effects of different variables on human reliability (for further details see Embrey and Lucas, this conference).

Estimates of absolute human error probabilities (type QUANT2)

The availability of absolute data on human error probabilities is a topic of considerable controversy. Briefly, the 'empirical' position on quantitative human error data holds that such data are only valid if they are ultimately based on observed frequencies of errors obtained from operational situations, simulations, or experiments which can be validly extrapolated to operational situations. The 'subjectist' position, whilst agreeing with the empirical approach for situations where such data can be validly collected, argues that:

- Such an approach cannot be applied to 'rare event' situations because it will never be possible to gather an adequate amount of data under conditions similar to those to which the data will be applied.

- Many crucial aspects of human performance in high risk situations, (e.g. decision making, diagnosis) do not involve externally observable processes that can be recorded in order to generate error rates and therefore probabilities.

- However imperfect the perceptions of the available experts may be regarding the likely HEPs for a given (rare) situation, these perceptions (where systematically elicited) represent the best available evidence and must therefore be regarded as being best estimates of HEPs.

- Strictly speaking error probability estimates derived from frequencies are only useful if the underlying causes are the same for every error in the sample. This assumption is rarely met.

Conclusion

Useful qualitative data can be collected from operational experience and from simulators. In the nuclear industry (particularly in the US, France and Scandinavia) the benefits of having structured and systematic human performance data gathering exercises are becoming clear. Other industries need to learn from this success. However, there remains the need to use cognitive models of error causation more extensively in the collection and analysis of human performance data. Future research should look to finding methods of assisting analysts to utilise such theoretical insights in a practical situation.

In our view the best approach to the collection of quantitative human error data within any industrial setting is a pragmatic one. Both empirical ("objective") and judgemental ("subjective") data should be utilised. Since only a small amount of empirical data can be collected (relative to the overall need), this should be used within expert judgement based techniques.

The best sources of data for these purposes are likely to be simulator studies (in the medium term) and a comprehensive system aimed at collecting data on operational errors and near misses in the longer term.

References

1. Ball, P. et al (1985) Guide to Reducing Human Error in Process Operations. SRD Report R347, UKAEA, Wigshaw Lane, Culcheth, Warrington.

2. Beare, A.N. et al (1984) A simulator-based study of human error in nuclear power plant control room tasks. NUREG/CR-3309, SAND 83-7095.

3. Embrey, D.E. (1986) SHERPA: A Systematic Human Error Reliability and Prediction Approach. Paper presented at the ANS/ENS international topical meeting on Advances in Human Factors in Nuclear Power Systems, Knoxville, Tennessee.

4. Embrey, D.E., Carroll, J.E. and De Montmollin, M. (1986) The INPO Human Performance Evaluation System: A review and proposals for further development. Proceedings of a conference organised by INPO and EDF, Lyons, France, June 1986.

5. Lucas, D.A. and Embrey, D.E. (1986) A pilot study of the root causes of human errors in dependent failures. Report prepared for EPRI, Palo Alto, California.

6. Lucas, D.A. (1987) Human performance data collection in the nuclear industry. Human Reliability in Nuclear Power. IBC Technical Services Ltd.

7. Norros, L. and Sammatti, P. (1986) Nuclear power plant operator errors during simulator training. Technical Research Centre of Finland, research reports 446.

8. Rasmussen, J. (1983) Skills, rules and knowledge: Signals, signs and symbols, and other distinctions in human performance models. IEEE **Transactions on Systems, Man and Cybernetics**, SMC-13 (3), 257-266.

9. Reason, J. (1987) Generic Error-Modelling System (GEMS): A Cognitive Framework for Locating Common Human Error Forms. In: Rasmussen, J., Duncan, K. and Leplat, J. (eds.) **New Technology and Human Error.** Chichester: Wiley.

10. Swain, A. and Guttman, H.E. (1983) Handbook of human reliability analysis with emphasis on nuclear power plant applications. NUREG/CR-1278.

11. Woods, D.D. (1984) Some results on operator performance in emergency events. Institute of Chemical Engineers Symposium Series, 90, 21-31.

Quantification of Human Reliability

Unni Nord Samdal & Jon Arne Grammeltvedt
Norsk Hydro Research Centre
N-3901 Porsgrunn
Norway

ABSTRACT
Human reliability has been brought increasingly into focus during the past years, and several quantitative models for human reliability have been proposed. In a man-machine system analysis of operator tasks in the control room of one of our PVC plants, four alternative models were used for quantification. These were: HEART, TESEO, PROF, and data from Swain & Guttmann's "Handbook of Human Reliability Analysis". Based on the experience from this study, it was decided to develop PROF further. This model uses a human reliability data base as a basis for estimating error rates. Most of the data in this data base were collected from open reports. As the amount of such data is limited, it was natural to concentrate on further data collection.

At Sandsli, Bergen, Norsk Hydro has developed a full-scale true-copy simulator for the control room at the Oseberg oil production platform in the North Sea. Operator training was first performed, using the simulator, in 1988. This opportunity was used to collect more data on human reliability.

1. COMPARISON OF QUANTITATIVE METHODS

Hydro started work in the field of human reliability in 1985. To gain experience in human reliability analysis and to compare some of the various methods for estimating human error probabilities, a case study was performed in 1987. The PVC factory at Porsgrunn Works had already been selected for a safety review project, and it was natural to couple the two activities.

1.1 Case study

The PVC production is based on a batch process, and is rather labour intensive. This phase of the safety review project focused on the work of one of the control room operators. Her tasks were analysed by means of the Man-Machine System Analysis technique as described by Swain and Guttmann /1/. This technique includes steps to quantify error probabilities. Though every task was analysed qualitatively, the quantitative steps were only performed for four of the tasks. These were selected by the PVC factory reference group, and were:

a) To set a specific temperature controller from manual to automatic mode, a routine task to be performed 10-15 times during a shift.

b) To check that another temperature controller has correct status (cascade coupling), a routine check done 10-15 times during a shift.

c) To have continuous control of the process parameters so that abnormal situations are detected, diagnosed and taken care of in time to handle undesirable events at an early stage. This requires good process knowledge and understanding, and continuous alertness.

d) To check that the process parameters are such that the termination of the batch process can start. This also requires good process understanding, and information must be sought from different sources. The task is performed 10-15 times during a shift.

Thus, of the four tasks quantified, one was a simple operation, and the three others were tasks of varying complexity involving check and assessment of process parameters and instrument status. None of the tasks are critical in a normal situation, but they are essential when abnormal events occur.

This paper will concentrate on the quantitative part of the analysis.

1.2 Models for quantification

Various models for quantification of human error probabilities have been developed. However, most of them lack verification.

For quantification of the four selected tasks given above, four methods which were easily accessible and applicable to the current problem were chosen. There are certainly others which could have been included. However, the four selected methods were:

1) Estimation from tables and directives in Swain and Guttmann /1/.

2) TESEO (Tecnica Empirica Stima Errori Operatori), as described in Bello and Colombari /2/.

3) HEART (Human Error Assessment and Reduction Technique), as described in Williams /3/.

4) PROF (PRediction of Operator Failure Rates), as described in Wright and Soma /4/ and Quasar /5/.

We will not describe the four methods in detail, but leave that to the originators (see the references). We will, however, point out the main differences and similarities between the models.

Method 1) (S&G) is based on a defined set of data, and directives for the use and modification of these. It is thus not possible to use this method for tasks which are not covered in the data set.

Methods 2), 3) and 4) are all based on mathematical models, with parameters which must be assessed and quantified. Though the mathematical formula and number of parameters vary for each method, the subjects to be assessed are much the same.

TESEO is a simple product of five parameters. This method is designed for control room tasks.

HEART has as starting point an error probability chosen from a table of task types, spanning from "Totally unfamiliar (task), performed at speed with no real idea of likely consequences" to "Respond incorrectly to system command even when there is an augmented or automated supervisory system providing accurate interpretation of system state". The figure chosen is modified according to a list of 17 unfavourable parameters which - if present - increase the probability with given factors.

PROF is based on the error probability being a logarithmic function of nine parameters, and also a function of the operation type. The nine parameters are weighted with weights calculated from information given in a data base, containing data from Swain and Guttmann /1/ and other sources. Additional information on PROF is given in the Appendix.

While TESEO and HEART give the probability of a final error, which means that error correction is not performed, S&G gives the probability of performing a first error, which might be corrected. PROF is not altogether clear on this point, but viewing the sources of information in the data base, the conclusion is that PROF also gives the probability of performing a first error.

S&G strictly defines the type of error that is considered (error of omission, error of commission, timing error etc.). The other methods do not distinguish clearly between these.

No correction for these differences is done in the results referred to below.

All the methods are based on more or less subjective evaluation of various factors. Using the methods with our best judgement gave the results in table 1.

Task	Method			
	S&G	TESEO	HEART	PROF
a)	0.002	0.002	0.04	0.0004
b)	0.006	0.005	1.0	0.007
c)	0.3	0.03	>1	0.09
d)	0.03	0.02	>1	0.05

Table 1: Results obtained through first experience with the methods[1]

As can be seen from table 1, the estimates obtained by use of HEART are rather unrealistic. Discussions with the originator made clear that our results were due to misinterpretation of the method. When using HEART, one must be careful not to account for unfavourable conditions twice, and cautious in use of the modification factors (normally only a few should be used). The HEART probabilities without accounting for any unfavourable conditions are of the same order of magnitude as the other results above.

Our experience with HEART shows how critical the user's perception and interpretation of each method are, and it demonstrates the need of extremely clear and accurate defininitions.

Except for the HEART estimates, table 1 shows that the results are in relatively good agreement.

1.3 Evaluation of the methods

It is clear that the conclusions that can be drawn from the study with respect to method evaluation are limited. Only four different tasks of much the same type were considered, and there were no empirical figures to which the results could be compared. But the study illustrated that it is essential to

---

1.)For all the methods, a probability greater than 1.0 is to be set equal to 1.0.

have a detailed knowledge and understanding of the method one wishes to use, and that this is more important than the mathematical formula of the model. Therefore, one method was chosen for further studies.

The method was selected using the following criteria:

a) The method should cover a great variety of operation types.

b) The method should be simple to use.

c) The method should make use of existing knowledge on human factors, and it must be prepared thoroughly and be consistent with respect to level of details and definitions of parameters and failures.

d) It must be easy to incorporate new knowledge and data in the method.

e) The method and the information and data on which it is based must be well documented.

Each of the methods has both strong and weak points according to these criteria. Which to prefer depends on which of the above criteria is considered the most important. In Hydro, we believe there is still a lot to be learned in human reliability. Criterion d) is therefore considered as principal. Also criterion e) is important to get a picture of the information which is the basis of the estimates.

On the basis of these considerations, it was decided to participate in further development of PROF. But at this stage of development, we cannot limit ourselves to use of PROF alone. This is mainly because of weakness with respect to criterion a). As for the other methods, there is a lot of valuable information in the HEART tables, information which for example can be useful when evaluating the parameters of other methods. TESEO is a method which is easy to use, and Swain and Guttmann's Handbook /1/ remains an outstanding piece of work in this field (as already mentioned, their data are also included in the PROF data base).

2. DATA ON HUMAN RELIABILITY

The PROF results are based on data in the PROF data base (as described in the appendix). Thus, an improvement and extension of this data base is essential to make the results more reliable and the method more flexible with respect to different operation types and conditions. When developing PROF further, it was therefore natural to focus on the data base.

Data on human reliability can be obtained from the following sources, listed by priority:

1) industrial plants
2) simulators
3) experiments using realistic industrial work tasks
4) experiments using artificial tasks

Unfortunately, data from the preferred source hardly exist. Simulator data are more easily obtainable. In the next chapter we describe how a control room simulator was used to collect operator failure data.

There is a continuous discussion on the meaningfulness of quantifying human reliability at all. Whatever opinion one might have on this question, it is clear that this is a complicated field with respect to data collection. To give a complete description of the conditions under which a mistake occurs is extremely difficult, if not impossible. Only the most important factors must therefore be defined and quantified in one way or another. All the methods described earlier more or less apply this approach. There will always be considerable uncertainty attached to the estimates, since, among other reasons, the defined factors might not actually be the most important in the current case, and the factor quantification will always be partly subjective. Also the modelling of task sequences is associated with great uncertainty. It is essential that this is recognized when dealing with human reliability.

Nevertheless, we believe that human failure data collection gives valuable information regarding, for example, design and job management. We also believe that the estimated probabilities obtained through use of the models give meaningful relative results. Whether the estimates are meaningful as absolute numbers, we don't yet know. Work concerning validation of the results is still needed.

3. THE OSEBERG TRAINING SIMULATOR

Hydro is operator for the Oseberg field, which is located in the North sea approximately 130 km north-west of Bergen, Norway. The reservoir contains a substantial amount of both oil and gas. At the Oseberg platforms the wellstream is separated and stabilized. The crude oil is then transported to shore in a pipeline, while the gas is reinjected into the reservoir to increase oil production.

An education and training centre has been developed to prepare the operators in safe and efficient platform operation. Based on the knowledge that human failures are the most common reason for shutdown and emergency situations in complex technical processes, a simulator was built. By using this full-scale true-copy simulator the operators gained experience in operating the platform before startup.

The simulator consists of a copy of the central control room, a room for the instructors, computer rooms and lecture rooms. The computer based simulator model is based on:

- Detailed mathematical models regarding pressure, temperature etc. which have been established for all important process and utility systems on the platforms. The models will react to the operating conditions and the control system in the same manner as the platforms.
- Models of the process control systems which have the same interface as on the platforms.
- An instructor system so that the instructors can lead and control the simulation sessions efficiently.

From the point of view of the central control room operators, the simulator cannot be differentiated from the real control room. The simulator has the same control equipment and communication systems and is operated in the same way as the platforms.

The instructors have full view of the control room from their room. They have screens and monitors to control the simulator and follow the operator actions. They also have the possibility of communicating with the operators, acting as the different persons that normally communicate with the control room operators. During the simulation session the instructors can introduce failures on various equipment, and the simulator model will respond to the failures as the real process does. The effect of over 1500 different failures have been modeled.

4. COLLECTING OPERATOR DATA FROM THE SIMULATOR

The purpose of the simulator training course was to give the control room operators the necessary knowledge to operate the Oseberg platforms. In the fourth and last part of the course, the operators operated the simulator like the real platform. They tried to optimize the production, diagnosed failures in different equipment, and trained in procedure operations (e.g. startup and shutdown of the oil-and gas trains, and other routine operations). It is from this part of the course that data on operator actions were collected.

There are different strategies for collecting data from a simulator. One possibility is to be present during the session and follow the operators very closely, so all the actions they take can be noted. A maximum amount of data can be collected in this way, but it is very resource-intensive.

Lacking the resources to collect data in this way, the project used the event log provided by the simulator. The event log records important process alarms and all instructor and operator actions carried out on the simulator. By going through the operator actions, it is possible to get good information about the operators' performance. The main disadvantage in this strategy is that the event log only includes direct operator actions (e.g. changing the set point of a controller, opening / closing a valve etc). An examination of the process parameter status will for example not be included in the event log.

To be able to collect data from the event log, specific types of operator actions had to be chosen from the session plans. The actions were sorted in groups so that the performance shaping factors had the same value for each group. Examples of the chosen type of operator actions are:

- Setting a valve in open/closed position.
- Setting a controller in auto and adjusting its set point.
- Setting a valve/controller in manual/auto.
- Adjusting a set point.
- Starting pumps.
- Testing of valve/pump after corrective maintenance.

Since the event log records information sequentially in time, it has been loaded into a data base and analysed with the help of a computer.

5. EXPERIENCE FROM THE DATA COLLECTION

The manner of data collection in the project restricted us to simple operator actions which are done according to procedures. But there is also a need for these data in the data base. In addition, collecting data in this way allows an easy analysis of any new simulator course which follow the same session plans. Thus the data base will increase with each course. Data from 6 courses have been collected so far.

We would also like to collect data on more complex operator actions, and therefore intend to do some more detailed studies on the simulator when it later is used for retraining and emergency handling. We would then like to collect data both through following the sessions and through analysing the event log. We also plan to vary the the performance shaping factors over a much wider range, which is a necessary next step in development of the PROF data base.

6. CONCLUSIONS

Although there exist several models for human reliability, a lot remains to be done with regards to verifying the methods and obtaining more reliable data. Also the uncertainties attached to the estimates need to be looked into further.

Our experience shows that the models are extremely vulnerable to the user's misinterpretation of the model's parameters and structure. It is therefore essential that the user is well acquainted with the method, and preferably experienced in its application.

Hydro has concentrated on further use and development of the PROF model, which is viewed as a useful and interesting method. The main weakness of PROF is the lack of reliable data in its data base. A program for data collection from an offshore control room simulator was therefore developed.

Collection and analysis of simulator data is a time consuming task, but results in useful information that can be used in design and job and organization management. Whether the validity of the results is universal is a topic for further investigation.

7. REFERENCES

/1/ A.D. Swain and H.E. Guttmann: 'Handbook of Human Reliability Analysis with Emphasis on Nuclear Power Plant Applications. Final Report'. NUREG/CR - 1278, 1983.

/2/ G.C. Bello and V. Colombari: 'The Human Factors in Risk Analysis of Process Plants. The Control Room Operator Model "TESEO"'. ENI, Roma, 1979.

/3/ J.C. Williams: 'HEART, a Proposed Method for Assessing and Reducing Human Error'. Proceedings from 9th Advances in Reliability Technology Symposium - 1986.

/4/ J. Wright and H. Soma: 'Prediksjon av operatørfeilrater. PROF'. Quasar Report no. F-05-86-1. (Norwegian.)

/5/ Quasar Consultants: 'Users Guide to PROF, Version 2'. Quasar, 1987.

APPENDIX

DESCRIPTION OF PROF

PROF (PRediction of Operator Failure rate) is a computerised method developed to estimate operator failure probabilities. The method was developed at Quasar Consultants in cooperation with Hydro and partly financed by The Royal Norwegian Council for Scientific and Industrial Research.

PROF is based on a mathematical model and on a data base containing data on human reliability.

The mathematical model is:

$$\log f = \sum_{k=1}^{m} W_k \ PSF_k + C \qquad (1)$$

where:

- f = the failure rate
- W_i = weight of PSF_i (numerical constant)
- PSF_i = numerical value of Performance Shaping Factor i
- C = numerical constant
- m = number of PSF's (in the present model, $m = 9$)

The Performance Shaping Factors (PSFs) used by the model are:

1. task complexity
2. information quality
3. feedback
4. breaking of stereotypes (standards)
5. distractions
6. time pressure
7. potential consequences (economic, risk or reputation related)
8. task relevant experience
9. personnel qualifications

W_i and C are constants which are calculated through regression analysis from selected data in the data base, typically all data which apply to a certain operation type.

The data base contains available data on human error probabilities, e.g. the data from Swain and Guttmann /1/. Since available data differ in credibility, they are given a weight (not to be confused with W_i) according to their origin. For example empirical data will be weighted higher than expert judgements. These weights are used in the regression model. Each piece of data which are included in the data base is evaluated for numerical values of PSF. In general the evaluation is done according to the scale given in table 1, but there are also some specific notes to most of the PSF values.

PSF subjective scale	PSF numerical scale
Optimal, much better than normal	1
Better than normal	2
Normal	3
Worse than normal	5
Far worse than normal	10

Table 1: PSF evaluation scale

When using PROF to make estimates, the appropriate operation type(s) must be selected, and the PSF values must be determined in the above manner. At this stage the following operation types are incorporated in PROF:

- reading of displays
- reading labels
- inspection on a change of shifts or plant walk through
- selection of manual controls
- actuation of manual controls
- responding to annunciated legend lights
- completing a set of multiple tasks
- communication tasks
- maintenance and calibration tasks

This list can of course be extended when data are obtained for other operation types.

Treatment of Uncertainties in Human Reliability Analysis

Pekka Pyy and Urho Pulkkinen
Technical Research Centre of Finland (VTT)
Electrical Engineering Laboratory

Abstract

The modelling of operator errors has been discussed in several contexts and also in various benchmark studies. The estimates of operator error probabilities have differed from each other several orders of magnitude depending on the modelling approach. In this paper, reasons for this phenomenon are discussed. Furthermore, the application of an approach based on influence diagrams in estimation of error probabilities is presented. The use of influence diagrams proved to be an effective tool in the modelling and the identification of the major uncertainties. However, some problems concerning data uncertainty discussed in the text still remain unresolved.

1 Introduction

Problems in assessing human reliability are among the main problems of PSA. The problems arise from the large variety of possible human actions in operation, maintenance, test and supervision of nuclear power plants. Thus, new accident sequences deviating from the design basis cases may occur and common cause failures (CCFs) may be initiated. This is mainly due to the fact that nuclear power plant tasks require more than the operators are capable to do i.e.to stay alert in routine operations or maintenance, to make decisions under uncertainty and conflicting goals, to win their reluctancy to take extreme emergency actions and , finally, to use procedures even if they are sure they know better solutions.

Under these circumstances it is interesting to note that the nuclear power plant operators actually perform their duties nearly without errors. PSA(Probabilistic Safety Analysis) usually represents their activities in too negative a light when trying to estimate human reliability in a conservative way. Besides the operating or maintenance errors engineers and designers commit errors as well. It has to be remarked that these upper level errors are in many cases direct causes of operator errors and should not be excluded from the event analyses [1].

The problems and uncertainties related with the human reliability analysis can be divided into two categories. The first group are the uncertainties due to the phenomenon itself e.g. degrees of freedom of the human behaviour, which means the characteristics of human actions make it difficult to create any model on them. The other group strongly coupled with PSA is the information on human actions. In reliability engineering this is known as human reliability data and quantification problem. The modelling and the data problem are discussed in the following. The discussion is, however, limited to post initiating event human actions.

2 The state of the human reliability modelling

The most common models used to illustrate the structure of human activities are conventional system reliability models event tree and fault tree. Here, Swain's HRA–tree [2] and the OAT–model [3] are also considered as event trees. Which one of these models is better for illustrating the event logic is up to the analyst.

The basic assumption behind the use of these models is that the failures or errors can be expressed as a logic function of decomposed basic events, especially in the case of manual activities. Here, the mutually exclusive basic events are generally assumed for the simplification of the quantification. Because the errors quantified, however, originate partly from the same mental processes, their quantification as exclusive events can induce a remarkable bias in the results. Besides,the decomposition of cognitive processes in probability assessment purposes is doubtful and can mislead rather than help the analyst [4].

In addition to fault tree and event tree modifications there is currently a number of models to be used in the human reliability analysis. Examples from the models are HCR–correlation [5] and SLIM–MAUD [6]. Also these models suffer from several uncertainty sources such as determining the time window, ability of the operators to correctly use the time available, relations of the performance shaping factors etc.

The result of the review of widely used human reliability models yields a conclusion that current models, especially event and fault trees, need foundation, which is more consistent with the human behaviour and thus decreases the modelling uncertainties inherent [7]. In the following, such a procedure is discussed.

3. An alternative human reliability modelling approach

3.1 Overview

In the development of new models, it has to be taken into account that the model created should also act as an identification means to indicate the important parts of an event sequence for corrective measures. The basic features of human behaviour have to be taken into account. Maybe the most important are the operators' goals and the plans the operators have for reaching these goals. At the same time, the model structure has to be simple to provide a possibility to be used in probability assessment.

In references [3], [7] and [9], principles for classification and modelling human actions have been presented. These principles form a good basis for the model construction. When thinking about the PSA quantification needs, the models [7] and [9] go, however, maybe too far. Our approach is to divide the operator behaviour during accident sequences into four phases on a coarse level. The phases are given in the following.

A. Situation perception and interpretation, where the operators'goal is to try to localise the cause of the alarms and the indications. The errors here are related with confusing or missing information (etc. lack of alarms, training, new event).

B. Selection of action programme, where the goal is to eliminate the event cause on the basis of the hypothesis formed in the previous phase. Here, the decision is influenced by the previous experience (training, operational experience, individual memories), team co–operation (social surroundings) and knowledge of the plant features (completely new sequences). The errors are related with the unsuitability of the implemented knowledge–base for the situation.

C. Execution of the tasks required, where the goal is to carry out necessary manual actions in the time available concurrently validating the hypothesis on the event cause. The errors here are small slips and deviations such as forgetting and mixing, but also deliberate violations are possible (as also in the selection of violating action programme) [11]. In some cases, the perception and action are simply highly integrated reacting (e.g. pushing a scram button), where the three phases cannot be separated.

D. Observation of the results, where the need to take additional measures or reanalyse the plant condition is under consideration. If the results do not satisfy the operators, they simply return to phase A if the time allows. The errors are related with the lack of ability to accurately enough observe changes in the plant state or with the lack of information needed.

The four phases form a control loop presented in the figure 1. The human behaviour in post initiating event conditions in the control room can be seen as a structure of these loops repeated until the desired state is roughly reached. Same kinds of ideas have also been presented in the references [9] and [10].

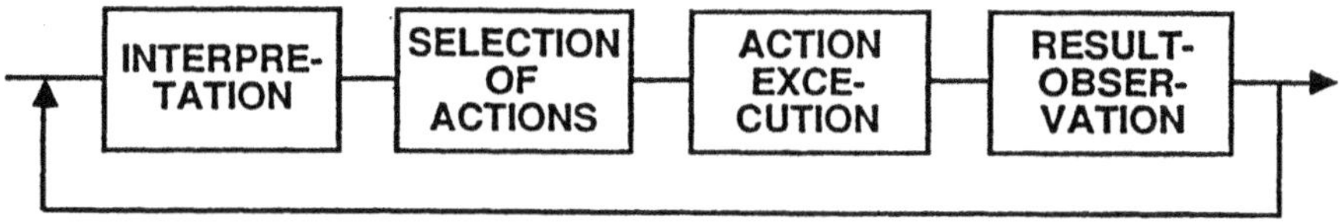

Figure 1. A loop of human behaviour functions in post initiating event sequences

This approach provides a frame, which can be used to validate current human reliability models and, if needed, a more sophisticated model can be created for psychological purposes. The other functions of the approach are:

- to serve as a qualitative basic model for human behaviour in accident sequences
- to give input to more detailled quantitative models such as influence networks.
- to help the identification of human errors and organisational pitfalls

3.2 Identification of operator errors

The identification of hazards due to operators is important, since here the model presented serves as a guideline to influence the operator behaviour. Through the identification of important operator task characteristics, the uncertainties related with the human action phenomenon itself are essentially decreased and the quantitative model structure concurrently simplified (See the example).

The contributing factors of the model nodes are important from several points of view, of which the quantification is not the least. The following classification for the contributors can be considered reasonable:

- Existence and validity of current instrumentation (I)
- Procedures, rules and instructions (P)
- Control room crew and its past experience (C)
- Task characteristics (complexity, time, reluctancy) (T)
- Scenario specific training (Tr)

In the identification, suitable questions for post initiating event conditions are e.g. (the contributing factors presented in the parentheses).

A. Perception and interpretation

- Is the information (I) required possible to see without moving around the control room and skimming alarms, what is the effect of a missing alarm etc?
- Are there indication patterns causing confusion (use of confusion matrix suggested if not clear) (I)
- What is the extent and quality of training given on this sequence (Tr), what is the real life experience (C)?
- Are there any procedures or operator aids to backup the interpretation of the indications (P), (I)?

B. Selection of action programmes

- Are there any action programmes, which may arise reluctancy in operators (avoid core melt, but cause extended outage) (T),(Tr),(C)?
- Which are the action programmes suitable for this particular scenario (procedures, rules–of–thumb) on the basis of phase 1 (Tr),(G)?
- Are there any temptative measures strictly forbidden (operating instructions, technical specifications), is there an easy–way–out solution(T),(G)?

C. Excecution of the tasks required

- What is the effect of time and consequence pressure(T),(G)?
- What is the number of subtasks performed(T), are the errors observable (T),(I), is the assumption of independent checking correct in a situation like this(T)?

What is the role of training and experience; are the procedures needed, what is the quality of the procedures (P),(Tr),(G)?

D. Observation of the results

- Does the plant response rapidly show the correctness or erroneousness of the actions(I)?
- What is the plant state, if an erroneous action programme has been selected, what is the time window(T)?

The questions above should reveal most of the major error possibilities arising from the operating crew behaviour. The organisational lability in respect of the information exchange in vertical direction (engineers – operators) and in the time axis (designers – operators – feedback) can also be identified, if the questions are made to different organisation stages and significantly different answers are received from different personnel groups.

3.3 Validation of the human reliability models

In the validation of the used event tree or related models, the procedure is to take the most probable path of the event tree into consideration. The similarity matching between this path and the presented model is then compared, and the result can be seen to be proportional to the validity of modelling human actions of the post accident scenario.

The operator crew behaviour depends on many factors such as training, procedures, crew characteristics etc. The modelling of this complicated dependence structure is a very difficult task. In many cases, it is not possible to describe the factors exactly although their effect on the crew behaviour is obvious. Here, we propose the use of the so called influence diagram or network as a modelling possibility. The influence diagrams have been discussed for instance by Pearl [12] ,Barlow and Pereira [13] and Howard and Matheson [14].

In influence diagrams, the dependencies are described as conditional probability distributions. A network created on the basis of our approach is shown in Fig. 2. The loop presented in Fig. 1 is reflected through changes in the influence parametres of the repeated attempts, if the time available allows. The nodes of the network represent different factors influencing other factors and thus finally affecting the manual actions, which are the only way man can interfere in a technical system i.e. induce errors. The model nodes are described further in the chapter 3.2.

The nodes are interpreted as random variables since we do not know their real value. The consequence arrows between the nodes represent the interdependence of the nodes. The degree of dependence is described with conditional probability distributions of the values of the node variable. The values of the node variables represent different states of nature, such as whether the instrumentation is available or whether the operators succeed in interpretating of a specified event. The following gives an example of how the modelling principle presented is utilized in the quantification of an operator error sequence (chapter 4).

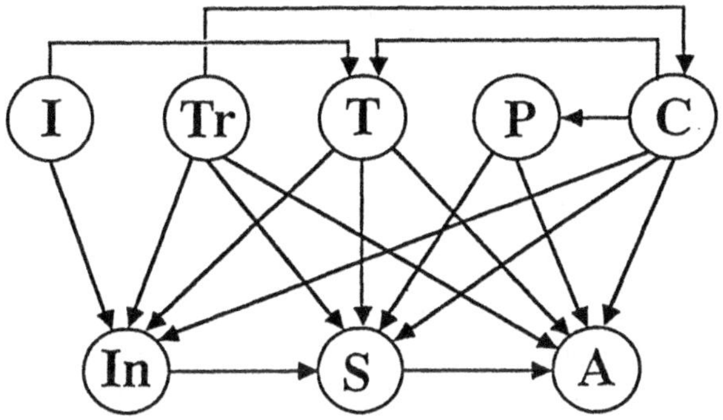

Fig. 2. An influence network for post initiating event sequences.

4 Application in the Nordic Reference Study on uncertainty and.sensitivity analysis

The nordic co-operation project NKA/RAS-470 financed partly by the Nordic Liaison Committee for Atomic Energy (NKA) concentrates on some critical aspects of level I PSA. The most important issues discussed in the project have been the estimation of CCF-probabilities and critical reviews of human error analyses of some recent PSAs. In the year 1988 the topic of the project has been the reference study on uncertainty and sensitivity analysis based on the most important accident sequence of the PSA of the Swedish nuclear power plant Forsmark 3. Nordic organisations (ABB-ATOM and Studsvik Energy from Sweden, Riso National Laboratory from Denmark and VTT from Finland) have participated in the reference study. The background of the reference study is described in detail in the reference [15].

The accident sequence, T_fUX_2, analysed within the reference study is composed of the loss of main feedwater as an initiating event (T_f), unavailability of the auxiliary feedwater (U) and the failure of manual depressurisation of the reactor (X_2). The sequence contributes about 90% to the total core melt frequency. The event U consists mainly of the quadruple CCFs of the auxiliary feedwater system. The analysis of manual depressurisation (X_2) is discussed further in the following.

The depressurisation is performed automatically, if the reactor level reaches about +0.5 m point and high pressure exists in the containment (as in the LOCA case). Because such pressure is impossible in the loss of feedwater transient, the high pressure condition has to be given to the logic by bringing in a certain key and turning it in the keyhole of the control panel. Then, after the level +0.5 is reached, the reactor blowdown is activated automatically.

Although the indications are quite clear and evidently there can be no doubt about what is going on, the operators may appear to be reluctant to activate the depressurisation. This is mainly due to the fact that a sudden pressure relief causes an extended outage with degraded pressure vessel lifetime. Therefore, it is obvious that the operating crew tries to restore the FWS and AFWS until it is necessary to launch the blowdown. Other features increasing uncertainty are that the place, where the key is to be found is not completely clear and the time available for the operating crew to perform their tasks is not exactly known. Thus, an appreciable extent of uncertainty is related with the sequence.

In the example, the task characteristics are fixed to be complex (as they really are) with remarkable stress and time pressure and the crew is supposed to be well experienced. These assumptions simplify the model interactions. The availability of the instrumentation, the use of the instructions and the quality of the training on the sequence are considered as stochastic parametres of the influence diagram describing the manual depressurisation. The network presented in Fig. 2 is thus simplified as shown in Fig. 3.

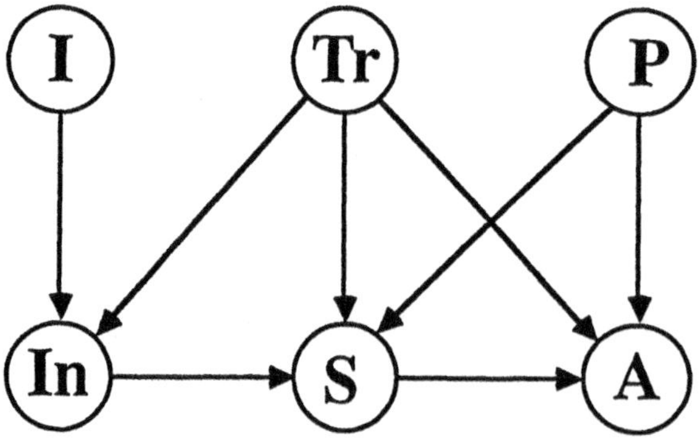

Figure 3. An influence network for the BWR blowdown case with fixed highly complex task and extremely good crew conditions.

The stochastic nodes of our influence network model are described in more detail in table 1.

Table 1. The nodes of the influence network model

Node	Description	Values of the variable
1 (I)	State of instru– mentation	I = 1: needed instrumentation available I = 2: instrumentation unavailable
2 (Tr)	Training	Tr = 1: specific simulator training Tr = 2: general sequence training Tr = 3: no training
3 (P)	Use of procedures	P = 1: procedures used P = 2: procedures not used
4 (IN)	Interpretation of the initiating event	In = 1: correct interpretation In = 2: incorrect interpretation
5 (S)	Selection of action	S = 1: correct actions selected S = 2: incorrect action selected
6 (A)	Action	A = 1: successful action A = 2: unsuccessful action

The nodes 1, 2 and 3 (I, Tr, and P) are assumed to be independent on other nodes. The other nodes depend of each other in the way presented in figure 3. For evaluating the error probability of the manual depressurisation, we have to determine the joint probability distribution of the node variables. Since we have postulated the dependencies between nodes, we may write the distribution in the form:

$$P(I,Tr,P,I,S,A) = P(A|Tr,P,S)\cdot P(S|Tr,P,In)\cdot P(In|I,Tr)\cdot P(I)\cdot P(Tr)\cdot P(P), \qquad 1$$

where $P(A|Tr,P,S)$, $P(S|Tr,P,In)$ and $P(In|I,Tr)$ denote the conditional probability distributions of the variables A, S, and In and P(I), P(Tr) and P(P) are the probability distributions of the variables I, Tr and P.

The quantity of interest is P(A) which represents the error probability of manual depressurisation. In order to find this probability we have to determine the marginal probability distribution of the variable A by integrating the joint probability distribution over the values of other variables:

$$P(A) = \sum_{I} \sum_{Tr} \sum_{P} \sum_{In} \sum_{S} P(I,Tr,P,I,S,A). \qquad 2$$

The probability distributions P(I), P(Tr) and P(P) and the conditional distributions P(A|Tr,P,S), P(S|Tr,P,In) and P(In|I,Tr) are needed in the evaluation of the distribution P(A). If the nodes of the influence network are defined properly, the above probabilities can be evaluated based on actual statistical evidence or judgement.

In principle, it is possible to develope the nodes further e.g. to include the effect of the time on the conditional probabilities explicitly, but it would lead to more complex network. Now, the time available is taken implicitly into account as one parameter affecting the task complexity. Furher decomposition of the network leads to increased number of independent nodes and to the need of more data.

In our case study we used judgemental estimates. Further, we assigned uncertainty distributions to the parameter of the model to describe our uncertainty about these parameter values. The data is described in appendix 1. Using our parameters estimates we obtained the failure probability of manual depressurisation $P(A) = 3.82 \cdot 10^{-2}$.

To express the sensitivity of the manual depressurisation failure probability to the parameter uncertainties we performed a Monte–Carlo–simulation where we propagated the parameter uncertainty distributions through the model. As the final result we obtained the uncertainty distribution for the manual depressurisation failure probability, P(A). The uncertainty distribution is shown in figure 4. Some of the most important characteristics of the distribution are presented in table 2.

Table 2. Characteristics of the uncertainty distribution

Expected value	$3.57 \cdot 10^{-2}$
Variance	$6.72 \cdot 10^{-3}$
Median	$1.61 \cdot 10^{-3}$
5% fractile	$5.03 \cdot 10^{-6}$
95% fractile	$2.19 \cdot 10^{-1}$

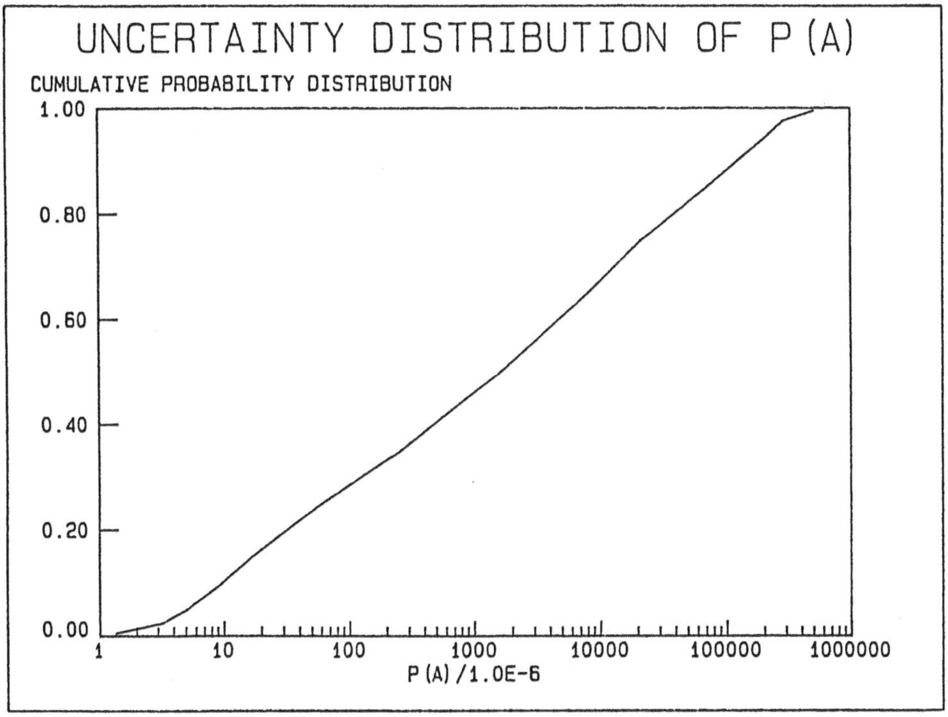

Figure 4. The uncertainty distribution of the error probability.

The uncertainty distribution is broad demonstrating the number of uncertain factors having effect on the manual depressurisation failure probability. This far, we have not yet analysed whether this kind of broad uncertainty distributions are typical of influence diagrams.

5. Uncertainties related with the data and quantification

The aim of the quantitative human reliability analysis is to evaluate probabilities of human errors in various situations subject to various environmental conditions or performance shaping factors. Due to the complexity of human behaviour and lack of direct statistical evidence the human error probabilities are often estimated by using the various models discussed earlier in the text.

The model construction is not complex and there are numerous ways of describing the interdependencies of various factors. Although the factors included in different models would be the same, the way in which the factors are related to each other may be different. Our conclusion is that based on the evidence we have on human errors we cannot say, which model is the correct one. Accordingly, we must understand also the stuctural dependence models as an extension of the conversation concerning the human error probability. When we describe the depencency of the human error probability on several factors we in a way extend our conversation to cover these, sometimes purely fictitious factors instead of discussing solely the human error probability. This approach is compatible with the principles of bayesian statistical modelling [18].

The recent benchmark studies have demonstrated that different approaches have led to results differing several orders of magnitude from each other [4]. The differencies are due to varying modelling approaches and varying data. Even in the case where similar approaches have been applied the differencies have been significant. From the bayesian view every probability estimate is equally good if there is no statistical evidence and the estimates used reflect the analysts degree of belief. However, if statistical evidence exists, the results should tend to equal figures (if the prior distributions are robust enough).

If the basic approaches are not similar, the analysts tend to interpret the statistical evidences in different ways. Further, the observations are dependent on the model in the sense that different models require different kind of information. This leads automatically to different probability estimates also in the cases where the statistical evidence is the same.

Our approach was the use of influence diagrams which help to express the various structural dependencies in a transparent way. Because we interpret the dependencies as conditional probability distributions, we need lot of parameters in our model. In other approaches the number of parameters may be smaller but, instead of having a large number of parameters, the models bring the information of complex dependencies within the parameter values. This may cause difficulties in understanding the results.

The model parameter values or their uncertainty distributions must in almost every case be chosen on the basis of subjective or expert judgements. Many authors have emphasised that experts tend to overestimate their expertise and they do not actually make the basis of their judgements explicit enough [19]. In many cases also the signifigance of mathematical methods is overemphasised.

Using the approach applied in this paper, the uncertainties of model parameters can be included in the model. This can be done by introducing extra nodes to the influence diagram and defining the possible dependencies between the model parameters. This, of course, increases the complexity of the model and makes the quantitative analyses difficult.

A interesting and important issue in statistical modelling is the scope of detail of the models. If the event (human error) is decomposed in a large number of variables, the statistical evidence concerning the basic variables weakens but the structure of the model becomes more explicit. On the other hand, if the decomposition is not detailled, there may be statistical evidence. However, due to an inexact model, the effect of different performance shaping factors cannot be taken into account, and further, the relevance of existing evidence with respect to the case under analysis cannot cannot be evaluated properly.

6. Conclusions

The framework for treating the uncertainties in the human reliability analysis gives new insights into the area. Against this background, it is obvious that the OAT–model [3] and the representation of the operating crew's mental states [10] are suitable for the PRA purposes. However, the influence network model presented in this paper helps to remark and study the related uncertainties in a more explicit way. Whether this is attempted or not, is up to the user. In many cases, greater uncertainty is related with the reluctancy of the evaluator to express the lack of knowledge in the case he is evaluating.

The propagation of uncertainties through the models is not necessarily the proper way to analyse uncertainty of probability estimates. In fact, probability is always one quantity without any uncertainty. Uncertainty distributions should be assigned only to the parameters of statistical or probabilistic models. Here, we have interpreted the probabilities as parameters, which is not correct according to the bayesian framework [16]–[17].

An interesting note is that the network model yields the more conservative estimates the more uncertainty about the parametres there exists. Even though the example shows the sensitivity of the result to the influence and parameter selection, the fact helps us only to revise our assumptions on the real situation. The model shows clearly, how complex even a simple network can be. Thus, also the secondary contributions are taken into account.

In addition to the uncertainty modelling and quantification questions, the paper gives a procedure for aiding the identification of hazardous stages of the operator behaviour in post initiating event sequences. It is emphasized that the questions presented should not be seen as the only alternative, but they provide a logical guideline. By using the guideline, the uncertainties related with a specific sequence can be explicitely analysed and also in many cases reduced.

References

1. Pyy, P., Wahlström, B.: Modelling the Human in PSA Studies. Reliability Engineering and System Safety 22(1988) 277–294.

2. Swain, A.D., Guttmann, H.E.: Handbook of Human Reliability Analysis with Emphasis on Nuclear Power Plant Applications. NUREG/CR–1278. Sandia National Laboratories, Albuquerque, New Mexico, USA. 1983.

3. Hall, R.E., Fragola, J., Wreathall, J.W.: Post Event Human Decision Errors: Operator Action Tree/Time Reliability Correlation. NUREG/CR–3010. Brookhaven National Laboratories, USNRC. 1982.

4. Poucet, A.: Survey of Methods Used to Assess Human Reliability in Human Factors Reliability Benchmark Exercise. Reliability Engineering and System Safety 22(1988) 257–268.

5. Hannaman, G.W., Spurgin, A.J., Lukic, Y.: A Model for Assessing Human Cognitive Reliability in PRA Studies. 1985 IEEE Third Conference on Human Factors and Power Plants, Monterey, CA. pp. 343–353.

6. Embrey, D.E.: SLIM–MAUD – A Computer Based Technique for Human Reliability Assessment. International Topical Meeting on Probabilistic Safety Methods and Applications, Volume 2. EPRI NP–3912–SR. 1985.

7. Rasmussen, J., Human Factors in High Risk Technology. Published in High Risk Technology, E.A. Green (Editor). John Wiley & Sons Ltd., London. 1982.

8. Apostolakis, G.E., Bier, V.M., Mosleh A.: A Critique of Recent Models for Human Error Rate Assessment. Reliability Engineering and System Safety 22(1988).

9. Bersini, U.,Cacciabue, P.C., Mancini G.: Cognitive Modelling: A Basic Complement of Human Reliability Analysis. Reliability Engineering and System Safety 22(1988) 107–128.

10. Worledge, D.H.: Some Useful Characteristics of
erformance Models. 1985 IEEE Third Conference on Human Factors and Power Plants, Monterey, CA. pp. 329–335.

11. Reason, J.: Errors and Violations: The Lesson of Chernobyl. 1988 IEEE Fourth Conference on Human Factors and Power Plants, Monterey, CA. 1988..

12. Pearl, J.: Fusion, Propagation, and Structuring in Belief Networks. 1986. Artificial Intelligence 29(1986) pp. 241–288.

13. Barlow, R.E., Pereira, C.A.de B.: The Bayesian operation and Probabilistic Influence Diagrams. 1987. University of California Engineering Systems research Center technical report ESRC 87–7.

14. Howard, R.A.,Matheson, J.E.: Influence Diagrams. In: Readings in the Principles and Applications of Decision Analysis. Two Volymes. 1984. Startegic Decision Group, Menlo Park, CA.

15. Hirschberg, S.: NKA–project Risk Analysis (RAS–470). Boundary Conditions for Reference Study on Uncertainty and Sensitivity Analysis. Report RAS–470(88)1 (ABB ATOM Report RPC 88–5), February 1988.

16. Clarotti, C.A., Spizzichino, F., Volta G.: Latest Advances in Bayesian Statistics: Theris Impact on Reliability and Probabilistic Risk Assessment. 1987. International Topical Conference on Probabilistic Safety Assessment & Risk Management, PSA'87. ETH, Zurich. pp. 137–142.

17. Clarotti, C.A.: Failure rate Estimation, A Dangerous Nonsense in a Bayesian View. 1987. Reliability Engineering and System Safety 20(1988) 117–126.

18. Singpurwalla, N.D.: Foundational Issues in Reliability and Risk Analysis. 1988. SIAM Review vol. 30, No. 2.pp 264–282.

19. Lichtenstein, S., Fischhoff, B., Phillips, L.D.: Calibration of Probabilities: The State of the Art to 1980. 1982. In: Judgement under Uncertainty: Heuristics and Biases, ed. Kahneman, D., Slovic, P. and Tversky, A.. Cambridge University Press, Cambridge, England.

APPENDIX 1. DATA USED IN THE ANALYSIS OF THE CASE STUDY

Probability parameter	Mean value of of the parameter	Uncertainty distribution of the parameter
P(I = 2)	$1.0 \cdot 10^{-5}$	Lognor($6.13 \cdot 10^{-6}$,5.0)[1)]
P(P = 1)	0.5	Uniform(0,1)
P(Tr)	0.7, 0.2, 0.1	Dirichlet(0.7, 0.2, 0.1)
P(In=1 | I=1,Tr=1)	1.0	Deterministic
P(In=1 | I=1,Tr=2)	0.9999	Beta(9.999,0.001)
P(In=1 | I=1,Tr=3)	0.999	Beta(9.99,0.01)
P(In=1 | I=2,Tr=1)	$1.0 \cdot 10^{-4}$	Beta(0.001,9.999)
P(In=1 | I=2,Tr=2)	$1.0 \cdot 10^{-5}$	Beta(0.0001,9.9999)
P(In=1 | I=2,Tr=3)	$1.0 \cdot 10^{-6}$	Beta(0.00001,9.99999)
P(S=1 | Tr=1,P=1,In=1)	0.9999	Beta(9.999,0.001)
P(S=1 | Tr=1,P=1,In=2)	0.0	Deterministic
P(S=1 | Tr=1,P=2,In=1)	0.9999	Beta(9.999,0.001)
P(S=1 | Tr=1,P=2,In=2)	0.0	Deterministic
P(S=1 | Tr=2,P=1,In=1)	0.995	Beta(9.95,0.05)
P(S=1 | Tr=2,P=1,In=2)	0.0	Deterministic
P(S=1 | Tr=2,P=2,In=1)	0.995	Beta(9.95,0.05)
P(S=1 | Tr=2,P=2,In=2)	0.0	Deterministic
P(S=1 | Tr=3,P=1,In=1)	0.95	Beta(9.5,0.5)
P(S=1 | Tr=3,P=1,In=2)	0.0	Deterministic
P(S=1 | Tr=3,P=2,In=1)	0.85	Beta(8.5,1.5)
P(S=1 | Tr=3,P=2,In=2)	0.0	Deterministic
P(A=1 | Tr=1,P=1,S=1)	0.9999	Beta(9.999,0.001)
P(A=1 | Tr=1,P=1,S=2)	0.0	Deterministic
P(A=1 | Tr=1,P=2,S=1)	0.99995	Beta(9.9995,0.0005)
P(A=1 | Tr=1,P=2,S=2)	0.0	Deterministic
P(A=1 | Tr=2,P=1,S=1)	0.995	Beta(9.95,0.05)
P(A=1 | Tr=2,P=1,S=2)	0.0	Deterministic
P(A=1 | Tr=2,P=2,S=1)	0.995	Beta(9.95,0.05)
P(A=1 | Tr=2,P=2,S=2)	0.0	Deterministic
P(A=1 | Tr=3,P=1,S=1)	0.90	Beta(9.0,1.0)
P(A=1 | Tr=3,P=1,S=2)	0.0	Deterministic
P(A=1 | Tr=3,P=2,S=1)	0.5	Beta(5.0,95,0.0005)
P(A=1 | Tr=3,P=2,S=2)	0.0	Deterministic

Information and Communication Technology in the British Coal Industry – Systems Management and the Organisation and Reliability of Production and Maintenance Work

Dr Stephen Heycock
School of Industrial Technology
University of Bradford
BRADFORD
Britain

SYNOPSIS

In the last ten years the British Coal Industry has introduced information and communication technology into the process of production in order to increase the reliability of its data collection, and to reorganise work and make it more flexible. In part this is to increase the machine operating time within machine available time by a speedier and more accurate identification of machine performance and likely breakdowns. This for example allows the reorganisation of both production and maintenance workers, and in particular the assessment and calculability of their availability in number, place, and time.

The focus of my paper is the interface between this new technology, the system of management and the organisation of work in coalmining.

The technology by which information concerning the production process and the reliability of machinery is gathered is itself a major factor in both how the data is collected and how the work of coalminers is organised. It is argued that it is not just the technology of production which determines human and machine reliability but how it is organised as a whole process of production and in what way information and communication technology is applied to this process.

My paper therefore analyses those interfaces between production technology, information and communication technology, the particular structure and system of management which has been developed and the organisation of work. It focuses upon those changes in the British Coal Industry since the 1970's and points forward to possible effects upon management, human reliability and the economic structure of the industry in the 1990's.

Introduction

The redesign of working time is not an effect of changes in the technology of production but a precondition for it. Technology is socially shaped and a major determinant of that is the division of labour required by capital. Nevertheless that division of labour may be prevented until a technology is designed which can accommodate it. However its conception may precede the technological advance necessary and therefore will already be part of its design, born of a managerial strategy and later applied to a process of production.

Coalmining is an example of an industry in which, after the beginning of large scale deep mining in the 1870's, the rate of technological change was very slow for a hundred years. However the rapidity of change after the 1970's has been dramatic in comparison. There are two questions here - why has the rate of technical change altered significantly, and what are its effects upon the organisation of work.

One effect, the re-design of working time, has only just begun, but it is potentially rapid and far reaching. However, working time was obviously designed before it was re-designed. But how and why - and how and why will it be re-designed?

It is necessary first to see how working time was designed in the coal industry before the introduction of new technology. We must know what working time was before we can understand how it can be made more flexible.

I intend to look at the history of working time in the coal industry and to attempt to answer the question, why it is being re-designed **now.**

The History of Coal Mining Technology and Working Time

Historically, methods of work and methods of payment under piecework ensured that the collier internalised control over his own labour power and designed his own working time. The extent to which he was able to do this, has however, been over-estimated in studies like that of Trist et al [1].

Although never regarded as a skilled worker in the traditional sense (Harrison [2]), he did up until the advent of early forms of mechanisation in the late 19th Century maintain some unity of mental and manual knowledge and skill. The division of labour did not reach that extreme point which produced a significant loss of control over the process of production in skilled occupations, as suggested by Bravermann [3] for example . However this point may now be within reach with the development of the microprocessor within automated mining techniques, [Fieckert [4].

Traditionally the collier was regarded as having a great deal of control over his own working time, and although it can be argued that this was in fact far more restricted than assumed [S Heycock [5]], writers as diverse as Marglin [6] and Trist assume that up until the mechanisation period of the 1970's, the British coal industry was one in which the division of labour never really took hold, and where leadership and supervision were internal to the work group which displayed the so called quality of "responsible autonomy", a term first used by Trist in their study of the coalminer and later taken up and used more generally by Friedman [7].

However coalminers never were able to design their own working time to the extent that has been generally accepted, and therefore the modern re-design of their working time is simply made seemingly more dramatic by new technology, but it is in fact only part of a much longer and slower process, speeded up and now made more visible and dramatic. [Heycock [8]]

It is necessary to look at how working time was controlled in pre-mechanised and then partially mechanised coalmining, before looking at the radical re-design of working time in the present period of full or automated mechanisation.

Pre-mechanised or Handgot Coalmining.

In **handgot shortwall mining** all the skills and knowledge necessary for the production of coal were contained within small work groups of usually 2-4 colliers and trammers. The division of labour therefore was minimal.

However a theoretical and practical knowledge of coalmining was also held by management. By law they had to hold certificates in mining theory and practice which included those techniques and methods of work adopted by skilled colliers.

The managers ability to conceptualise the miners work was augmented by the fact that this was only part of a wider knowledge of the whole mine, and was therefore a more holistic understanding than that of the individual miner. This was a "systematic" knowledge which preceded the systems theory to be adopted by the National Coal Board under full mechanisation and new technology in the 1980's.

The skill and knowledge of miners was delimited by the fact that not only did they work in separate workplaces but also in different seams, each with their own peculiarities of coal conditions.

Generally therefore, during this technological phase of production, management were not dependent upon the manual skill and conceptual knowledge of the miner. Skill does not always create responsible autonomy and technological change does not always destroy it.

In coalmining the general problem for management at all levels of technology has been how to control the work of miners hundreds of metres underground and perhaps many miles from the pit bottom. In handgot shortwall mining they were working in separate workplaces. In the next stage of mining technology - **handgot longwall mining** - they were still working in separate work **groups,** but less spatially separate along a whole coalface. Longwall faces were open and miners became more visible to mangement.

An important indicator of how the organisation of work had been changed was the introduction of shiftwork. This was now possible in longwall faces whose working conditions were more uniform and therefore more than one set of colliers could work in the same place.

At this point miners began to ask for a shiftwork allowance on the price paid for the coal they produced to compensate them for their worktime being restricted by the double shift system.

An allowance was asked for because miners were not able to design their own working time. They were no longer able to work as long as the conditions of work directed they must to earn a "living wage". Later, Acts like the 1908 Eight Hours Act, further prevented the miner designing his worktime and resulted in him asking for allowances and for a Minimum Wage in recompense.

Allowances therefore are an important indication of lack of control over working time.

Colliers could not, as Roy [9] noted pieceworkers in other industries could, "fiddle" their work to bring a notion of effort and wages back into line, when conditions of work caused fluctuations in them. This was because colliers could not hold back stored up work or "time" as coal. Coal had to be removed from the workplace to allow work to continue.

Contrary to what has been generally accepted up to now miners in handgot mining had very little flexibility in their work time, because owners took control over and asserted their ownership of the product at the pit head and were able to reach this control back down the pit to remotely control the miners working time in their workplace.

Allowances were an attempt by the miner to counteract this inflexible control of their working time by making payment more flexible.

Partially mechanised or machine cut longwall mining

The process of production in machine cut longwall mining became a more social process with 3 shifts, each with a different function [preparation, under cutting the coal by machine, hand-filling and hauling, later done by mechanical conveyor] and employing miners with different jobs and skills on each shift, though each dependent on the work done in the previous shift.

The whole process was sequential and cyclical. However while the need for cycle completion by the miners might limit managerial control it did not necessarily increase the control of miners over their part of the cycle by an equal amount. The whole process of production became even more opened up to the direct observation and greater potential control over working time by management. Where separate groups of miners performed simpler more repetitive tasks on separate shifts, the calculability of their different labour costs was facilitated.

The real importance of the change from handgot to machine cut mining was not the technological change represented by mechanisation but the change in the organisation of work associated with it.

The whole point of machine cut mining in terms of the organisation of work was that it was a more continuous process of production which allowed less flexibility of worktime to the discretion of the miner. Management attempted to maintain flexible methods of payment whilst at the same time decreasing the flexibility of work and worktimes. Management still wanted to pay them individually for clearing a "stint" or yardage of face. The fillers wanted to be paid as a team on an "all throw in" basis. This allowed them to utilise their work time more flexibly, and this notion of their ability to control the flexibility of their work was opposed by the attempt to reduce the flexibility of worktime by increasing the division of labour with increasing mechanisation. Paradoxically an increasing inflexibility occured independent of any change in the technology of mining, and as a result of the miner's struggle in the 19th Century for the 8 Hour Day and the Minimum Wage, gained in 1908 and 1911. Mineowners could only then extract higher levels of surplus value by intensifying the rate of exploitation of labour power. In large part they attempted this by increasing their control over the flexibility of miners work by reorganising and redesigning worktime through the introduction of mechanisation.

There were of course even wider influences upon decisions taken by both miners and management as to the flexibility of working time. For example absenteeism, irregular attendance and short time working, have always been a major feature of coalmining. They can also be seen as both an attempt to control their own flexibility of working time by workers and as a response through withdrawal to increased inflexibility of working time.

Coalmining is also almost unique in that it is impossible for the miner to leave his place of work underground without the permission of management. This is a form of control over the flexibility of worktime, at the workplace, which is almost absolute.

New Technology and Flexible Working

Machine cut longwall mining lasted from the end of the 19th Century right up until the 1960's. The major technical advance during the 1960's and 1970's was to introduce power loaded coal cutting machines which did not undercut the coal but sheared it off the coal face in one action by scraping a toothed drum along it. The coal ripped off the face was then automatically and continuously loaded onto a movable conveyor as the coal cutter moved up and down the face. The obvious change to the organisation of work was that a cycle of 3 shifts was no longer necessary or appropriate. There was now a single coal production shift and a repair, maintenance, and preparation shift - usually at night.

Dramatic though this was technologically, it was in a sense only an intermediate stage in the development of the organisation of work in the coal industry. Its most dramatic effect was the introduction of the National Power Loading Agreement [NPLA]. In 1966 this brought significant and far reaching changes in the flexibility of work, worktime, and methods of payment.

Secondly, the NPLA brought in hourly or daywage payment for the first time ever in coalmining history for the key faceworkers. This was something the miner's union had been struggling for, for generations. However it was initiated now by the National Coal Board not the National Union of Mineworkers because of the difficulties of paying faceworkers by the ton when miners were now manning machines which produced so much coal. The problem for the NCB however was how to motivate the faceworkers to run this expensive machinery as long and efficiently as possible. Their first solution was to impose a more direct discipline upon faceworkers by changing the role of Deputies. Previously they were mainly responsible for safety and workplace negotiations over piece rates and allowances. Now they were to act more as foremen, or first line supervisors. They became the agents of management discipline, particularly charged with keeping the powerloading machines and conveyors running.

A number of significant changes with regard to the flexibility of work had already occurred in the 1950's prior to the NPLA of the 1960's and 1970's and well before the introduction of new technology to coal production machinery and methods in the 1980's. First, the flexibility of methods of payment of hourly paid workers was greatly reduced. That is, daywage workers had the flexibility of their payment by time greatly reduced by the simplification of the grade structure, which at the same time increased the flexibility of their deployment by management. Secondly the NPLA reduced the flexibility of payment to pieceworkers by removing piecework and paying them by time also.

A greater inflexibility in miner's earnings and work was therefore introduced. However in the late 1970's the advent of new technology was to embark the NCB upon a search for greater flexibility of payments, deployment of labour, and working time. Basically the problem and the constant factor throughout all these preliminary attempts to increase flexibility was that actual hours of work in coalmining were limited by Act of Parliament to 7½. Powerloading technology held out a promise of increased production which was frustrated by an inability to radically reorganise the work of miners. The new technology of the microprocessor held out the promise of a solution through greater flexibility of labour.

Basically no change took place in the technology of coal cutting machinery after the introduction of powerloaders in the 1960's. What changed was the organisation of labour surrounding them. What made this change possible was the successful implantation of the microprocessor into the already existing machinery which in turn allowed a development of managerial control. It must not be forgotten that changes in technology are not simply bound up with changes in the organisation of work but with changes in managerial strategy.

The major problem for management in the 1960's and 1970's was how to control the running time of machinery. Typically it was only running for $^1/_3$ of its potential worktime according to management. One third was due to so called "lost time" or delays caused by miners, and $^1/_3$ due to delays caused by maintenance and repair called by management "operational and ancillary time". [Burns et al [10]] The Mine Operation System [MINOS] developed by the NCB was an attempt to design around the microprocessor new systems of controlling the working time of miners.

One part of this system, FIDO [Face Information Digested on Line] is designed to redesign and restore "lost time" by gathering information about coal cutting machinery as it operates from inbuilt transducers, feeding the information into a control room where the causes and duration of such delays are noted and acted upon.

The identification of such delays is linked directly to pay and the Area Incentive Scheme [AIS] under which delays of less than 20 minutes result in a loss of bonus, and for unavoidable delays of more than 20 minutes bonus is paid at an average rate.

The flexibility of workingtime for miners is thereby considerably reduced with their reduced ability to control their own pace of work. It is also associated with an intensification of work and a potential reduction of jobs, due to increased productivity if there is a static demand for coal.

Another microprocessor controlled system, IMPACT, [In Built Machine Performance and Condition Testing] monitors diagnoses, and even carries out standard inspections of machines through an implanted transducer. Principally this is designed to reduce "operational and ancillary time", but it also redesigns the working time of skilled fitters and electricians [Burns et al]. Because of its ability to forecast and diagnose breakdowns it removes the need for fitters and electricians to be on standby and therefore a smaller number of them can be directed centrally from a control room to deal with already identified breakdowns.

Thus the flexibility of this skilled part of the labour force is greatly increased for management - though decreased for the miners - as their work is intensified, and their numbers decreased. These skilled miners were an expanding group of workers in the coal industry after the introduction of power loaded coal mining, but are now a declining occupational group.

All the information so far described is fed into the Management Information System [MIS] in which there are programmes designed to allow managers to analyse and collect data on productivity, absenteeism, machine performance by shift etc. which leads some to conclude that :-

> "..... it does not allow the workforce to use the information to make any decisions about more flexible patterns of work organisation" [Burns et al [11] p.104].

The National Union of Mineworkers Industrial Relations Department estimated that 52,000 jobs would be lost by 1987/8 through the introduction of such new technology and that estimate has been substantially correct. [NUM [12]]

Throughout the 1980's and into the 1990's they believe that computer technology will result in the reduction of jobs at high technology collieries, together with the continued deskilling of the labour force, and the closure of older collieries with lower technology.

Flexibility must be seen as only part of a much wider managerial strategy in the coal industry. The NUM sees this strategy as also comprising the new commercial orientations of the NCB possibly with a view to privatisation, the problems of motivating the workforce, and the emergence of the Union of Democratic Mineworkers. However they see flexibility of work as a major component of the NCB's plans for the 1980's and 1990's. The key elements of this plan, they see as :-

1. Coal production over six days a week.
2. 4 x 9 hour shifts per day underground on a continental shift basis.
3. Winding shifts to work for 23 hours per day at full capacity.
4. Fitters and electricians to be merged into one multi-skilled electro-mechanic grade, with some maintenance tasks transferred to machine operators.
5. The revision of the Mines Health and Safety Legislation to permit the use of new working practices and reduce the influence of the Deputy, as a health and safety monitor.
6. The increased use of roof bolts, trackless vehicles and other highly flexible production techniques used in America.
7. The increased use of outside contractors.

The NUM calculates that if the NCB's plan in total was successful then current output could be maintained with 70,000 miners instead of the present 100,000. [NUM [13]]

If just a 6 day week of 4 day 9 hour shifts was worked this in itself they believe would lose thousands of jobs. They further claim that in the 1980's and 1990's computer technology in coalmining will result in the reduction of jobs, to deskilling in high technology pits, and to the closure of other pits. The goals of the NCB are increased productivity of labour and capital, and increased control over the industry through MINOS which

> "...has a much greater flexibility because it is defined in terms of sub-systems which can be implemented in a modular basis or on their own without requiring the whole of the system for their functioning". [NUM [14]p.12].

The NUM have been very concerned with flexibility of work and see present trends linked to new technology as part of a longer history of change in the labour process. They identify four major types of flexibility.

Functional flexibility they define as both horizontal, (for example the integration of different skilled workers) and vertical, (for example between production workers.) Their conclusion is that -

> "....genuine functional flexibility has long been a part of the process of deep coal mining. The Board's new strategy would appear to be, therefore, designed to intensify the work process and prepare again for massive job cuts". [NUM [14]p.12].

Numerical flexibility they define as the ability to alter the size of the workforce by the use of sub-contracted labour and temporary or part-time workers. The latter is only really possible in clerical work, but sub-contracting of skilled ancillary work like that of electricians and fitters they claim is on the increase.

Temporal flexibility they define as flexible shift patterns, annual hours, length of the working week and structure of the working day. They see the strategy of the NCB as itself "flexible" in their overall aim to increase pit working time. They could either increase "within shift efficiency" by increasing machine operating time within machine available time, or they could change the current statutory limit on underground hours and work longer shifts on a 4 day week or 9 day fortnight basis.

The NUM claim that the NCB is :-

> ".... most desperate to change working practices in the most capital intensive pits, but because those pits are generally the most profitable, they have less leverage over the workforce. Consequently, they have targetted vulnerable pits in the North East, threatening no more investment and closure if longer shifts are not agreed and have placed similar pressure in South Wales at Margam [NUM [14]p.13].

According to the NUM, as far back as 1981 the NCB conducted a survey of shiftwork amongst its mine managers and in over 80% of collieries, managers thought that underground shifts should be between 8 and 9 hours to achieve maximum machine utilisation.

In 1985 the strategy was outlined as taking high cost (low technology) pits out of production and working low cost (high technology) pits at higher rates of intensity per hour for more hours per day, and more days in the year. However, the NUM do not believe that flexible working will be restricted to high technology, low cost pits (the new "super pits"), but will spread from them to all "surviving" collieries. The word "survival" is used because of the effect they think even only one aspect of flexible working would have, i.e. a 6 day working week. The NUM calculate that this alone would be able to maintain the present deep mined output of approximately 90 m. tonnes with 31 fewer collieries, i.e. 31 collieries could close, with an actual employment of 27,000 miners. However as more miners would be needed for a 6 day week in the remaining collieries the potential loss of jobs would be less than this figure.

However, if what they term the "full strategy" of the NCB was implemented, as already outlined earlier, then the present output of 90 m. tonnes could be produced by 55 of the low cost, high technology collieries. This would result in the closure of 55 of the present 110 collieries, with a total job loss taking into account additional miners needed at the high technology collieries of 40,000. That is the number of collieries would be halved, and the labour force reduced from 100,000 to 70,000 miners.

The question of flexible working in the coal industry has been linked closely to the new "super pits" like Selby in North Yorkshire and the proposed Margam pit in South Wales in particular, because the NCB laid it down as a condition for going ahead with it that the union accept 6 day working. The national NUM leadership are opposed to 6 day working on principle, and their policy is to reduce shift times and the hours worked per week.

Paradoxically present coalmining technology as it has been developed is relatively "inflexible". It requires good geological conditions of straight, thick seams for high productivity, and less miners and higher tonnages over longer periods of coal production. Thus this technology derived from a particular managerial strategy requires "flexible" working as defined by the NCB. This conflicts with a long historical struggle on the part of the miners to achieve a shorter working week, now made even more relevant to them since the decline of the coal industry in 1957, and the present general recession and increasing unemployment since the late 1970's and 1980's.

The NUM has continued its policy of attempting to gain a shorter working week. The NUM concludes that the major reason the NCB wants a 6 day week is to reduce the number of pits and the number of miners employed in the industry.

Central to this strategy are flexible working patterns which stem from a managerial strategy influenced by an increasingly uncertain commercial market for coal as Government policy opens it up to foreign and domestic competition from coal and other fuel sources.

REFERENCES

1. Trist E.L. et al. "Organisational Choice", London, Tavistock 1963.

2. Harrison R. (ed) "Independent Collier : The Coal Miner as Archetypal Proletarian Reconsidered" Sussex, The Harvester Press 1978.

3. Braverman H. "Labour and Monopoly Capital", New York, Monthly Review Press, 1974.

4. Fieckert, D. "Of men and MINOS", Computing Europe 22 November 1979.

5. Heycock S. "Effects of Collective Bargaining Structures on The Nature of Industrial Conflict", PhD Thesis (unpublished) School of Economic Studies, University of Leeds, May 1986.

6. Marglin S. "What do Bosses Do? The Origins and Functions of Hierarchy in Capital Production" in Gorz A (ed) "The Division of Labour : The Labour Process and Class Struggle in Modern Capitalism", Sussex, Harvester Press, 1976.

7. Friedman, A.L. "Industry and Labour. Class Struggle at Work and Monopoly Capitalism". London, Macmillan, 1977.

8. Heycock S. "The Effect of Changes in Technology and Techniques of Production upon the Organisation of Labour and Work in Coalmining", British Sociological Association Annual Conference, "Science, Technology and Society", University of Leeds, April 1987.

9. Roy D. "Quota Restriction and Goldbricking in a Machine Shop", American Journal of Sociology, Vol. 57 1952.

10. Burns et al. "The Restructuring of the British Coal Industry" Cambridge Journal of Economics, Vol. 9, 93-110, 1985.

11. Burns et al. "The Miners and New Technology" Industrial Relations Journal, Vol. 14, No.4, 7-20, 1983.

12. NUM Industrial Relations Research Department "Current Industrial Reorganisation and New Technology in the British Coalfields", 1987.

13. NUM Industrial Relations Research Department "New proposed Working Practices in the British Deep Coal Mining Industry", June 1987.

14. NUM Industrial Relations Research Department "Flexibility in the British Deep Coal Mining Industry" 1987.

ESFAS: An Information System on Worldwide Nuclear Power Stations

M. Melis, B. Lisanti

Edra
Milano, Italy

1. Introduction

The idea which was at the basis of the development of the data base "ESFAS-Engineered Safety Features and Auxialiary Systems" was to create a centralized information system in which nuclear power stations should be described from the standpoint of their safety features.
Existing "pedigree" data-bases in the nuclear field are mainly addressed to nuclear steam system and plant target data (e.g., IAEA file), or to commercial data (on NSSS, suppliers of main components, operation dates - e.g. ELECNUC/CEA).
No organized collection of information existed on nuclear power plants safety strategies and philosophies, with adopted solutions from the standpoint of plant layout.
This is presently the aim of ESFAS, which is not an alternative but a complement to the abovesaid data bases.
Such a data-base is mainly addressed to all those involved in nuclear power plants safety and availability, in the phases of design, licensing, construction and operation of nuclear installations; ESFAS potential users can thus be identified in Licensing Authorithies, Utilities, Research Centers, Decision Makers in the field of nuclear safety.
It is important to notice that ESFAS is a "dynamic" system, in continuous evolution: the number of nuclear power plants included will be continuously increased, as well as the software-tools supporting data analysis and data base inquiry.

2. General Description and Applications of Esfas

ESFAS is an organized collection of information on nuclear power plants engineered safety features and auxiliary systems. It aims at providing an overall description of existing nuclear power stations, mainly focused on engineering characteristics of safety related systems.

Available information are stored both in descriptive and coded formats.

Present version of data base includes Pressurized Water Reactors (PWRs) and Boiling Water Reactors (BWRs).

Plant-systems on which information are collected are presented in table 1.I.

All the main safety-related systems are included, together with relevant support systems/auxiliaries serving the nuclear island and engineered safety features.

Each plant is identified by "address" information (IAEA-code;plant-type; NSSS; criticality date; loops number; MWth, MWel; number of PORVs and safety valves; containment type) (fig. 1.1).

A general description of plant is also presented, indicating main design characteristics, safety topics and plant layout, main remarks arising from safety studies performed on plant. (fig. 1.2).

Furthermore, for each of the systems indicated in table 1.I the main engineering characteristics are presented (pumps number; system layout description; design flow and head; suction sources and support systems). (fig. 1.3).

The resulting plant-description is rather exhaustive, and allows an overall set of analyses on the real design characteristics of nuclear power stations.

For what concerns the possible applications of ESFAS , the following items can be focused:

- ESFAS allows a comparison of safety philosophies in various countries and of adopted layout of safety systems.
- ESFAS allows a general insight on real protection levels of plants.
- ESFAS allows an analysis of evolution/development of PWR and BWR plant types for different NSSS.

- ESFAS allows a comparison of plant design, for different NSSS in the same country or in different countries.
- ESFAS allows focusing of links between safety system and identification of multiphase systems.
- ESFAS is a source of engineering characteristics data for plant-components in safety and auxiliary systems.
- ESFAS allows the access/creation of subsets of data, subdivided according to the desired inquiry criteria, on which to perform elaborations (e.g., NSSS, criticality date, containment type, etc.).
- ESFAS draws attention of the user on particular systems configurations.

Furthermore, linking of ESFAS with data-bases containing records on events affecting plant safety and availability allows a wide spectrum of analyses (e.g., effect of failure events on overall systems function, on the basis of the various systems-configurations adopted in different plants); trend and pattern analyses based on homogeneous samples of data; etc.).

- PWR

  Containment System

  Auxiliary Feedwater System (AFWS)

  High Pressure Safety Injection System (HPSI)

  Low Pressure Safety Injection System (LPSI)

  Chemical and Volume Control System (CVCS)

  Residual Heat Removal System (RHR)

  Containment Spray System (CSS)

- BWR

  Containment System

  Isolation Condenser (IC) or Reactor Core Isolation Cooling (RCIC) System

  High Pressure Core Spray, High Pressure Safety Injection (HPSI)

  Feedwater Coolant Injection (FWCI)

  Automatic Depressurization System (ADS)

  Low Pressure Safety Injection (LPSI)

  Low Pressure Core Spray (LPCS)

  Reactor Shutdown Cooling (Residual Heat Removal) (RHR)

  Reactor Vessel Head Spray Cooling (RVHSC)

  Suppression Pool Cooling (SPC)

  Containment Spray System (CSS)

  Standby Liquid Control System (SBLCS)

- Support Systems

  Primary Component Cooling Water System (PCCWS)

  Essential Service Water System (ESWS)

  Emergency Power Supply (EPS)

  Reactor Building Component Cooling Water System (RBCCWS)

Table 1.I: Systems Considered in ESFAS

GENERAL DATA

IAEA CODE	US 417
PLANT	GRAND GULF 2
TYPE	BWR
NSSS	
CODE	GENE
DESCRIPTION	GENERAL ELECTRIC
CRITICALITY YR	
UNIT NO.	1
LOOPS NO.	
MW-TH	3833
MW-EL	1225
CONTAINMENT TYPE	MK3
SAFETY VALVES NO.	0
RELIEF VALVES NO.	22

Fig. 1.1 : Plant General Data

GENERAL INFORMATION

PLANT IS A BWR-6 TYPE, WITH 24 JET PUMPS INSTALLED;
IT HAS REACTOR CORE ISOLATION COOLING - RCIC AND HIGH PRESSURE COOLANT SPRAY-HPCS INSTALLED;
RHR PROVIDES FOR LOW PRESSURE SAFETY INJECTION-LPSI, SHUTDOWN COOLING AND SUPPRESSION POOL COOLING;
STEAM CONDENSING MODE OF RHR/RCIC IS A MODE OF OPERATION AT THIS PLANT;
HPCS (MOTOR DRIVEN PUMP) REPLACES HPSI (TURBINE DRIVEN PUMP) , AND HAS A DEDICATED DIESEL GENERATOR
PLANT DHRS-RELATED CORE MELTS ARE SLIGHTLY HIGHER THAN SUGGESTED CRITERION; ANYWAY, BLACKOUT-RELATED CORE MELTS ARE REDUCED

Fig. 1.2 : Plant General Description

LPSI

PUMPS NO 3 LAYOUT & TYPE A
3 MOTOR DRIVEN PUMPS
CAPACITY

DESCRIPTION	FLOW(KG/S)	HEAD(MPA)	%
3 MOTOR DRIVEN PUMPS	0470	00.1	100

SUCTION
POOL SUPPRESSION POOL
SUPPORT SYSTEMS
ESWS ESSENTIAL SERVICE WATER SYSTEM
REMARKS
DUAL FUNCTION RHR;
3 REDUNDANT 100% LOOPS;
ESWS COOLS HEAT EXCHANGERS AND PUMP COOLERS

Fig. 1.3 : System Engineering Data

3. Present Status of Esfas Data Base

Present version of data-base includes units from USA, Sweden, France, Italy, Belgium, The Netherlands.

201 nuclear power plants (PWR and BWR type) are included.

ESFAS is anyway a "dynamic" system; population of covered plants will increase to cover all existing nuclear power units.

4. Hardware/Software

Data base management system is "ADABAS"; data base enquiry language is NATURAL.

PC version of data base are under development. An overall set of menu-driver enquiry possibilities are associated to the data base.

Graphic displays and computerized maps for the geographical localization of plants are also being developed.

Availability Assessment of a Technological Railway Information System

S. HANŽEL

Institute of Transportation Ljubljana

A. HUDOKLIN

University of Maribor,
Faculty of Organizational Sciences Kranj

Summary

In the paper an assesment of availability and reliability of the computer supported railway technological information system in Slovenia is described. Two versions of a Markov model comprising hardware failures and software and human errors have been used to calculate the system steady state availability and mean time to failure. The forms for the collection of field data on failures and repairs of elements of the central and communication part of the system are presented.

1. Introduction

The Railway Enterprise of Slovenia has 1200 km of track lenght, with 150 stations capable to manipulate freight. About 1000 trains operate daily in this system. An average yearly performed work is about 4 millons ntkm. The freight and passenger traffic is supported by about 15000 employees.

Since the technology process within the railway becomes more and more complex, it can not be adequately managed in the classical "manual" way any more. Consequently, a computer supported technological information system was introduced.

Technology includes the activities which enable the transport of goods and passengers in the railway system. Therefore, it represents the principle part of the whole system. Accordingly, a railway technological information system provides the basic information for decision taking on strategic, tactical and operational management level. In order to assure an effective operation of this system an appropriate level of its reliability and availability must be achieved and continuously monitored.

In the following paper a computer supported technological information system of Railway Enterprise Ljubljana is presented. Different failures of the system are defined and the forms for failure and repair data collection are described. These data are used to calculate reliability and availability parameters of the information system in the framework of two versions of a Markov model, first for the whole system, and the second for the central part of the system.

2. Description of the system

Railway technological information system provides data entry, control and management of waggons, locomotives and trains. It gives at any time actual information on vehicles, trains and consignment inventories, as well as status within the covered area. On the basis of these data different reports and statistics for different levels of management and decision taking are produced. This system is shown schematically in Fig. 1.

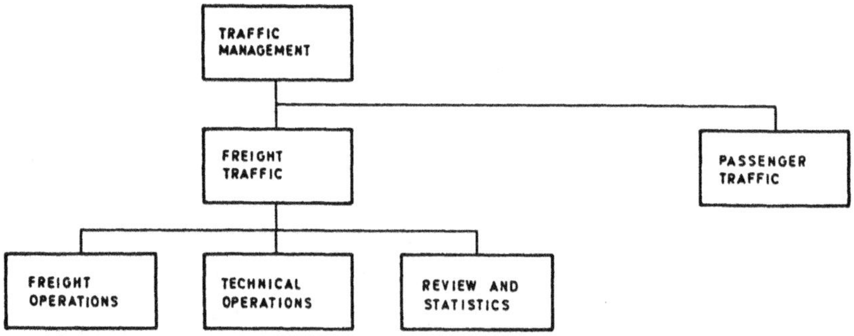

Figure 1: Schematical presentation of a railway technological information system

A continuous operation in real time of the technological information system of Railway Enterprise Ljubljana is enabled by introducing a Sperry 1100/73H1 computer [1]. A simplified functional scheme of this system is shown in Fig. 2.

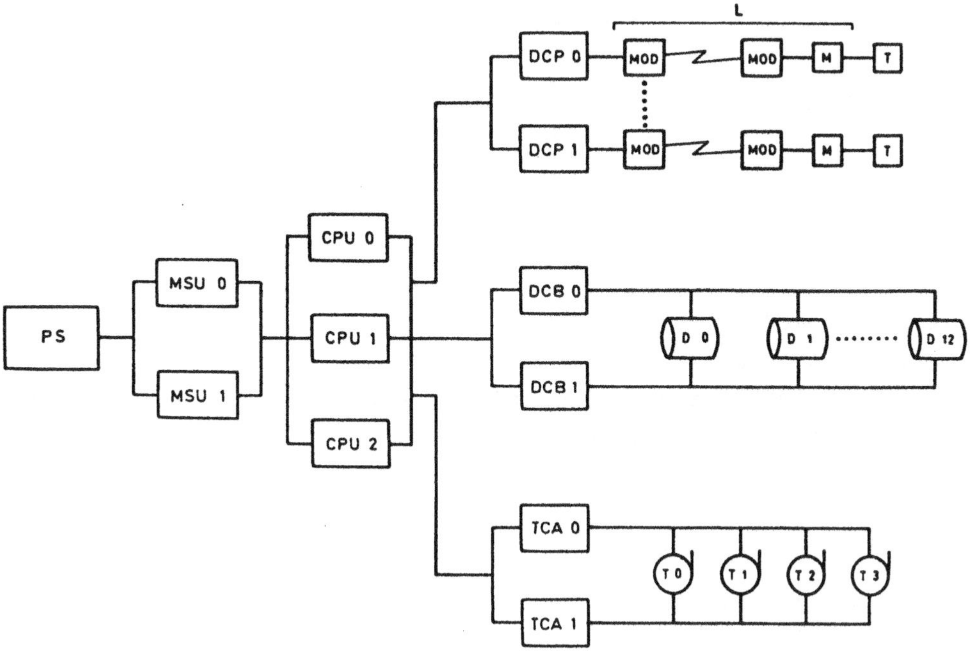

Figure 2: Functional scheme of the computer supported technological information system of Railway Enterprise Ljubljana

Legend:

PS - Power supply unit
MS - Main storage units
CPU - Processor units including two I/O units
DCP - Distributed communications processors
MOD - Modems
M - Multiplexers
L - Communication Lines (27)
T - Terminals (150)
DCB - Disc control units
D - Disc storage units
TCA - Tape control units
T - Tape storage units

3. Definition and description of system failures

The types of system malfunctions observed are: catastrophic and degradation failures. Catastrophic failures are:

- breakdown of operation
- incorrect output
- loss of substantial amount of data

Degradation failures are:

- slow response time
- loss of small amount of data from a voluminous output
- partial loss of services

Both types of system failures are caused either by catastrophic hardware failures either by software or human errors. Hardware failures are failures of all functional units shown in Fig. 2 including I/O units. Software errors are devided into system and application software errors. System software errors are errors of:

- Operation System EXEC
- Communication Management System CMS
- Communication Delivery Software Base TELCON
- TIP Utilities Programs TIPUTIL
- System Support Processor Supervisor SUPVR

The only application software error is an error in the Mapper. Human errors represent malfunctions of the operator.

4. Data collection

The forms for the collection of failure and repair data of the central and communication part of the information system have been prepared. These two forms are shown in Fig. 3 and Fig. 4.

Page

CENTRAL SUBSYSTEM

Month Year

Time of failure			Unit	Failure cause			Subsystem up			Time of repair			Service resp. time	Subsystem start	Down time	Comments
Day	Hour	Min		HW	SW	Env.	YES	NO	Degraded	Day	Hour	Min	Min	Min	Min	
1	2		3	4	5	6	7	8	9	10			11	12	13	14

Figure 3: The form for the collection of data on failures and repairs of the central part of the system

Page

COMMUNICATION SUBSYSTEM

Month Year

Time of failure			Location	Unit	Failure cause				Terminal up			Time of repair			Report time	Service resp. time	Down time	Comments
Day	Hour	Min			HW	SW	Line	Env.	YES	NO	Degraded	Day	Hour	Min	Min	Min	Min	
1	2		3	4	5	6	7	8	9	10	11	12			13	14	15	16

Figure 4: The form for the collection of the data on failures and repairs of the communication part of the system

Failures of system elements are detected either automatically by comprehensive circuitry contained in the main storage unit, or by the user. Collected failure and repair data enter a personal computer where a data base is formed.

5. System availability and reliability assessment

Steady state availability and mean time to failure have been chosen as measures of the system availability and reliability. For the calculation of these two characteristics two versions of a Markov model including hardware failures and software and human errors were developed. The following assumptions were made:

(i) Failures are statistically independent.

(ii) Hardware units failures rates and software and human error rates are constant.

(iii) Repair rates of hardware units and of the system are constant.

The steady state availability A of the system with N states was calculated according to the relation

$$A = \sum_{i\text{-up states}} p_i \qquad (1)$$

where p_i are components of the vector $[p_1, p_2, \dots, p_N]$ calculated from the equation

$$[\, p_1, p_2, \dots, p_N \,] = [\, 0, 0, \dots, 0, 1 \,] \begin{bmatrix} a_{11} & a_{12} \cdots & a_{1,N-1} & 1 \\ a_{21} & a_{22} \cdots & a_{2,N-1} & 1 \\ \cdot & \cdot \;\cdots & & \cdot \\ \cdot & \cdot \;\cdots & & \cdot \\ \cdot & \cdot \;\cdots & & \cdot \\ a_{N1} & a_{N2} \cdots & a_{N,N-1} & 1 \end{bmatrix}^{-1} \qquad (2)$$

Elements a_{ij} of the above matrix are the elements of the matrix $[\underline{P} - \underline{I}]$, $\underline{P}$ being the stochastic transitional probability matrix, and $\underline{I}$ identity matrix.

Mean time to failure MTTF for the system with n up states was calculated from the relation

$$MTTF = m_{11} + m_{12} + \dots + m_{1N} \qquad (3)$$

where m_{1j} are the elements of the first row of the matrix $[\underline{I} - \underline{Q}]^{-1}$, $\underline{Q}$ being the truncated probability matrix in which the row and column associated with the absorbing state is deleted from the stochastic transitional probability matrix.

5.1 Markov model - version I

In version I of the model the hardware of the central and communication part of the system was devided into functional units, while different modules of the system software and application software were treated as a single subsystem. Human errors include only those operator errors leading to a catastrophic failure of the system. The system failures were devided into following groups:

- degradation failures due to failures of hardware units
- catastrophic failures due to failures of hardware units
- catastrophic failures due to software errors
- catastrophic failures due to human errors

The state space diagram for the version I of the model is shown in Fig. 5.

The states shown in Figure 5 are defined as follows:

1 - normal operation (hardware and software operate within specifications, response time is less than 3 sec)
2 - failure of one terminal
3 - failure of one distributed communications processor
4 - failure of one processor unit
5 - failures of two terminals connected to different communication lines
6 - failure of one processor unit and one distributed communications processor
7 - failure of one terminal and one distributed communications processor
8 - failure of one terminal and one processor unit
9 - failure of one I/O unit
10 - failure of one tape control unit
11 - failure of one distributed communications processor and two terminals connected to different communication lines
12 - failure of one processor unit and two terminals connected to different communication lines
13 - failure of one processor unit and, afterwards, one I/O unit
14 - failure of one processor unit and, afterwards, one tape control unit
15 - failures of two terminals connected to different communication lines and, afterwards, one I/O unit
16 - failures of two terminals connected to different communication lines and, afterwards, one tape control unit
17 - failure of one terminal and, afterwards, one I/O unit

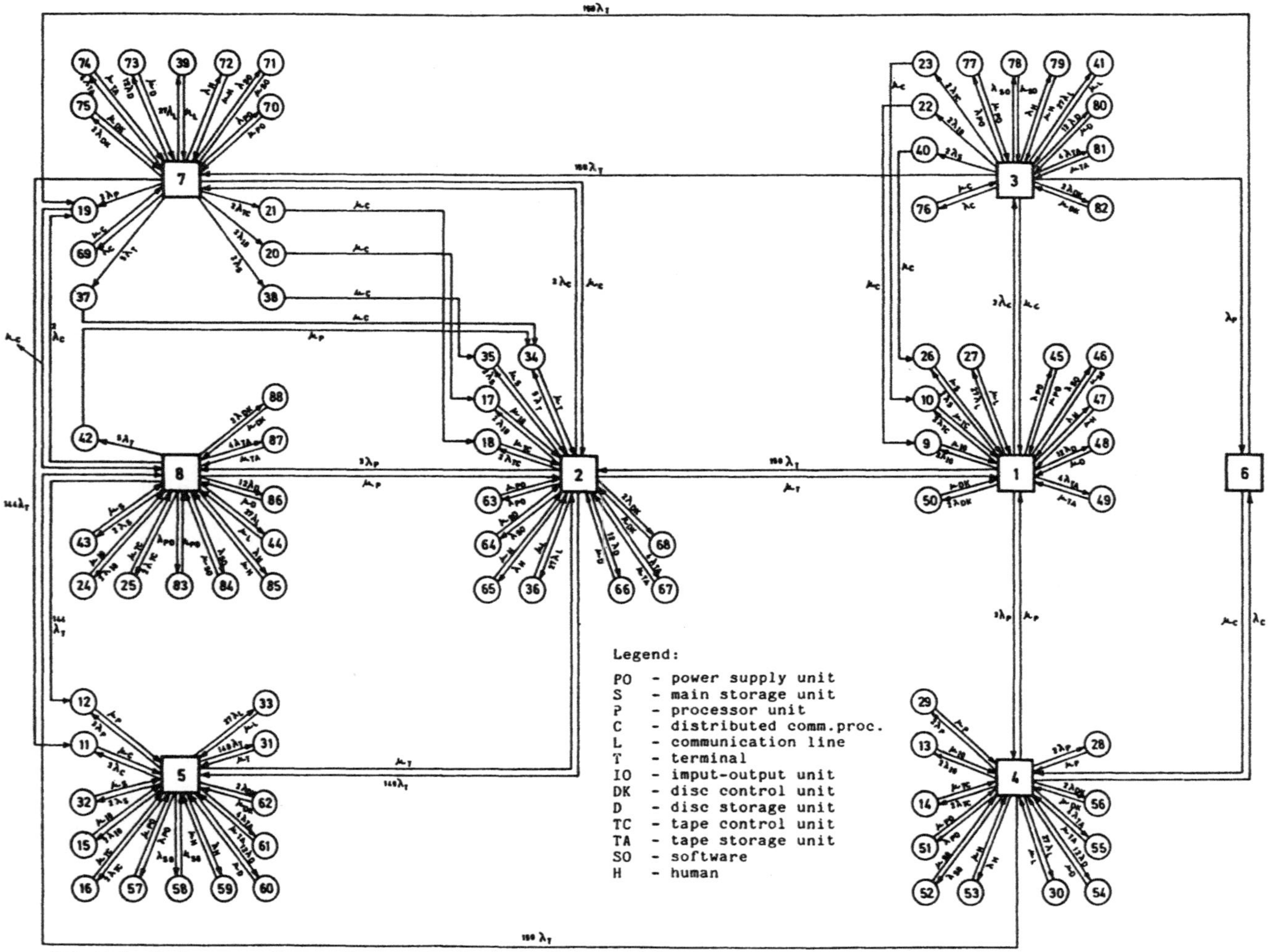

Figure 5: State space diagram for version I of the Markov model

18 - failure of one terminal and, afterwards, one tape control unit

19 - failure of one distributed communications processor, one processor unit and one terminal

20 - failure of one distributed communication processor and one terminal and, afterwards, one I/O unit

21 - failure of one distributed communications processor, one terminal and, afterwards, one tape control unit

22 - failure of one distributed communications processor and, afterwards, one I/O unit

23 - failure of one distributed communications processor and, afterwards, one tape control unit

24 - failure of one processor unit, one terminal and, afterwards, one I/O unit

25 - failure of one processor unit, one terminal and, afterwards, one tape control unit

26 - failure of one main storage unit

27 - failure of one communication line

28 - failures of two processor units

29 - failure of one processor unit and, afterwards, one main storage unit

30 - failure of one processor unit and, afterwards, one communication line

31 - failures of three terminals

32 - failures of two terminals connected to different communication lines and, afterwards, one main storage unit

33 - failures of two terminals connected to different communication lines and, afterwards, one communication line

34 - failures of two terminals connected to the same communication line

35 - failure of one terminal and, afterwards, one main storage unit

36 - failure of one terminal and, afterwards, one communication line

37 - failure of one distributed communications processor, one terminal and, afterwards, second terminal, connected to the same communication line

38 - failure of one distributed communications processor, one terminal and, afterwards, one main storage unit

39 - failure of one distributed communications processor, one terminal and, afterwards, one communication line
40 - failure of one distributed communications processor, and, afterwards, one main storage unit
41 - failure of one distributed communications processor and, afterwards, one communication line
42 - failure of one processor unit, one terminal and, afterwards, second terminal connected to the same communication line
43 - failure of one processor unit, one terminal and, afterwards, one main storage unit
44 - failure of one processor unit, one terminal and, afterwards, one communication line
45 - failure of power supply unit
46 - software error
47 - human error
48 - failure of one disc storage unit
49 - failure of one tape storage unit
50 - failure of one disc control unit
51 - failure of one processor unit and, afterwards, power supply unit
52 - failure of one processor unit and, afterwards, software error
53 - failure of one processor unit and, afterwards, human error
54 - failure of one processor unit and, afterwards, one disc storage unit
55 - failure of one processor unit and, afterwards, one tape storage unit
56 - failure of one processor unit and, afterwards, one disc control unit
57 - failures of two terminals connected to different communication lines and, afterwards, power supply unit
58 - failures of two terminals connected to different communication lines and, afterwards, software error
59 - failures of two terminals connected to different communication lines and, afterwards, human error
60 - failures of two terminals connected to different communication lines and, afterwards, one disc storage unit
61 - failures of two terminals connected to different communication lines and, afterwards, one tape storage unit

62 - failures of two terminals connected to different communication lines and, afterwards, one disc control unit
63 - failure of one terminal and, afterwards, power supply unit
64 - failure of one terminal and, afterwards, software error
65 - failure of one terminal and, afterwards, human error
66 - failure of one terminal and, afterwards, one disc storage unit
67 - failure of one terminal and, afterwards, one tape storage unit
68 - failure of one terminal and, afterwards, one disc control unit
69 - failure of one distributed communications processor, one terminal and, afterwards, second distributed communications processor
70 - failure of one distributed communications processor, one terminal and, afterwards, power supply unit
71 - failure of one distributed communications processor, one terminal and, afterwards, software error
72 - failure of one distribtued communications processor, one terminal and, afterwards, human error
73 - failure of one distributed communications processor, one terminal and, afterwards, one disc storage unit
74 - failure of one distributed communications processor, one terminal and, afterwards, one tape storage unit
75 - failure of one distributed communications processor, one terminal and, afterwards, one disc control unit
76 - failure of one distributed communications processor and afterwards, second distributed communications processor
77 - failure of one distributed communications processor and, afterwards, power supply unit
78 - failure of one distributed communications processor and, afterwards, software error
79 - failure of one distributed communications processor and, afterwards, human error
80 - failure of one distributed communications processor and, afterwards, one disc storage unit
81 - failure of one distributed communications processor and, afterwards, one tape storage unit
82 - failure of one distributed communications procesor and, afterwards, one disc control unit

83 - failure of one processor unit, one terminal and, afterwards, power supply unit
84 - failure of one processor unit, one terminal and, afterwards, software error
85 - failure of one processor unit, one terminal and, afterwards, human error
86 - failure of one processor unit, one terminal and, afterwards, one disc storage unit
87 - failure of one processor unit, one terminal and, afterwards, one tape storage unit
88 - failure of one processor unit, one terminal and, afterwards, one disc control unit.

States 1 - 8 are up states. States 9 - 25, 50, 56, 62, 68, 75, 82 and 88 are in fact up states but were treated as down states due to presumed low occupation probability. States 26 - 44 represent a degraded mode of system operation (e.g. increased response time). States 45 - 49, 51 - 55, 57 - 61, 63 - 67, 69 - 74, 76 - 81, 83 - 87 represent breakdown of system operation.

Failure and repair rates of the system elements were collected in a period of one year. The repair rate was defined as a reciprocal value of the system down time due to the failure of an element. Values of failure rate and repair rate for system elements are given in Table 1.

Table 1: List of failure and repair rates of system elements

System element	$\lambda\,[10^{-3}h^{-1}]$	$\mu\,[h^{-1}]$
main storage unit	0,257	0,40
processor unit	0,305	0,10
I/O unit	0,076	1,00
tape storage unit	0,903	0,03
tape control unit	0,076	1,00
disc storage unit	0,482	0,53
disc control unit	0,038	0,50
distributed communications		

processor	0,684	4,50
communication line	0,559	0,26
terminal	0,200	0,28
power supply unit	0,152	0,50
software	1,110	1,03*
operator	0,017	1,00*

* Repair rate of the system after a failure due to a software or human error.

Using failure and repair rate data from Table 1 and equations (1) and (2), the system steady state availability was calculated to be

$$A = 0,839$$

The system unavailability due to degradation failures was found to be

$$(1 - A)_D = 0,050$$

and the system unavailability due to catastrophic failures was found to be

$$(1 - A)_C = 0,111$$

The relatively low value of the system availability is ascribed to relatively long repair times because of the dislocated service station.

Using (3) mean time to failure was calculated to be

$$MTTF = 37 \text{ h}$$

5.2 Markov model - version II

Version II of the model was adapted for the calculation of availability and reliability characteristics of the central part of the system. It was derived from the version I by omission of the states belonging to the communication part of the system.

Besides, the state representing the system failure due to a software error was branched out to 6 states corresponding to errors of different modules of the system and application software.

The calculation according to equations (1), (2) and (3) using the data of Table 1 gave the following results:

$A = 0{,}882$
$(1 - A)_D = 0{,}001$
$(1 - A)_C = 0{,}117$
$MTTF = 86\ h$

In both versions of our model the calculation of availability and reliability characteristics of the system was performed using a personal computer.

6. Conclusion

An assesment of availability and reliability of a computer supported railway technological information system was performed using a Markov model. The main elements of the information system, hardware, software and human factor were taken into account. Two versions of this model were developed because of the complexity of the system. In the first version, the system hardware was considered in detail while all the software modules were treated as a single unit. A more detailed analysis of the software modules was performed using the second version of the model in which only the central part of the information system is included.

In no version the number of states exceeds 100, therefore our model is very suitable for the calculations by means of a personal computer. The data base needed for these calculations was also built up in a personal computer using appropriate field data.

The system steady state availability and mean time fo failure have been calculated according to both versions of the model. A comparison of both MTTF values indicates a big contribution of the communication subsystem failures to the unreliability of the information system. However, the difference between the two calculated values of the system steady state availability is relatively small. This could be ascribed to the long repair times of hardware units in the central part of the system.

Our approach enables a relatively simple monitoring of the availability and reliability of a computer based information system and its parts. The results could be used for the improvement and development of computer supported information systems.

References

1. Hudoklin A.; Hanžel S.; Rozman V.: Reliability and Safety Parameters of a Railway Traffic System, Report to Research Community of Slovenia, University of Maribor, Faculty of Organizational Sciences Kranj, Kranj 1987, pp. 75.

Assessment of Failure Probabilities of Launch and Orbital Manoeuvres of Satellites

R. Leicht and H.J. Wingender

NUKEM GmbH, P.O. Box 1313

D-8755 Alzenau, FRG

Abstract

The question of failures leading to abortions of space missions is of importance in terms of financial and scientific expenditures. However, if nuclear devices such as power sources of satellites are involved and re-entries are to be envisaged, these failures will become a matter of safety, too. The results found are that abortive mission failures lay in the range of 10% probability. A break down of accessible event data provides a sound basis for these results. The more complex question of safety can only be touched here.

1 Introduction

The use of space flight technology, in particular the deployment of satellites, requires devices and equipment functioning reliably with respect to their launch and flight manoeuvres and their mission performance, since substantial expenditures are on stake in commercial and scientific terms, as well. Thus it is of some importance to know the frequencies of failures resulting in mission abortion and early re-entry of space vehicles. More recently, an other aspect of satellite failures has drawn the attention of the public: the re-entry of military satellites using nuclear power devices and the contamination and radiation exposure in consequence of such events. The following paragraphs deal with the scenario considered, the parameters affecting the lifetime of satellite orbits, the derivation of failure probabilities, and with some conclusions.

2 Scenario

Space missions to the outer regions of the solar system or still further require an electricity supply independent from the radiation of the sun. In addition, earth orbiting satellites for particular military purposes also require such a power supply. Such space vehicles are usually equipped with nuclear power devices - i.e. radionuclide batteries or fission reactors - and converters transforming the heat provided by the nuclear device into electricity.

Radionuclide batteries - more specific radioisotope thermoelectric generators (RTG's) - usually make use of Pu238 because of its heat output (0.56 kW/kg) and half life (87 years). The very source of energy is the radioactive decay of the nuclides. Thus, there is no technical means to stop the decay; the radioactivity is highest on the launch pad and gradually decreasing with time.

Fission reactors make use of the neutron-induced chain reaction at U235. The reaction can deliberately be switched on or off according to power requirements. Before the reactor has reached its mission orbit,it contains only the rather small radioactivity of the fresh fuel. Operating, the reactor rapidly increases its radioactivity inventory by generating considerable amounts of fission products. After shut down, the radioactivity decreases at first quickly - due to the decay of the short-lived products - and then gradually, as the long-lived nuclides become dominant. Some data of nuclear power sources (NPS) are compiled in table 1 from references /1/, /2/, and /3/.

After shut down of a reactor, the activity decreases by roughly two orders of magnitude within two months, and by additional four to five orders of magnitude within a further 300 years. Thus, it has become common practice either to use NPS-satellites at high orbits (above 1000 km altitude) and in outer space (no return) or to deploy them at low earth orbits (around 250 km altitude) for a limited time of three months - roughly the lifetime of such orbits - and then separate the nuclear reactor and transfer it to a 800 to 1000 km altitude orbit with a 300 to 500 year lifetime, in order to give the radioactive inventory time for sufficient decay. These orbits are so-called nuclear safe orbits (NSO) as recommended by the UNO /4/. In this way 33 reactors have been launched from 1967 to 1987, of which 31 were stored in NSO, and two devices failed and re-entered the atmosphere. However, all material disposed of in NSO will return sooner or later. Thus, it is important to know of the amount of radioactive material to be expected to return. A rough estimate is shown in figure 1, based upon calculations with the ORIGEN-2 code /5/. In addition the failure induced early re-entries are also of interest.

Type of NPS:	RTG	Reactors		
		1	2	3
Name of NPS	SNAP19	SNAP10A	TOPAZ	ROMASHKA
Vehicle	NimbusIII	Snapshot	Cosmos	?
Type of fuel	Pu238	U235	U235	U235
Quantity of fuel (kg)	~2	~5	20	5
Enrichment of U235	-	93%	90%	10%
Thermal power (kW)	1	40	100	40
Activity at ground (Bq)	1.3E15	4.E8	1.6E9	1.E9
" in operation (Bq)	1.3E15	7.E15	1.E16	7.E15

Table 1: Nuclear power source data

3 Orbit lifetime parameters

The lifetimes of satellite orbits are affected by a great variety of parameters, of which only a few can be discussed here and only to a very limited degree of detail.

Low earth orbits - i.e. those used for reactor supplied reconnaissance satellites in the range of 250 to 350 km altitude - are stable for only a few days and decay to re-entry within 1 to 3 months, depending on the initial altitude, the atmospheric drag and the "ballistic coefficient" taking into account the mass and the friction surface (kg/m²) of the device. The atmospheric drag depends on the solar activity (11 year cycle), the season in which the device is operating, and the inclination of the orbit, and the orbit altitude, of course.

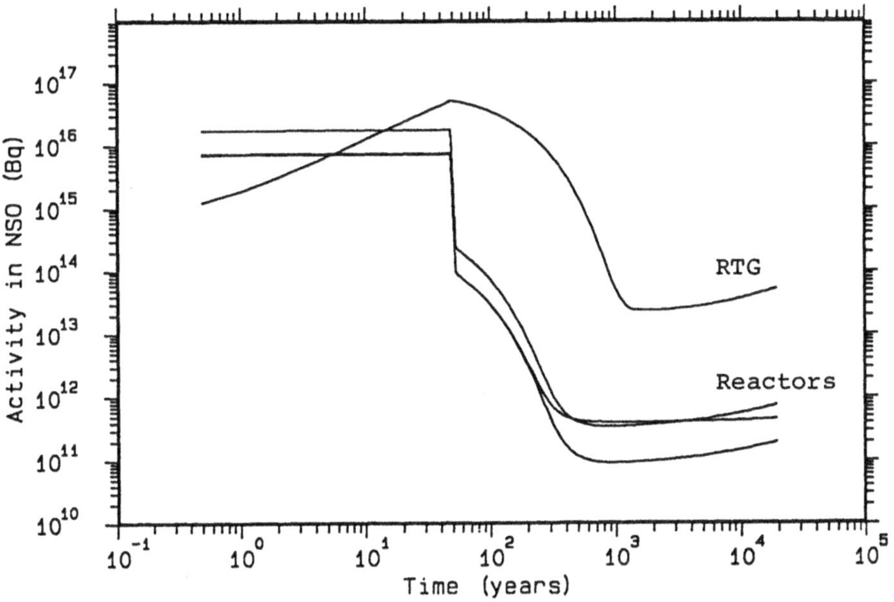

Fig. 1: Estimates of NSO radioactivity inventories for 3 types of reactors or 1 kg Pu238 RTG's, each launched at a constant rate of 2 per year over 50 years

Lifetimes of transfer orbits from low earth orbit to nuclear safe orbit (LEO to NSO) depend on the same parameters. For instance, the lifetime of a transfer orbit between 300 km (perigee) and 950 km (apogee) can vary from 2.6 years to 5.5 years for devices with 50 kg/m² to 200 kg/m² ballistic coefficient, if they are launched at the solar minimum. These lifetimes will change to 0.8 years (5.2 years), if the launch takes place at a solar maximum, and to 1.3 years (4 years), if launched 3 years after a solar minimum.

For a NSO of 800 km altitude and a worst case solar activity model, the lifetimes range from 100 years (50 kg/m² ballistic coefficient) to 400 years (200 kg/m²), whereas the respective figures are 800 years and 3000 years for a 1000 km altitude NSO. These figures hold as long as other effects, for instance collision with space debris, can be neglected. However, there is strong evidence /6/ that just in the altitudes between 500 and 1000 km the maximum of debris flux density is expected, leading to an impact frequency of roughly 10^{-5} impacts per m² and year. And these data are growing. The impacts themselves will increase the debris density, although more than 99% of the impacted mass will re-enter within 50 to 60 years after the collision.

4 NPS Incident Statistics

There are two possible ways of evaluating expected failure or event frequencies :

- derivation from equipment reliability and availability data, and
- derivation from event experience.

It has been found preferable to use event experience here, since this data base is available and in addition reliable, whereas the equipment data are not generally accessible. Three possible experience based data bases could have been used:

- USSR-NPS experience
- USA -NPS experience,and
- total western launch and space manoeuvre experience.

The USSR-NPS experience does not qualify as a total probability data base, since it is questionable whether it comprises all launch attempts or not. However, one can be sure that all attempts to put the device into an NSO are included. A second flaw of those data is its poor statistics. The USA-NPS experience is completely communicated, however of a poor statistics, too.

The total western experience provides complete information and sufficient quality of statistics. It has been choosen as the reference data base for the probability evaluation. This is justified, because the manoeuvre failures of interest are independent of the type of the electric power source of the space craft.

Table 2 shows a compilation of the experience data /7/ of western satellite deployment attempts. Non-geosynchronous orbits have been entered directly. This is a data set referring to the probabilities of launch and ascent failures. Geosynchronous orbits have been entered via interim steps: initial orbit, transfer orbit, final orbit. This is a type of manoeuvre quite close to that used for USSR-NPS: mission orbit (LEO), transfer orbit, NSO (800 - 1000 km). Moreover, the degree of detail of the data is such that the first stage data - i.e. the attempts to enter the initial orbit - can be used as an additional contribution to the former set: enter non-geo-synchronous orbit directly.

The probability of failure of a task such as a launch or orbit manoeuvre is estimated by assuming this probability to be constant. In this case, the only necessary data to be collected are:

X = number of failures observed

n = total number of attempts

X is a random variable with a binomial distribution. The probability of observing exactly x failures during a total of n attempts is therefore given by:

$$P(X=x) = \binom{n}{x} p^x (1-p)^{n-x} \quad \text{for } x = 0,1,2,\dots,n$$

The best estimate for p is the mean value:

$$\hat{p} = \frac{X}{n} = \frac{\text{number of failures}}{\text{total number of attempts}}$$

launch vehicle (period)	-non-geosynchr.-- launch attempts	total failures	------- geosynchronous ---------- launch attempts	-- failures to enter: -- initial orbit	transfer orbit	geosyn. orbit
ATLAS (1963-1987)	22	2	37	3	5	5
TITAN (1965-1986)	10	3	34	2	3	5
DELTA (1960-1987)	100	6	81	3	5	10
N + H (1975-1987)	7	0	9	0	0	1
ARIANE (1979-1987)	4	1	15	3	3	3
total	143	12	176	11	16	24

Table 2: Mission success statistics of geosynchronous and non-geosynchronous satellite launches /7/ (US, Europe, and Japan only)

The lower 5% limit $p_{0.05}$ and the upper 95% limit $p_{0.95}$ for the 90% confidence interval are given by:

$$p_{0.05} = \frac{x}{x + (n-x+1)\, f_{0.95,2(n-x+1),2x}}$$

$$p_{0.95} = \frac{(x+1)\, f_{0.95,2(x+1),2(n-x)}}{n - x + (x+1)\, f_{0.95,2(x+1),2(n-x)}}$$

with $f_{a,v1,v2}$ = Fisher distribution
a = 100a% percentile
v1, v2 = degrees of freedeom

Two points are still to be discussed in more detail:

(1) Assuming a constant failure probability is a good approximation for well established launch vehicle systems with proven reliability in many launches or manoeuvres over a period of many years. As can be seen from fig. 2, there is a significantly increased failure probability for a new system during the first few launch attempts. However, after a certain number of launches, the failure probability decreases and maintains an almost constant level. Thus, assuming a constant failure probability is justified.

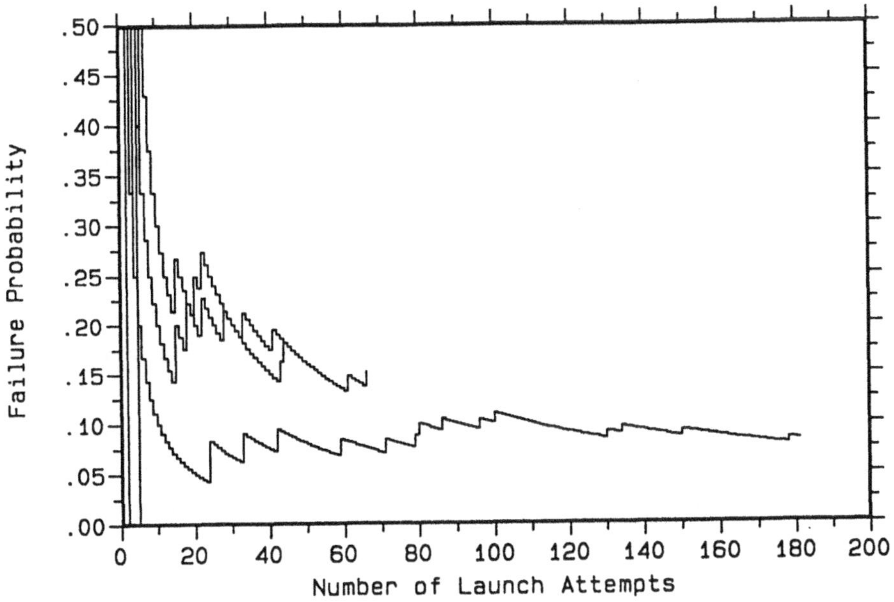

Figure 2: Failure probabilities versus launch attempts (particular carrier systems)

(2) One way of improving the poor failure statistics is to combine the data of different launch vehicle systems. This can be justified only, if the behaviour of these particular "populations" is not too much different. Otherwise, one would introduce additional systematic errors which might cancel the gain of statistical improvement. Although comparison of different types of missions and systems has shown that there are some evident differences, all available failure information of the various launch vehicles systems has been combined. The systematic error introduced by that seems to be smaller than the resulting statistical errors of the combined failure probabilities, and these probabilities are used for a general judgement only, independent of the particular launch vehicle.

From the data all the required information can be derived (see table 3):

a) probabilities and frequencies
   - of launch and ascent failures preventing the satellite to enter LEO or NSO, if these are the initial orbits,
   - of failures to enter the transfer orbit, when LEO was successfully reached,

b) uncertainties of those probabilities for a required confidence level of the statistical mean value,

c) comparisons with USA-NPS and USSR-NPS experience, the latter under the assumption that the available information refers to attempts to put NPS into NSO from the mission orbit(LEO).

The return probability from NSO is equal to one, of course, and the frequency almost that of the charging rate to NSO, provided that debris impact effects can be neglected. This question is investigated in an other study /6/.

Type	manoeuvre	attempts/ failures	failure probabilities % mean	90% confid. limits low	high
A 1	ground-orbit	143/12	8.4	5.0	13.0
B 1	ground-initial	176/11	6.3	3.6	10.
B 2	initial-transfer	165/5	3.	1.2	6.2
B 3	transfer-final	160/8	5.	2.5	8.7
B 1+2+3	ground-final	176/24	13.6	9.8	18.3
B 1+2	ground-transfer	176/16	9.1	5.8	13.2
B 2+3	initial-final	165/13	7.9	4.8	11.9
C 2+3	initial-final (USSR-NPS)	33/2	6.1	1.1	17.9
A1+B1	ground-initial	319/23	7.2	3.0	9.9

A : non-geosynchronous data from table 2

B : geosynchronous data from table 2

C : USSR-NPS data

Table 3: NPS failure probabilities (%)

5 Consequences

Even with respect to financial expenditures, the failure frequencies appear to be considerable, the more so with regard to safety in the case of nuclear power devices deployed in space. The question of radiation impact and environment contamination expected from the delayed however inescapable re-entries of those devices "stored" in so-called nuclear safe orbits remains unanswered. Together with the space debris effect on the lifetimes of these orbits, the radiation exposure should be investigated now, and considerable effort should be made in finding a means of actually keeping the material off the atmosphere, and cleaning the space environment, if possible.

6 Acknowledgements

The findings reported here have been evaluated in the course of a study financed by the Bundesminister für Forschung und Technologie. The considerations concerning orbital mechanics and orbital lifetimes were contributed to the study by Professor D. Rex and Dipl. Ing. J. Bendisch of the Technische Universität Braunschweig. The authors wish to thank both of them for their kind permit to mention these findings.

7 References

/1/ Bennet, G.L.,
Overview of the U.S. Flight Safety Process for Space Nuclear Power, Nuclear Safety, Vol. 22, No. 4, p423-434, July/August 1981

/2/ Loeb, H.W.,
Nuclear Energy in Space, Earth Oriented Appl. Space Technology, Vol. 2, No. 1, p1-16, 1082

/3/ Angelo Jr. A.J., Buden D.,
A 21st Century Nuclear Power Strategy for Mars, 37th Congress of the International Astronautical Federation, Innsbruck, Austria, Oct.4-11,1986

/4/ UN Committee on the Peaceful Uses of Outer Space, Scientific and Technical Subcommittee, Working Group on the Use of Nuclear Power Sources in Outer Space, Report 1981

/5/ Croff A.G.,
ORIGEN-2, A Revised and Updated Version of the Oak Ridge Isotope Generation and Depletion Code, ORNL-5621, Oak Ridge, July 1980

/6/ Prof. Dr. D. Rex, Universität Braunschweig, FRG
private communication

/7/ Space Statistics Review (until April 1987)
Aviation Information Services Ltd., England

Investigation of the Safety of a Ship Propulsion System by Monte Carlo Technique

A. BRANDOWSKI

Maritime Academy, Gdynia, Poland

Abstract

The simulation model of safety/risk to investigate ship's systems,taking into consideration their usage and maintenance process ,is presented in the paper. The study of a ship's propulsion system is presented and discussed as the example of safety/risk analysis in the area of sea going ships.The numerical simulation methods have been used there. The emphasis in the following paper is laid on the presentation of the analysis over the safety/risk and reliability of system. It may be useful for ship's design.

Basic concepts and assumptions

ACCIDENT means an uncontrolled series of events which run at extremely short time in relation to potential preventive action, producing serious damages of persons and/or material property and/or environment.

RISK is related to serious and negative consequences whose frequency of apperence is rather small. In this case, it is related to accidents and it means the ability of a system to produce them.

SAFETY means the system's ability not to produce accidents. It is generally treated as an opposition of risk.

The usage and maintenence process /U-MP/ of a ship's system is the subject of modelling. The process depends on environmental states which are created by a ship and the waters she is going on. The system consists of a finite number of elements and possesses well-defined usage structure, reliability structure and maintenence structure. The element can be repairable or unrepairable at sea. Their failures can produce different results to the system and its environment. In an extreme situation a sea accident may be produced.

The assumption of probability independence of all the times describing the behaviour of a ship, a system and its elements has been accepted generally. The cycles between ship class renewal /CBCR/ are studied. These times intervals last until five years.

The usage and maintenance states /U-MS/ of system are divided into absorbing states /e.g. the loss of propulsion ability which isn't convertable at sea/ and non-absorbing states. In the first case the simulation of CBCR is finished and the calculation goes to the simulation of the next one.

The aim of simulation study is to calculate the safety/risk measures of a system. The idea of the safety model is shown in the figure below. It is designed for project analyses of not only the variants of a system but the variants of its U-MP as well.

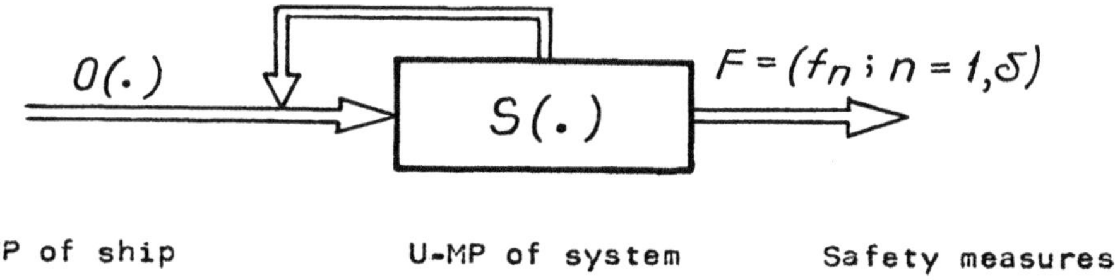

Fig. 1. Scheme of model

Model of ship usage and maintenence process [4]

Let's denote:

$O = \{o_i \ : \ i = 1,\alpha\}$ - set of names of ships U-MS; (1)

$O(\cdot): [0, T) \rightarrow O$ - U-MP of a ship, when: (2)

$[0, T)$ - time interval.

One can assign a set of the kinds of round voyages over a U-MP of a liner ship. The voyages of well-defined kind possess a deterministic repartoire of U-MS appearing one after the other according to exact order and stochastic times $T_{i/l/f}$ of ships existance in states, when i is numebr of state, l is number of voyage segment and f is number of repertoire $/i=\overline{1,\alpha}, l=\overline{1,\beta_f}, f=\overline{1,\varphi}/$. From natural investigation it has been known that these times have cut-normal distribution [4].

The U-MP of a ship is broken by plan repairs /repair II/, when a ship goes to a repair yard. This is connected with the special kind of round voyage /repertoire number $\varphi+1$/. It has been assumed that such voyages take place after all Δ normal round voyages.

Model of system's element [1],[2],[3]

A system is considered as consisting of elements with numbers $m=\overline{1,\nu}$. Each element can change its usage, reliability and maintenence states. The state of an element is described by ordered three $\langle u, n, o\rangle$, when:

$$u = \begin{cases} 1 - \text{active usage/operation/,} \\ 0 - \text{other form of usage or absence of usage,} \end{cases} \quad (3)$$

$$n = \begin{cases} 4 - \text{satisfactory state,} \\ 2 - \text{unsatisfactory state, failure repairable at sea,} \\ 0 - \text{unsatisfactory state, failure unrepairable at sea,} \end{cases} \quad (4)$$

$$o = \begin{cases} 4 - \text{emergency repair /I/} \\ 2 - \text{plan repair /II/} \\ 0 - \text{absence of repair.} \end{cases} \quad (5)$$

The following kinds of element states are interesting in our safety study:

$\langle 4, 4, 0\rangle$ - active usage with satisfactory state,
$\langle 0, 2, 4\rangle$ - unsatisfactory state with repair I,
$\langle 0, 0, 0\rangle$ - unsatisfactory state unrepairable at sea,
$\langle 0, 4, 2\rangle$, $\langle 0, 2, 2\rangle$, $\langle 0, 0, 2\rangle$ - repair II of element existing in different reliability states
$\langle 0, 4, 0\rangle$ - stoppage /stand-by or turning off/.

Notion of times discribing the behaviour of elements has been assumed as follows: $T_{m/2}$ - calender time between failures repairable at sea, $T_{m/0}$ - calender time between failure unrepairable at sea, $T_{m/u/2}$ - time of active usage between failures repairable at sea, $T_{m/u/0}$ - time of active usage between failures unrepairable at sea, $T_{m/04}$ - time of repair I, $T_{m/02}$ - time of repair II,when: m is the element's number,$m \in \{1,2,\ldots,\vartheta\}$. These times have been treated as stochastic variables with exponential distribution. The reliability of an element is also described by one of the four above mentioned times between failures and times of repairs I and II.

The elements reserved are used actively in the convertible way. E.g. if in the system there are two elements with numbers m and m+1 which make reserve one another,turning on the first one and turning off the second one foolows at moments assigned by $\Delta t_{m/u} = \Delta t_{m+1/u}$-deterministic time intervals of active usage.

The repairs II of elements can be carried out in the frames of ship's plan repairs /in repair yard/ or independently after $\Delta t_{m/02}$ - deterministic calender time interval between repaire II.

Model of system [2],[2],[3]

Let's denote:

$$S = \{S_i \; ; \; i=\overline{1,\alpha}\} \text{ - the set of the subsets of system's U-MS,} \quad (6)$$

$$S_i = \{s_i \; ; \; i=\overline{1,\alpha} \; ; \; k=\overline{1,\gamma_i}\} \text{ -the subset of system's U-MS which can take place in ship's state } o_i, \quad (7)$$

$$S(.): [0, T) \rightarrow S \text{ - the U-MP of system.} \quad (8)$$

System's states depend on the states of all their elements and ship's states too. During system's U-MP there take place interstate changes among specified states. These result from the logic of system's U-MP. The times of system's existence in its states are not only assigned by times of usage, between failures and lasting of repairs of system's elements but the times of ship's existence in its states as well.In each state $s_{ik} \in S$

a system possesses certain potential capacity which is naturally taken from the safety/risk point of view.

Safety/risk measures

The safety or risk measure of a system may be 1/ the frequency of system's accidents and 2/ the frequency of ship and its environment accidents which are dependent on system's accidents. In the first case the safety/risk vector

$$F' = \left(f'_n \; ; \; n \in \{1,2,\ldots,\delta\}\right) \text{ , when:} \qquad (9)$$

$$f'_n = \sum_{i=1}^{\alpha} \sum_{i=1}^{\gamma_\alpha} \Phi_{ik} \, z'_{iku}$$ - the frequency of system's accidents of n-th kind, 1/yr,

Φ_{ik} - the frequency of system's enters into the state s_{ik}, 1/yr,

$$z'_{ikn} = \begin{cases} 1 - \text{when the accident of n-th kind can appear in the state } s_{ik}, \\ 0 - \text{when it cannot appear.} \end{cases}$$

In the second case the safety/risk vector

$$F'' = \left(f''_n \; ; \; n \in \{1,2,\ldots,\delta\}\right) \text{ , when} \qquad (10)$$

$$f''_n = \sum_{i=1}^{\alpha} \sum_{k=1}^{\gamma_\alpha} \Phi_{ik} \, z''_{ikn}$$ - the frequency of ship's and its environmental accidents, 1/yr,

$$z''_{ikn} = p\left(A_n / S(.) \to s_{ik}\right)$$ - the probabity of event A_n that the accident of n-th kind appears when system is being in the state s_{ik}.

Let's pay attention that z'_{ikn} and z''_{ikn} constitute the potential capacity of a system from the safety point of view.

Study of the safety/risk of a ship's propulsion system

The safety/risk of the propulsion system with Diesel engine and fixed blade propeller will be studied. The scheme and reliability structure, of this system is pictured in fig.2.

In this structure some auxiliary installations /colling, lubrication, fuel, compressed air and electrical energy installations/ haven't been taken into account. This abbreviation will not result in essential errors of the assessment of system's safety because their level of reliability is always very high

/they have got the redundancy of structure and functional kind and their elements are always repairable at sea/. The safety study of auxiliary instalations is also very labour-consuming by reason of their complex structures and complex U-MP.

The catastrophic failures of propulsion system,unrepairable at sea, are the most essential in regard to ship's safety. These result in the loss of ability to propel a ship. When it has happened to a ship it is left at the mercy of the forces of nature. The loses of propulsion ability are the accidents of a system and can lead to the dependent accident of a ship and its environment. It is probable the ship with cargo and crew may be lost.

The study was carried out over the system installed on board of a container ship operating between Gdynia and the ports of East Coast of the USA. Eight kinds of round voyages, which are combinations of some number of visited ports were, identified there.

It has been assumed the reliability of propulsion system elements is dependent on their active usage time.The failures of these elements only can occur when they are in active usage states. The graph of the U-MP under consideration is pictured in fig.3. The process is the result of the composition of two process: the usage process of a ship and the failure process of a propulsion system.

The investigation plan is shown in the table below. The failure rate λ_1 is corresponding with the following mean times to failures: 10,20,25,30,40 and 60 years of active usage time. The failure rates $\lambda_2 = \lambda_3$ however, are corresponding with time 50 years.

λ_1 [1/h]	$\lambda_2 = \lambda_3$ [1/h]	Δ
$1{,}9\ 10^{-7}$, $9{,}5\ 10^{-8}$, $7{,}6\ 10^{-8}$ $6{,}3\ 10^{-8}$, $4{,}7\ 10^{-8}$, $3{,}2\ 10^{-8}$	$3{,}8\ 10^{-8}$	20

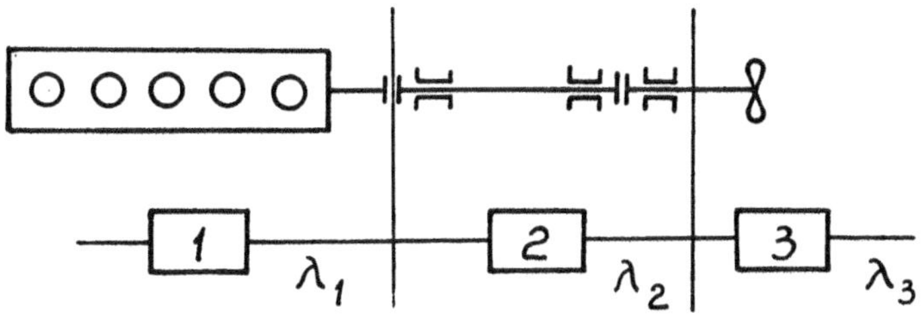

Fig.2. Scheme and reliability structure of studied propulsion system,when:1-Diesel engine,2-shafting,3-propeller.

The results of simulation experiment are shown in fig.4. The frequency of propulsion losses f_1' was calculated according to the formula (9), where potential capacity $z_{ik1}' = 1$ for the absorbing states and $z_{ik1}'' = 0$ for the non-absorbing states.In fig.4 there is simulationsly the frequency of ship's losses f_2'' calculated according to the formula (10), where potential capacity was assumed $z_{ik2}'' = p$ (ship's loss/S(.)$\rightarrow s_{ik}$)$= 10^{-3}$ for the absorbing states and $z_{ik2}'' = 0$ for the not absorbing states.

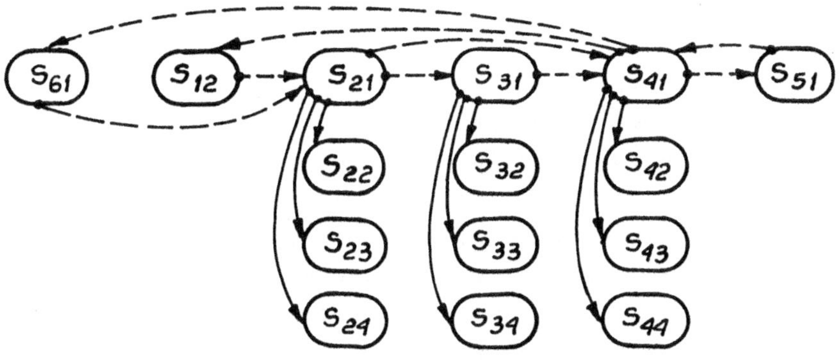

Fig.3. Graph of the U-MP of propulsion system

<u>Findings</u>

The problem presented above can be only studied effectively by means of Monte Carlo techniques. It is the result of complicated character of the safety/risk model. The stochastic times of system's being in states are associated variable /the same U-MP, the same operator/. That is why the measures f_1' and f_2'' should

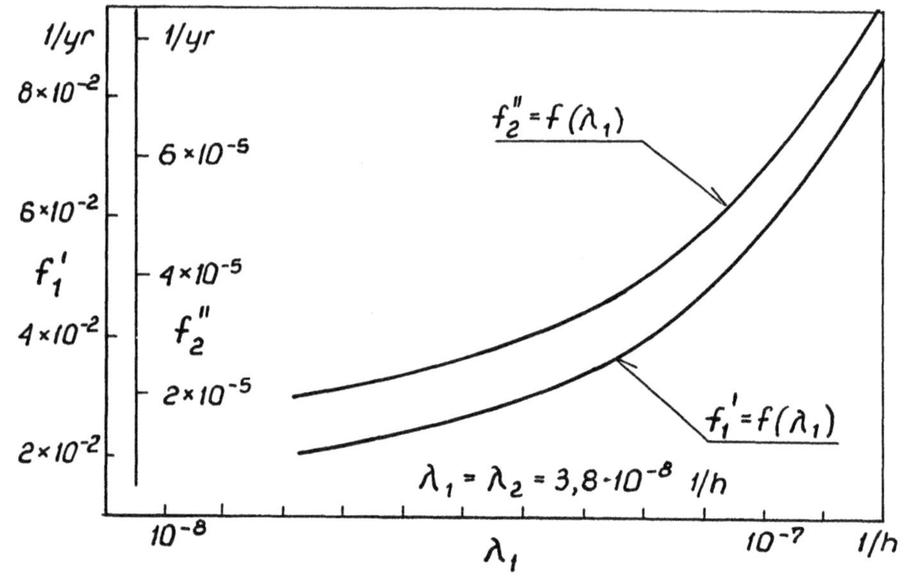

Fig.4. Frequency of the losses of propulsion ability and frequency of the losses of ships

be treated as upper bound of safety/risk assessment. If the reliability of auxiliary installations is taken into account these bound will be moved upwards a little.

The contents of this paper should be treated as some example of the methodology of ship system's safety assessment. Generally, the system's reliability is dependent not only on system's reliability but their U-MP and potential capacity /from the safety point of view/ as well. If the safety/risk criterions for ships are implemented /e.g. acceptable level of measure f_2''/. it will be possible to assign the requirements of a system's reliability for fixed U-MP.

Reference

1. Brandowski,A.:Efektywnościowo-niezawodnościowy model systemu technicznego.Zagadnienia Eksploatacji Maszyn,No 1-2.Warszawa 1987.
2. Brandowski,A.:Reliability Effectiveness Models of Ships Engineering Systems. ICMES'84 Symposium.Tiriest 1984.
3. Brandowski,A.:Relationsship among the Effectiveness of Ship System and its Reliability.IV IMAEM Congress. Varna 1987.
4. Kowalczyk,J.: Metoda prognozowania użytkowania wybranego, złożonego systemu mechanicznego. Doctoral thesis, Kielce 1980.

From the Plant Risk Analysis to the Area Risk Management: Problems and Perspectives

G. Uguccioni, A. Crotti, S. De Sanctis

Snamprogetti SpA
Safety and Reliability Department
San Donato Milanese (Milano)
Italy

Abstract

The application of risk analysis techniques to major hazard plants has increased in the recent years, and these studies are now common practice in the design and authorization procedures for major hazard plants. A natural development of this approach is the consideration of groups of plants with the associated infrastructures to assess the risk at an area level.

This approach is not so consolidated and Authorities are now starting experimenting its benefits.

The paper presents the Author's point of view of the problem based on their experience in developing and testing risk assessment for industrial areas.

Some different meanings of 'Risk analysis'

The application of probabilistic techniques to the design and the assessment of major hazard installations is widely accepted in the industrial community.

Among the Authorities, the way indicated by the European Community for the assessment of the acceptability of major hazard installations has been followed with more or less delay in all the EEC countries. In some cases the fact that the Directive is in practice requiring a quantified assessment of the risk posed by the plant as a whole (flammable, explosive and toxic hazards) has been accepted time ago while in other instances, like in Italy, this has required a longer debate: only recently the Directive has been accepted globally /1/ but, at the moment of writing the paper, the technical specifications on the techniques and procedures to be adopted to issue a safety report have not been formalized.

Both among operators and authorities areas of 'resistance' to the application of a probabilistic approach still exist, but no one is now objecting (at least in principle) to the need of a systematic and quantified approach to the safety of plants.

This situation however has given the lack of a commonly accepted interpretation of the contents and techniques for the development of a Risk Analysis Report, in spite of the consensus that has been reached among the European specialists involved.

As a result, the terms 'Risk analysis', or 'Hazard analysis' or 'Safety report' are common, but have a different meaning in different countries, and in different organizations in the same country.

The main differences that arise are related to the methodology adopted for the hazard identification (analytical methods like HAZOP and Operability Analysis versus synthetic methods like historical experience and engineering judgement) and to the application of quantified techniques for the study of the probability of the accidents.

The term 'Risk analysis' could therefore indicate the elaboration of an overall index, or plots of the consequences of selected accident (those indicated by the historical experience as plausible), without indication of the credibility of such events, or individual risk contour plots and F-N curves for the plant under investigation.

These different approaches corresponds from one hand to the attitude to secrecy of some organizations, that believe that the less information is given, the lower the probability that part of the public can argue against the safety of their plants. It should also be considered that the attitude of part of the public toward the industrial risk, based on the myth of the 'no risk plant', poses a strong temptation to some operators to try to demonstrate that the 'no risk' situation has been attained. On the other hand, the Authorities are not so happy to have to decide on the ground of probabilistic judgements; a 'credible-not credible' approach poses much less problems in taking a decision.

Some problems with Risk Analysis reports

The situation described has consequences on the Authorities' attitude towards the use of risk analysis: the fact that the content of the reports issued can vary so widely has not helped in developing a consolidated procedure for the management of major hazard installations.

In some cases, there could be the mistake to use the Risk Analysis results in a deterministic decision framework. The Authority could be tempted to adopt, instead of the check of compliance with safety factors, a check of compliance with 'accident probabilities' or 'risk indexes'.

However, the tendency is to require a probabilistic assessment of the risk at least for the 'major hazard' plants, where 'major hazard' indicates a plant classified according to some 'objective' criteria like the quantity of hazardous material or the value of an index that takes into account the engineering characteristics of the plant.

This approach has advantages that have been widely described and recognized; it however focuses on the installation 'alone' without consideration to the interaction with other hazardous installations in the area.

It allows to simulate the behaviour of a plant taking into account its actual engineering characteristics and the performances of the operators and can give to the Authorities the technical elements to develop emergency plans for that installation.

There are however problems that a Risk Analysis report alone does not address: accidents induced by other installations or systems in the same area; additional risk caused by modifications in other systems (e.g. transportation network); modifications caused in the environment; concentration of risk over some areas due to the simultaneous presence of more installations (not necessarily classified as 'major hazard' ones..).

These problems have been recognized and partial answers have been given, from different points of view: the integration of Risk Analysis into EIA studies tend to include in the assessment any aspect that could be modified by the installation /2/; the development of integrated Risk Analysis computer codes that start from the description of the plant to obtain complete isorisk plot and F-N curves can give to the assessors an homogeneous set of analyses of all the plants in the area. All these proposals however do not give, in the Authors' opinion, a complete answer to the need of integration of the analyses from the point of view of the area involved.

<u>The need for a step ahead</u>

A characteristic of innovations is that, after the period of time necessary to 'impose' the new way of thinking, they induce an acceleration due to the development of techniques available for their application, the consideration of the usefulness of the results, the new needs induced etc.; this has happened for the probabilistic approach to the industrial installation safety. It took about 15 years from the earliest applications in the nuclear field to the pioneering applications in the chemical industries, but it took less than 5 years to become accepted as a common practice in the legislation (EEC Directive 501/82). In addition, the availability of integrated computer codes to deal with the probabilistic analysis and the consequence analysis now available for a few thousands dollars has led to a 'mass production' of Risk Analysis reports that has required to the Authorities an effort to be able to control and manage the results; on the other hand, it has led to new questions and problems to be answered.

The necessity to deal with reports based on comparable

methods has led to the proposal of technical specifications for the development of Risk Analysis studies. These specifications however, if issued without a clear understanding of the possible uses of these studies, could be found superseded at the very moment of their issuing.

It has now become clearer that a report focused on the aspect of the engineering control of risk is not enough to manage the risk of a complex area.

It is interesting to note that some pioneering applications of the Risk Analysis techniques in the chemical industry were devoted to the consideration of a whole area (Canvey and Rijnmond studies /3,4/). The complexity of the problem and the attention posed on the engineering aspects of the control has then moved the attention to single installations; many signals show that now the time has come to move the attention again on the territory.

Some Authorities, either pressed by public concern over the concentration of hazardous activities or by the necessity to understand the intersection of many different Risk Analysis reports, have highlighted the need for a methodological approach to the Risk focused not on individual installations, but on the area that receives the effects of a complex system (plants, storages, distribution and transportation networks etc.).

This has led to projects aimed to develop methods to determine the risk at an area level. This has been done mainly in the UK (with the development of the Riskat procedure at HSE /5/) and in Italy, with the ARIPAR project that is under progress for the risk management of the Ravenna industrial area /6/.

The main characteristics of an area risk assessment

This new approach is based on the following considerations:

- the Authorities need to 'recompose' the information that are dispersed over a number of Risk Analysis reports;
- the main use of this is to assess the risk to be attributed to a specific area with the purpose to give to the Authorities the techniques and methods to manage the risk;that is to reduce it through appropriate measures, to develop emergency plans and to plan the future development of the area.

The theoretical and methodological choices that can be done have consequences on the content of Risk Analysis reports and on the definition of the plants that need to be analyzed.

The authors are involved in the development of the methodological and analytical tools for the ARIPAR project and their views and opinions that will be given in the following are based on this experience.

It is the Authors' view that the only variable that can correctly represent the state of an area is the risk in its most complete meaning: a single 'global' numerical index (like the R=F*M definition) is not enough to give a clear understanding of the risk (as already noted for the case of a single plant by other authors /7/).

This is even more true if we move to an area assessment where we do not need to have a generic figure to make a 'still photograph' of the acceptability, but a complex figure to handle the future developments (residential and industrial) of the area.

The definition of 'risk'

A discussion on the different ways to express the concept of risk can be found in other works /7,8/; the most appropriate ways to express risk are the isorisk contours, the F-N curves and the I-N curves (suggested in /7/).
These representations are linked to the situation of the present demographical aspects of the area; it is therefore advisable to adopt a definition not dependent on contingent factors, to allow to simulate the variation induced by planning options.

A suggestion has been given in the Rijnmond study, where the individual risk has been defined as a 'property' of the space, not linked to the presence or not of a real individual in that point. This parameter must not be confused with the individual risk (as defined in /8/) and could be appropriate to give to it a different name; in the frame of the ARIPAR project we have adopted the definition of 'Local' Risk (indicated by R).

It can be defined as:

$$R(x,y) = \sum_a F_a * p(e|a) * p(c|e)$$

where

F_a is the frequency of occurrence of accident (a);

p(e|a) is the conditional probability that the accident (a) will cause the physical effect (e) at location (x,y). The term p(e|a) includes all the considerations related to the environmental conditions on the area (wind speed and direction, atmospheric stability etc).

p(c|e) is the conditional probability that the physical effect (e) will cause a specific consequence (c) over an human being, without consideration of the exposure of any real individual in the area.

The summation is extended to the whole set of accidents (a) (mutually exclusives).

The local risk R can then be evaluated, on the basis of the results of individual Risk Analysis reports, for each point of the area; the present and future demographic characteristics of the area can then be superimposed on it, giving the individual risk I, defined as:

$$I_n = \sum_a R_a(x,y) * p_o * p_m$$

where

R_a is the local risk for a specific accident (a);

n is one of the partition of the population of a discrete area centered around (x,y) with homogeneous characteristics with respect to the risk exposure:

p_o is the probability that an individual of the group n will be present at (x,y) at the accident;

p_m is the probability that an individual of the group n will be present indoor or outdoor at the accident.

The partition of the population into groups with homogeneous characteristics (same exposure time, same vulnerability characteristics etc) allows to overcome the problems posed by the definition of I for the 'average individual' or the 'most exposed individual', discussed for example in /7/.

The evaluation of the F-N curves is straightforward, on the basis of the same results adopted to calculate R and I.

To adopt these indicators for the risk at (x,y) imposes certain requirements to the content of Risk Analysis reports. They have to be probabilistic, to give an estimation of the frequency of the accidents (F_a); the physical effects of the accidents must be coupled with an estimation of p(e|a); this implies that they must be assessed for a series of weather conditions and not for 'average' or 'worst' conditions alone; the use of global indexes as an overall measure of the plant safety (useful tool to give a judgement on the engineering characteristics of the plant) is of no use in assessing the risk of the area.

The domino effects

To represent correctly the risk in (x,y) as a summation of risks posed by different events it has to be ensured that the events are mutually exclusive. This in practice means that for any initiating event (defined as Top Event in the Risk Analysis report of a specific installation) it has to be developed an event tree to indicate all the possible outcomes of the event, depending on mitigating measures, atmospheric conditions etc.

This shall not be done for all the Top Events; a suitable criterion to discriminate between initiators of domino effects and other events is the consideration of the 'interaction distance', i.e. the maximum distance where a given physical effect able to cause damage to equipments is verified.

This implies to perform consequence calculations for extreme environmental conditions and to define a threshold for the physical effect over which the interaction is considered possible. Reference can be made to the threshold values given in the literature for heat radiation and overpressure effects. /9/.

Definition of hazardous installation

It shall be noted that the plants to be studied can not be defined once and for all on the basis of the quantity of material alone or of some numerical index. The criteria to adopt for the classification need to be modulated on the peculiar area to be studied.

The interaction of a number of plants that are classified as 'medium' or even 'low' risk adopting the method currently proposed can give a concentration of risk over certain areas that need to be considered. Moreover, the effect of a plant can be transferred far away from its fence by the transportation system adopted; transportation systems do not generally fall under the legislation based over the 501/82 EEC Directive.

This highlights the necessity not to rely upon classification indexes developed strictly for a single installation alone, but indicates the need for a calibration of the criteria presently adopted on the basis of the results of Area risk analysis experiences.

The criteria to be developed will require consideration not only of the engineering characteristics but of the 'density' of plants, the vicinity to vulnerable areas etc. as well.

Procedural aspects

The characteristics of an area Risk Assessment are such that they should be managed by the local Authorities, with the development of coordinating structures to ensure that all the parties involved (industrialists, population, authorities, experts) are involved in the development of the project.

From a technical point of view the decision to perform an area Risk assessment has consequences on the Risk Analysis reports to be issued for the individual installations (as discussed before).

In addition, it poses the need for the implementation of a system of collection of the territorial data to be adopted in the assessment (transportation fluxes, demographical aspects etc). This is an effort that could require a significant amount of the resources devoted to an area risk assessment exercise.

Conclusions

It is clear to all that although the PRA techniques applied to chemical complexes need refinements and research work in some areas, they are accepted by designers, operators and Authorities as an useful tool to decide about the engineering quality of a plant.

The development of such studies has induced the demand for their application at complex installations like industrial areas.

This moves the attention from the plant to the area, with the purpose to give to the Authorities decision aids in support to the development of planning or design modifications of the present situation. The most appropriate index, on the basis of the Authors' experience, is the Local risk calculated on the basis of the effects of accidents in all the industrial installations and transportation systems in the area. from this figure the individual risk contours, F-N and I-N curves can be derived to decide on the acceptability of the present situation.

The area assessment can be based upon the Risk Analysis reports that the operators have to develop under the 'major hazard' legislations in the EEC countries. However, this use of the Risk Analysis reports poses certain conditions on the techniques and criteria adopted for their elaboration, including the definition of 'major hazard installation' itself.

Area risk assessment exercises poses certain requirements not only at a technical level but at administrative levels as well: they require the cooperation of Authorities, experts and the other parties interested; this can take the form of Committees or management groups to control the development of the exercise; in addition, the necessity to make available data on the characteristics of the area (transport system, included the statistics of accidents; demographic characteristics etc) require the implementation of data collection system that could not be readily available in all situations.

The approach to the plant safety with a view not only of its engineering aspects but of its effect over a wider system as well should be considered in the regulations and standards that have to be issued to regulate the matter, to avoid that legislations that are just now discovering the Risk Analysis for individual plants alone are superseded at the moment of their approval.

References

/1/ DPR 17/5/88 n. 175

/2/ G.C. Bello;
Esperienze di gestione dei rischi in grandi aree industriali di alcuni paesi latini (in Italian)
3ASI International Symposium - Milano november 1988

/3/ Canvey - An investigation of potential hazards from operation in the Canvey Island - Thurrock area
HMSO

/4/ Risk Analysis of six potentially hazardous industrial objects in the Rijnmond area, a pilot study
D Reidel Publ. Company - 1982

/5/ G A Clay; N W Thurst; C Nussey
Quantified riskassessment; HSE's Risk Assessment Tool, RISKAT
Workshop on the major accidents hazards - london, Dec 1987

/6/ ARIPAR - Progetto di fattibilita' (in Italian)
Bologna, luglio 1987

/7/ P L Holden
What are we to make of estimated risk?
4th Nat. Rel. Conference Reliability 83

/8/ M Considine
The assessment of individual and societal risk
SRD Report SRD R 310 - 1984

/9/ F P Lees
Loss prevention in the process industries
Butterworths 1980

Risk Analysis of Dangerous Goods Transport in Spain

C. GONZALEZ, J.A. DOMENECH, A. LAZARO, J.M. RENAU, A. TASIAS.

Territorio y Medio Ambiente, S.A.
Barcelona, España.

Summary

Between December 1986 and October 1988, three projects on risk analysis of dangerous goods transport by road and train, have been developed in Spain.
The following products have been studied: chlorine, butadiene, styrene, propylene oxide, anhydric ammonia, anhydric fluorhydric acid, acrylonitrile, propane, butane, gasoline, gas-oil, toluene, methyl methacrylate, chlorhydric acid solution and caustic soda solution.

Introduction

The transport of dangerous goods, originates a variety of operations that can be summarized in the following steps:

1. Product load at origin
2. Transport to the destination point, through the selected itinerary.
3. Vehicle, or train stops at intermediate points of the trip.
4. Unloading of the product at the destination point.

Each one of these steps bears a risk, with its own characteristics. In this study we have determined the risk corresponding to the transport step (point 2).
A risk analysis study of dangerous goods transport starts with the identification of the traffic of the analysed products, where we establish:

origin and destination points, alternative routes, volume transported, and way of transportation.

The object of this study is to fix the risk value, expressed as a number of casualties/year ratio caused by the presence of dangerous goods on the population at risk.

This population is constituted by the people living in the vicinity of the transport pathway, and by the drivers and passengers of the vehicles, that use the same transport roads.

The study allows to establish a risk classification for each one of the itineraries and alternative transport ways, which constitutes an additional tool for the management of the transport of dangerous goods, and adds also to the savings and flexibility of the process.

Methods

In this study, we have analysed individually the accident frecuency and its consecuences.

First we divided each route into sections, which are studied individually.

The frecuency of incidents with leakage of material, expressed as the ratio leaks/year along a section of the route, is a function of:

1. Total frequency of incidents (accidents/km.vehicle) FI (I)
   I is the reference of the sections with a different value of frecuency of incidents. In our case we have used different values of frecuency of incidents for highway, and two-lanes road traffic.
2. Itinerary length L (km)
3. Number of trips per year (N)
4. Damage probability to the recipient PD (K)
   K shows the likelihood of different types of damage:
   K=0 Recipient not perforated.
   K=1 Recipient or connection crack.
   K=2 Recipient or connection break.
   K=3 Recipient collapse.
5. If there is a leakage, the probability of causing a particular type of consequence (BLEVE, UVCE, Poolfire) is ex-

pressed as a function of the type of damage to the recipient.
PBLE (K), PUVC (K), PPOOL (K)

6. Net amount of product transported per vehicle. CANT
7. Ignition probability PIG (P). This is related to the type of product, and the area considered.
   P=1 rural type area
   P=2 semirural type area
   P=3 urban type area

The frequency of material leakage could be expressed as:

F(I,K,P)=FI(I)*L*N*PD(K)

The Bleve's frequency could be expressed:

FB(I,K,P)=FI(I)*L*N*PD(K)*PBLE(K)*PIG(P)

In this way all the different frequencies, for each type of consequence are calculated.

The consequences of an accident with product leakage, expressed as the ratio casualties/year are a function of:

1. The product's own characteristics
2. Exposition area for the different types of leakage, weather, type of land, and landscape A(I,J,K)
3. Population density in the area DP(P)
   P=1 rural population
   P=2 semirural population
   P=3 urban population
4. Population escape factor FF(K). Expressed as a function of the type of leakage.
5. Exposition factor FEXP(P). Expressed as a function of the type of area.

The number of casualties in an accident, might be expressed as:

N(I,J,K,P)=A(I,J,K)*DP(P)*(1-FF(K))*FEXP(P)

The risk is a function of the frecuency of incidents, and of the number of casualties in the accident.

R(I,J,K,P)=F(I,K,P)*N(I,J,K,P)

The calculation of these algorithms can be solved by means of multivectorial matrix procedures. This is obtained by means of the TRANSIT calculation code, originally developed by TEMA, Italy, and complemented in TEMA, Barcelona. When the transport is by road, we should introduce a second factor which is accumulated to the calculated one, which is the risk of the passengers and drivers,using the road simultaneously with the tanker. This factor is accumulated to the first one. This is calculated by substitution of the population density for the medium daily intensity (MDI) expressed as vehicles/24 h and assuming an occupation of 2 people per vehicle.

The values used were the following:

1. Frequency of incidents

   a) ROAD

Frecuency of incidents in the general traffic

Spain	13E-7
Italy	6.7E-7 - 7E-7
Holland	6.0E-7 - 8E-7
France	8.5E-7 - 9E-7
Great Britain	9.0E-7 - 11E-7
German Federal Republic	10.0E-7 - 12E-7
Belgium	13.0E-7 - 15E-7

Frequency of incidents in the transport of dangerous goods in Spain

Source: General Department of traffic

Total	3.6E-7
Road	4.3E-7
Highway	2.0E-7

b) RAILROAD

Country	Source	acc/train.trip	acc/train.km	Notes
Spain	Ineco		2E-6	1
USA	AAR	2.3E-5		2
USA	DOT		1.3E-6	3
UK	HSE		1.0E-6	3
UK	ICI		2.4E-6	4
Finland	VTT		2.2E-9	5
France	EDF		5.2E-8	5
Switzerland	Smith		3.0E-6	
Holland	TNO		1.4E-7	4

Notes

1. For TDG (Transport of Dangerous goods)
2. Only for general merchandises
3. Only for derailments
4. Only significant incidents
5. Only serious collisions

The applied values for incidental frequency in this study were:

	Rail	Road	Highway
acc/km.veh.	2.0E-6	4.3E-7	2.0E-7
acc/km.t (mean)	3.3E-9	2.2E-8	1E-8

2. Leakage Probability

Source: A mean value is obtained from the General Dept. of traffic and it is inversely proportional to the wall thickness of the recipients.

	Rail	Road	Highway
Chlorine	0.08	0.24	
Ammonia	0.05	0.13	0.19
Fluorhydric acid	0.07	0.18	0.25
Acrylonitrile	0.07	0.18	0.25
Propane	0.06	0.16	0.23
Butane	0.06	0.16	0.23
Gasoline	0.11	0.29	0.41
Gas-Oil	0.11	0.29	0.41
Toluene	0.11	0.29	0.41
Methyl methacrylate	0.11	0.29	0.41
Chlorhydric acid	0.11	0.29	0.41
Caustic soda	0.11	0.29	0.41
Butadiene	0.06	0.16	0.23
Styrene	0.11	0.29	0.41
Propylene oxide	0.06	0.16	0.23
Mean Value	0.085	0.21	0.30

The leak probability is distributed into probability of crack, break and collapse of recipient, with the following coefficients.

Source: TNO

Fissure 0.90

Break 0.07

Collapse 0.03

Total 1.00

3. Source: Sonata data bank. (LPG transport)

	Prob. BLEVE PBLE %	Prob. UVCE PUVCE				
		ACN	$C_3^{\bullet}$	$C_4^{\bullet}$	Gasol.	Tol.
Crack	1	1	10	10	10	1
Break	5	2	20	20	20	2
Collapse	15	3	30	30	30	3

4. Source: Guesstimate

	Run-off factor (People) FF
Crack	0.9
Break	0.5
Collapse	0.2

5. Source: Sonata data bank. (LPG(transport)

	Ignition Prob. PIG						
	ACN	$C_3^{\bullet}$	$C_4^{\bullet}$	Gasol.	GasOil	Tol	MMA
Urban area	5	20	20	10	2	2	5
Semirural	3.5	15	15	7	1.5	1.5	3.5
Rural area	2.5	10	10	5	1	1	2.5

6. Source: Guesstimate

	Exposition factor (People) FEXP			
	BLEVE	UVCE	Tox. Real.	Pool F.
Urban area	0.4	0.6	0.4	0.4
Semirural area	0.6	0.4	0.75	0.6
Rural area	0.8	0.2	0.95	0.8

Table 1 lists the types of consequences considered for each product.

Results

This method of calculation uses several statistical data such a frequency of railway, road or highway incidents and probability of recipient leakage.
There are nevertheless, other types of data such as the probabilities of each type of consequence (BLEVE, UVCE, POOLFIRE) or ignition probabilities, that have to be calculated or estimated from the available data. This may question the accuracy of the end results. Nevetheless, if we accept this group of working data, we will be able to establish a series of comparisons between alternative routes, different products or ways of transport, that are of great interest in the management of the risks of the transport of hazardous materials.

Leakage frequency comparison between road an railway

	ROAD	HIGHWAY	RAILWAY
Frequency of incidents (casualties/vehic.km)	4.3E-7	2E-7	2E-6
Frecuency of incidents(*) (casualties/t.km)	2.2E-8	1E-8	3.3E-9
Leakage probability (leakages/t.km)	21.3	29.9	8.5
Frequency of leakage	4.7E-9	3E-9	2.8E-10
Leakage frequency ratio	16	11	1

(*) Assuming that the transported net amounts by road and railroad are 20t and 600t, respectively.

Absolute risk results

These studies have been extended to 15 different compounds, in 138 different routes, representing a transport of 3827,6 kt/year. Nowadays, 2684,3 kt/year are been transported by road and the rest by railway.

The total risk value is 7,02E-2 casualties/year in road transport, considering an hypothetical total transport of 3827.6 kt/year and 5.14E-3 casualties/year for the transport via railway, assuming the same total hypothetical figures.
See Table 2 for details on absolute risks expressed as the ratio casualties/year.

Risk by products

If we observe the specific risk expressed as casualties/t km instead of the absolute risk , we will see that the ranking from higher to lower risk of Table 1 is altered, and we obtain a progressive risk scale of each analysed product. We must nevertheless take into account, that this specific risk value is influenced by the different routes assumed for each product. See the table 3.

Conclusions

1. The products with an inherent greater risk are the toxic materials, with chlorine first in this list.
   Immediately after it is the group of LPG, with monomers like butadiene leading this group. Finally, there are the inflammable liquids ranked according to their degree of inflammability, and lastly, the aggresive chemicals.
2. The highest risk found for the road transport was in the section Alicante-Cartagena by N-332 with 7.2 cas./year due to the ammonia and butane traffic accumulation with 115 and 54 kt/year respectively, and the population characteristics.
3. For the rail transport the highest risk was in the section Tarragona-Vinaroz with 3.6E-4 cas./year due to the ammonia, gas-oil and chlorhydric acid traffic accumulation and the population characteristics.
4. Globally we found a risk ratio between road and rail transport of 17.

Glossary

BLEVE: Boiling Liquid Expanding Vapour Explosion.
UVCE : Unconfined Vapour Cloud Explosion.
LPG : Liquefied Petroleum Gas.

Aknowledgements

To Eidos, s.p.a. specially to J.C. Bello and E. Gallatolla, for their work in the preparation of risk model calculation TRANSIT.

Table 1

	Toxic Release	Bleve	UVCE	Poolfire	Chemical Agression
Chlorine	X	–	–	–	–
Ammonia	X	–	–	–	–
Fluorhydric acid	X	–	–	–	–
Acrylonitrile	X	X	X	X	–
Butadiene	–	X	X	–	–
Propylene oxide	–	X	X	–	–
Styrene	–	X	–	–	–
Gasoline	–	–	X	X	–
Gas-Oil	–	–	–	X	–
Toluene	–	–	X	X	–
Methyl metacrylate	–	X	–	X	–
Chlorhydric acid	X	–	–	–	–
Caustic soda	–	–	–	–	X
Propane	–	X	X	X	–
Butane	–	X	X	X	–

Table 2

Absolute Risk	railway (casualties/year)	road (casualties/year)	(kt/year)
Ammonia	2.95E-3	5.21E-2	356.5
Chlorine	1.13E-3	7.02E-3	16
Acrylonitrile	4.76E-4	3.78E-3	99.5
Butane	1.27E-4	1.72E-3	284.4
Propane	1.73E-4	1.44E-3	169.5
Chlorhydric acid	2.95E-5	1.31E-3	105.3
Propilene oxide	6.63E-5	9.26E-4	52.3
Butadiene	1.00E-4	8.26E-4	46.5
Fluorhydric	8.30E-5	5.34E-5	13.1
Gasoline	1.55E-6	2.89E-4	526.2
Gas-Oil/Fuel	4.27E-7	1.80E-4	1896.5
Caustic soda	6.54E-7	5.93E-5	216.5
Methyl methacrylate	6.30E-8	1.11E-5	27.3
Styrene	9.70E-7	8.57E-6	8.8
Toluene	7.23E-9	3.75E-6	9.2
Total	5.14E-3	7.02E-2	3827.6

Table 3

Specific Risk

	Casualties/t.km (road)	Risk index
Chlorine	7.35E-10	1000
Ammonia	3.50E-10	476
Acrylonitrile	1.51E-10	205
Fluorhydric acid	9.46E-11	129
Chlorhydric acid	4.95E-11	67
Butadiene	3.03E-11	41
Propylene oxide	2.48E-11	34
Propane	1.68E-11	23
Butane	1.30E-11	18
Gasoline	2.59E-12	3.5
Methyl methacrilate	1.23E-12	1.7
Toluene	4.83E-12	0.7
Gas-Oil/Fuel	4.43E-13	0.6
Caustic soda	4.05E-13	0.5

Criteria for the Brazilian Directive on High Risk Plants Recent Developments

Tania Maria Amorim (*), Gian Carlo Bello (**)
Rosana Santalucia (*)

(*) Snamprojetos - Praca Floriano, 19/23° andar
Rio de Janeiro, Brasil
(**) Consultant

SUMMARY
A directive for the classification of hazardous installations and for the specification of the content of the required risk assessment will be issued by the State of Sao Paulo and subsequently by the Federal Government of Brazil. A critical review of the Seveso Directive suggested the development of new criteria to improve the consistency of the specified masses with the intrinsic hazard of the substances, to include consider special physical conditions into the process that can modify the hazard, to deal with plants using many different hazardous substances, to consider the characteristics of the scenario and to specify the contents of the risk analysis and the criteria for emergency planning.

1) INTRODUCTION

The objective of the work is to define criteria for the classification in process plants, storage and pipeline transportation. A critical revision of the Seveso Directives was carried out. This work was developed in two parts:

1) Classification of the installations and pipelines according to the danger of the substances. This depends on: type of substance involved, the quantity and the conditions of process / storage / pipeline transportation.
2) Analysis of the surrounding area considering the different parameters (Sensitive Area), such as :
   - density of population
   - ecological sensibility
   - presence of public services, etc.

The classification of installations was made possible, taking into account the risks to the population and environment (fig.1).

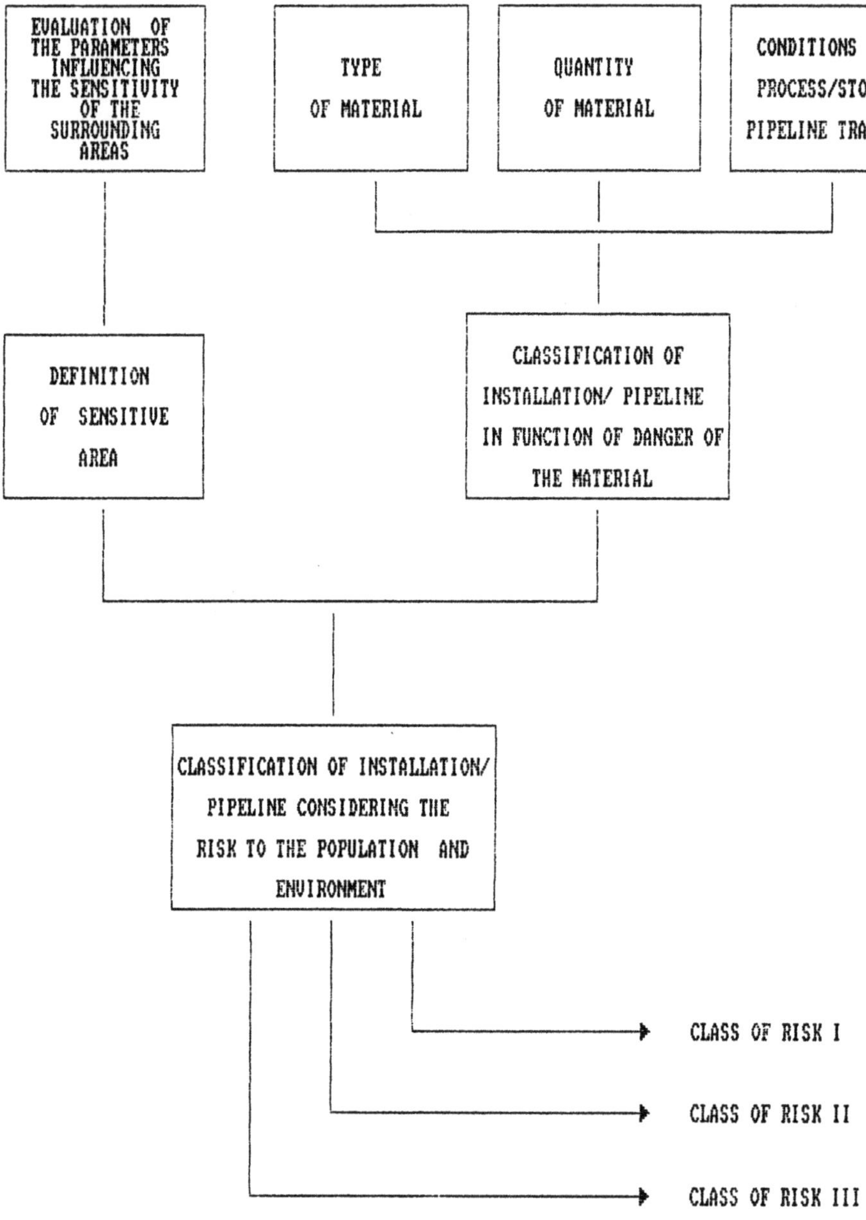
EVALUATION OF THE PARAMETERS INFLUENCING THE SENSITIVITY OF THE SURROUNDING AREAS
TYPE OF MATERIAL
QUANTITY OF MATERIAL
CONDITIONS OF PROCESS/STORAGE/ PIPELINE TRANSFER
DEFINITION OF SENSITIVE AREA
CLASSIFICATION OF INSTALLATION/ PIPELINE IN FUNCTION OF DANGER OF THE MATERIAL
CLASSIFICATION OF INSTALLATION/ PIPELINE CONSIDERING THE RISK TO THE POPULATION AND ENVIRONMENT
CLASS OF RISK I
CLASS OF RISK II
CLASS OF RISK III

FIGURE 1

2) CRITERIA FOR THE CLASSIFICATION OF SUBSTANCES AND DETERMINATION OF CRITICAL AMOUNTS FOR INSTALLATIONS OF PROCESS AND STORAGE

The hazardous materials have been divided into toxic, explosive and inflammable. Toxic materials have been further divided into four levels of danger in function of DL50 oral in rats (quantity of substance in mg/kg of body weight, ingested, causing death in 50% of the rats) and/or the inhaled dosage causing the death of 50% of exposed subjects (calculated as product of CL50 for the relative duration of inhalation, expressed as ppm and hours).

The higher levels for the classification in the four classes have been selected respectively at 50, 500, 5000, 15000 mg/kg for DL50 and at 500, 5000, 50000, 150000 ppm*hour for the inhaled dosage. The choice of a double criteria of classification was made necessary since, the inhaled dosage is more significant in the assessment for high industrial risks, whilst the CL50 values of many substances are unknown, data on DL50 are more easily found. The choice of parameters for rats is necessary due to lack of knowledge for the same parameters referring to humans. Each substance has been classified in function of DL50 and DOSAGE in four categories of danger.

For toxic substances:

CATDL50: Danger level obtained according to the classification of DL50

CATDOS: Danger level obtained according to the classification of DOSAGE

CATBRAS: Danger level of substances in function of toxicity

CATBRAS = CATDL50 IF CATDL50 >= CATDOS
CATBRAS = CATDOS IF CATDOS > CATDL50

and:

CATBRAS 4 : Highly toxic substance
CATBRAS 3 : Toxic substance
CATBRAS 2 : Slightly toxic substance
CATBRAS 1 : Practically non-toxic

74% of the materials are within the same category, both when classified according to DL50 and when classified according to DOSAGE, 24% show a variation of a level, and 3% a variation of two levels.

For materials where the DL50 and CL50 levels are not available, a TLV-TWA based classification is allowed (with the levels defined as 1, 30, 400, >400 ppm). The classification obtained using TLV must be interpreted as a "temporary classification" due to the fact that the TLV index is used in cases of chronic intoxication, and not an index of acute intoxication. It can be used until the values of DL50 and DOSAGE are known.

The definition of Critical Amounts, quantity of material which represents a potential risk to the process plant or storage of each toxic substance is based on a series of theoretical simulations of the effects of a release of toxic material. Chlorine has been used as a characteristic reference.

It has been observed that in case of an instantaneous release of a toxic substance, a concentration in the air of 1000 mg/m3 per ton, at a distance of one kilometer, will remain for approximately 5 minutes; in other words the relative dosage is equal to 100 mg/m3*hour for each ton of toxic substance released. The material in reference, chlorine, has a dosage equal to 35 ppm*hour per ton released. Using the lethal dosage for 50% of the exposed population as 325 ppm*hour, we observe that this concentration is verified at a distance of one kilometer leeward for a release of 9 tons of chlorine.

The hypothesis of a total release in an installation where there is a Mt mass of chlorine is unrealistic; the product is infact contained in a series of different vessels and tubes, separated by block valves and the release would be gradual and continual and certainly not instaneous. According to the Incident Data Bases, except in exceptional cases, the amount released has not been superior to 10-15% of the total contents (a domino effect with multiple rupture of vessels of toxic material is not assumed generally). A released mass of 9 tons corresponds to a total mass in the installation of 50-100 tons. On the other hand, the concentration at one kilometer equal to a lethal dosage for 50% of exposed subjects would produce terrifying effects,it is reasonable to introduce a coefficient of reduction.

We reach the conclusion that "an installation with more than 5-10 tons of chlorine present is a high risk installation".

The decision to assume a 10 ton mass as critical mass Mc for chlorine has been made. The critical mass for toxic materials other than chlorine have been determined, using the equivalence of toxic effects, or in relationship to respective DL50 and DOSAGE of chlorine.

A table has been compiled with several hundred substances, each with its critical mass Mc. The Mc values start with a minimum 683 kg for methyl-isocyanate and go to a maximum of several hundred tons for moderately toxic materials. The central values for the classes of toxicity are 10 tons for the class 4 toxic materials, 140 tons for class 3 materials, 600 tons for class 2; class 1 materials do not have Mc mass values as their toxic characteristics do not represent a high risk.

The table sometimes differs greatly to that of the CEE Directive known as Seveso Directive.

As for the explosive substances, we started by defining the three classes of danger (characterized by nitroglycerine, TNT and ammonium nitrate). A standard was developed using the "classification of affinity" of explosives already known, considering the sensitiveness to heat and shock as well as the maximum detonation rate.

For explosive materials:

CATBRAS 3: Very explosive material
CATBRAS 2: Quite explosive material
CATBRAS 1: Slightly explosive material

The critical mass used for reference was one which provocated overpressure at 6200 N/m2 at as distance of 0,5 Km (the distance for reference of explosive materials is equal to half of that for toxic materials, to take into account that the distribution of damage does not have preferential directions, but stretches, evenly over the surrounding area). Using these standards critical masses of 2.5, 25 and 2500 tons for the three classes of materials have been derived.

Inflammable materials have been divided into 4 classes, in function to their flashpoint and boiling point. To determine their critical mass, eventual UVCE of different materials if referred to and not the simple thermal radiation due to fire, which generally does not create great danger to the population.

Using this procedure we determined a critical mass of 250 tons for class 4 inflammable materials (flash point and boiling point < 37.8 C), 2500 tons for class 3 (flash point <37.8 C and boiling point > 37.8 C). For class 2 and 1 materials we defined a mass of 250 tons, to consider only under certain conditons (in other words conditions of overheating in a way that would produce a vapour cloud which could be potentially deflagratore).

For inflammable materials :

CATBRAS 4: Extremely flammable liquids or flammable gas
CATBRAS 3: Easily flammable liquids
CATBRAS 2: Flammable liquids
CATBRAS 1: Slightly flammable liquids

Critical masses calculated with these standards are applied to materials in process installations. For normal separate storage, the values are multiplied by a factor of 5; a study of actual accidents determined this factor; the study shows how the frequency of accidents is 10 to 100 times less when the materials are stored rather than processed, and that it exists a non linear correspondence between the mass involved and the damage extension.

3) APPLICATION OF CRITICAL MASSES IN INDUSTRIAL INSTALLATIONS

The principle is to consider a potentially high risk installation, where there is an amount of hazardous material greater than the characteristic critical mass of the material itself. To calculate the mass present in the installation, the "Criterion of 500 m" defines the mass M as quantity of material in process or in storage within the installation, plus the quantity of material present in the radius of 500 m from the baricenter of the installation.

In application of these standards, important corrections are needed for particular conditions in process which could alter the danger of the plant. A series of correction factors have been defined to consider in particular, the thermal conditions of the material (overheating) and the density of vapours given off (problem of heavy gasses). It is obvious that for materials with equal potential toxicity, the risk associated to a product with vapour density higher than that of air is greater than that of a material with light vapours; it is also clear that overheated conditions can greatly increase the danger of a material which in normal conditions would be in a liquid state and therefore not excessively hazardous.

Taking this into account , the actual mass of product present on the plant is multiplied by a series of correction factors before it is compared with the relative critical mass to decide on the existence of potential high risks. The correction factors are shown in figures 2 and 3. We can obtain the corrected mass (M') present on the plant by using these correction factors.

$$M' = M * f$$

The quantity of material in an installation represents a potentially high risk if:

$$\frac{M'}{Mc} = \frac{M * f}{Mc} \geq 1$$

As an example let us consider a storage of chlorine (critical mass: 5*10 = 50 ton). If the clorine is stored as a gas under pressure f=1 and a major hazards occurs with stored quantity above 50 tons, whilst if chlorine is stored in criogenic conditions f= 0.2 and a major hazards occurs with stored quantity above 250 tons.

Standards have been developed to consider concurent presence of different hazardous materials on the same plant. In these conditions the plant is considered high risk when the summation of the ratio between the mass of each material and the relative critical mass is superior to 1. By doing this we evitate the risk of not considering high risk those installations where there are many differents hazardous materials each with a mass equal to 99% of the relative critical mass.

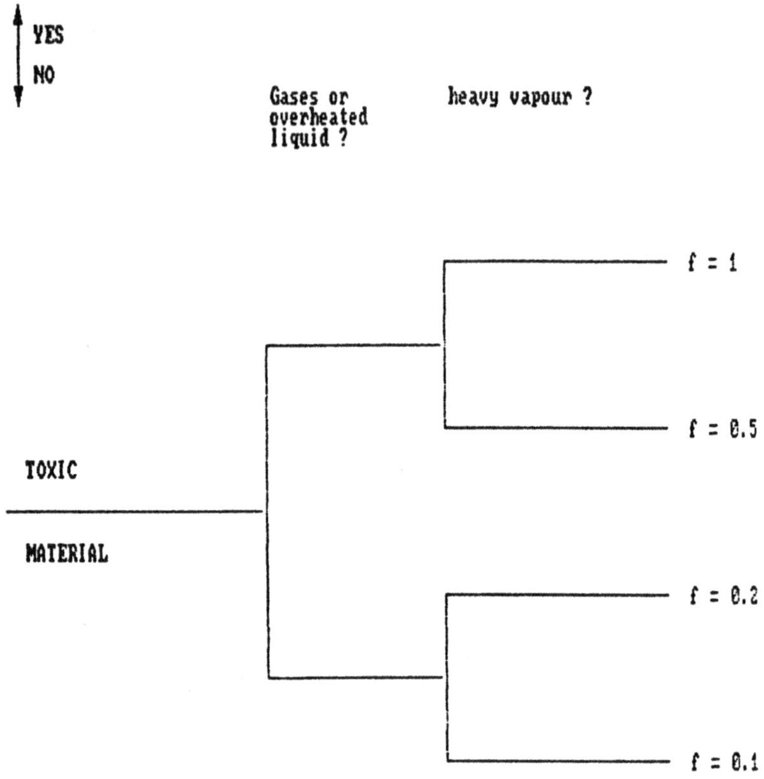
YES
NO
Gases or overheated liquid ?
heavy vapour ?
TOXIC
MATERIAL
f = 1
f = 0.5
f = 0.2
f = 0.1

FIGURE 2

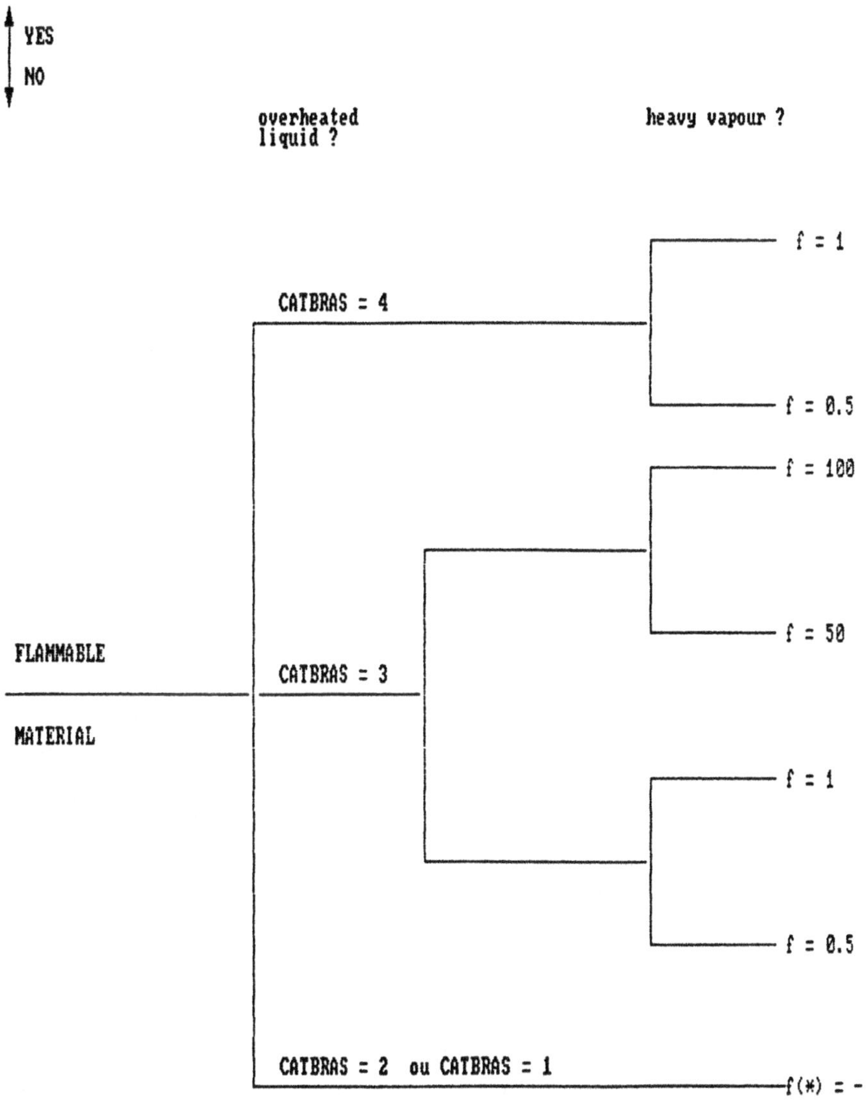

(*) Critical Mass defined in particular conditions

FIGURE 3

4) STANDARDS TO CLASSIFY MATERIALS AND DETERMINE CRITICAL LENGTHS FOR TRANSPORT IN PIPELINES.

This classification was developed for transport of liquids and gas normally transported as toxic and flammable gas and flammmable liquids.

The levels of danger of materials (toxic and inflammable) are the same as those for process.

The critical masses are multiplied by a factor of 0.1 to consider major vulnerability to people due to difficulty in intervention, being the pipeline localized in open unguarded areas.

Starting with Critical Mass (M_c) we developed a Critical Length (L_c) criteria to improve the characterization of hazards of the pipeline.

The determination of the Critical Length considers the relationship of Critical Mass and internal volume of the pipe and density under working conditions. This length is called L_{c1}. A Critical Length L_{c2} has been defined also, which represents the upper value for the length of the pipeline, above which the pipeline is considered a major hazard anyway.

For toxic materials with the danger level the same as chlorine L_{c2} = 2Km. For materials with levels as ammonia, L_{c2} = 20 Km. For inflammable gas L_{c2} = 30 Km. For the highest flammable level of liquids (level 4), L_{c2} = 10 Km. For level 3, L_{c2} = 100 Km.

The Critical Length (L_c) of the pipeline has been defined as:

$L_c = L_{c1}$ if $L_{c1} \le L_{c2}$

$L_c = L_{c2}$ if $L_{c2} < L_{c1}$.

5) APPLICATION OF CRITICAL LENGTHS IN PIPELINE TRANSPORTATION

We must consider the potential high risk in transport when the length of the pipe is greater than the calculated Critical Length, L_c.

6) CONTENTS OF THE SAFETY REPORT

The level of detail of the safety report must be specified taking care of both the intrinsic danger of the plant or pipeline (determined by the corrected mass present in them) and the sensibility of the environment where the pipeline or plant is situated.

For this reason the installations are classified in 3 classes of risk: installations with a mass superior to the Critical Mass, installations with a mass inferior to the Mc but superior to 0.1*Mc and installations with hazardous materials in quantities less than 0.1*Mc. The same procedure is carried out for materials in pipelines using Critical Length Lc. The contents of a safety report vary from a simple notification of less serious cases to an exhaustively quantitative analysis of risks for the more critical cases. In any case, a minimum safety audit should be carried out and an analysis of historical accident evidence in similar installations.

When defining the completeness and depth of a risk analysis, the sensibility of surrounding environment must be carefully considered. For sensibility to damage , we mean the presence of a highly populated area surrounding the plant up to a distance defined in function of the type of possible dangerous event, or critical public services (schools, hospitals, etc.), main roads or railways, ecologically protected areas, other high risk installations.

This way we avoid considering a plant with a 10 ton mass of chlorine in a semi-deserted area as a high risk, while not considering high risk an installation where there is a 9.5 ton mass of chlorine situated in a densely populated area. The class of risk for an installation or pipeline is increased by 1 level if it is situated in a sensitive area.

The contents of the risk analysis are specified clearly and in detail in the proposed standards and a guide to application of each method (from safety audit to FMEA, HAZOP event/fault tree, calculations of the consequences, estimation of the vulnerability, etc.). Each risk analysis must finish with the preparation of a relative emergency plan for both internal and external emergencies. Standards have been developed to classify the surrounding areas in terms expected impact; for example, when if the hazards is caused by toxic releases the surronding scenario is classified as areas where the dosage is lethal, where the dosage is toxic and areas where the dosage is irritating. The area surrounding the installation must be mapped, pointing out the risk area and specifying the emergency plans for each area.

The contents of the safety report are specified according to the hazard classes: for class 1, high risk installation must carry out a detailed risk analysis; if it belongs to class 2, then the analysis will be less detailed (DOW/MOND method). A class 3 installation needs only a notification and description of the process.

7) CONCLUSIONS

The Brazilian directive for high risk installations has been developed under the initiative of CETESB (San Paulo State Organization for Environment and Safety) by Snamprojetos Consultoria Tecnica ed Informatica, the Brazilian society of the ENIDATA holding. The directive, with relative technical annexes and a complete example referring to Cubatao industrial area, will be active firstly in San Paulo State and the rest of the country will follow.

Design Optimization of New Plants by Reliability Engineering Methodologies: Application to a Subsea Pumping Station

Giovanni Uguccioni (*)
Fausto Zani (+)
Simberto Senni (*)

(*) SNAMPROGETTI S.p.A. - San Donato Milanese (MI)
(+) SYRECO S.r.l. - Gavirate (VA)

Summary

Reliability Engineering methodologies may be successfully applied to the analysis of new plants for which no direct experiences exist. For these plants the risk analysis represents the only possibility to optimize the design in alternative to the "a posteriori" experience.

In this context the conventional methods for the risk analysis have to be calibrated on the particular needs of the case to be analyzed.

The paper shows the main results of an application carried on in SNAMPROGETTI in the framework of SBS (Subsea Booster System) project involving the design of a remotely controlled pumping station to be installed in deep sea (up to -1000 m) for oil/gas reservoirs development. The installation and maintenance costs of the system require to ensure the system reliability; the design solutions will therefore be defined on the basis of a reliability analysis.
The procedure that has been applied starts from the reliability data calibration to enable, by FMEA, HAZOP and Fault Tree methods, the system availability evaluation and the identification of critical components.

On the basis of the reliability model, indications are provided about the expected reliability level for each sub-system and the optimum design configuration in order to achieve the economical feasibility targets.
A sensitivity analysis is also applied to evaluate the influence of the parameters' estimation on the final results and the benefits due to increased component quality level, different functional solutions or component redundancies.

1. System description

The project objective is the development of a sub-sea pumping station designed for the development of marginal oil/gas field in deep sea; the pumping station will be characterized by:

- two-phase fluid pumping capability
- allowance for completely remote operation.

The entire plant is constituted by a sub-sea well head system, connected by flow-lines to a central manifold, from which it is possible to drive the products of each single well toward the test system or (mixed with the production of the other wells) towards the pumping system.

The pumping system consists of pumps designed to handle two-phase fluid in a wide range of void fraction, with a developed head sufficient to boost the products up to the on-shore receiving station.
The electric power required is fed from an on-shore main station by a sub-sea high voltage feed cable; the same bundle will transport the communication signals from/to the on-shore control room to/from the sub-sea control system.

The sub-sea transformers and the switchgears are located in a dry module in inert atmosphere while all the process units of the plant will be installed in direct contact to the external marine environment.
The hydraulic power to the valves' actuators and to the connectors will be given by an electrically fed hydraulic system.

The plant is constituted by separated modules, properly located on a central unit and three identical peripheral pumping units, in order to allow for the retrieval of a single module up to the surface in case of maintenance and repair intervention.

2. General methodology outlines for system availability analysis

During the project basic design, special attention has been focused on the prediction of the system availability.
For technological and economic reasons the remote maintenance and repair of each single component of the system has not been considered practicable.

Repair each interventions are performed on the surface, after the disconnection and retrieval of the plant section (module) of interest. This implies that the system shall be capable to operate without the need of retrieval operation for long periods, in the order of two years.

The main purpose of the system availability analysis has been the identification of the measures to be adopted (in terms of design solutions and component quality and reliability) to guarantee a low failure probability within two years of continuous production.

The availability analysis has been performed in two phases; a first phase aiming to compare five different design alternative solutions in order to identify the most promising one for further development of the design, a second phase focused on the availability evaluation of the chosen solution, aiming to identify the measures to be adopted to ensure the required performance.

The block diagrams of the activities of Phase I and II are shown on Figs. 1 and 2 respectively and shortly presented in the following.

2.1 Phase I - Analysis of process alternatives and comparison with the availability goals.

The five process alternative solutions have been compared in the preliminary phase of the SBS basic design. They differ mainly for the characteristics of the pumping units and the type of machines to pump the oil/water/gas two-phase mixture.

The process flow scheme is presented in Fig. 3, where the main functional sub-systems and auxiliaries are presented.

As a first step, a preliminary analysis of the consequences of single failure has been performed for each design solutions by a Failure Mode and Effects Analysis (FMEA), enlightening redundancies and protections, in order to obtain events combinations (Minimal Cut Sets, MCS) that may lead to the loss of production of the pumping unit (Top Event).

Attention has been focused on MCSs of 1st order, that is single failures that can lead to the Top Event occurrence. For each one of them preventive and protective measures have been identified in order to eliminate single failures and hence increase the total system availability.

Furthermore, the feasibility of the system, from the point of view of the availability targets derived on the basis of economic considerations, has been verified, starting from the most critical components' failure rate evaluation.

For this purpose, the main system's process functions have been identified, assuming a simple "series" reliability block diagram with which the system may be modeled being any function essential to ensure the system full production.

For a "series" system, the mean time to failure (MTTF), for a generic process solution, is given by

$$MTTF = 1 / \Lambda \tag{1}$$

where

$$\Lambda = \sum_{1}^{N} {}_J \, \lambda_J \tag{2}$$

λ_J = being the failure rate of the j-th process function

The system target failure rate, Λ_t, is related to the admissible failure probability (target availability of the system) in two years of continuous operation, by the following expression:

$$P\,[T = 2\text{ years}] = 1 - \exp\,[-\Lambda_t\, T] \tag{3}$$

Assuming a uniform distribution of the system target failure rate among all the N process functions, the target to be achieved for any process function, in terms of mean time to failure, will be:

$$MTTF_J = N / \Lambda_t \tag{4}$$

The availability targets taken into account in the analysis (10%, 20%, 40% failure probability within two operating years without any programmed maintenance of the system) have been distributed among the process function on the basis of eq. (4) and, for each one of them, the achievement of the partial target has been verified on the basis of the MTTF of the corresponding MCSs. When the proposed target cannot be fulfilled, the critical (less reliable) functions have been identified and another distribution of the target has been verified, on the basis of equation:

$$MTTF'_J = 1 / \;\; [\Lambda_t - (1 / MTTF*_K)] / (N-1)$$

where

$MTTF*_K$ = maximum admissible MTTF for function k-th (unless greater than $1 / \Lambda_t$)

$MTTF'_K$ = new target MTTF for all the remaining process functions $j \neq k$.

Assuming standard components' failure rates, provided by literature or Data Bases, the results of the analysis are summarized in the recommendations of Tab. 1, where the required design improvements and component quality standards are provided.

The adoption of 'conventional' components' failure rates implies that the quantitative analysis results can only be considered as a first estimation of the system's performance. As far as the comparison between the alternatives is concerned however the relative merits of the different alternatives can be considered to be modeled with sufficient accuracy for decision purposes. Furthermore, the quantitative results can be viewed as an index of the gap between the proposed application and the conventional ones.

From those recommendations and their severity rankings, from a qualitative point of view, process solutions # 2 and 5 result to be the most promising ones to assure the system availability target of 20% failure probability within two continuous operating years.

2.2 Phase II - Availability evaluation of the selected process alternative

On the basis of the results of the previous phase, also including other design and technological considerations, process solution # 2 has been taken into account for further development of the design.

A suitable procedure has been set up to identify all the Top Events with the related MCSs that could have significant effects on the system production continuity. The application of standard reliability techniques, based on functional FMEA and detailed Operability Analysis, to each sub-systems constituting the plant (central unit, production unit, electrical and hydraulic distribution, telecontrol and remote operation) has allowed the development of Fault Trees for the following Top Events:

TOP 1 Complete loss of SBS system production

TOP 2 Loss of a single pumping unit production (1/3 of the total)

TOP 3 Loss of 50% of a pumping unit (1/6 of the total)

TOP 4 Loss of 1/3 of a pumping unit (1/9 of the total)

TOP 5 Loss of a single flow-line or well (1/12 of the total)

For each one of them a logical-probabilistic analysis has been performed, aiming to the identification of critical components and Top Event probability evaluation.

Quantitative analysis has been executed assuming selected failure rates for each involved component and calibrating them according to the SBS design characteristics.

A sensitivity analysis has been also performed to evaluate the effects of uncertainty, assumptions and possible design improvements on the final results.

3. Some specific aspects of the analysis

The methodology adopted follows the current system reliability analysis techniques.
Nevertheless, some unconventional aspects of the SBS project have required the development of suitable procedures and approaches to apply standard methodology, taking into account the existing uncertainty on components' characteristics and behavior according to the actual operating conditions.

In the following paragraphs the procedures developed for selection and calibration of reliability data and sensitivity analysis are described in detail.

3.1 Reliability data selection and calibration

The lacking of data that could be applied to the SBS components operating and environmental conditions has required to develop a procedure to attribute the reliability parameters (failure rates or component unavailability) to the basic events of the Fault Trees.

Moreover, in some instances SBS design requirements can only be satisfied by new components for which no reference data exist in literature; in these cases the designer's judgment or laboratory test results (if any) should be assumed.

For all these reasons the procedure of data selection, combination and calibration outlined in Fig. 4, has been set up, in which special emphasis is given to the expert judgment elicitation for the calibration of existing data or the attribution of reliability figures for new components.

Data selection has been performed by attributing a "score" to each one of the following attributes:

- component engineering characteristics
- component failure mode
- component environmental and operating characteristics

- data source.

This partial scores have been combined to obtain a final "applicability index" for the data available. The data characterized by the higher score have been selected as the most reliable ones from the point of view of their application in the SBS project context.

The pooling of the selected data has been performed according to a simple geometrical average, in order to emphasize the importance of the exponent of the reliability figures to be combined w.r.t. their mantissa, keeping in mind the uncertainty with which the data are available.

The "applicability index" has allowed to define criteria (Table 2) for the applicability of the data obtained from the previous phases; from these criteria the actions to be performed for further investigations and improvements of the final data to be introduced in the Fault Tree analysis.

In the procedure, a special emphasis has been given to the experts elicitation and their combination, by means of an inquiry procedure derived from a standard Delphi methodology. The information required is mainly related to the evaluation of a corrective factor to be applied to the available data or, alternatively, to direct estimation of a MTTF (with related uncertainty bounds) in case of new components for which no data are available from literature or Data Base.

In Table 3 the module for data collection from experts is presented both for existing and new components.

3.2 Sensitivity analysis

The results of the logical and probabilistic analysis of Fault Tree enlighten the criticality of some components or process functions for each Top Event of interest.

Examples of distributions of initiating causes leading to the loss of a single pumping unit or control unit are presented respectively in Figs. 5 and 6, while in Fig. 7 is included the criticality index of the process component of a pumping unit, defined as:

$$I_J = \Sigma_i \; Q(MCS_J)_i \; / \; Q(TOP)$$

where

$Q(MCS_J)_i$ is the probability of MCS i-th that involves the j-th event

$Q(TOP)$ is the total Top Event probability.

For these critical components a sensitivity analysis has been performed to verify the benefit (in terms of higher expected reliability) of:

i) best component quality and hence greater availability (longer associated MTTF)
ii) introduction of suitable redundancies for the most critical components.

An example of the sensitivity analysis, for each type of the above suggestions i) or ii), is reported respectively in Figs. 8 and 9.

In these figures the Top Event probability at the mission time of 2 years is reported as a function of a reduction coefficient (ranging from 0.1 to 1.0) to be applied to the failure rate available from Data Base or literature for standard components.

The benefits of the interventions aiming to improve the component quality have been verified for valves, pumps, pressure sensors, safety valves and RTU of the process control unit in Fig. 8, while the effects of the introduction of one or more redundancies have been studied for the communication modem in Fig. 9. The latter shows that the modem redundancy has a greater effect on the communication system than the possible improvement on the RTU quality.

4. Main results

Intermediate results of the SBS availability analysis are presented in Tab. 2 that includes the required intervention to satisfy the reliability targets, as far as the Phase 1 of the study is concerned.

The phase II analysis has provided a system availability evaluation, that, under the hypothesis of standard components, does not meet the prefixed targets.

Therefore, some interventions both at a design level (electronical components redundancy and process functions simplification) and at a component quality requirements level have been identified.

The set of all the possible improvements, identified by subsequent design review (Rev. 1.0, 1.1 and 1.2) with respect to the basic design (Rev. 0.0) leads to the final results of Fig. 10, from which the effects on SBS system availability can be easily evaluated as a function of time up to the mission time of 2 years.

5. Conclusions

The application of system reliability analysis techniques (FMEA, Operability Analysis, Fault Tree) allows for useful information for feasibility study and basic design of plants, for the comparison of different alternatives and the definition of the interventions to improve the system performances, other than its safety operation.

In case of new plants, for which no applicable statistical data are available, special attention has to be paid to the selection, calibration and evaluation of the reliability parameters to be applied in quantitative analysis. For this purpose expert judgment can be the only applicable way to be undertaken.

The extensive application of sensitivity analysis permits to evaluate the effects of improvements in components' quality and design; a cost-benefit analysis can then be operated to take decisions during the project development.

Special emphasis is given to the reliability characteristic requirements for the most critical components; they have to be assured from the design phase to the construction activities, in the framework of a suitable integrated quality assurance system.

The methodology applied has not allowed to take into account some important factors, like those related to the intervention operations. For this purpose, a new improved methodology, set up in SNAMPROGETTI, based on decision tree (ADMIRA) will be applied in the next future.

References

1. Uguccioni, G.; Zani, F.:
Analisi di disponibilità dell' impianto SBS.
FMEA, Analisi di Operabilità, DSI ed identificazione dei punti critici di impianto
Snamprogetti Internal Report - Agosto 1987

2. Uguccioni, G.; Zani, F.:
Procedura di selezione, combinazione e calibrazione dei parametri di affidabilità
Snamprogetti Internal Report - Ottobre 1987

3. Apostolakis, G.:
Expert judgement in Porbabilistic Safety Assessment
Societa' Italiana di Fisica - Corso Lerici August 1986

4. Linskore, H.A.; Turoff, M.
The Delphi method: technique and application
Addison Wesley Publishing Company, 1975

5. OREDA
Offshore reliability data handbook
Ist edition, VERITEC, Oslo 1985

Tab. I Requirements for reliability targets obtainment

SOLUTION	FLOW CONTROL	WELL PERIODICAL TEST	SYSTEM START-UP	GAS/LIQUID PHASES SEPARAT.	VOID FRACTION EVALUATION	GAS EXTRACTION/ COMPRESSION	LIQUID PHASE FILTERING	HYDRAULIC POWER DISTRIBUTION	BIPHASE PUMPING
1.	- (A)	(A),Q,C (D) (F),(S)		Q,C		Q,C,D (F)	Q,(A)		Q,C,D,F (S)
2.	- (A)	(A),C,Q (D) (F),(S)			Q,C,D		Q,(A)		Q,C,F (D)
3.	- (A)	(A),C,Q (D) (F)						(A),Q,C,D	Q,C,D,F (S)
4.	- (A)	(A),C,Q (D) (F)						(A),Q,C,D	Q,C,F (D)
5.	- (A)	(A),C,Q (D) (F),(S)			C,Q	(A) (C)	Q,(A)	(A),Q,C,D (F)	(A),(Q)

Legend:
- \- no intervention required
- A further investigation required
- Q component quality improvement
- C component redundancy
- D process function target reduction
- F total function redundancy
- S function segregation in a separate module to allow for most frequent interventions

Between branches, those interventions that should require further investigations.

Tab. 2 Data acceptability criteria and actions to be performed.

ACTION	ENGINEERING CHARACTERISTICS	FAILURE MODE	OPERATING ENVIRONMENTAL CONDITION	SOURCE
a) Applicable data	A	A	A	A
	A	A	A	S
	A	S	A	A
	A	S	A	S
b) Data calibration required. (Delphi method)	A	A	S	-
	A	S	N	-
	A	A	A	N
	A	N	A	-
	A	N	S	-
	A	A	A	-
	A	S	S	-
c) Independent experts estimation required.	A	N	N	-
	S	-	N	-
	S	N	A	-
	S	N	S	-
	N	-	-	-

A Applicable
S Partially applicable
N Not applicable
\- Whichever A or S or N

Tab. 3 <u>Module for data gathering by expert inquiry</u>.

COMPONENT INFORMATION

Component

Application

Main Engineering Characteristics

Operating and Environmental Conditions

Failure Mode of Interest

EXISTING COMPONENTS

ESTIMATION OF RELIABILITY PARAMETERS FOR COMPONENTS FOR WHICH DATA ARE AVAILABLE

The mean time to failure (expressed in hours) obtained from the available information for similar components is estimated as:

MTTF =

The upper and lower values are:

$MTTF_{min}$ =
$MTTF_{max}$ =

These values are not directly applicable to this case for the following reasons:

It is requested to give an erstimation for a multiplicative corrective factor to these MTTF data to make it more suitable to the above descriptions.

NEW COMPONENTS

ESTIMATION OF RELIABILITY PARAMETERS

On the basis of the information provided, an estimation of the reliability parameters of the component is requested; this estimation can be based on the knowledge of information related on similar components or on the engineering experience of the expert.

It is requested an estimation of:

- <u>most probable MTTF</u>:

  MTTF = (hours)

- <u>upper MTTF</u> (maximum credible operation time without failures):

  $MTTF_{max}$ = (hours)

- <u>lower MTTF</u> (minimum operation time without failures):

  $MTTF_{min}$ = (hours)

<u>Additional information</u>:

Is the estimation based on experimental data, and, if affirmative, can these data be specified? (tests, literature etc.)?

Which are the main assumptions that have lead to this reliability parameters ?

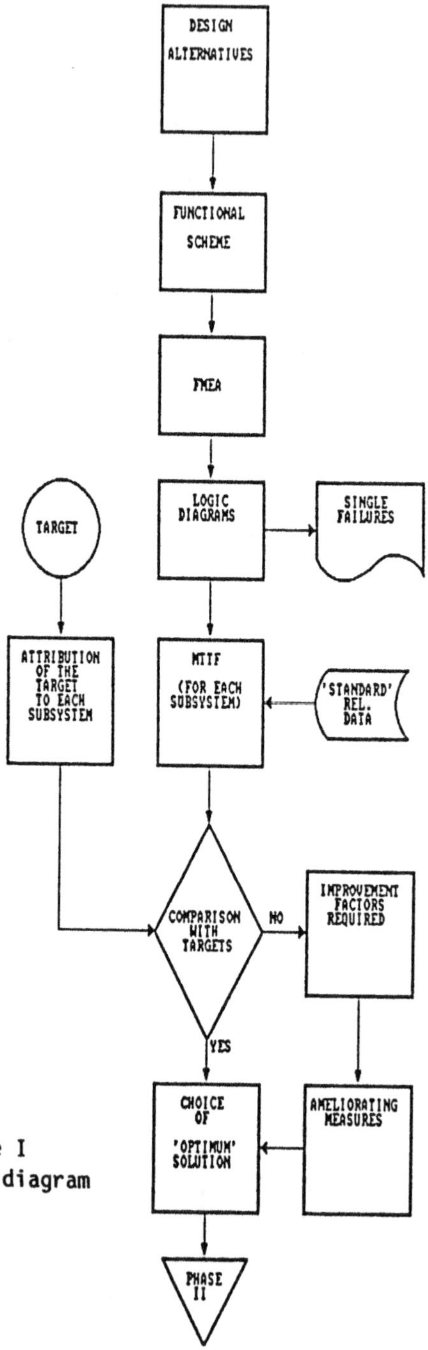

Fig.1 Phase I flow diagram

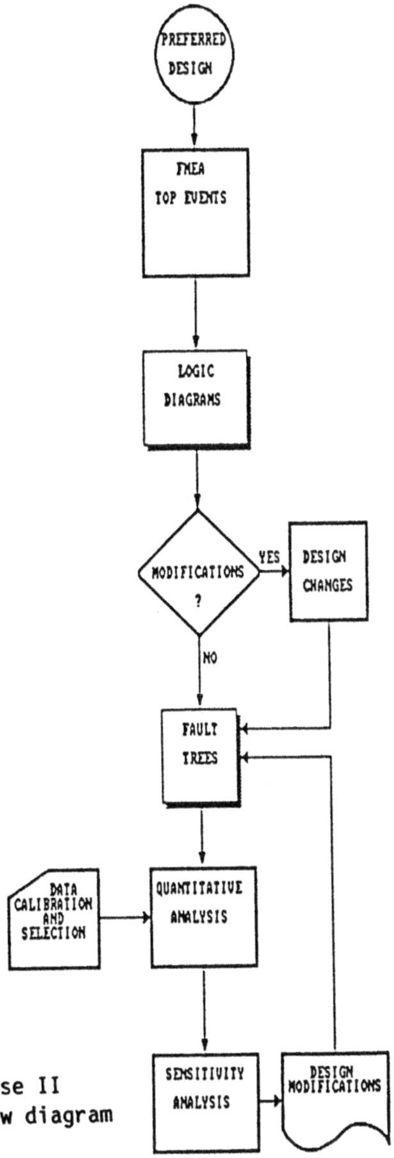

Fig.2 Phase II flow diagram

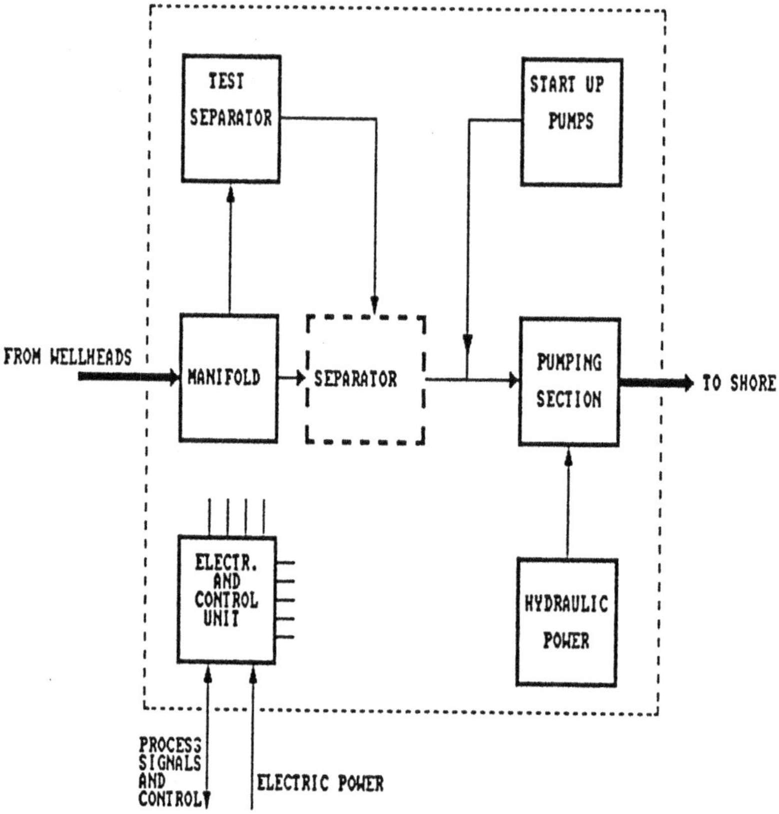

Fig. 3 SBS system block diagram

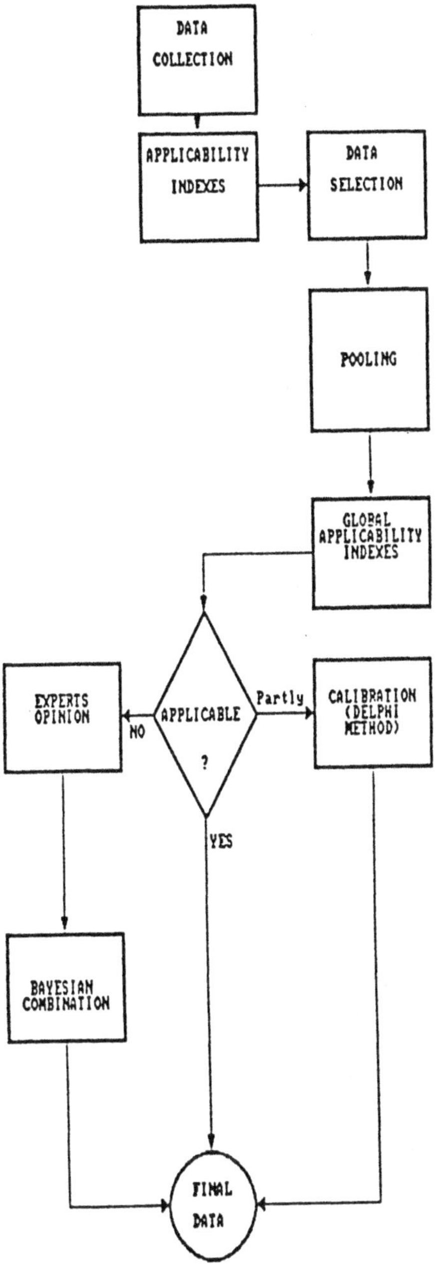

Fig. 4 Data selection and calibration procedure

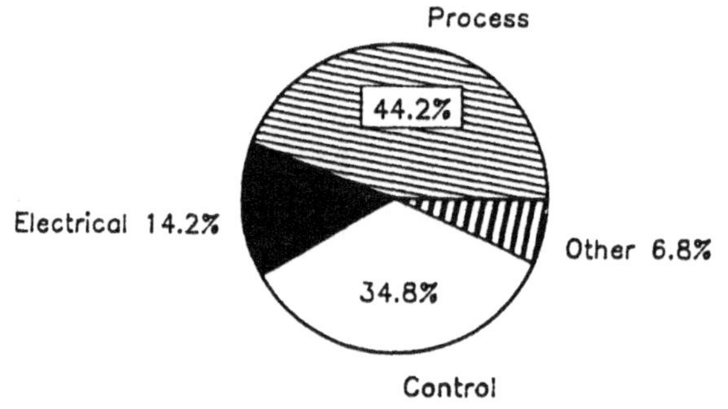

Fig. 5 TOP 2 — Initiating causes
Loss of production of 1 process module
Mission time T = 2 years

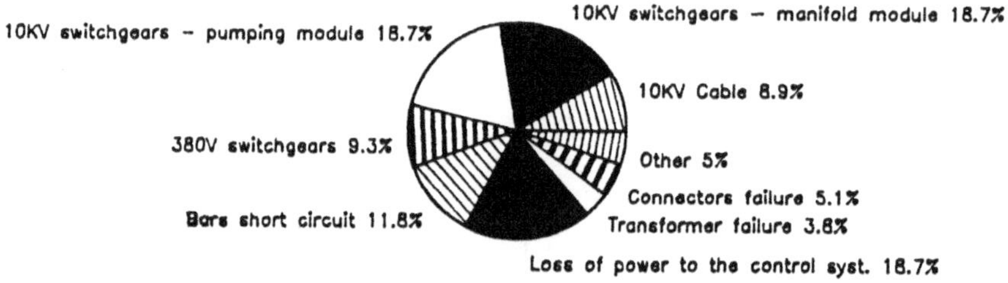

Fig. 6 TOP 2C — Failure causes
Loss of production due to control failure
Mission time T = 2 years

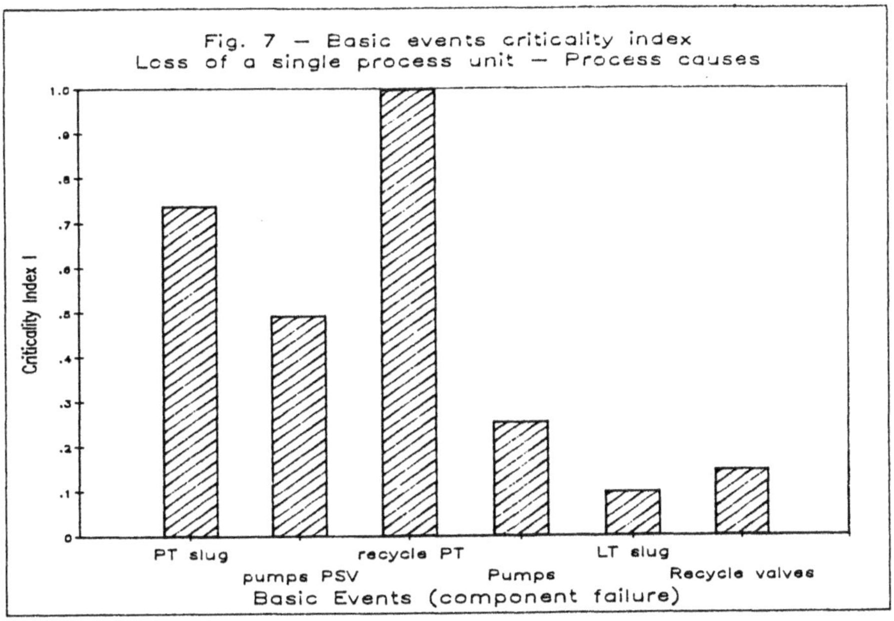

Fig. 7 — Basic events criticality index
Loss of a single process unit — Process causes
Criticality Index I
1.0
.9
.8
.7
.6
.5
.4
.3
.2
.1
0
PT slug
pumps PSV
recycle PT
Pumps
LT slug
Recycle valves
Basic Events (component failure)

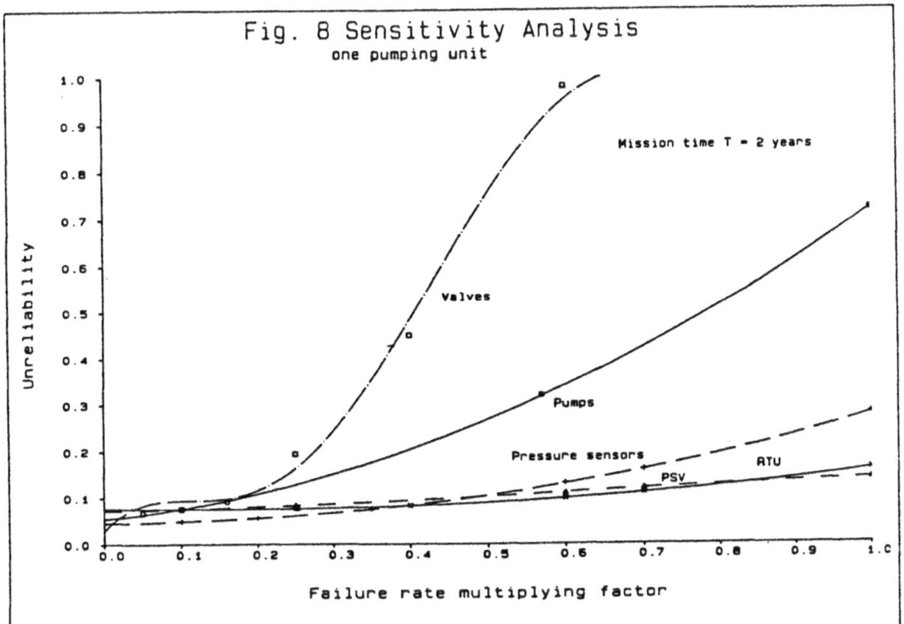

Fig. 8 Sensitivity Analysis
one pumping unit
Mission time T = 2 years
Unreliability
Valves
Pumps
Pressure sensors
RTU
PSV
0.0
0.1
0.2
0.3
0.4
0.5
0.6
0.7
0.8
0.9
1.0
Failure rate multiplying factor

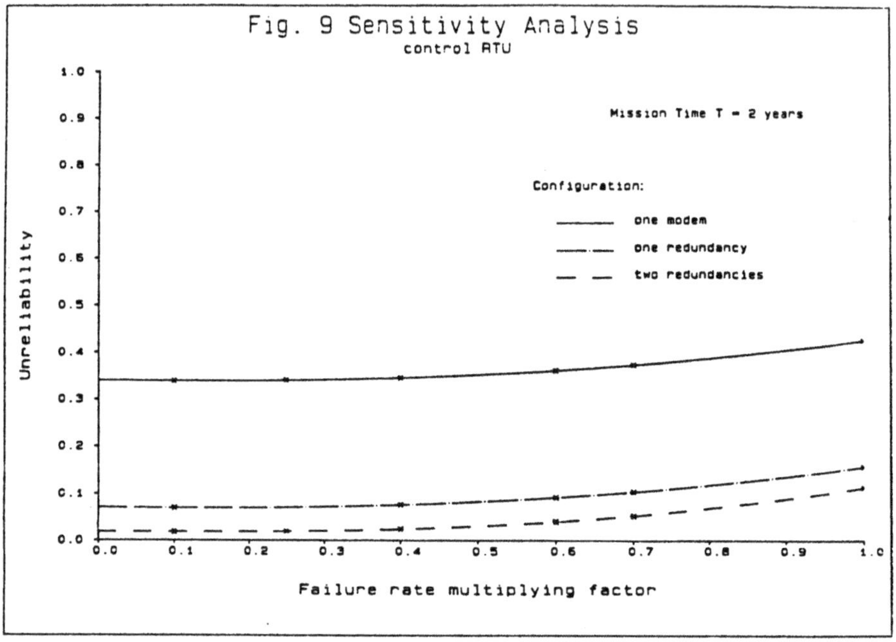
Fig. 9 Sensitivity Analysis
control RTU
Mission Time T = 2 years
Configuration:
one modem
one redundancy
two redundancies
Unreliability
Failure rate multiplying factor
0.0
0.1
0.2
0.3
0.4
0.5
0.6
0.7
0.8
0.9
1.0

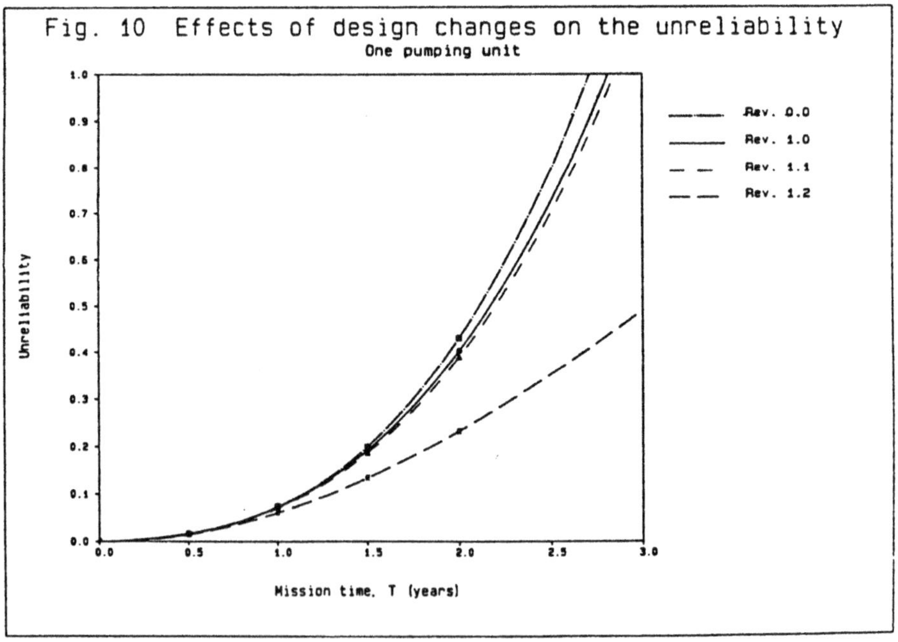
Fig. 10 Effects of design changes on the unreliability
One pumping unit
Rev. 0.0
Rev. 1.0
Rev. 1.1
Rev. 1.2
Unreliability
Mission time, T (years)
0.0
0.5
1.0
1.5
2.0
2.5
3.0

Risk Assessment of Operational Aspects in Offshore Drilling

R. Østebø and T. Gjerstad

Technica a.s
Norsea Base, Dusavik, P.O. Box 138
N-4001 STAVANGER, NORWAY

ABSTRACTS

A new and different type of risk assessment of the operational aspects of offshore drilling is discussed. Main emphasis is given to a shallow gas risk analysis which were performed to evaluate different risk-reducing measures. Consideration is also given to a relatively new hazard identification technique and its use in offshore drilling, as well as the importance of risk management.

1. INTRODUCTION

Offshore drilling may be associated with several potential hazards and operational problems, and it is one of the most critical operations being performed in connection with offshore oil and gas exploration and field development. Drilling related accidents such as the West Vanguard shallow gas blowout (Nov. 1985), the Enchova blowout (Apr. 1988) and the Ocean Odyssey blowout (Sept. 1988) remind us that blowout is a significant hazard which should be subject to safety and risk analysis, in order to optimize safety and regularity.

The purpose of this paper is to outline a new and different type of risk assessment of drilling operations which have proved to be successful in recent studies performed by Technica. The study was undertaken to establish the historical blowout risk profile associated with shallow gas, and to evaluate on a common format and framework how and to what extent the risk would be reduced if specific measures were implemented. The use of risk management and qualitative & quantitative analytical techniques to control and prevent drilling accidents are discussed.

2. SHALLOW GAS RISK ANALYSIS

2.1 Background

A shallow gas blowout is characterized by drilling into a high-pressure gas zone in the tophole section of a well, with both primary and secondary control measures failing. The hydrostatic pressure of the mud in the wellbore constitutes the primary barrier, while subsea or surface valves typically make up the second barrier. The special concern with the blowout risk during top hole drilling as compared with reservoir blowouts, is that shut-in of the well may not be possible (BOP not installed) or derirable at shallow depths (formation brekdown risk), as well the experienced problems with surface diverter arrnagement.

A shallow gas blowout in the North Sea initiated a major program by a Norwegian oil company to identify and recommend corrective measures within certain selected problem areas; refer to Table 1. This project series was subject to overall analysis by Technica to evaluate the risk reduction benefit and the safety merits of the recommendations being put forward.

1) Tophole drilling procedures
2) Drilling without marine riser (no mud return to rig)
3) Cementing with silica (to avoid gas migration)
4) Formation blocking (to avoid gas influx in wellbore)
5) Rig ventilation shutdown (to avoid ignition of escaping gas)
6) Anchor release systems (to move off location)
7) Diverter system (to avoid erosion)
8) Seismic methods (to improve shallow gas predictions)

TABLE 1: EXAMPLE OF SHALLOW GAS RISK-REDUCING MEASURES

2.2 Shallow Gas Risk Management

The above-mentioned areas illustrate the variety of measures which could reduce the shallow gas risk during drilling from a mobile rig. These risk-reducing measures include operational procedures and equipment modification relevant to both normal and emergency operations, as well as planning requirements prior to drilling commencement. Risk analysis provides a useful tool in the risk management as part of the decision-making process when selecting measures to be implemented.

Proposed measures for risk reduction may sometimes be presented individually as a basis for selection, with the result that no optimum decision with respect to safety may be achieved. However, by the development and use of an integrated approach in this study, it was possible to assess the safety and economic implications of each individual measure on a common format according to a baseline risk level. The underlying main assumption was drilling in an area where shallow gas could be expected.

2.3 Database Establishment

A common problem in risk analysis is that accident statistics are too general to be of much value in a specific context. This may sometimes lead to many assumptions having to be made in the analysis, which again could affect the robustness of the results. The assumptions may of course be justified by engineering judgement and sound industry practice, but ideally more experience data should be used if the study scope allows for it.

The intention of the shallow gas study was to make extensive use of detailed top hole well history data, as well as performing a state-of-the-art survey of worldwide shallow gas blowout events. The database established provided a unique, detailed and complete risk picture for the type of operations under study.

2.3.1 Worldwide Shallow Gas Events

A thorough investigation of 55 worldwide shallow gas blowouts during exploration and development drilling was performed. Cross-checking of different data sources was necessary to make the new database as complete and correct as possible. This data quality review identified weaknesses in external databases. As incident reporting is rather poor before the 1970's, the worldwide drilling activity in the period 1974-85 was used to predict the following basic blowout frequencies:

- P(shallow gas blowout while drilling an exploration well) = 2.0×10^{-3}

- P(shallow gas blowout while drilling a development well) = 7.2×10^{-4}

Considering all blowouts where the type of well is known (exploration or development), 67% of the shallow gas blowouts occurred during exploration well drilling. The lower frequency can generally be explained by more knowledge about formation, location and type of well control equipment etc.

The shallow gas database make it possible to investigate and quantify safety and operational aspects such as:

- Type of installation
- Operation in progress when blowout occurred
- Use of diverter and BOP to control blowout
- Emergency disconnection and move-off of mobile rig
- Fire and explosion hazards

2.3.2 Well History Handbook

An in-depth data collection and processing of drilling experience from 61 exploration wells on the Norwegian Continental Shelf was performed. The database established made it possible to quantify risk aspects in a way which had not been done before. A key feature of the data excercise was the wide coverage of statistics which gave input to the overall risk analysis, i.e.:

a) Shallow gas predictions (seismic surveys prior to drilling)
b) Warning signals during drilling
c) Normal and abnormal well control actions
d) Equipment failures
e) Operational problems

2.4 Human Factors and Drilling Safety

The nature of offshore drilling operations requires human interpretations and the response/role of operating crew to have the same importance as equipment performance and reliability. A rather new approach to this important issue was developed as part of the shallow gas risk analysis, providing new type of data related to drilling safety:

i) The human element in seismic interpretations prior to drilling

ii) Comparison between perceived and actual reliability of shallow gas warning signals during drilling

2.4.1 Seismic Predictions

By knowing in advance the exact depth where shallow gas is present top hole drilling operations can adequately take shallow gas risk into account at the planning stage. In this study the shallow seismic surveys were subject to probabilistic assessment to quantify the successfulness of these predictions. The statistical comparison of the shallow gas predictions versus actual experience while drilling, were based on 61 wells in areas where shallow gas generally could be expected. Predicted gas zones were compared against any encountered gas zones.

A human factors analysis of the decision problems associated with shallow gas predictions was performed using socalled signal detection theory. Evidence of shallow gas ("signal"), on a seismic map, could be misinterpreted with something that looks like shallow gas ("noise"). This confusion is influenced by two factors. One is the extent to which the signal and noise distribution overlap each other, i.e. how easily they can be distinguished by the person doing the task, also called "sensitivity". The second factor, is the individual choice of criterion of when to call an indication a signal or noise, also called "response bias".

A quantitative calculation of sensitivity and response bias was performed for various geographical areas, using the following parameters (see Figure 1):

- False alarm: A noise event is treated as a signal
- Correct accepts: A noise event is correctly predicted as lack of shallow gas
- Misses: A signal is treated as noise
- Hits: A signal is correctly predicted as shallow gas

FIGURE 1. A VISUAL DESCRIPTION OF THE DEFINITION OF HITS, CORRECT ACCEPT, FALSE ALARMS AND MISSES

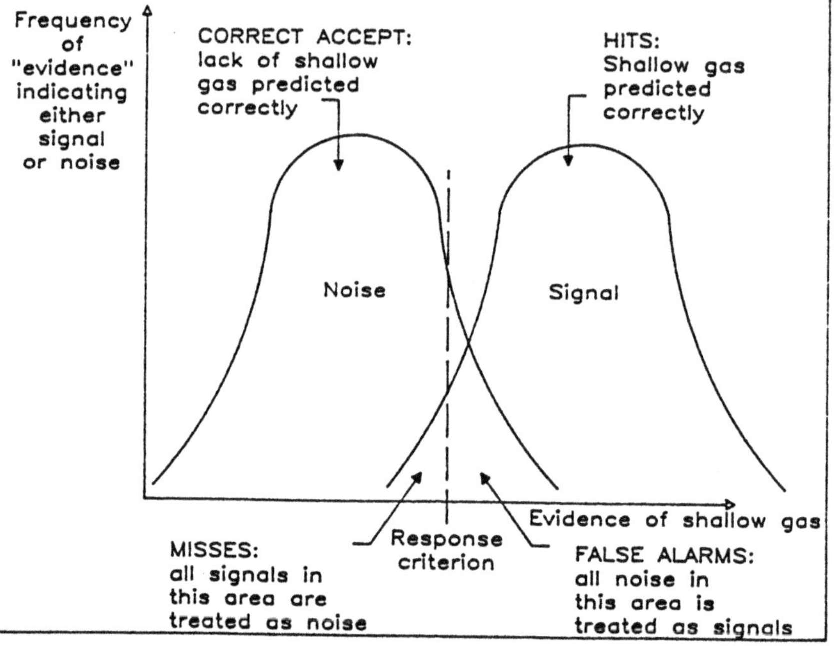

FIGURE 2
PROBABILITY OF HITS VERSUS PROBABILITY OF FALSE ALARM

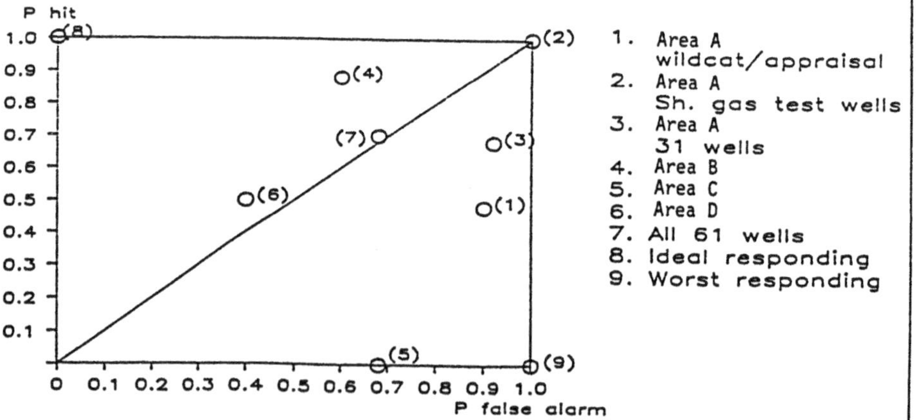

The results for the various geographical areas are shown in Figure 2. The diagonal line represent predictions which are chance responding, i.e. values above this line are better than chance (=tossing a coin), and vice versa. It is generally evident that the chance of making a correct prediction based on seismic analyses is poor: in the order of 50/50 per well. One may argue that these predictions have an adverse effect on safety, since, if taken seriously by the drill crew, they may result in lack of attention to shallow gas warning signals at depth intervals predicted as safe. This study indicates therefore that the drill crew should pay very little attention to non-shallow gas predictions, and be constantly alert during top hole drilling, using warning signals.

2.4.2 Warning Signals

Various parameters can be used during drilling operations to detect high pressure zones and/or gas bearing sand zones in the shallow part of the well; ref. Table 2. Totally, more than 400 drilling situations were analysed with respect to warning signals and well control actions. This did also include lost circulation/returns, mud loss, swabbing events, as well as serious shallow gas events such as blowouts, severe gas kicks and gas migration.

Drilling parameters	**MWD parameters**
1. Rate of penetration (ROP)	8. Gamma ray
2. Weight on bit (WOB)	9. Resistivity
3. Torque	
Drilling mud parameters	**Electric logging**
4. Gas trends in return mud	10. Gamma ray
5. Mud level in pit	11. Resistivity
6. Mud return flow	12. Interval transit time (sonic)
7. Sand in shale shaker samples	

TABLE 2: SHALLOW GAS DETECTION PARAMETERS

A questionnaire survey was carried out among on- and offshore drill crew, engineers and supervisors to evaluate human confidence in the above-mentioned shallow gas warning signals. A paired comparison technique was used: The 12 indicators were presented in pairs and respondents were asked to identify which of the indicators in each pair they had most confidence in, ignoring other factors. The combined results showed a clear preference for the following ranked warning signals:

1. Increasing mud return flow
2. Increasing mud level in pit (pit gain)
3. Increasing gas in return mud (gas peak)
4. Increase in ROP (drilling break)
5. Increase in resistivity (MWD)

It should be noted that the two first indicators will appear very late in a gas kick sequence, and hence their importance in an effort to avoid a serious incident to develop is marginal. In nearly half of serious shallow gas events these signals were experienced before remedial well control action was taken. The actual reliability of the three other indicators was estimated as follows:

Gas peak:

1 out of 3 reported gas peaks indicates a true gas zone. Furthermore, only 1 out of 3 gas zones generates a gas peak. Flow check response was taken in 62% of the reported gas peaks.

Drilling break:

1 out of 8 reported drilling breaks indicates a true gas zone. Furthermore, only 1 out of 8 gas zones results in a drilling break. Reported drilling breaks were acted upon in 96% of the cases.

Increased resistivity (MWD):

Measurement while drilling (MWD) is a relatively new method which proved to give positive response in 3 out of 4 cases when penetrating a shallow gas sand. When MWD worked, it detected gas in 87% of the cases where electric logging showed gas.

Gas peaks and drilling breaks seem to be less reliable than MWD, which is somewhat different from the perceived reliability opinion by the drilling personnel.

Although the combined questionnaire results indicates a clear ranking of the most critical indicators, this analysis showed significant differences in the individual responses. Some personnel seem to have very poor confidence in some of the above indicators. This may explain why, in retrospect, many serious drilling incidents seem to take place despite clear indications of downhole problems: Personnel with different points of view and biases as to what constitutes a significant warning may not be able to communicate adequately, resulting in filtering and bypass of important information. This is perhaps one of the most important safety problems in offshore drilling operations.

2.5 Quantitative Risk Analysis of Shallow Gas Project Measures

In addition to an individual qualitative assessment of each shallow gas measure, the study objective was also to evaluate on a common format the effect on risk level by implementing these. The combination of the established database, human factors methodology and risk analytical techniques made it possible to perform a unique and overall probabilistic risk assessment.

A Cause-Consequence Diagram (CCD) model was the developed for this purpose. An example of such an approach is shown in Figure 3. The CCD model generates two risk parameters: the probability of experiencing a critical shallow gas incident, and the expected average cost arising from shallow gas problems. Comparison was made with a base case, which represented the situation before the shallow gas blowout. The model indicated a 9% probability (per well in a shallow gas area) of experiencing a critical incident. Such incidents would include gas releases on rig surface or underneath the rig, both ignited and unignited cases. The expected average cost arising from shallow gas problems was predicted to be 12 million NOK per well when drilling in an area with shallow gas sand zones. Note that this is an average. Most wells will cause little or no extra costs due to shallow gas problems, whereas any accident could generate significant costs well in excess of 12 million NOK.

The overall results are presented on the graphs in Figure 4 and 5.

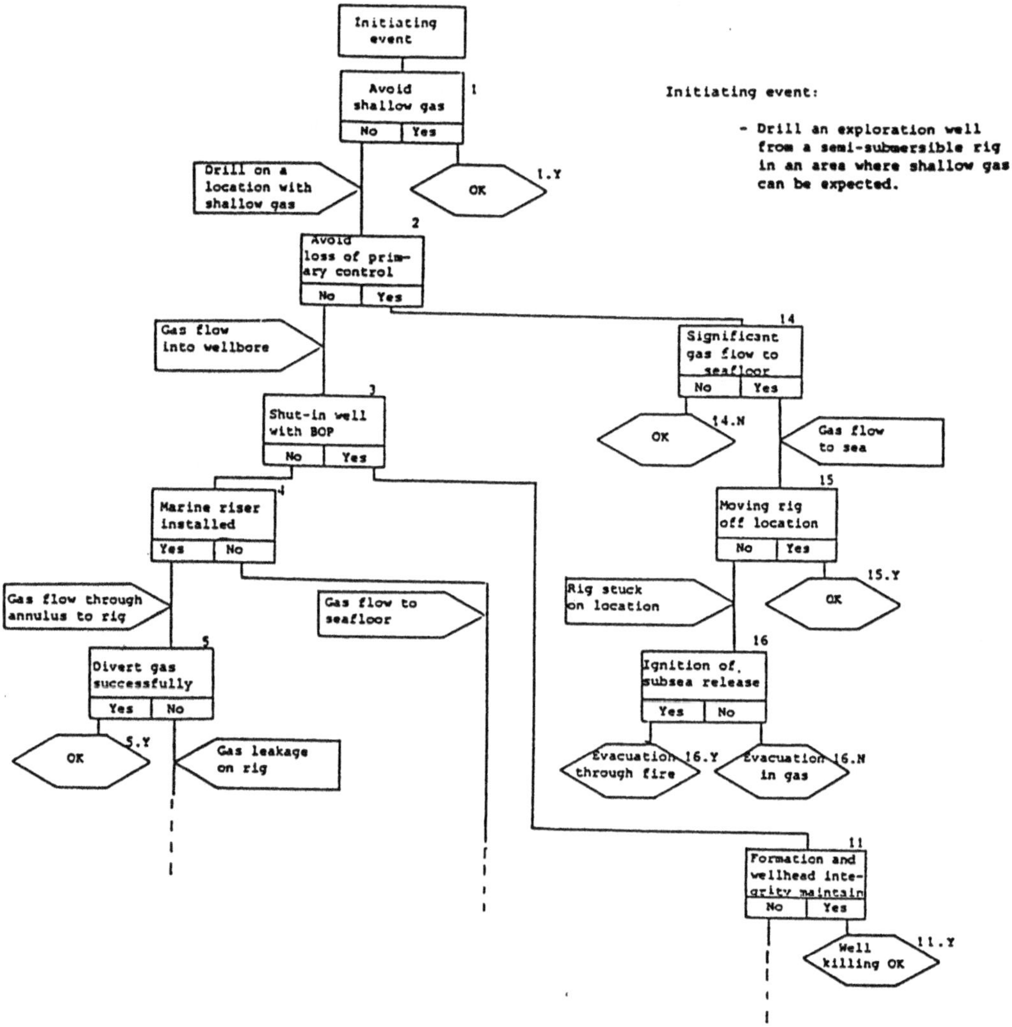

FIGURE 3 : SHALLOW GAS CAUSE-CONSEQUENCE DIAGRAM (CCD). (PARTIAL)

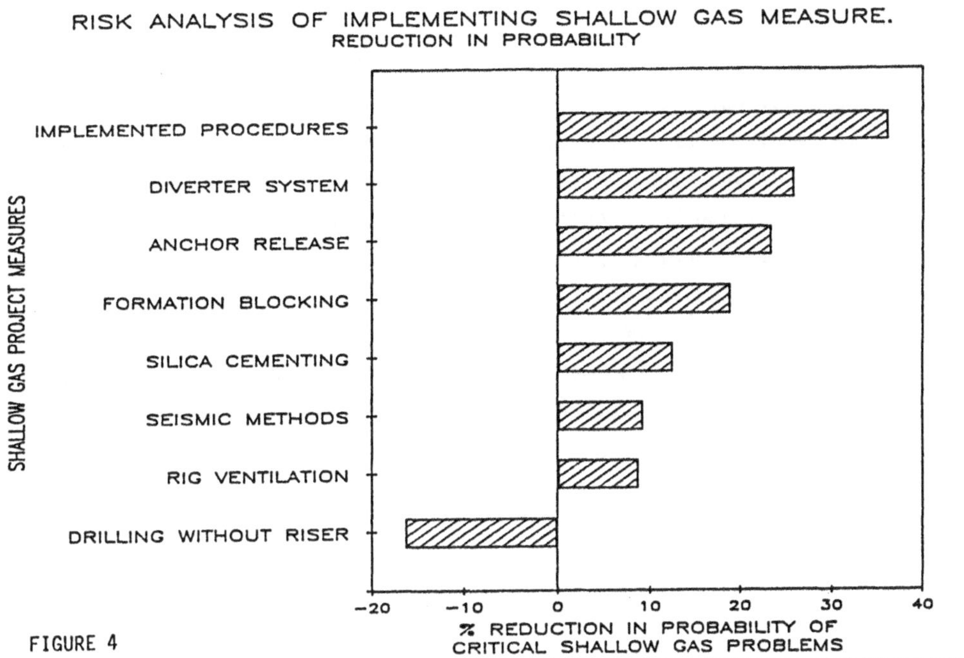

FIGURE 4

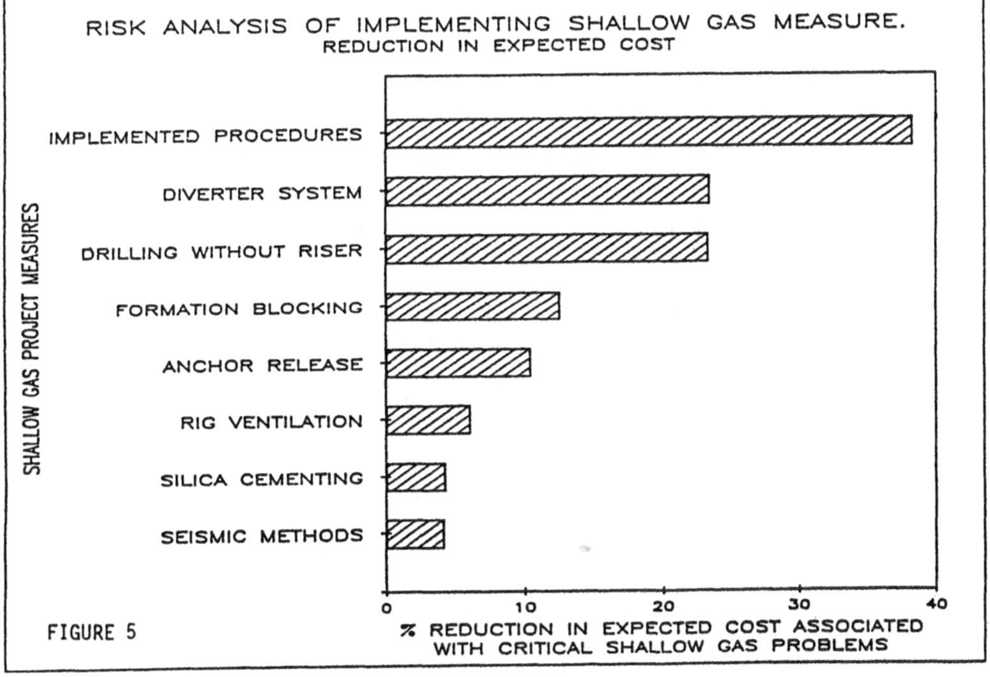

FIGURE 5

3. SYSTEM FOR DRILLING OPERATIONS BREAKDOWN AND ANALYSIS

The experience gained from the shallow gas risk analysis and recent safety and reliability work for other drilling related issues, demonstrate generally the importance of evaluating the operational aspects in addition to equipment design alone. A method has been developed which make it possible to evaluate how proposed equipment design matches the operational requirements, e.g. during operations such as drilling, completion, workover and other well interventions. Technica has developed a framework for systematic breakdown of operations, problem identification, documentation and follow-up which can be used as a tool for risk management in drilling, as well as in other applications (e.g. diving, lifting). Two main elements are:

- Operational Breakdown (OBD) System
- Driller's HAZOP

3.1 OBD System

The OBD system provides a tool for controlling the safety level of a rig and the associated drilling operations from the design stage to the operational phase. This risk management system was developed in connection with a subsea field development and the associated rig interventions. The general purpose is to provide a "database" system which contains:

- Breakdown of operations to take place
- Associated potential problems which could occur
- Proposed measures to reduce risk
- Documentation of follow-up actions by rig owner/operator etc.

The operational breakdown involves defining the operating procedures in terms of phases, operations, activities and tasks in a suitable and systematic format. Depending on the degree of information available and study scope, all or selected parts of this can be subject to identification of hazards and operational problems (e.g. by using the Driller's HAZOP approach).

The critical situations (e.g. operation or task) and potential problems which are identified should subsequently be reviewed to identify corrective and risk-reducing measures which could be taken. Possible improvements could be done on equipment hardware (subsea or surface), organisation of work and procedures and operating limits. Even if this does not seem to represent anything new in risk analysis, the important issue is the systematic recording of study results and follow-up requirements.

Traceability is a key word in this context, to provide a clear overview of the evaluation process and selections made for implementation. The reasoning behind important safety philosophies and equipment design are sometimes "lost" in the course of project development from conceptual design to the operational stages. In the worst case, sound built-in safety functions and design principles can be "removed" without realizing their existence. As different personnel will be involved at the various stages, lack of documentation is generally a main problem. The above-mentioned system allows for "bookkeeping" of critical areas related to drilling operations when updatings, redesign etc. are made.

Cost factors may be assigned to the recommendations, while probability estimates for significant critical areas can be used to allow for cost/-benefit evaluations when deciding follow-up actions.

The follow-up of recommendations put forward in a safety study would generally depend on economic implications, other implemented safety measures etc. However, a system for good coverage of potential problem areas will often be worthwhile having for safety and risk management purposes. It will also be useful for demonstrating the compliance with regulatory requirements as well as documenting the safety philosophies. At the same time, it will be important to document the reasons for any rejected recommendations, if the potential problems are controlled by other means, not cost-beneficial etc.

3.2 Driller's HAZOP

The Drillers' HAZOP is a newly developed method for reviewing drilling equipment and drilling operational procedures in order to improve safety and operability. "HAZOP" is an acronym for "Hazard and Operability Study" which traditionally has been related to the socalled Process HAZOP's, which for many years have been used with great success in the processing industries.

While the Process HAZOP is based on a review of Process and Instrumentation Diagrams (P&ID), the systematic breakdown of operating procedures (OBD tables) provides the basis for a Drillers' HAZOP. A more complete risk picture is obtained by this supplementary technique. Another key element of the method is the systematic use of the operators expert knowledge (in session meetings) applied to existing or new design and operations. This multi-disciplinary approach provides a method of evaluating how equipment design matches operational requirements.

In the HAZOP session itself, the master document (i.e. OBD tables) is examined task by task. The systematic procedure involves the HAZOP chairman to ask a series of questions in order to identify deviations from the norm, and subsequently causes, consequences and recommendations when necessary.

Recent applications of the method in the offshore industry shows that the method represents an efficient way of identifying hazards and operational problems. Unsatisfactory operational specifications which can limit human and equipment performance, can be detected at a planning stage. Both normal and emergency procedures can be verified by the method. Standards for procedural design can be achieved by integration of Human Factors techniques.

Potential problem areas will generally relate to equipment design, procedures or organizational aspects. Key issues may be fed into the design process, development of procedures or evaluated further by other type of safety, reliability or risk analyses.

4. CONCLUDING REMARKS

Risk analysis and safety assessment methodology provides useful tools to carry out a critical and systematic review of drilling equipment and the associated operations involved. The development and use of an integrated approach has made it possible to evaluate and quantitatively rank on a common framwork the particular potential problems and possible risk-reducing measures associated with shallow gas blowouts. To optimize offshore drilling safety generally, it is particularly important to review operational procedures in combination with equipment design and the man-machine interface.

5. ACKNOWLEDGEMENT

The authors gratefully acknowledge the support and assistance provided by Statoil in the execution of the shallow gas risk analysis project, as well as the sponsors for the Driller's HAZOP project which were NTNF, Statoil, Norsk Hydro and Saga Petroleum in addition to drilling expertise knowledge provided by the drilling contractors Wilh. Wilhelmsen and Norcem Drilling.

Availability Assessment of Complex Distributed Control System of a Petro Chemical Plant

G. PICCIOLO

EniChem Anic, Safety/Reliability Dept. S. Donato Milanese and University of Pavia, Electrical Engineering Dept, Italy

Summary

Particular application of reliability techniques is developing to evalu ate dependability (availability, integrity and security) of large proce ss distributed control system (DCS) for a Ethylene Plant. Each critic al plant section is associated to own control subsystem which is analyz ed taking into account major field unavailabilities contributing to glo bal and section outages figures. Control system availability,(hardware and software) is concerned, evaluating section fault tolerant approach by Stochastic Petri Nets Theory application on computerized control sub systems over a prior extensive standard Fault-Tree analysis conducted all over the plant. Cost-benefit balance is then highligthted as decis-ion-making tool, where reveals important performance in final control derated conditions.

Introduction

Complex Control systems play a major role in modern large chemical and petrochemical plants. Recently turn-over instrumentation and revamping of existing plants, point out a Cost-benefit approach where large contr ol systems are to be installed, and being compared different Manufactu ring design and architectures. Furthermore new projects provide partic-ular features that DCS will perform as: process optimization, advanced controls, safety enhancement. Plant Users normally require, other than above performances, high plant availability targets. Global availabili-ty (dependability) evaluation is a complex task to achieve, due to the relevant and diversified elements concerned: hardware and organizative. The former are fundamentally related to the equipments, the latter to the plant Operations and Management. Each one of these should be caref-ully analyzed in order to assess plant availability correctly.
Reliability Engineering, normally applied in other industrial areas,si-nce over the last 15 years, reveals a key basis to perform full assess-ment; infact simple subsystem and functional relationships connected to the plant performances, may be so complex that its understanding requi-res the knowledge of high specialized profession and advanced technolo-gies, normally not available and beyond the capabilities of single Engi neer. In particular, referring to control system, the over all dependab ility evaluation of an integrated instrumentation system merged into the plant, requires the experience of a reliability Engineer backed-up by advice from specialized disciplines. The reliability Engineer will base the technology on past experience failure plant data compelled to stati stical and logical techniques in order to provide plant operation upti-me and downtime. Reasonably comparison might be experienced from

field data and so performance may be allowed to accomplish correct pre predictions and conclusions which can be drawn to quantify the reliability of ne wly designed systems taking into account hw and sw alternatives in a co stbenefit conflict. As the integrated control system belongs to the whole plant management, it is essential to point out that instrumentation availability must be analized not only in a steady state condition, but also in the transient context related to the dynamical operation of each subsystem into which the plant should be operatively organized and divided. Control system Manufacturers normally apply reliability techniques to define own optimum hw and sw arrangement; many standard are ava ilable ranging from MIL, international (IEC, SA-SAMA,etc.) to internal ones and they carry out sub system and system reliability (dependability) figures for quality tar-get assessment.
Even if the control system complexity and quality of the final product requires carefully application of the reliability techniques and methodologies, on the other hand the problem the User encounters is how reliability figures are representative in order to assess the present control system. Furthermore and depending on the fact that the User requires plant availability figure approaching unity, he tends to avoid human intervention to control plant in up-set conditions to minimize shutdown and maintain rated production and quality levels. Owing to Manufacturers' ranges and relatively vast reliability feedback experience, User tends to increase control availability itself by qualitative and intuitive application of redundancies against high plant investment and operation costs in order to achieve illusory high plant availabilit y targets. On ther other hand, a correct reliability analysis should be comprehensive not only of the control system itself, but also of the whole plant configuration and dynamical behaviour. This criterion is being set-up to verify the integrated control system threri described where an advanced reliability technique-Stochastic Petri Net Theory, is implemented to handle the derating problems.

Plant availability evaluation approach.

It is of paramount importance that the objectives, scope and bounds of the system being assessed, be clearly established at the onset of the analysis. Where peripheral services exist which cut across the bounds, these require to be identified so the influence on the system has to be not overlooked; control system is a block of the plant (figure 1).
The porpouse of a control system is to control the plant in a optimum way, complying with the availability target. In our case the availlability target was set 120 h per year (average). It is here that the DCS's may provide selective and structural design of the control availability and reliability due to inherent characterics and capabilities of these systems. It is important to point out that the process characteristics itself are essential to evaluate the optimum control hw and sw for actual organization. An indipendent treatment for the DCS alone will be me aningless and could lead to incorrect conclusions, resulting in mismatched performances and high costs. From a general equipments failure ra te overview it is interesting to realize that whole plant availability figure is by far less than electronic systems which provide efficient selfdiagnostic features and operator interface, even if subsystems are arranged in different configurational architecture where redundancy levels are engineered in distributed control organization. Fur thermore plant provides general intrinsic availability owing to the good enginee ring and practice codes complying to Client specifications: unit sections are sometimes decoupled, equipments (pumps, filters etc.) are redundant, intermediate hold-up inventories installed. In this sense opti-

mum control system should not only match plant availability (comprehensive of field equipments only), but it is necessary to define selectivity of the DCS devices. Should the control system performances be evaluated according to functional and reliability standard, integration fault and relative correlation should to be built-up with plant sections, taking into account operator actions (man-machine interface). Due to high complexity of the plant an hybrid technique has been adopted to evaluate the plant availability also in derated conditions. A standard Fault-tree technique was applied to extract plant sections availability figures (figure 1, table a) showing critical items and connected control system failure figures (control room equipments) to be further analyzed. Taking into account fault tolerant general features of the control systems, operator intervention has been considered in derated control conditions (remote manual control) where section "tolerance" is still allowed. This particular dependency problem is being handled by Stochastic Petri Nets with Phase Type distributed transition times methodology. The stochastic realization of the net throughout its reachability set is evaluated by mean s of the isomorphic Markov process: in this way SPN's offer a powerful tool for a compact representation of the dynamic system, providing an evaluations of the various probabilistic measures complying with the analytical tractability of Markov processes where they require a larger number of states. Application of the above technique has been applied in previous work (1) and actually is being primarily implemented on Propylene purification section of the plant, where safety is of rilevant concern (Run away reaction). The general problem may be classed in realtime control process systems. Realtime systems fall into two classes, failure critical and non failure critical. In the former class of systems no downtime can be tolerated, that is, any failure during a specified interval (mission time) is constructed as mission failure. In the latter class of systems, some downtime may be tolerated. Failures are allowed to occur as long as some constraints are met, i.e, the time spent in the down states does not exceed a specified limit, or, the number of systems downtimes does not exceed a specified figure. As soon as such a constraint is violated, the system fails. The probability that the system does not fail over a specified interval is an appropriate measure of reliability for this class of real-time systems. That is the system can be down (not perfor-ming) but not in failed condition. In this way operator intervention reveals relevant task. The threshold value-the allowance interval of the section is sometimes well represented by a random variable distributed across a centered value(mean value), where the distribution should be Erlang shaped, commonly applied in stochastic practical models. Erlang distribution (m stages) could be defined as the time to the abs-orbing state of Markov process, where the states from 1 to m are trans-ient and the m+1 is the final absorbing state, that is, where total section failure occurs. The system, for seek of calculation semplicity, is schematically modelled by a markovian two state graph (figure 4). If failure is removed in a k-th transition, the system fallsback into the initial good state and the m state failure chain has to be routed to reach the final full failure (absorbing) state.

The Ethylene process.

The Ethylene production from virgin naphta and gasoil steam cracking is a well known process. From feed tanks the raw material enter the cracking furnaces; the cracked gas leaves the furnaces at low pressure and high temperature (700 C). The remainder of the gas is concerned with separating the cracked gas into its various purified components, some of which form the plant products, while others are recycled for further processing. This separation and purification is achieved by various stages including:

- cooling and chilling, reaching nearly -100 C in some stages, to condense liquids out of vapour streams,
- compression,
- fractionation in distillation towers to separate components with different boiling points,
- reaction in catalyst beds to remove impurities such acetylene, allene

Product vessels are provided for intermediate storage.

All process control is through DCS activating pneumatically operated control valves. Special optimization techniques and adavanced controls are implemented. Trips are actuated by indipendent measurement and electromechanical devices actuate safety functions; the majority of trips have pre-alarms in order to allow the panel operator some time to take remedial action. A very few critical trips use three primary process measurements with a two out three voting for trip action.

Preliminary Availability analysis

Object of the study was first to identify, by Fault Tree technique, critical plant sections or equipments leading to plant shutdown; then providing hw and operative recommendations to follow up. Availabilty (average value: one year) was set 120 hours. In table 1 are summarized the results carried out. From data analysis it is interesting to highlight that critical items should be classed in three main arrangements:

- utilities and feedstocks losses (a,g,h, table 1),
- field equipments failure (comprehensive of primary instrumental devices: transmitters, control valves, t/c, etc.) (b,d,e,m, table 1),
- control room instrumentation (f,l,n, table 1).

At same time, electromechanical spurious trips (two out three- majority voting logic) provides a low frequency rate as 7.10-5 (1/y), not comparable with remaining failure figures. This result (c, table 1) justified, on the other hand, by EniChem field test survey which has been conducted since some years ago, (FDT = 0.7.10-2) confirms the validity of electromechanical choice for interlock system vs.electronics or PLC, when economically comparable. Standard Availability analysis (Block diagram technique, not taking into account operator intervention) conducted by Manufacturers on large DCS control systems for a EniChem plant show global MTBF figures ranging from 1.86 to 2.86 years and an average Availability of .999878 and .999920 considering a MTTR of 2 hours, corrisponding to an unavailability of 1.07 and .7 hours respectively.

These general data compare almost favourably with equipments unavailability listed on Table 1, bearing in mind homogeneity of the analysis conducted on cases above mentioned. First of all it is interesting to point out the slight difference between the above results, assumed the 2 h MTTR on both calculations (no assumptions have been made on software and hu man reliability). In any case this relative result highlights that genneral control system features comply together, despite the Manufacturer philosophy design, secondly MTTR plays a fundamental role where failure autodiagnosis is essential system feature.

Furthermore repair rate associate to failure of b,d,e,m, (table 1) events reveal that instrumentation play no significant role. The restricted list to be further analyzed on the above context, requires a deeper examination only of few events such: f,l. Attention must be paid to repair times, compared to frequency rate, as cost-benefit balance of improved instrumentation redundancy vs, cost of production losses is emphatized.

In figure 2 a detailed analysis is outlined on plant subsystem in order to match specialized reliability analysis to justify DCS (subsytem) redundancies where panel operator might demonstrate recovery fault capabilities, object of SPN's being carried out.

Conclusion

Availability analysis (dependability) reveals a powerful tool to identify and quantify weak points in a complex plant, even if relative comparison is concerned. Control systems (DCS) play a fundamental role where service continuity and safety requirements have to be fulfilled. Splitting control system into iterconnected plant sections points out that standard availability analysis techniques (Fault-trees, Markov), are unsuitable to treat reliability systems where exceeding a specified continous interval of time (interval reliability problem) reveals criticality due to intrinsic plant "tolerance" abnormal condition should occur before total failure. In this sense SPN's phased time application seems to be an important technique to allow dynamical control to be handled, where subsystems performances are strictly related to plant sections outages and failure and where operator intervention could be relevant element to restore up-set conditions.

TABLE 1.

Event	Frequency rate (1/y)	Repair time (h)	Unavailability (h/y)	% unav. (87 h)
a-Main power supply fail.	0.2	24	4.6	5.3
b-Turbines TK-11A/B	5.10-2	360	17.28	19.87
c-Compressors K-11A/B trip spurious	1.13.10-2	24	.27	.3
d-Compressors K-11A/B fail.	1.4.10-2	360	5	5.75
e-Compressors K-81A/B fail.	6.10-2	240	14.4	16.55
f-No flow process gas	.38	24	9.12	10.5
g-Low pressure F.G. from B.L.	2.10-1	24	3	3.5
h-Feedstock V.N. from B.L.	6.42.10-2	48	3	3.5
i-Q.O. Pumps P-01A/B cavitation	1.6.10-2	24	.38	.44
l-Transfer line coke formation	5.10-2	24	1.2	1.4
m-Exchangers tube failure	10-2	480	4.6	5.3
n-Propylene loss to B.D.	2.4-2	24	.57	.65
o-General plant shutdown for relay spurious trip	7.10-5	24	.0017	.002
CALCULATE TOTAL UNAVAILABILITY			87	

Remark:-Failure data have been provided from EniChem and Snam Progetti.
-Repair time data have been provided from site EniChem Maint.Dep
-Frequency failure rate less than 10-2 have not been considered.

A	B	D	E	F	G	H	L	M	AVAIL.
4.6	17.28	5	14.4	9.12	3	3	1.2	4.6	87

Fig. 1. Ethylene production Block Diagram affecting primarily the overall availability figure (87 h)

```
                        TOP EVENT F
                             .
                    NO FLOW PROCESS GAS
                             .
FIELD EQUIPMENTS------- OR------------------ AND
FAILURES              .                       .
                      .                       .
SPURIOUS TRIP...... .                         ....... NO OPERATOR
                      .                       .       INTERVENTION
FULL DCS FAILURE .. .                         .
      .                                       ....... DCS SUBSYSTEM
     OR                                               DERATED
      .
HW  SW FAILURE
LOCAL & COMMON
```

Fig. 2. Subsection main Fault-tree where DCS partial and full failure is considered.

```
                    NO OPERATOR INTERVENTION
                                              NO TIME ALLOWED
NO PANEL ALARM-----------OR---------------TO MANUAL OPERATOR
OF DERATED                 .                  INTERVENTION TO SECTION
SUBSECTION          OPERATOR NOT              RESTORE
CONTROL             ATTENDANT
```

Fig. 3. Subfault-tree relative to operator intervention restoring plant section up-set, should partial dcs subystem failure occur.

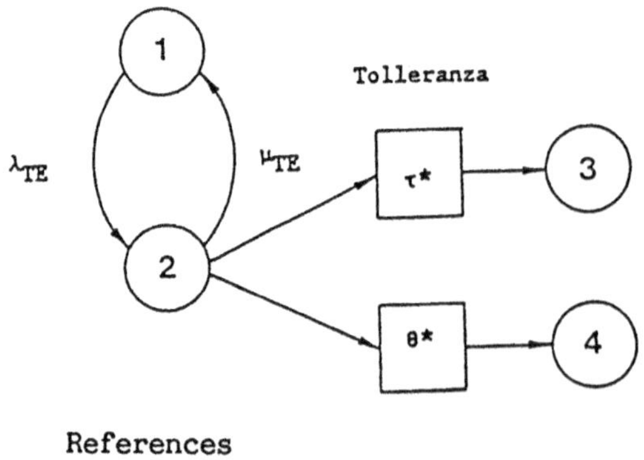

Fig. 4. SPN model to simulate operator restoring probability to exceed plant section allowable time.

References

1. Bobbio, A.; Piccinini, N.; Picciolo, G.; Scarrone, M.; Probabilita' della durata di disservizio in un impianto di abbattimento cloro. XV Animp Inter. Conf. on Industrial Hazards. Florence, 3/5 nov.1988.

Reliability Assessment of a Charging Machine for Radioactive Waste Disposal

R. Leicht, B.Puttke and H.J. Wingender

NUKEM GmbH, P.O. Box 1313

D-8755 Alzenau, FRG

Abstract

The charging machine designed for use in an experimental programme for disposal of vitrified high level radioactive waste canisters in the ASSE salt mine was the subject of a design state reliability assessment. The aim was to allow a judgement of the general reliability and availability of parts of the machine as well as of the system as a whole. The assessment was performed according to the fault tree method. The results are presented in the form of a reliability profile. It was found that those subsystems depending on the hydraulic unit contribute most to the overall failure frequency. No additional weak points were identified. The total failure frequency per year of operating time was estimated to be 29. However, the corresponding availability of the system amounts to more than 95%, thus meeting the requirements for such a machine.

1 Introduction

According to the present German concept for the back end of the nuclear fuel cycle, the high level radioactive waste (HLW) resulting from the reprocessing of spent fuel elements is to be vitrified, sealed in waste canisters, and disposed of in bore-holes in a salt dome. In order to demonstrate the feasibility of this procedure, an experimental disposal will be performed on the -800 m level of the ASSE 2 salt mine, testing both the technical components and the required handling. At the end of the test period, the waste canisters have to be removed from the bore-holes and transported back to above ground - a procedure which also is necessary and always must be available in the case of an unpredictable intrusion of water.

One of these newly designed components is the charging machine, which is to be used for filling the waste canisters into the bore-holes, and the related bore-hole shutter. The reliability and availability of these systems had to be judged in the design stage because of reasons concerning safety: insufficient reliability of safety-relevant components can lead to accidents, and frequent repairs can cause increased radiation exposure of the personnel.

The aim of the preliminary reliability assessment of the charging machine system was to obtain information on the general reliability of the system and of some of its essential parts, and to identify weak points, which then could be eliminated in a re-design process.

2 Description of the System

On the -800 m level of the salt mine, the vitrified waste canisters are transported to the bore-holes in shielded transport casks. For charging or discharging, the transport cask is set down on the bore-hole shutter which is located on top of the lined bore-hole. The charging machine takes the canister out of the transport cask into a shielding cover and then inserts it into the bore-hole. The bore-hole can be as deep as several hundred meters.

The charging machine is a vehicle which can be driven to and positioned at a bore-hole. Its major subsystems are (fig. 1):

- undercarriage,
- jib,
- cable winch with grab, and
- lifting device for the shielding cover.

In addition, the bore-hole shutter is considered here.

The undercarriage is equipped with a hydraulic mechanism consisting of a separate hydraulic motor drive on each of the four wheels. Each of the wheels has its own hydraulic steering and a 180° angle of steering lock. After an approximate positioning of the charging machine at the bore-hole, it is located and held horizontally by means of four hydraulically extendable supports.

The undercarriage carries a winder table with a jib which can be fine-positioned by a slewing gear and by axial adjustment. The cable winch has a steel cable which contains electrical wires for electromagnetic operation of a coupling grab which is attached to the cable by means of a special cable clamp. This coupling grab is used for attachment to the canister grab which is located inside the transport cask, and for electromagnetically opening and closing the cask for lifting or putting down the canister, respectively.

The cable clamp and the coupling grab are located within the shielding cover which is lowered onto the transport cask by means of a separate lifting device after having positioned the charging machine.

For shielding and containment reasons, the bore-hole is equipped with a bore-hole shutter, which is electrically opened at the beginning of the charging or discharging, and closed and locked at the end.

The only medium required by the charging system is the 500 V three-phase current power supply available in the mine.

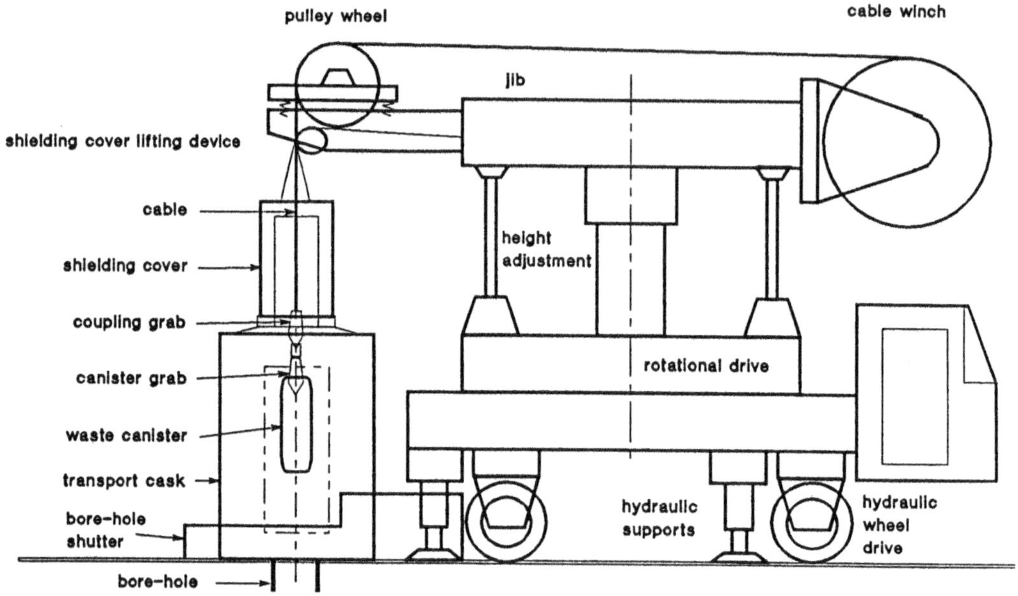

Figure 1: Schematic sketch of the charging system

3 Description of the Fault Tree

The fault tree describes the conditions of failures which lead to the unavailability or interruption of the charging or discharging procedure. It is structured in such a way as to identify subsystems which can then be separately analysed. Because of interdependencies between these subsystems, the behaviour of the total system is not given by the sum of the particular contributions of the subsystems, but must additionally be evaluated. The essential parts of the fault tree are discussed in the following.

The top event of the fault tree is defined to be

"HLW charging system is unavailable"

This occurs, if one of the following conditions is fulfilled:
- driving not possible,
- positioning not possible, or
- charging/discharging not possible.

The charging machine must be able to drive from one bore-hole to the other. This is not possible or possible only to a limited degree if:
- the hydraulic drive fails,
- the hydraulic steering fails, or
- the brakes fail.

The positioning of the charging machine at the bore-hole is not possible if one of the following events occurs:
- the hydraulic drive fails,
- the fine positioning fails,
- the horizontal level control fails, or
- the jib-positioning fails.

Charging or discharging of canisters into or from the bore-hole is not possible in the case of
- failure of the winch,
- failure of the coupling or canister grab,
- failure of the shielding cover lifting device, or
- failure of the bore-hole shutter.

The corresponding fault tree is shown in fig. 2.

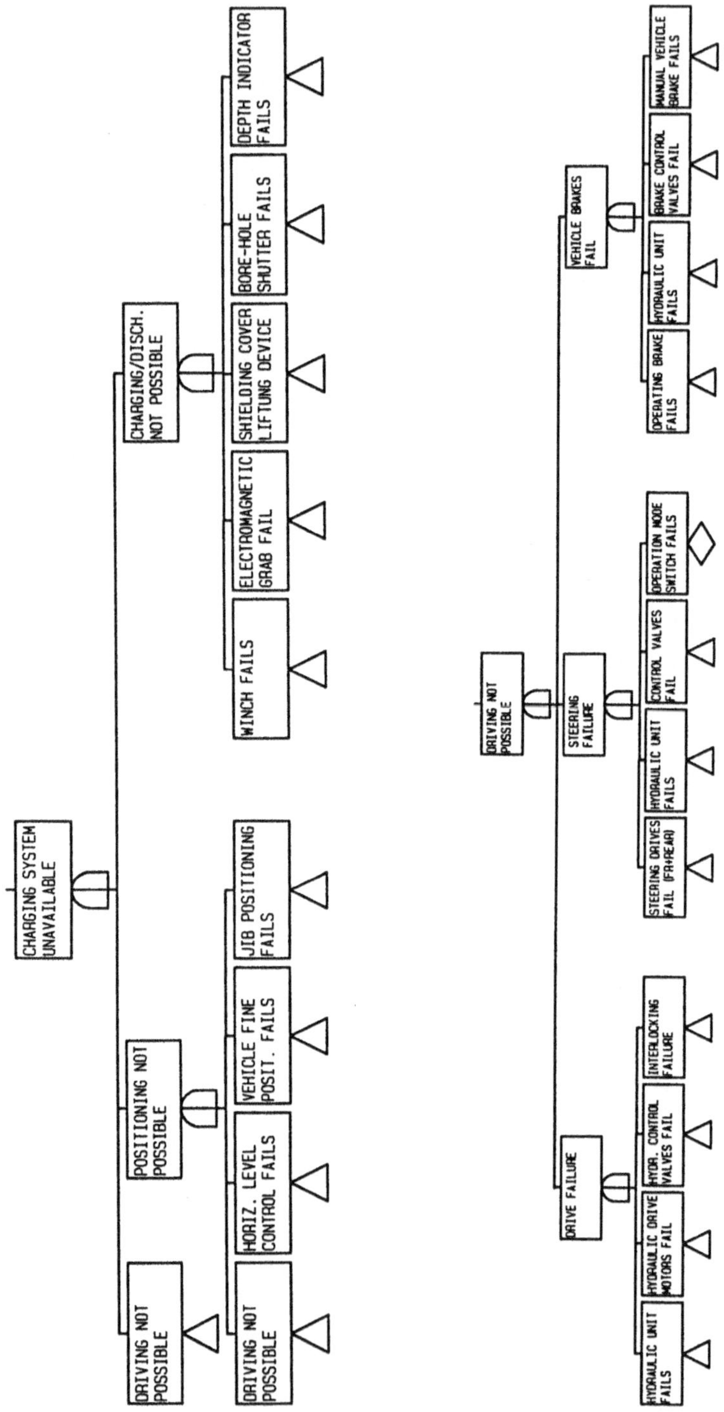

Figure 2: Upper part of the fault tree of the charging system

4 Evaluation of the Fault Tree

The quantitative evaluation of the fault tree requires failure data for the basic events, mainly failure rates and mean repair times. Since no specific failure data were available for this newly designed charging system, data from comparable fields have been used, and estimates had to be made in some cases. Hence, the results of this reliability assessment are considered preliminary and prognostic, and test experience should be used for updating and improving the failure data base as soon as available.

The main sources for the failure data used are the OREDA Handbook /1/, the Systems Reliability Service /2/, an NCSR report /3/ and failure data recommendations /4/. The mean repair times (MTTR), however, had to be estimated without exception, and were based on a classification into 3 groups as recommended in /4/:

MTTR = 6 h: repair by changing parts or by prepared work
MTTR = 12 h: in-situ repair by disassembling components with only little resources required
MTTR = 24 h: repair requiring special resources and longer set-up time

The question whether or not additional delays may arise due to unavailability of spare parts for the prototype charging machine remains presently unanswered.

The fault tree which contains 142 basic events and 74 gates has been quantitatively evaluated with the fault tree code RISA /5/. The results are given in table 1.

System or sub-system	failure frequency (op. year)	unavaila-bility (%)
charging system	29	4.0
driving	12.2	1.7
- hydraulic drive	8.6	1.2
- hydraulic steering	7.4	1.1
- brakes	7.6	1.1
positioning	24.0	4.1
- fine positioning	1.6	0.1
- horizontal level control	14.8	2.0
- jib-positioning	2.6	0.3
charging/discharging	11.8	1.6
- winch	9.6	1.3
- coupling or canister grab	1.6	0.2
- shielding cover lifting device	1.4	0.1
- bore-hole shutter	1.6	0.2
hydraulic unit	4.0	0.6

Table 1: Failure frequencies per year operating time and unavailabilities of the charging system and essential sub-systems

5 Results and Discussion

The results presented in table 1 can be understood as the "reliability profile" of the charging system and its major sub-systems. The evaluation is based on an operating time of 1 year, which is presumed to correspond to a calendar time of much more than 1 year. At present, the time required for disposing of one canister is not known, otherwise it would be reasonable to evaluate the probability of failure per charging procedure of one canister.

According to the results of the evaluation, 29 failures of the charging system are to be expected during one year operating time, with an unavailability of 4% and a mean down time of 12 h per failure. This includes also minor failures, which may not necessarily lead to complete unavailability or cause safety problems. On the other hand, the mean down time and the unavailability essentially depend on the repair times used, which in case of unavailable spare parts for the prototype system might be considered too optimistic.

These considerations, however, do not really affect a relative comparison of the failure contributions of the subsystems. The highest contribution is given by the positioning, followed by the drive and the charging/discharging which both are better by a factor of 2. It can be seen from table 1 that those subsystems which depend on the hydraulic unit have the highest failure frequencies and unavailabilities. Among those are the drive, the steering and the brakes of the vehicle as well as the horizontal level control. Also the relatively high failure frequency of the winch is caused by its hydraulically operated brakes.

Specific weak points cannot be identified, and the reliability profile can in general be judged as well-balanced. In particular there is no component recognizable for which adding a redundant component would result in a significant increase of the system availability.

6 References

/1/ OREDA - Offshore Reliability Data Handbook
published by the OREDA Participants
ISBN 82 515 0087 7, 1984

/2/ Systems Reliability Services Data Bank, UK
Failure data for brakes and valves

/3/ A.J. Duke
Reliability techniques used in the assessment of cranes
NCSR/GR/64, March 1985

/4/ J. Weymann
Empfehlung von Zuverlässigkeitskenngrößen für Fehlerbaumrechnungen im Projekt Sicherheitsstudien Entsorgung
PSE internal report, Technische Universität Berlin, 1983

/5/ A. Becker, L. Camarinopoulos
Programmbeschreibung RISA, Version 3.8
Institut für Kerntechnik
Technische Universität Berlin, 1981

Acknowledgement

The work reported was performed under a contract with the Gesellschaft für Strahlen- und Umweltforschung (GSF) mbH, München, for a charging machine designed by SIEMAG TRANSPLAN, Netphen. The authors thank GSF and SIEMAG for their kind approval to report the item.

The Availability of a Set of Inhomogeneous Machines

B. D. BUNDAY and E. KHORRAM

Department of Mathematics,
University of Bradford, U.K.

Summary

This paper presents a solution of the machine availability problem in which N different machines are looked after by a team of r operatives.

The run time of each machine is assumed to have a general distribution, different for each machine, and the repair times are assumed to have a negative exponential distribution with different means for the different machines. An explicit expression for the probability that a particular set of machines is running in the steady state is given. From this other useful measures for the system can be obtained. It is shown that these quantities depend on the run time distributions only through the means of these distributions.

Introduction

The machine availability problem has been considered by a number of authors. The main aim has been to predict the availability and operative utilisation for a group of identical machines looked after by one or more operatives.

Ashcroft [1] obtained a solution to the M/G/1 model. Benson and Cox [2] dealt with the M/M/1 model and extended their solution to the M/M/r case. Extensive tables for work study applications for this M/M/r case were calculated by Peck and Hazelwood [10]. Later Mack and Stoodley [8] gave a solution for the M/D/2 case and Maritas and Xirokostas [9] showed how to compute numerical results for the $M/E_k/r$ model. An important advance was made by Bunday and Scraton [4] who showed that the solution to the G/M/r homogeneous case was the same as that for the corresponding M/M/r solution.

In recent years a number of authors have tackled the problem for non identical sets of machines. Among these are Elsayed [5], Gross and Ince [6], Kameda [7] and Sztrik [11].

This paper develops the work of Bunday and Scraton [4] to deal with the case of N heterogeneous machines looked after by a team of r operatives. The repair time of machine i has a negative exponential distribution with mean $1/\mu_i$.

Running times for each machine are assumed to be independently distributed and for machine i, if that machine is started at time zero, we let $G_i(t)$ be the probability that it is still running at time t later.

Thus

$$G_i(0) = 1 \quad \text{and} \quad G_i(\infty) = 0 \quad \text{for all } i. \tag{1}$$

The mean running time of machine i is

$$\frac{1}{\lambda_i} = \int_0^\infty t[-G_i'(t)]dt = \int_0^\infty G_i(t)dt. \tag{2}$$

The repair discipline is SIRO (service in random order), in which stopped machines are randomly allocated to the r operatives whenever a new breakdown occurs or a repair is completed. Although this differs from the more usual FIFO (first in, first out) discipline it will in many practical situations closely approximate it. This form of service discipline, frequently referred to as processor sharing, occurs widely in computer systems models.

The Solution for the Steady-State Situation

Bunday and Khorram [3] have developed a mathematical model for the system just described. They let

$$Q_{i_1 i_2 \ldots i_n}(t_{i_1}, t_{i_2}, \ldots, t_{i_n}) \tag{3}$$

be the probability density that in the steady state a particular set $i_1, i_2, \ldots, i_n$ of the N machines are running, one of them for time t_{i_1}, another for time t_{i_2} etc.

Of course it is only in the case of negative exponential repair times that $n, t_{i_1}, t_{i_2}, \ldots, t_{i_n}$ provide an adequate description of the system. We need not specify the state of each repair, since this will not influence the future behaviour. Indeed we need not even specify which particular stopped machines are being repaired since the residual repair time has the same distribution whether or not the repair has started.

Bunday and Khorram [3] show that

$$Q_{i_1 i_2 \ldots i_n}(t_{i_1}, t_{i_2}, \ldots, t_{i_n}) = R_{i_1 i_2 \ldots i_n} \cdot G_{i_1}(t_{i_1})\, G_{i_2}(t_{i_2}) \ldots G_{i_n}(t_{i_n}) \tag{4}$$

where $R_{i_1 i_2 \ldots i_n}$ is independent of $t_{i_1}, t_{i_2}, \ldots, t_{i_n}$.

It is convenient to denote $\{i_1 i_2 \ldots i_n\}$, the set of running machines, by A_n while $B_n = A_n^c$ denotes the set of stopped machines. Then it is shown that if

$$R_{i_1 i_2 \ldots i_N} = \kappa \tag{5}$$

then

$$R_{i_1 \ldots i_n} = \frac{\kappa}{\prod_{k=1}^{N-n} \mu_{j_k}} \tag{6}$$

where $j_k \in B_n$ and $N - r \leq n \leq N - 1$,

$$R_{i_1 \ldots i_n} = \frac{(N-n)!\,\kappa}{r^{N-n-r} r! \prod_{k=1}^{N-n} \mu_{j_k}} \tag{7}$$

where $j_k \in B_n$ and $0 \leq n \leq N - r$.

Thus the probability that a particular group $i_1 i_2 \ldots i_n$ of machines are running and the rest are stopped is given by

$$Q_{i_1 i_2 \ldots i_n} = \int_0^\infty \ldots \int_0^\infty Q_{i_1 i_2 \ldots i_n}(t_{i_1}, t_{i_2}, \ldots, t_{i_n})\, dt_{i_1} \ldots dt_{i_n}$$

so that from (2)

$$Q_{i_1 \ldots i_n} = \frac{(N-n)!\,\kappa}{r^{N-n-r} r! \prod_{k=1}^{N-n} \mu_{j_k}} \prod_{j=1}^{n} \lambda_{i_j}^{-1} \tag{8}$$

for all groups with $0 \leq n \leq N - r$.

$$Q_{i_1 \ldots i_n} = \frac{\kappa \prod_{j=1}^{n} \lambda_{i_j}^{-1}}{\prod_{k=1}^{N-n} \mu_{j_k}} \tag{9}$$

for all groups with $N - r \leq n \leq N$.

Thus the probability that n machines are running is

$$q_n = \sum_{\{i_1,\ldots i_n\}} \frac{(N-n)!\,\kappa}{r^{N-n-r}r!\prod_{k=1}^{N-n}\mu_{j_k}} \prod_{j=1}^{n}\lambda_{i_j}^{-1} \tag{10}$$

for $0 < n < N - r$.

$$q_n = \sum_{\{i_1,\ldots i_n\}} \frac{\kappa}{\prod_{k=1}^{N-n}\mu_{j_k}} \prod_{j=1}^{n}\lambda_{i_j}^{-1} \tag{11}$$

for $N - r < n < N$.

κ is determined by the condition

$$\sum_{n=0}^{N} q_n = 1. \tag{12}$$

Useful Measures for the System

For a particular machine i, if $A^{(i)}$ denotes the long run proportion of time that machine i is running, essentially the availability of machine i, then

$$A^{(i)} = \sum_{n=1}^{N} \sum_{i\epsilon\{i_1,\ldots i_n\}} Q_{i_1,\ldots i_n}. \tag{13}$$

Then using a result first derived by Tomkó [12] we have

$$A^{(i)} = \frac{1/\lambda_i}{1/\lambda_i + W_i + 1/\mu_i} \tag{14}$$

where W_i is the mean time that machine i has to wait whilst not actually being repaired.

Thus

$$W_i = \frac{1 - A^{(i)}}{\lambda_i A^{(i)}} - \frac{1}{\mu_i}. \tag{15}$$

The mean time that machine i remains out of action and hence non productive in its cycle is thus

$$T_i = W_i + 1/\mu_i = \frac{1 - A^{(i)}}{\lambda_i A^{(i)}}. \tag{16}$$

The mean number of stopped machines is given by

$$\bar{N} = \sum_{i=1}^{N} \left[1 - A^{(i)}\right] = \sum_{i=1}^{N} \lambda_i T_i A^{(i)}. \tag{17}$$

Numerical Results

It is not difficult to evaluate the closed form results given at (10) and (11). The program uses N, r, $\lambda_1, \ldots, \lambda_N$, $\mu_1, \ldots, \mu_N$ as data. The output from a particular run is given. In the case of identical machines we simply reproduce the results for the M/M/r case. The output given illustrates the case of 8 machines and 2 operatives. Here, and this conclusion appears to be generally true, those machines with the same "λ/μ" have the same overall availability. Their mean time out of action and mean waiting time are calculated from (15) and (16).

```
                    Number of Machines = 8
                    Number of Operatives = 2

Breakdown rate of mach. 1 = .0100     Repair rate of mach. 1 = .1000
Breakdown rate of mach. 2 = .0200     Repair rate of mach. 2 = .2000
Breakdown rate of mach. 3 = .0200     Repair rate of mach. 3 = .4000
Breakdown rate of mach. 4 = .0300     Repair rate of mach. 4 = .6000
Breakdown rate of mach. 5 = .0300     Repair rate of mach. 5 = .3000
Breakdown rate of mach. 6 = .0400     Repair rate of mach. 6 = .4000
Breakdown rate of mach. 7 = .0400     Repair rate of mach. 7 = .2000
Breakdown rate of mach. 8 = .0500     Repair rate of mach. 8 = .2500

         Probability that 0 mach. are stopped = .41942884
         Probability that 1 mach. are stopped = .37748596
         Probability that 2 mach. are stopped = .14365438
         Probability that 3 mach. are stopped = .04529831
         Probability that 4 mach. are stopped = .01151332
         Probability that 5 mach. are stopped = .00226492
         Probability that 6 mach. are stopped = .00032322
         Probability that 7 mach. are stopped = .00002973
         Probability that 8 mach. are stopped = .00000132

                 Availability of mach. 1 = .9005
                 Availability of mach. 2 = .9005
                 Availability of mach. 3 = .9470
                 Availability of mach. 4 = .9470
                 Availability of mach. 5 = .9005
                 Availability of mach. 6 = .9005
                 Availability of mach. 7 = .8219
                 Availability of mach. 8 = .8219

Mean time 1 waits = 1.04738       Mean time 1 is stopped = 11.04738
Mean time 2 waits =  .52369       Mean time 2 is stopped =  5.52369
Mean time 3 waits =  .29981       Mean time 3 is stopped =  2.79981
Mean time 4 waits =  .19987       Mean time 4 is stopped =  1.86654
Mean time 5 waits =  .34913       Mean time 5 is stopped =  3.68246
Mean time 6 waits =  .26185       Mean time 6 is stopped =  2.76185
Mean time 7 waits =  .41802       Mean time 7 is stopped =  5.41802
Mean time 8 waits =  .33442       Mean time 8 is stopped =  4.33442
```

Particular Cases

It is possible to specialise these general results to the case where the operatives are in charge of two groups of machines. There are N_1 machines in the first group which each have mean run time $1/\lambda_1$ and mean repair time $1/\mu_1$. In the second group there are N_2 machines each with mean run time $1/\lambda_2$ and mean repair time $1/\mu_2$.

Then if $q_{n_1 n_2}$ denotes the probability that n_1 machines of the first type are running and n_2 machines of the second type are running then

$$q_{n_1 n_2} = \begin{bmatrix} N_1 \\ n_1 \end{bmatrix} \begin{bmatrix} N_2 \\ n_2 \end{bmatrix} \left[\frac{1}{\lambda_1}\right]^{n_1} \left[\frac{1}{\lambda_2}\right]^{n_2} \frac{\kappa}{\mu_1^{N_1-n_1} \mu_2^{N_2-n_2}} \tag{18}$$

if $N_1 + N_2 - n_1 - n_2 < r$.

$$q_{n_1 n_2} = \begin{bmatrix} N_1 \\ n_1 \end{bmatrix} \begin{bmatrix} N_2 \\ n_2 \end{bmatrix} \left[\frac{1}{\lambda_1}\right]^{n_1} \left[\frac{1}{\lambda_2}\right]^{n_2} \frac{\kappa(N_1 + N_2 - n_1 - n_2)!}{r!\; r^{N_1+N_2-n_1-n_2-r} \mu_1^{N_1-n_1} \mu_2^{N_2-n_2}} \tag{19}$$

if $N_1 + N_2 - n_1 - n_2 > r$.

κ is determined by the requirement

$$\sum_{n_1=0}^{N_1} \sum_{n_2=0}^{N_2} q_{n_1 n_2} = 1. \tag{20}$$

These results enable us to compute availability, waiting times, etc. for the machines of each type.

Numerical results from the application of these formulae to a case where $N_1 = 10$, $N_2 = 8$ and $r = 3$ are given.

Number of Machines of Type 1 = 10, Number of Machines of Type 2 = 8
Number of Operatives = 3

Type 1

Breakdown rate of machines = .030
Repair rate of machines = .100
Availability of machines = .5950
Mean time that a machine is stopped = 22.6877
Mean time that a machine waits = 12.6877
Mean number of machines running = 5.9501

Type 2

Breakdown rate of machines = .060
Repair rate of machines = .300
Availability of machines = .6823
Mean time that a machine is stopped = 7.7590
Mean time that a machine waits = 4.4256
Mean number of machines running = 5.4587

Mean number of machines running = 11.4089
Mean number of machines stopped = 6.5911
Operative utilisation of each operator = .9589

References

1. Ashcroft, H.: The productivity of several machines under the care of one operator. Journal of the Royal Statistical Society B12 (1) (1950) 145-151.

2. Benson, F.; Cox, D.R.: The productivity of machines requiring attention at random intervals. Journal of the Royal Statistical Society B13 (1951) 65-82.

3. Bunday, B.D.; Khorram, E.: A closed form solution for the $\vec{G}/\vec{M}/r$ machine interference model. International Journal of Production Research. To appear 1989.

4. Bunday, B.D.; Scraton, R.E.: The G/M/r machine interference model. European Journal of Operational Research 4 (1980) 399-402.

5. Elsayed, E.A.: An optimum repair policy for the machine interference problem. Journal of the Operational Research Society 32 (1981) 793-801.

6. Gross, D.; Ince, J.F.: The machine repair problem with heterogeneous populations. Operations Research 29 (1981) 532-549.

7. Kameda, H.: A finite-source queue with different customers. Journal ACM 29 (1982) 478-491.

8. Mack, C.; Stoodley, K.D.C.: Machine interference with two repairmen when repair time is constant. New Journal of Statistics and Operational Research 4 (1968) 1-7.

9. Maritas, D.G.; Xirokostas, D.A.: The $M/E_k/r$ machine interference model. European Journal of Operational Research 1 (1977) 112-123.

10. Peck, L.G.; Hazelwood, R.N.: Finite queueing tables. ORSA Publications in Operations Research 2. Wiley, New York, 1958.

11. Sztrik, J.: On the finite-source $\vec{G}/M/r$ queue. European Journal of Operational Research 20 (1985) 261-268.

12. Tomkó, J.: On sojourn times for semi-Markov processes. 14th European Meeting of Statisticians, Wroclaw, 1981.

Availability of the Pressure Relief Panel System at Pickering "A" Nuclear Generating Station

R. Rajagopalan
Ontario Hydro
Ontario, Canada

F. Camacho
Ontario Hydro
Ontario, Canada.

Abstract

Based on the overall containment system reliability requirements, targets have been apportioned for the new pressure relief panel system at Pickering NGS A. The availability targets for the rupture panels and the valves were derived from the overall containment system requirements.

A method has been developed to estimate, with a given confidence, the availability of the rupture panels. An acceptance sample plan for the rupture panels has been developed, to ensure that the availability targets are met. The reliability of the bypass valves is determined by using the fault tree analysis techniques.

Introduction

The containment system is recognized as the last barrier to mitigate the effects of a potential loss of coolant accident (LOCA). Hence, a specific reliability target has been designated for the containment system in order to meet the regulatory requirements. The containment system consists of a number of specific independent subsystems. Therefore, to meet the overall reliability of the containment system, the target is appropriated between these independent subsystems.

This paper gives a brief overview of the containment system at Pickering NGS and describes the method used for arriving at the reliability targets for the new pressure relief panel system at Pickering A. The paper also describes the methodology used for determining the reliability of the components of the system.

Background

The CANDU containment design employed by Ontario Hydro is in the generic category of "negative pressure containment" (reference 1). This design was first applied to the four unit station at Pickering NGS A which came in to service in the period 1971 - 1973. Subsequently, another four units, designated as Pickering NGS B, have been built and are connected to the same pressure relief duct (PRD) and vacuum building (VB) (Figure 1).

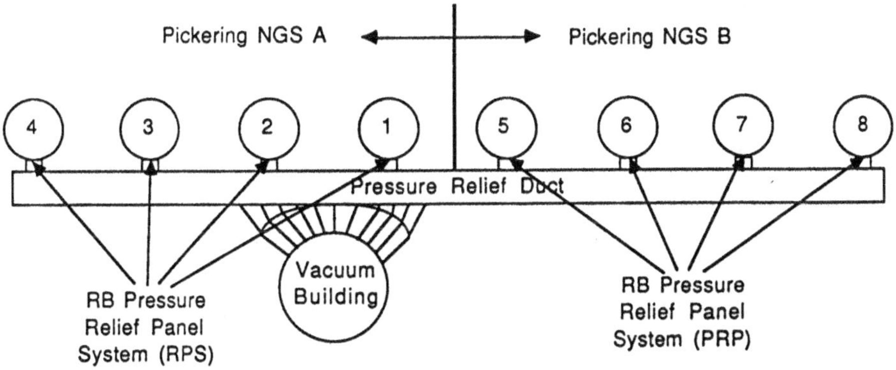

Figure 1. PICKERING NUCLEAR GENERATING STATIONS

The containment system at Pickering NGS consists of eight reactor buildings (RB), connected to a PRD. Each RB is isolated from the PRD via a reactor building pressure relief system, louvres and vapour barrier panels at Pickering NGS A and pressure relief panel (PRP) system at Pickering NGS B. The PRD is in turn connected to a VB through sixteen (16) pressure relief valves. The reactor building pressure during normal operation is maintained at a slightly negative pressure whereas the PRD is at atmospheric pressure. The VB is maintained at a pressure of approximately 7 kPa(a). Access to containment is provided via six airlocks in each of the RBs and two airlocks in the PRD.

In the event of a loss of coolant accident (LOCA) or a main steam line break (MSLB) inside containment, the pressure inside the accident RB rises. In the original design of Pickering NGS A, when the pressure differential between the RB and PRD reaches approximately 3 kPa, the louvres and vapour barrier panels relieve pressure into the PRD. The increase in PRD pressure in turn opens the self-actuating pressure relief valves (PRVs) and relieves pressure into the vacuum building. This initiates a water spray system which condenses the steam and reduces the pressure in the VB. In addition to the PRVs, there are three instrumented pressure relief valves (IPRVs) which maintain the pressure in the PRD at a slightly negative pressure (of approximately -2 kPag). Steaming from the break may continue in the accident RB but containment pressure will be maintained at a subatmospheric pressure by the operation of the IPRVs.

In order to provide effective post LOCA isolation between the non-accident Pickering A RBs and the PRD, a new pressure relief system called the rupture panel system (RPS) has been designed to overcome the inadequacies of the previous system.

The RPS consists of twenty nine (29) rupture panels, two bypass valves and an emergency exit door, all mounted on a bulkhead separating each RB from the PRD (Figure 2).

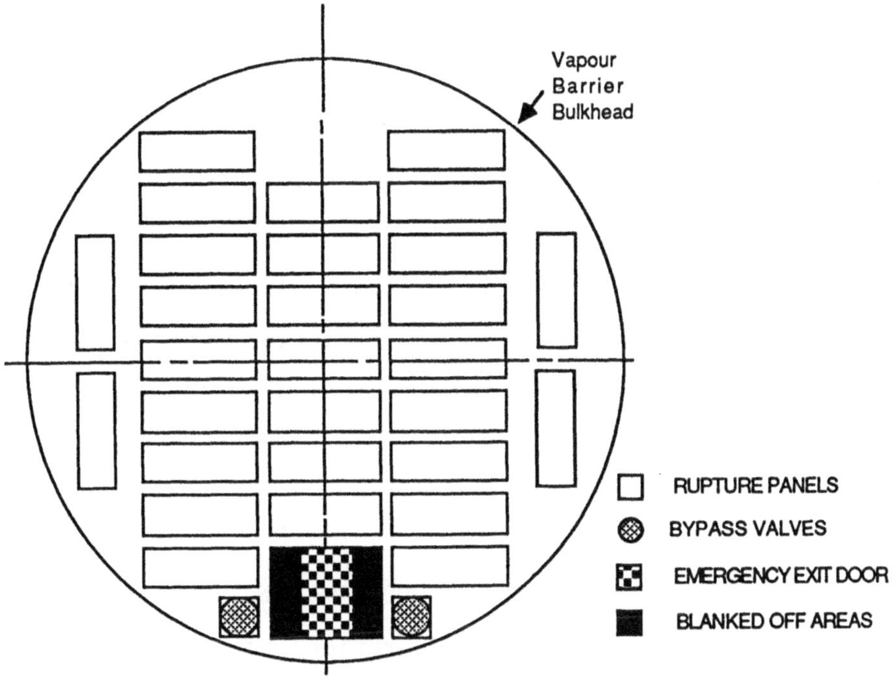

Figure 2. RUPTURE PANEL SYSTEM

Functional requirements of RPS

The following requirements have been imposed on the rupture panel system:

a) Following a main steam line break, a minimum of 13 m^2 of flow area is required to limit the peak pressure in the RB to the commissioning test pressure.

b) A flow area of 6 m^2 is required to cater for a large loss of coolant accident.

c) A flow area of 0.3 m^2 must be available to cater for very small LOCAs and to cater for under pressure transients in the PRD.

d) In case of an accident with personnel working in the PRD, each RPS bulkhead must provide access into a non-accident reactor building.

Component description

Rupture Panels

The rupture panel assembly consists of two stainless steel plates, sandwiching a polymeric sealing membrane. The panels are designed to burst when the pressure differential between the RB and the PRD reaches approximately 15 kPad. Each panel measures 161 cm by 54 cm; thus the total flow area provided by the panels is 17.4 m^2.

Bypass Valves

The bypass valves are 61 cm butterfly valves, pneumatically operated. During normal plant operation, both bypass valves are closed. If the pressure in the RB increases to 3.5 kPa, then the valves are signalled to open.

Emergency Exit Door

The emergency exit door at each RPS bulkhead provides access into a non-accident reactor building. Since the door is designed for emergency access into the RB, there are no pressure relief requirements.

Reliability

Methodology

The containment functions are carried out through a number of independent subsystems. The overall containment unavailability is the sum of the unavailabilities of all the containment subsystems, required to function following a postulated break inside the RB.

Some of the subsystems are shared by both Pickering NGS A and B stations. Hence, the unavailability of these common subsystems are also considered in deriving the target for the unavailability of the RPS.

<u>Assumptions</u>

The break is assumed to be a large LOCA or MSLB in one of the Pickering NGS A units.

It is assumed that the bypass valves in the Pickering NGS B pressure relief system open. Thus the containment envelope is assumed to consist of the accident unit, Pickering NGS B units, PRD and VB.

Since the Pickering B units are assumed to be connected through the bypass valves in the pressure relief system, the probability of breach of containment through Pickering B air locks is included.

The reliability of isolation of Pickering NGS B, isolation of the accident unit in Pickering NGS A and the flow path to the common vacuum building have been assessed. This is termed as the containment isolation in table 1.

Derivation of target for RPS

Table 1 lists the containment subsystems and their respective calculated unavailabilities:

Table 1. UNAVAILABILITIES OF CONTAINMENT SUBSYSTEMS

Subsystem	Common to A & B	Unavailability
a) Containment isolation	yes	1.7×10^{-4}
b) RB/PRD Airlocks	no/yes	1.2×10^{-5}
c) Pressure Relief Valves	yes	8.1×10^{-7}
d) Instrumented PRVs	yes	9.9×10^{-5}
e) Vacuum & Dousing Systems	yes	9.2×10^{-5}
f) Filtered Air Discharge System	yes	4.0×10^{-4}
g) RB Rupture Panel System	no	2.0×10^{-4} (LOCA)
		1.0×10^{-3} (MSLB)
TOTAL		9.7×10^{-4} (LOCA)

The total of the unavailabilities of the containment subsystems (not including RPS) for a LOCA is 7.7×10^{-4}. Since the overall unavailability target for Pickering A containment system is 3×10^{-3}, the maximum unavailability that can be assigned for the RPS is 2.23×10^{-3}; however a conservative target of 2×10^{-4} has been assigned for the RPS. For the main steam line break scenario, a target of 1×10^{-3} has been assigned to be consistent with Pickering NGS B.

The target of 2×10^{-4} for the LOCA scenario has been equally divided between the bypass valves and the rupture panels. With a target of 1×10^{-4} assigned for the rupture panels to cater to a LOCA, the flow area required for the LOCA is approximately 6 m^2.

Since the operation of the bypass valves is the same for LOCA and MSLB, the target unavailability for the bypass valves is the same for both cases. Thus, for the rupture panels, a target of 9×10^{-4} has been derived to cater to a MSLB with a maximum flow area requirement of 13 m^2.

Reliability of RPS components

Failure Criteria

The following failure criteria are established for each of the components of the rupture panel system:

(a) The area of each rupture panel is 0.6 m^2. Hence, for a postulated LOCA, if 9 panels or less open, then the required area of 6 m^2 is not available. Therefore, the system will be considered unavailable if 20 or more panels (out of 29) do not open for a LOCA.
For a postulated MSLB scenario, if less than 20 panels open, the total area from the rupture panels is a maximum of 12 m^2. The system will be considered unavailable if 9 or more panels (out of 29) do not open.

(b) The bypass valves are provided to cater for the small break LOCA scenario. The area provided by one valve is 0.3 m^2 and hence is sufficient to relieve the pressure in the RB. Hence, the failure of both valves to open is considered as the failure of the system.

<u>Rupture Panels</u>

The panels are designed to open when the pressure differential between the RB and PRD reaches approximately 15 kPa. Sufficient controls have been imposed during the material procurement, manufacture and quality assurance testing of the panels that it is reasonable to assume that the panels are indistinguishable and that all panels have the same probability of failure (ie. not opening) when the pressure differential reaches 15 kPa.

If the number of panels installed in each reactor building is denoted by n, out of which m panels are required to open, then the unavailability of the panels for an accident is given by the probability that less than m out of n panels burst. This unavailability is a function of the failure p and can be given by

$$UNAV(p) = \sum_{k=0}^{m-k} \binom{n}{k} p^{n-k} (1-p)^{k}.$$

The complement of this probability is the probability that at least m out of n panels burst and is given by

$$AV(p) = 1 - UNAV(p)$$

Figure 3 presents the plot of this function for n=29 and m=21 where it can be observed that this is a decreasing function of p.

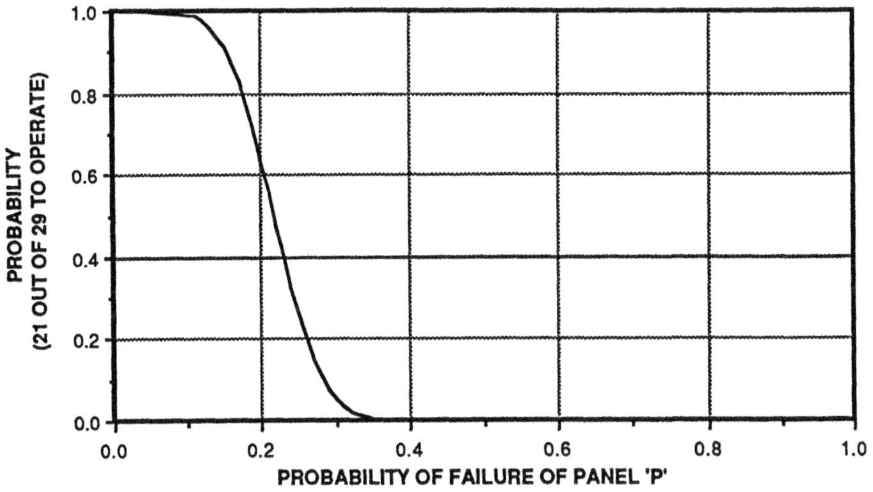

Figure 3. PANEL AVAILABILITY AS A FUNCTION OF 'P'

In general the probability of failure, p, is not known. One way of determining this failure is to manufacture a number of panels, N, and test them. If a number of panels, F, do not burst within the acceptable range of pressure difference, then they can be considered failures, and the probability p can be estimated as p = F/N. Since the test is destructive in nature, the panels cannot be reused.

It is possible that no failures may be observed during the tests. The estimate for p will be zero and the corresponding estimate for AV(p) will be one. This however, may not be realistic and better estimates may be obtained by using confidence intervals. Statistical analysis techniques can be used to arrive at the number of panels that must be tested in order to establish a failure rate with sufficient confidence in the failure rate.

A specific upper confidence interval for p (for example 99%) may be constructed so that it contains the true value of the parameter with a pre-established confidence or probability. For example, a 99% upper confidence interval for p will be of the form $(0,p_u)$ where p_u is chosen so that, if repeated samples were taken, the interval would contain the true value of p an average of 99 out of 100 times.

The value of p_u can be evaluated in each sample tested as the solution to the following equation

$$1 - c = \sum_{j=0}^{F} \binom{n}{j} p_u^j \ (1-p_u)^{n-j}$$

where c is the confidence value.

Based on the reliability requirements, an acceptance sampling plan can be designed to indicate the sample size N, and the number of defective panels, F, that can be tolerated. Table 2 gives the values of UNAV(p_u) for several values of N and F when n = 29 and m = 21 (for the MSLB scenario) and table 3 gives the values for n = 29 and m = 10 (for the LOCA).

It can be seen from table 2 that the smallest sample size at which the unavailability of the system (ie. 1×10^{-3}) is met, is with N = 48 panels and F=0. If there are any failures during the test, then additional panels must be tested to retain the 99% confidence level.

Table 2. UNAVAILABILITY FOR MAIN STEAM LINE BREAKS

Number of Panels installed in the RB = 29
Number of Panels required to open per RB = 21
Number of panels tested = N
Number of panels failed to open within pressure range = F

N	F=0	F=1	F=2
10	.80E+00	.99E+00	.1E+01
15	.35E+00	.80E+00	.96E+00
20	.12E+00	.47E+00	.77E+00
25	.44E-01	.24E+00	.50E+00
30	.16E-01	.11E+00	.30E+00
32	.11E-01	.84E-01	.24E+00
34	.77E-02	.63E-01	.19E+00
36	.55E-02	.47E-01	.15E+00
38	.39E-02	.36E-01	.12E+00
40	.28E-02	.27E-01	.93E-01
42	.20E-02	.21E-01	.74E-01
44	.15E-02	.16E-01	.59E-01
46	.11E-02	.12E-01	.47E-01
48	.82E-03	.94E-02	.38E-01
50	.62E-03	.73E-02	.30E-01
52	.47E-03	.57E-02	.24E-01
54	.36E-03	.45E-02	.20E-01
56	.28E-03	.36E-02	.16E-01
58	.22E-03	.28E-02	.13E-01
60	.17E-03	.23E-02	.11E-01

Table 3. Unavailability for Large LOCA

Number of Panels installed per RB = 29
Number of Panels required to open = 10
Number of panels tested = N
Number of panels failed to open within pressure range = F

N	F=0	F=1	F=2
10	.46E-03	.34E-01	.25E+00
15	.21E-05	.44E-03	.87E-02
20	.00E+00	.91E-05	.29E-03
25	.00E+00	.30E-06	.14E-04
30	.00E+00	.00E+00	.89E-06
32	.00E+00	.00E+00	.60E-07
34	.00E+00	.00E+00	.00E+00
36	.00E+00	.00E+00	.00E+00
38	.00E+00	.00E+00	.00E+00
40	.00E+00	.00E+00	.00E+00

Bypass Valves

Fault tree analysis techniques are used to determine the unavailability of the bypass valves.

Each of the two valves is fed from a different high pressure instrument air station. In addition, each valve is equipped with its own air tank with sufficient capacity for at least two valve cycles, in case of failure of instrument air supply.

The valves are designed to fail closed on loss of instrument air or power supply. Power for one valve is supplied from the odd bus whereas the other valve is supplied from the even bus.The bypass valves and the associated controls are tested periodically to ensure their operation.

Three channelized pressure transmitters are mounted on three separate local panels in the steam generator area to monitor the RB pressure. During normal operations, the bypass valves are closed and the hand switches are set at "auto" position. The common logic failures and independent failure of each valve (including the failure of instrument air tanks) have been evaluated. Figure 4 gives simplified fault trees.

The fault tree analysis demonstrates that the valves and the associated control logic must be tested on a monthly basis. The calculated unavailability of the valves is 7.2×10^{-5}.

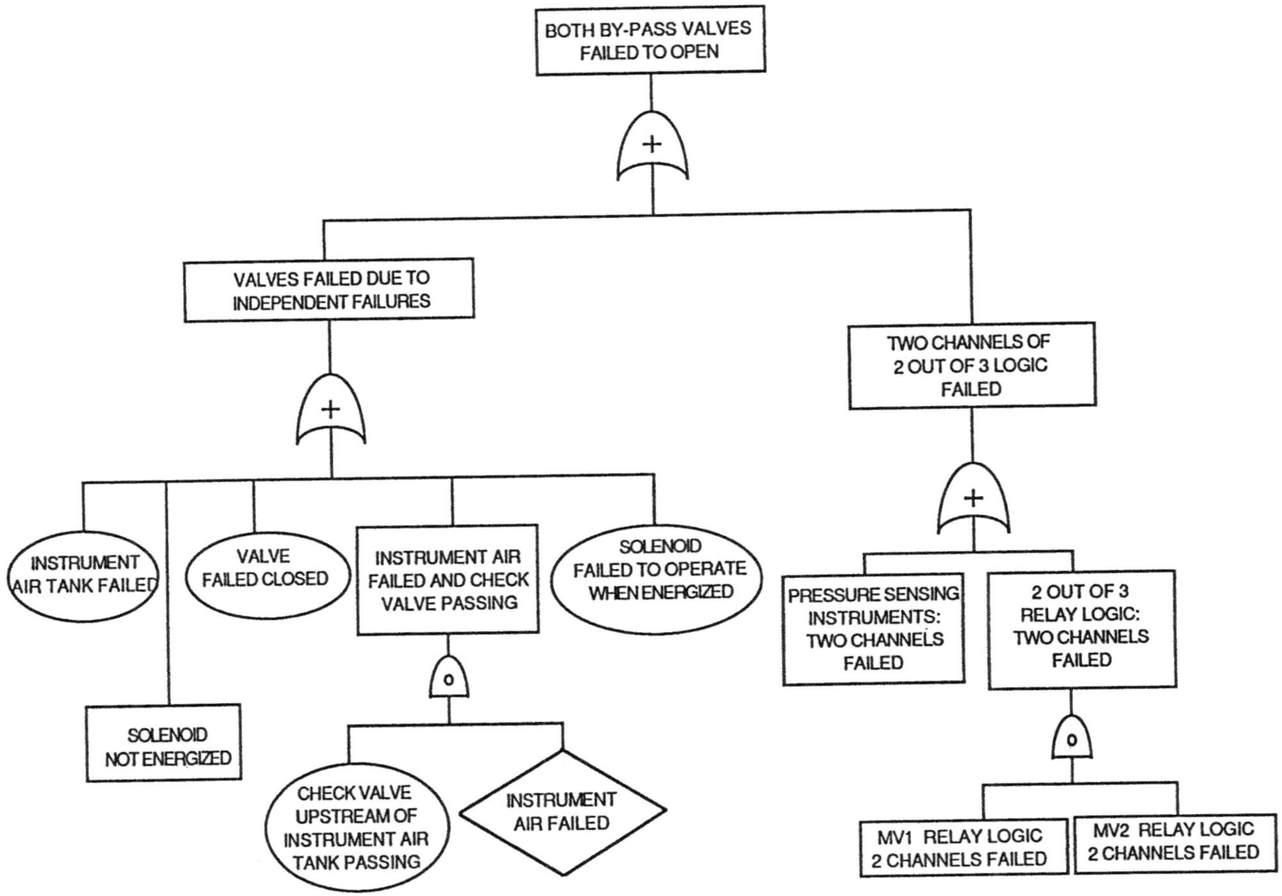

Figure 4. VALVE FAULT TREE

Conclusions

The reliability requirements for the new pressure relief panel system have been determined. These requirements have been apportioned between the two main components of the system, namely the rupture panels and the bypass valves. The statistical analysis techniques have proved to be a very useful tool to determine the number of panels that must be tested in order to have sufficient confidence in the reliability of pressure relief through the rupture panels. Since all panels passed the rupture tests, without this statistical analytical technique it would not have been possible to estimate the reliability of the rupture panels with any confidence. The fault tree analytical technique has been used to determine the reliability of the bypass valves. The valves and their associated control logic will be tested at a frequency of once per month.

References

1. Blanik, C., McKean, D.W., Meneley. D.A., Skears, J. and Yousef, N., "Principles of operation of CANDU multi-unit containment systems", International Conference on Containment Design, CNS Conference, June 1984.

2. Lloyd, D.K. and Lipow, M., (1962) "Reliability: Management, Methods and Mathematics", Prentice Hall, Englewood Cliffs, New Jersey.

Comparative Availability and Reliability Assessment of Design Options for the Secondary Sodium Loops of the EFR

Hartmut Pamme

Interatom GmbH

D-5060 Bergisch-Gladbach FRG

ABSTRACT

The EFR (European Fast Reactor) project has entered a conceptual study period where different design alternatives are compared concerning feasibility, safety and economic aspects.

This paper describes a comparative probabilistic availability and reliability assessment of alternative design options for the secondary sodium loops.

These loops will provide heat transfer from the reactor pool to the water-steam (power generating) side. So a high operational availability of the secondary loops during plant lifetime is essential for economic power generation. Additionally a high reliability is required to fulfill the operational decay heat removal function in case of a reactor trip.

Availabilities and reliabilities of the different options were assessed using failure mode and effect analysis and the fault tree method.

1. INTRODUCTION

Two different technical secondary system concepts and 3- or 6-loops versions are options considered in the project and thus must be compared concerning their availability and reliability performance.

Under operational aspects the availability of secondary loops as quotient between available time for power generation and operation period of the plant was assessed. Furthermore the ability and reliability of secondary loops for operational decay heat removal (DHR) (to reduce demands on the safety systems) under forced and natural convection were examined.

2. TECHNICAL DESCRIPTIONS

Figure 1 shows a simplified system configuration of the "REGAIN- concept" for a 6-loops version (REGAIN = réservoir d'expansion, génerateur de vapeur, auxilieures integrés). The main aspects of this concept are:

- The heat transfer from the reactor pool sodium to the secondary sodium takes place in the intermediate heat exchanger (IHX). The secondary sodium heat is transferred to the water-steam side within the steam generator (SG). The forced circulation in the main loop is achieved by the secondary pump (SP).

- A constant sodium level is achieved in the steam generator (SG) and mechanical secondary pump (SP) by provision of overflow pipework. This pipework is connected to the secondary expansion tank (SET), which is the lowest point of the system.
- The sodium is pumped back to the main loop (suction side of SP) via the level holding system (consisting of level holding electromagnetic (EM) pump (LHP) and associated pipework and valves).
- The Argon cover gas system (CGS) is connected with the SET. The SET, SG, SP have a common gas pipe system to achieve a constant gas pressure ("communicating systems"). One important task of the gas system is to provide sufficient pressure on the suction side of the level holding EM pump (LHP, placed above the SET).
- One sodium purification system (PS) is connected with the SETs of two secondary loops. The EM purification pump (PP) is able to fulfill the level holding function for one loop if the corresponding EM pump (LHP) has failed.

Figure 2 shows the simplified configuration of the "conventional concept" (for a 6-loops version). The main differences to the REGAIN-concept are:

- It is a specific "inherent" property of the conventional concept that the main loop operates as a "closed circuit", i.e. no active means are needed to keep a constant mass of sodium in the loop (if no failures in the purification system occur, which might lead to sodium loss into the secondary dump tank (SDT)).

- Each loop is provided with an secondary expansion vessel (SEV) in the cold leg of the main loop to cope with temperature induced sodium expansion.

Figures 1 and 2 show the secondary loops of the two concepts in a 6-loops version. Two loops have to share one purification system (PS). Existing dependencies between the cover gas systems of two loops (via the purification system) are not shown in the figures but were taken into account in the system analysis.
Figure 3 shows the system structure for a 3-loops version. Each of three secondary pumps (SP1-SP3) has to feed into two IHX-SG subcircuits. So especially the number of secondary pumps and associated valves and pipework is reduced compared to the 6-loops version.

3. QUALITATIVE ANALYSIS

Boundary conditions and assumptions
Due the the preliminary stage of design some "basic" assumptions and conditions had to be defined to clarify success requirements for the system in different states. The term "basic" means that there are many other requirements necessary to guarantee the technical function of the secondary loops but they do not contribute significantly to the derivation of reliability or availability parameters. Some main assumptions and conditions which influenced the construction of the fault trees shall be mentioned here.

Operational Aspects
It is assumed that operation (= power generation) with the secondary loops is possible

- if not more than two loops are unavailable with a 6-loops design,

- if not more than one loop is unavailable with a 3-loops design.

Furthermore "normal" operation requires

- availability of Argon cover gas pressure (e.g. to avoid pump cavitation),
- forced convection using the SP
- availability of the level holding system (REGAIN-specific requirement).

DHR Aspects

Due to preliminary thermohydraulic calculations it is assumed that the DHR-function can be fulfilled with one loop. DHR within a time range of 400 hours can be performed without sodium purification (in case of failure of the purification system).

The design of all versions of secondary loops has to guarantee that the minimum sodium level for natural convection (depending on sodium contraction with decreasing temperatures) is preserved during the DHR-phase.

The "operational" DHR is possible

- with forced convection using the SP,
- if no sodium-losses in the main loops occur (REGAIN-concept specific: no active level holding necessary) and
- with reduced Ar pressure or even total pressure loss.

DHR under natural convection only demands that no sodium losses in the main loops occur.

Based on these assumptions and conditions a failure mode and effect analysis down to component level was performed. Fault trees were constructed individually for both concepts in both redundancy levels with the TOP-events "no operation with secondary system" and "no DHR with secondary system". Additionally subtrees were modelled for the analysis of partial loop operation.

4. DATA BASE

General problems

It is essential in any reliability study that realistic and

relevant data are used. Failure data banks mostly providing generic data must therefore be compared with available data from operating experience to achieve justification for the intended use of a data base.

The derivation of failure data for sodium components is aggravated by problems of data adaption from generic "non sodium" data banks and the transformation of operating experience to an expected future component reliability.

Derivation of failure rates

An important data source within this study was the Interatom data bank of KNKII sodium-components. Additionally a literature search was performed to find comparable component data sets to reduce the "subjective uncertainties" concerning the use of only one data source.

Especially for large sodium components the available amount of data was found to be very spare so that the derivation of "generic" failure data became necessary.

The derivation of failure rates of some large sodium components shall be described briefly.

- Steam Generator

Generic failure data for tubing and welds based on broad operating experience can be used for an assessment with

$$\lambda(\text{tube}) = 10^{-11} \,/\, (\text{hr ft tube})$$

$$\lambda(\text{weld}) = 10^{-5} \,/\, (\text{year weld}).$$

As no detailed design for the EFR-SG is available up to now the design option for the straight-tube SNR2-SG (for an 8-loops concept) was used for further assessment. The relevant technical SG data are:

Number of tubes	:	1513
Length of one tube	:	28.3 m
Wall thickness	:	1.9 mm
Tube diameter	:	15 mm

It is assumed that only two welds are needed to connect one tube to the upper and lower bundle plate.

Using the generic rates from above the resulting SG failure rate is

$$\lambda(\text{SG}) = 4.3 \;\; 10^{-2} \; [1/\text{yr}].$$

An alternative assessment was performed using generic failure rates for

piping : $\lambda(\text{pipe}) = 6.7\ 10^{-11}$ / (year Q)
with the factor
Q = (length * diameter) / (wall thickness)^2

and welds: $\lambda(\text{weld}) = 3.25\ 10^{-5}$ / (year weld) .

The result is

$$\lambda(SG) = 1.01\ 10^{-1}\ [1/yr]$$

These two generic failure rates for a SG (based on an 8-loops SNR2 concept) are not too pessimistic as the future EFR SGs in a 6-loops or 3-loops design (with 6 SGs) will not have smaller dimensions. Additionally the SG was modelled as a simple tube bundle and failure modes other than pipe or weld rupture or leakage are not quantified here.

The assessment of a range for breeder-SG failure rates based on international operational experience was also performed by Interatom. Examining the available informations about tube ruptures with sodium-water reactions and corresponding operating times the resulting range of failure rates was

$$7.2\ 10^{-2} \leq \lambda(SG) \leq 1.2\ 10^{-1}\ [1/yr].$$

For this study an average failure rate based on the above assessments of

$$\lambda(SG) = 9\ 10^{-2}\ [1/yr]$$

was used.

- Intermediate Heat Exchanger

In a similar way a generic failure rate for the IHX was calculated. The results show that both IHX- and SG-failure rates must be expected within the same order of magnitude with approximately the same numerical size.

- Mechanical Secondary Pumps

The range of failures rates for large centrifugal pumps is scattering over nearly one order of magnitude. For a large breeder a failure rate of 0.82 [1/yr] was estimated for the secondary sodium pump. Operational data from a prototype breeder show a range of 1.1 up to 2.4 [1/yr] for large centrifugal pumps. Another data summary provides a failure rate of 3.5

[1/yr] and 0.4 [1/yr] for secondary pumps of two other prototype breeders. A direct comparison of these data is inadmissible because the failure modes and technical details which influenced the published failure rates are unknown (especially a problem of codified data aggregation in data banks). However the data show clearly that approximately one failure per year must be expected.

- EM Pumps

All EM pump data show a large scattering in failure rates. From KNKII-experience a rate of 0.011 [1/yr] can be derived. In other data sources rates between 0.51 [1/yr] and 0.07 [1/yr] are published. Again no detailed information about failure modes or technical details are given. For this study a rate of

λ(EM level holding pump) = 0.2 [1/yr] and

λ(EM purification pump) = 0.1 [1/yr] is assumed.

Derivation of repair data

A survey of repair times for small sodium components (e.g. valves) showed that the repairs were mostly finished within some hours or a few days. The range of repair times however increases significantly for main sodium components. For these components not only the repair time itself but also "logistic" problems (e.g. spare parts) or accessibility and the provision of suitable conditions for repairs (e.g. low radiation levels) contributed to the component unavailability. Thus "overall repair times" of large components between one week and more than a half year can be observed.

To quantify the impact of repair times for large sodium components on the operational availability the repair times of the SGs and IHXs were used as variable parameters in the study.

5. QUANTITATIVE RESULTS

DHR reliability targets

Preliminary safety considerations indicated that a reliability target of $P \leq 10^{-4}$ (probability of failure per demand) should be reached for the operational DHR using the secondary system. Thus the secondary system shall contribute significantly to the reduction of demands on the safety DHR system. It must be noticed that the target does not yet take into account common mode failures.

Achieved results

Based on the fault tree construction and the data base the probabilities of DHR failure within 400 hours after a demand were calculated. In a conservative way it was assumed that no repairs are performed in a DHR phase. That means that the components with high failure rates (especially the large sodium components) contribute significantly to the DHR failure probability.

Table 1 shows the numeric results of the DHR analysis for both concepts in the 3- and 6-loops version.

EVENT PROBABILITIES on demand		
Event	6 loops	3 loops
------------- R E G A I N C O N C E P T -----------		
1. DHR (forced convection) fails	5.86E-8	4.31E-4
2. DHR over sec. system fails	5.80E-12	1.90E-5
-------- C O N V E N T I O N A L C O N C E P T -------		
3. DHR (forced convection) fails	5.74E-8	4.27E-4
4. DHR over sec. system fails	5.10E-12	1.84E-5

- Events 2 and 4 explicitly describe the loss of DHR under forced and natural convection

Table 1

From table 1 (events 1 and 3) it can be seen that the 6-loops version fulfills the target reliability of 10^{-4} already under forced convection (operational DHR) whereas the 3-loops version slightly misses the target. For natural convection both

version meet the reliability requirements.
The results also show very clearly that the 6-loops version with 6 nearly redundant loops performs much better in achieving the target reliability than the 3-loops version with 2 identical sub-circuits within each loop. The reason is the fact that most identical component failures in both versions provoke the loss of 1/3 of DHR-capacity in the 3-loops design and only a loss of 1/6 of DHR-capacity in the 6-loops design.

Operational availabilities

Figure 4 shows the operational availabilities of both concepts for a 3 loops and 6 loops design in a bar chart. These results are derived from the calculation of the fault trees TOP-event probabilities. To show the influence of a variation of repair times two alternative calculations are shown in this paper. The terms "long (short) repair" in figure 4 mean assessed repair times for SG and IHX of 6 (1) months.
The unavailabilities of one secondary loop with respect to power operation are:

6 loops design:	REGAIN,	long repair:	P = 0.145
	"	short repair:	P = 0.074
	Conventional,	long repair:	P = 0.136
	"	short repair:	P = 0.065
3 loops design:	REGAIN,	long repair:	P = 0.232
	"	short repair:	P = 0.089
	Conventional,	long repair:	P = 0.230
	"	short repair:	P = 0.088

Figure 4 shows the overall availabilities for operation of all design combinations (see left bars for 6 loops at label "≥ 4/6" and 3 loops at "≥ 2/3"). The sign "≥" means that operation is possible if 4 (2) or more loops are available in the 6 (3) loops version.
The sum of the availabilities for full and partial load operation (see bars at labels "6/6", "5/6", "4/6" and "3/3", "2/3") provides the overall availabilities.
The right bars (labels "0") show the unavailabilities of a 6

or 3 loops design. So the sum of the values of left and right bars in each row is 1.

A comparison of REGAIN and conventional availabilities shows approximately the same results for both concepts with a slightly better tendency for the conventional concept. The reason is that the "qualitative" dependency of the REGAIN concept on the active level holding system has only a minor influence on the "quantitative" results. The main contributions to unavailabilities come from the large sodium components which are common to both concepts.
Significant differences can be found for availabilities between the 6 loops and 3 loops version. In the 3 loops case failures of components lead to the unavailability of 1/3 of power generation capacity and thus have a "double" effect in comparison to the 6 loops version. So the "load factor" as percentage of the maximum possible amount of power production (100% would express full load operation provided by secondary loops over the whole lifetime of the plant) is in general lower for 3 loops than for 6 loops . This result is shown in figure 5. The conventional concept again has slightly better results than the REGAIN alternative.
Figures 4 and 5 also show the sensitivity of availability calculations to variations of repair times. It can be seen very easily that the "short repair" case leads to increasing availabilities for full load operation and consequently to increasing load factors. Additionally figure 5 shows the overproportional effect of increasing repair times comparing the 6 loops and 3 loops load factors. For the short repair case the difference in load factors between 6 and 3 loops version is approximately 2 % and increases to more than 10 % in the long repair case.

6. GENERAL CONCLUSIONS

The assessments of both operational availability and DHR reliability show that the REGAIN concept and the conventional concept yield approximately identical results for a given

version of 3 or 6 loops. The reason is that the large sodium components which are common to both concepts are assessed to be the main contributors to unavailibility and unreliability (due to high failure rates for the DHR case and additionally long repair times in case of operation).
Significant differences can be found in a comparison of reliability and availability parameters of the 3 and 6 loops versions. The 3 loops version must be expected to have a lower operational availability than the 6 loops alternative. The unavailability margin between 3 and 6 loops version grows up with increasing repair times for components. The reason is again the "double" impact of a component failure in a 3 loops design.
The DHR unreliability of a 3 loops concept is several orders of magnitude higher compared with a 6 loops concept. A 6 loops concept is also assumed to have reliability advantages within a still necessary common mode failure discussion.

It is admitted that the used data base has a partly generic character. However a comparison of generic data with the few existing failure data especially for sodium components shows that the numeric range for failure rates within an order of magnitude must be acknowledged as "established" data. That means for example that based on the experience of prototype breeders at least one failure per SG or IHX must be expected in the lifetime of a plant. This fact reduces reliability and availability but the consequences of component failures can be weakened by minimizing repair times. Especially short repair times for the large sodium components have a significant effect on the availability of secondary loops for power generation in full or partial load operation (see e.g. [1]). This is one important result of the study. To achieve short repair and unavailability times for components keywords like logistics of spare parts, good accessibility and suitable repair conditions must be considered more intensively in future designs. As a perfectly reliable component will never exist the repair time becomes an important reliability parameter which is however often treated in a perfunctory manner.

The summary of the assessment indicates that the discussion of "REGAIN- or conventional-concept ?" should be extended or even turn to a discussion of "3 loops or 6 loops ?". Apart from economic aspects which give favour to the REGAIN-concept a safety and mainly availability orientated point of view must demand a 6 loops option no matter whether a REGAIN or conventional concept will be realized.

7. LITERATURE

[1] Guidez, Jolly
Assessment of the availability and viability of the French PHENIX Fast Breeder after 12 years' operation in proceedings of "Experience Gained and Path to Economical Power Generation", Richland, USA, Sept. 1987

REGAIN-concept

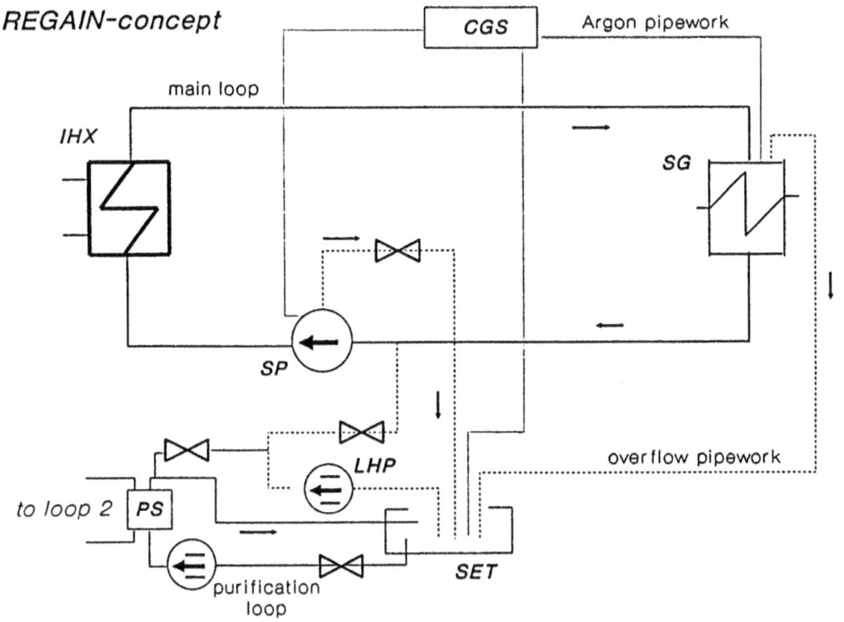

Figure 1

Conventional concept

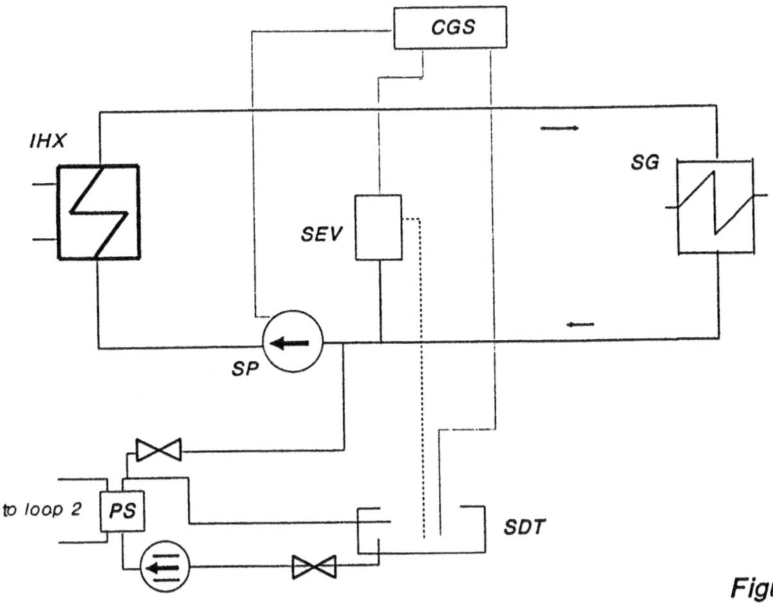

Figure 2

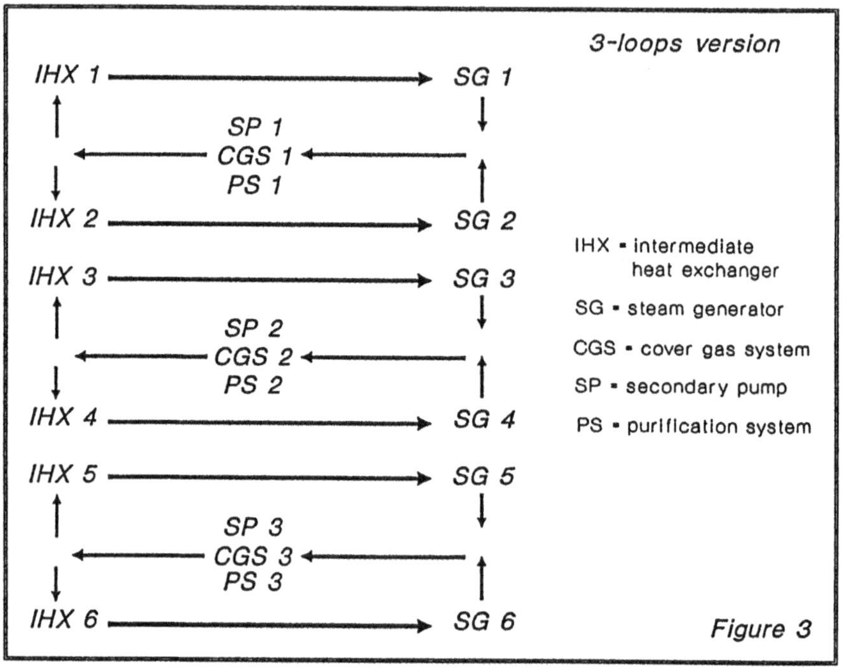

Figure 3

EFR Secondary Loops
Availabilities for Operation

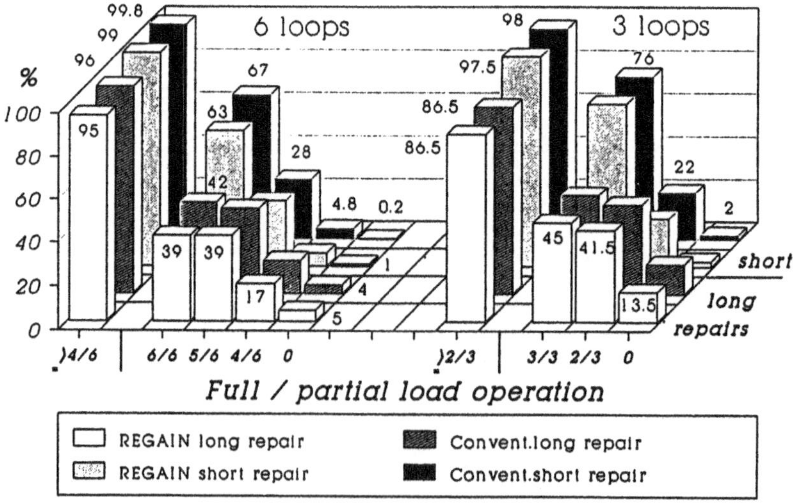

Figure 4

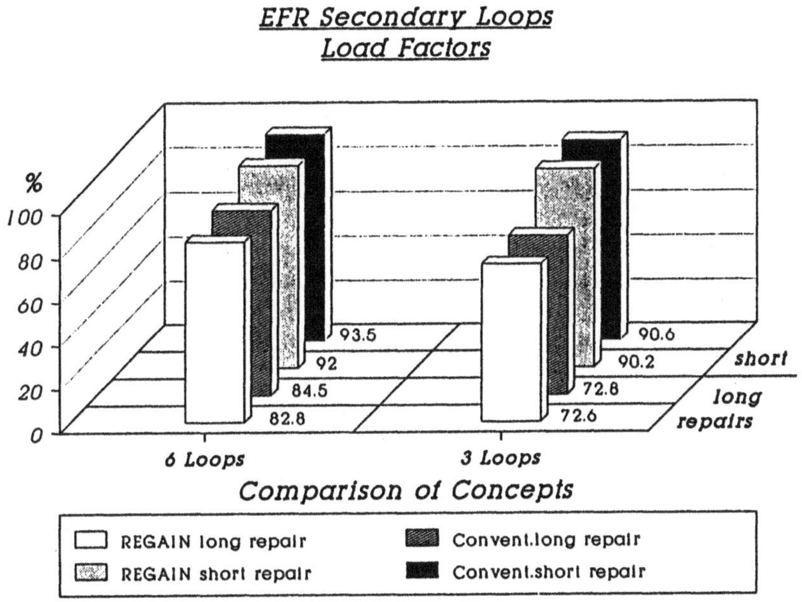
EFR Secondary Loops
Load Factors
%
100
80
60
40
20
0
93.5
92
84.5
82.8
90.6
90.2
72.8
72.6
short
long
repairs
6 Loops
3 Loops
Comparison of Concepts
REGAIN long repair
REGAIN short repair
Convent.long repair
Convent.short repair

Figure 5

Risk Analysis of Areas with High Concentration of Industrial Activities: Methodology Used for Priolo and Naples Areas (Italy)

R. Fox

Fiat Engineering
Torino, Italy

M. Melis

Edra
Milano, Italy

1. Introduction

The paper deals with the methodology adopted for the execution of Risk analyses of two areas with high concentration of industrial activities.

Risk analysis was a part of more complex environmental analyses for the areas of Naples and Priolo (Sicily).

For Naples-study, not only industrial risk, but also seismic/volcanic risks were considered.

For Priolo-study, industrial, seismic and hydro-geological risks have been analysed.

2. Risk Analysis for Naples Area

Analysis was executed upon committment of Ministry of Environment to a group of Italian firms (Fiat Engineering, Infrasud, Snamprogetti); aim of the study was the preparation of a plan for the reduction of pollution and environment improving.

The following typologies of risks were analysed:

- seismic risk
- volcanic risk (Vesuvio and Campi Flegrei)
- industrial risk (analysed by Edra; according to Health Ministry and Civil Protection Ministry Census, 14 Class-A industries and 20 Class-B industries are present in the area; classification is according to CEE-Seveso Directive).
- risk deriving from transportation of dangerous substances (e.g., about 2.5 Millions of tons/year of oil-product

circulating in the area).
Results of the study will not be presented in the following; attention will be focused only on the adopted methodologies.

3. Risk Analysis for Priolo Area

Analysis was committed to Fiat Engineering by Priolo Local Authorities and covered the following types of risks:

- seismic risk
- hydrogeological risk
- industrial risk (analysis performed by EDRA)

As for the Naples study, only the adopted methodologies are presented in the following pages.

4. Seismic and Volcanic Risk Analysis

Methodologies adopted are the ones commonly used by the Civil Protection Ministry.
For each area, with the contribution of Italeco, the following earthquake data were identified:

- max. intensity corresponding to a return-time of 500, 200, 100, 50 years.
- max. historical intensity
- max. foreseable intensity

Information were derived from the "Seismic Handbook" developed by CNR-Italy; seismic simulation models were utilized for consequence-analyses of various types of earthquake.
For volcanic risk, the following information sources were utilized:

- Campi Flegrei area: analyses developed by Civil Protection Ministry during Pozzuoli volcanic crisis (1983-1984)
- Vesuvio: analyses performed by Osservatorio Vesuviano and University of Pisa

On the basis of the abovesaid information, it was possible to identify scenarios of catastrophic events and to analyse possible consequences on industrial areas.

5. Hydrogeological Risk Analysis

The hydrogeological risk analysis covered phenomena of natural (e.g., alluvion, floading, landslides) or non-natural (e.g., dam overflow/failure) origin.

For Naples area the event which was investigated was landslide due to collapse of underground cavities.

For Priolo area, the consequences of a catastrophic failure of a dam (380 m. over sea-level, 5.7 millions of cubic-meters volume) were analysed.

6. Industrial Risk Analysis

Methodology adopted by EDRA for the analysis of industrial risk was the following one:

1. Collection of information on analysed plants (identification of processes, substances, plant layout, safety features, etc.)
2. Plant inspection/interviews with plant personnel
3. Preparation of "data-sheets" for each one of the substances handled in the plant
4. Identification of major accidents (explosion/fire/toxic release) and definition of risk-contours

Detail of obtained results was related to the information available from each plant.

6.1 Material Safety Data Sheets

For each substance characterizing the processes of the analysed plants, a data-sheet was prepared, in order to provide identification of physical-chemical characteristics of substances and of risks deriving from their use.

Pedigree-data (name, formula, CAS-CEE code, etc.), threshold values (according to CEE Seveso Directive), explosion/toxicity limits were identified together with flammability/self ignition data, TLV-ACGIH, DL50, CL50, etc.

6.2 Consequence Analysis of Accidental Events

Consequence analysis was performed using EDRA computer codes and evaluation models in order to identify risk-contours.
The following events were analysed:

a) Fireballs
b) Fire
c) Explosion
d) Dispersion (continuous/instantaneous discharge)
e) Evaporation (including "flash" phenomena)

7. Risks Deriving from Transportation of Dangerous Substances

The following accident-scenarios were considered

- dispersion of oil-products on sea-surface (and related environmental inpact)
- fire/explosion on ships/vessels in port areas
- accidental release of toxic substances from trucks and/or tank-cars
- explosion of GPL trucks/tank cars

Results of analyses were used to provide indications for re-routing or main communication lines modifications.

8. Emergency Plans and Conclusions

Information resulting from the abovesaid analyses represents a powerful input for the preparation of emergency plans for high-risk industrial areas.
Nevertheless, indications resulting from studies where different types of risks are considered (e.g., seismic,

volcanic, hydro-geological, industrial) must be carefully evaluated, in order to ascertain possible synergisms, identify accident scenarios and define topics for which more detailed analyses are required.

Risk Assessment for Transportation of Dangerous Materials – A Comparative Study

D. DIAMANTIDIS*, A. GUIDUCCI*, C. BUTTASI**, G. PERETTI***, F. BATTISTA***

* D'Appolonia S.p.A.

** Tema S.p.A.

*** Istituto Sperimentale Ferrovie dello Stato

Summary

A general methodology for the risk assessment of transportation of dangerous materials is presented in this contribution and is illustrated in a comparative study dealing with the rail and road transport of chlorine in an important North Italian route.

Introduction

Transportation risk of dangerous materials has been a cause of serious concern in the last decades due to the large number of accidents occurred and their high consequences for the society and the environment. Even more striking are the so-called "near accidents", which are accidents with catastrophic effect on areas fortunately unpopulated, and that could have caused great damage in less favourable circumstances.

Investigations have shown that the transportantion of dangerous substances has an important effect on the overall threat to the environment /1/. The risk inherent in the transportation of dangerous substances should be considered equally with the risks involved in their production and storage.

The procedure developed herein combines two different types of models. Firstly, physical models for the dispersion of chlorine after a release are selected to compute the areas affected by such an event. Secondly, the results are implemented in a general and transparent probabilistic consequence model together with other influencing parameters - such as metereological, environmental and population parameters - which are considered random. The proposed concept can be programmed in a personal computer.

Dispersion Models

In order to describe the dispersion of chlorine in atmosphere, mathematical models have been used to identify and simulate the basic processes of dilution of the cloud and to determine its space-time evolution under specified weather conditions. Available models are summarized in /2/ and /3/.

The following cases of accidental scenarios are considered:

- o chlorine dispersion due to cracks in the tank (leakage);
- o chlorine dispersion due to tank failure.

The leakage model used to simulate the chlorine dispersion considers the following conditions of emission:

- o two-phase flow (liquid and vapour) at high rate;
- o stationary emission.

The model uses the equations of mass-balance, momentum and energy. The air entering the cloud is described by the horizontal shifting and spreading velocity due to gravity. The energy balance takes into account the entrainment of liquid drops in the initial jet. The transition from the gravitational dispersion phase to the gaussian is given by the difference of density between the jet and the air: when this difference is 10^{-3} kg/m^3, the gravitational dispersion becomes not significant and the resulting dispersion is assessed by introducing the Pasquill model. The model computes the chlorine concentration behavior, the quantity of spilled liquid which forms the pool, the evaporation rate from the pool and the subsequent dispersion of the chlorine.

To simulate the chlorine dispersion due to tank failure a "box" model has been used; it assumes that the initial gas cloud is described by a cylinder, with given height and radius. The height was taken equal to the radius, since the gravitational slumping velocity is assumed to occur as in a liquid column. The model computes dimensions, temperature and concentration of the cloud during its evolution to the neutrality phase: at this point the Pasquill model is applied to describe the gaussian dispersion. The transition point from the gravitational dispersion to the gaussian is again determined by the difference between cloud and air densities.

These dispersion models are used to estimate the areas in which irritation, toxicity and letality thresholds are reached. Figure 1 illustrates the areas for the letality thresholds in case of tank failure.

Risk Model

The risk associated with the transportation of a dangerous substance like chlorine is a function of the accident probability p_i (respective to the i-th event) and the probable damage M_i following the accident:

$$R = \sum_i p_i M_i \tag{1}$$

Three risk levels are considered here: risk of irritation, intoxication and lethality. The accident probability is affected by the atmospheric conditions, by the type of losses and by the population characteristics. The atmospheric stability conditions are divided into the six categories of Pasquill, as indicated in Figure 1. The loss of contents is classified into three different types, i.e. crack in containers, broken flanges or manholes and total collapse of the container. The population density is classified into rural, semi-rural or urban areas.

The total route to be analysed is divided into homogenous part routes of lenght l_j. Based on the above considerations, the following occurrence frequencies can be defined: f^a_{kj} for the k-type category of atmospheric stability, f^l_{hj} for the h-type of loss and f^p_{mj} for the n-type of population area. If ν_{oj} is the annual accidental rate (per unit length), then the occurrence rate of accident in the i-th condition is given as:

$$\nu_{ij} = \nu_{khmj} = \nu'_{oj}\, f^a_{kj}\, f^l_{hj}\, f^p_{mj}\, l_j \tag{2}$$

By assuming that the accidents follow a Poisson process, and that ν_{ij} is a small number, the annual accident probability p_{ij} may be approximated as:

$$p_{ij} = \nu_{ij} \qquad (3)$$

The magnitude of the expected consequences M_{qij} relative to a risk level q, for a condition i (e.g. atmospheric stability conditions k, type of loss h and population area m) and in a part route j may be computed as:

$$M_{qij} = M_{qkhmj} = A_{khq} \; d_{mj} \; e_{mj} \; (1-V_{hj}) \qquad (4)$$

where:

A_{khq} = area exposed to concentration of type q, because of a h-type release, which occurs in a h atmospheric-stability class

d_{mj} = average number of inhabitants in a m-type of area

e_{mj} = average exposure factor in a m-type of area

V_{hj} = escape factor from a h-type of release

The area A_{khq} is obtained by the dispersion model. The exposure factor e_{mj} reflects the fact that a part of the population present will evade exposure (for example because they are on high floors of buildings and the release is heavy gas); the material escape factor represents the ratio of population which is able to move in time.

By combining Equations (1) - (4) the total risk is obtained as:

$$R_q = \nu'_{oj} \; f^{1}_{hj} \; f^{P}_{mj} \; A_{khq} \; d_{mj} \; e_{mj} \; (1-V_{hj}) \qquad (5)$$

For the automatic computation of equation (5) appropriate algorithms are described in /2/.

Example Application

The procedure developed and the corresponding computer code have been applied to compute the risks for road and rail transport of chlorine in the important North Italian route Mantova-Venezia. Road and rail routes are illustrated in Figure 2.

Accidental frequencies for road and rail transport have been derived by applying a Bayesian stochastic model able to combine literature with actual data as $\nu'_{oj} = 5\times10^{-7}$ and 5×10^{-8} (per tanker/kilometer/year) respectively. The probabilities of release from a road or rail tanker following an accident were taken as given in /4/. With respect to the other input data, as atmospheric conditions and surrounding scenario distribution, information have been taken from an extensive study /2/.

The results obtained in the study are illustrated in Figures 3 and 4.

Figure 3 compares road and rail transport risks. It is concluded that for an equal amount of transported chorine the ratio between the railway risk and the road risk is of 1/10.

Figure 4 illustrates the influence of the accidental rate on the risk ratio. The high accidental rates for the road transport are the main reason for the risk difference.

Conclusions

A general concept to assess transportation risks of dangerous materials is presented in this contribution. The concept is based on dispersion models for gases and on a simple probabilistic consequence model.

This procedure has been applied to evaluate the transportation risk of chlorine between Mantova and Venezia and especially to compare the rail with the road route, indicating the benefits in terms of risk by using a rail transport. The proposed methodology and the corresponding computer package give optimum possibilities for the analyst to introduce specific data, such as different types of dangerous substances, different routes, different means of transportation. The fundamental scope of the analysis is to optimize the transportation procedure in order to minimize the risk.

REFERENCES

1. Simmons, J. A.; Erdmann R. C.; Naft N.: The Risk of Catastrofic Spills of Toxic Chemicals. UCLA, Los Angeles, CA 1974.

2. D'Appolonia: Comparative Risk Analysis Chlorine Transport by Road and Railway, Report. Genoa 1988.

3. Lees, F. P.: Loss Prevention in the Process Industries. Butterworths, London 1980.

4. TNO: Methods for the Calculation of the Physical Effects on the Escape of Dangerous Materials. Voorburg, the Netherlands 1979.

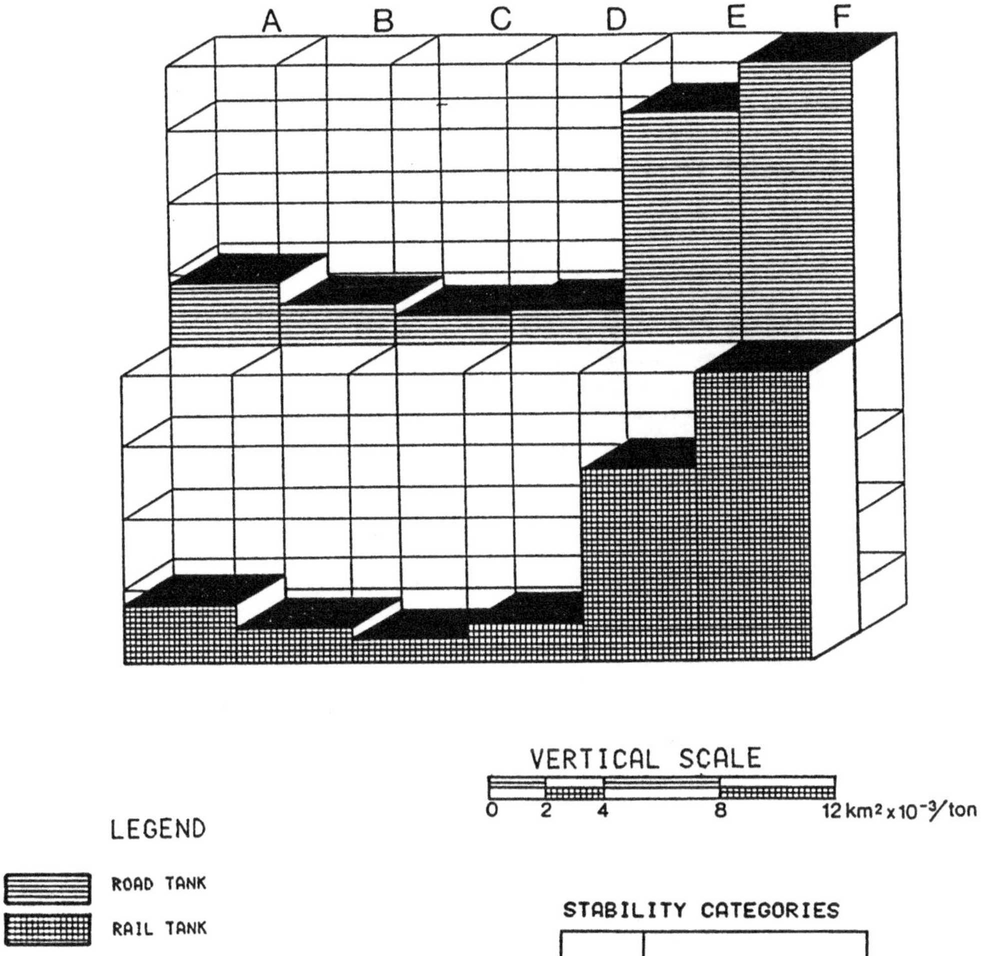

STABILITY CATEGORIES

A	VERY UNSTABLE
B	UNSTABLE
C	LESS UNSTABLE
D	NEUTRAL
E	STABLE
F	VERY STABLE

Fig. 1. Areas for lethality thresholds in case of tank failure

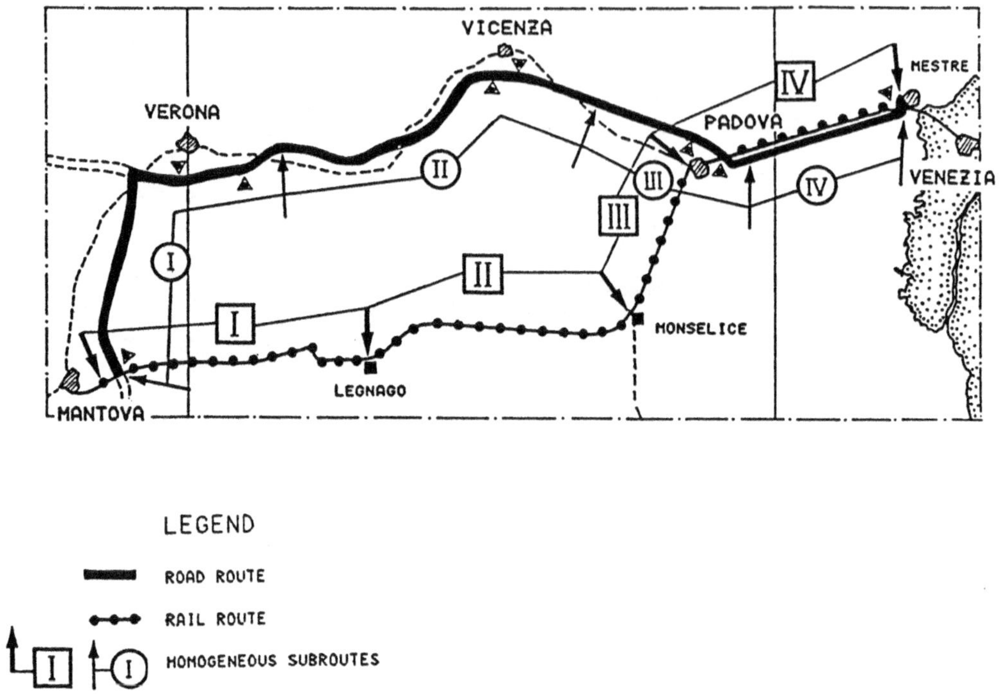

Fig. 2. Representation of road and rail route

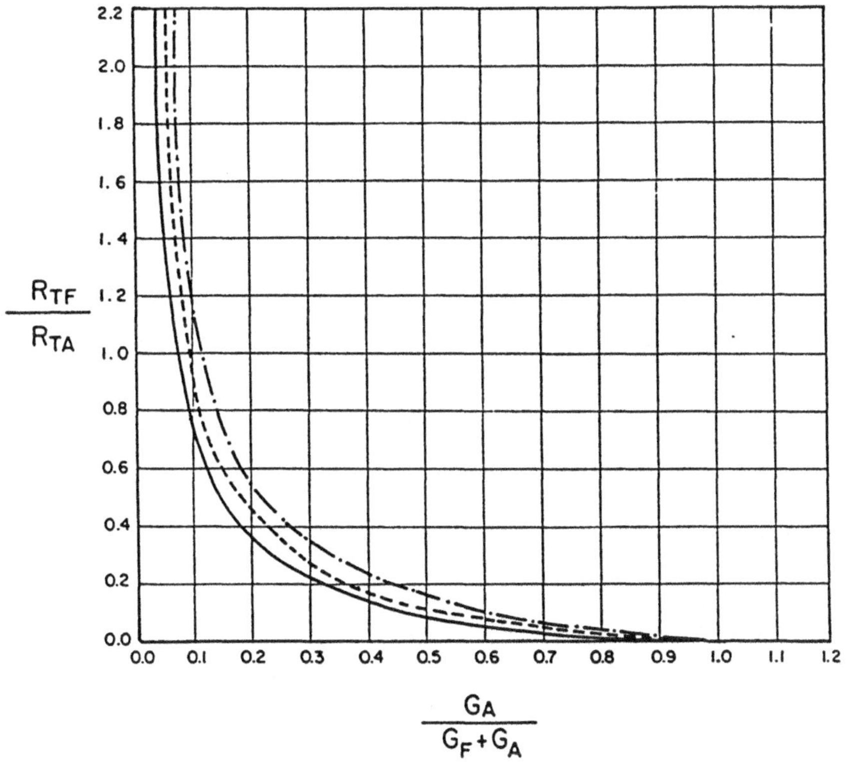

LEGEND

R_{TF} RAIL RISK

R_{TA} ROAD RISK

G_A CHLORINE QUANTITY TRANSPORTED BY ROAD TANK

G_F CHLORINE QUANTITY TRANSPORTED BY RAIL TANK

SYMBOL	RISK EFFECT	VALUE [ppm]	[mg/m³]
------	IRRITATION	15	45
———	INTOXICATION	50	150
—·——	LETHALITY	1000	3000

Fig. 3. Comparison between rail and road transportation risk

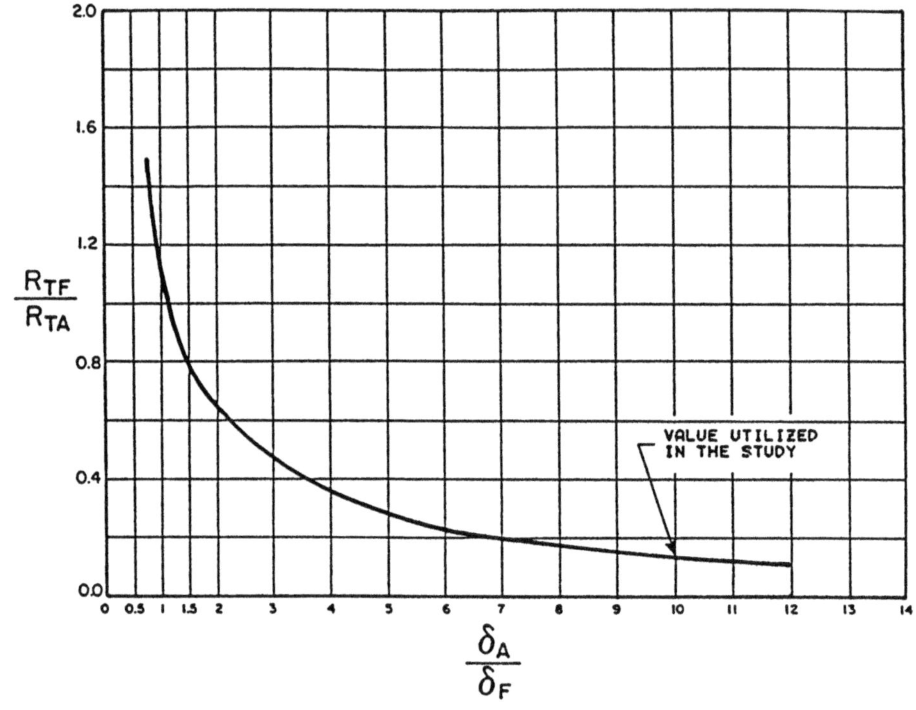

LEGEND

δ_F ACCIDENTAL FREQUENCY FOR RAIL TRANSPORT

δ_A ACCIDENTAL FREQUENCY FOR ROAD TRANSPORT

R_{TF} LETHALITY RISK IN RAIL TRANSPORT

R_{TA} LETHALITY RISK IN ROAD TRANSPORT

Fig. 4. Lethality risk ratio between rail and road transport as a function of the accidental frequency

Effect of Maintenance Factors on Risk and Adequacy of Power Generation

Farag Ali El-Sheikhi
Arabian Gulf Oil Co.

PO Box 296
Benghazi
Libya

Roy Billinton
Power System Research Group
University of Saskatchewan
Saskatoon
Saskatchewan
Canada

ABSTRACT

The overall availability assessment of a generating unit involves an appreciation of the forced outage parameters for the unit and its maintenance reqirements.

This paper introduces an algorithm for performing maintenance scheduling of generation systems in terms of levelizing the system risk throughout the year. The moment-cumulant method is used in developing the technique of this paper. The paper investigates the effects of the system forecast peak load and forced outage rate of a generating unit on the maintenance schedule of the generation system. Applications to the standard IEEE Reliability Test System (RTS) have been used to illustrate these studies.

INTRODUCTION

Generating units require a significant amount of maintenance time in order to maximize their operating reliability and to enhance the overall system adequacy. Scheduling of generating equipment for preventive maintenance plays an important role in long term generating planning. Improper generating unit preventive maintenance schedules can force a generating system expansion plan into the installation of capacity to cover maintenance at a time when economic and environmental concerns dictate that electric power utilities conserve their resources. The available installed capacity of the system is decreased during a unit maintenance period which results in an increase in the system risk. This risk should not exceed the system acceptable planning value at any time during the annual load cycle. It is important, therefore [1] to levelize the/...

the system adequacy throughout the year by appropriate preventive maintenance scheduling. The technique utilized in this paper for evaluating the system risk is the Loss of Load Expectation (LOLE) method [2]. Maintenance requirements can have a substantial effect on the reserve required to meet a desire LOLE level. Maintenance requirements increase the contribution of off-peak periods to the annual LOLE as most maintenance operations, especially of large units, are usually scheduled at relatively low system loads.

This paper presents a new technique based on the moment and cumulant of the system generating unit outage distribution and the hourly load distribution together for maintenance scheduling of generating units.

Sensitivity studies of the effect on maintenance scheduling of generating facilities with unit forced outage rate and system hourly peak demand variations are presented in this paper using the risk-levelization approach. The standard IEEE Reliability Test System is used to illustrate the practical significance of the concepts proposed.

PROPOSED TECHNIQUE

This technique is based on the Gram-Charlier expansion of a distribution. It can recognize the probabilistic nature of the generating unit forced outage rates together with the probabilistic nature of the load cycle.

Description of notation

m_{ji} = jth moment about the origin of the ith unit.

C_i = capacity of unit i.

q_i = forced outage rate of unit i.

n = number of generation units in the system.

$l^j{}_i$ = jth moment of the hour i load about the origin.

L_j = jth moment of the load distribution about the origin.

KL_j = jth cumulant of the hourly load distribution.

KE_j = jth cumulant of the equivalent load curve.

EL(x) = the value of the equivalent load corresponding to any cumulative prob.

ELCC = effective load carrying capability of an unit.

FOR = forced outage rate of an unit.

ALGORITHM STEPS

1. Calculate the first six cumulants of each unit [3]:
   a. Calculate the first six moments about the origin of each unit:
   $$m_{ji} = c_i^{\,j} q_i \qquad j = 1,2,\ldots,6 \qquad i = 1,2,\ldots,n$$
   b. Calculate the first six cumulants of each unit [3].

2. Calculate the cumulants of the system generating unit outage distribution:
$$KG_j = \sum_{i=1}^{n} K_{ji} \qquad j = 1,2,\ldots,6$$

3. Calculate the first six cumulants of the hourly load distribution:
   a. Calculate the first six moments about the origin of the hourly load distribution
   $$L_j = \frac{1}{T}\sum_{i=1}^{T} l^j{}_i \qquad j = 1,2,\ldots,6$$
   b. Calculate the first six cumulants of the hourly load distribution.

4./...

4. Coefficients of the continuous approximation of the equivalent load curve can be obtained from the cumulants of the system generating unit outage and hourly load distribution cumulants:

   a. Calculate the cumulants of the equivalent load curve (load + outages):

   $$KE_j = KG_j + KL_j \qquad j = 1,2,\ldots,6$$

   b. Calculate the coefficients of the equivalent load curve:

   $\mu = KE_1$ mean
   $6^2 = KE_2$ variance

5. The equivalent load value (load + outages) corresponding to any cumulative prob. can be obtained as follows [4]:

   $$EL(x) = \mu - 6.\phi(z,G_j)$$

6. Maintenance scheduling procedure can proceed using the unit ELCC as calculated from the EL(x) [4].

SYSTEM STUDIES

The proposed technique described in this paper, has been implemented in a computer program which was then used to perform different studies on the IEEE Reliability Test System. This system has 32 units, ranging from 400 MW nuclear to 12 MW fossil-oil Units [5]. The installed capacity is 3405 MW and the system annual peak demand is 2850 MW. The following cases were performed on the IEEE-RTS:

Case 1 - Base Case Study

The original system data given in Reference [5] has been used. Plan 1 in Table 1 shows the IEEE-RTS maintenance schedule using system reliability levelization idea. This/..
This plan represents the base case maintenance schedule of the system to which the other cases will be compared.

Case 2 - Effect of the Unit FOR on the System Maintenance Plan

The influence of the unit FOR on the maintenance schedule can be illustrated by changing the FOR of a 400 MW unit which is the largest unit in the system. The FOR of one 400 MW unit has been increased by 50% to examine this effect on the system maintenance plan. Plan 2 of Table 1 shows the maintenance schedule in this case. The rest of the system data is unchanged. An examination of Plan 2 and the base case plan, i.e. Plan 1, reveals that the maintenance schedule is different in 26 weeks. It is evident from Plan 2 that changing the value of the FOR of just one unit can result in a totally different system maintenance schedule.

Case 3 - Impact of the System Forecast Peak Load on the Preventive Maintenance Plan

The load model of the IEEE-RTS [5] gives 8736 hourly loads for one year on a per unit basis, expressed in chronological fashion. In this study, hourly as well as weekly peak demands are used. The effect of the system forecast peak load on the maintenance plan was investigated by increasing each hourly value by 5%. The results are shown in Plan 3 of Table 1.

The planning criterion was taken as 0.1 days/year in order to compare this plan with the base case plan, i.e. plan 2. It can be seen that the revised plan has different maintenance activities in the weeks 3,4,5,40,41 and 42. Although the difference is not great in this case, our experience with other real power systems indicates that changing the system forecast peak demand can influence the maintenance plan very much.

CONCLUSIONS

Preventive maintenance of generating facilities plays an important role in the long term planning decisions regarding the size and type of units to be installed. Scheduling outages for planned maintenance is relatively difficult because of the probabilistic nature of the overall problem. Maintenance scheduling will become much more serious if the load growth picks up and system reserves shrink in the future. A preventive maintenance schedule should permit the required work to be accomplished but not introduce excessive risk to the system.

A new approach for scheduling preventive maintenance of the system generating faciliteis based on the Gram-Charlier expansion of distribution on the risk-levelization criterion has been introduced. The new approach can recognise the different probabilistic aspects of the maintenance scheduling problem and levelizes the system reliability throughout the maintenance period as well. It was shown that the maintenance schedule of the generation system is sensitive to the system forecast peak load and any variations in the unit forced outage rate can seriously affect the preventive maintenance plan.

REFERENCES

1) F A El-Sheikhi and R Billinton, "Power Generation Systems Availability Maintainability Assessment", 10th Advances in Reliability Technology Symposium, University of Bradford, UK. 6-8 April, 1988 pp. 197-210.

2) R Billinton and R N Allan, "Reliability Evaluation of Power Systems", Longman, London (England)/Plenum Press, New York, 1983

3)/...

3) F A El-Sheikhi, "Maintenance Scheduling of Generating Facilities in Single and Interconnected Power Systems", Doctor of Philosophy in Electrical Engineering, University of Saskatchawan, Saskatoon, Canada, 1983.

4) F A El-Sheikhi, R Billinton, "Probabilistic Simulation of Power System Operation for Maintenance Scheduling of Generating Facilities", Africon `83, 1st IEEE Conference in Africa, December 7-9, 1983, Nairobi, Kenya.

5) "IEEE Reliability Test System", A Report prepared by the Reliability Test System Task Force of the IEEE Trans; Vol. PAS-98, No 6, Nov/Dec 1979, PP 2047-2054.

Table 1

IEEE MAINTENANCE SCHEDULE

Week No.	Level											
	1	2	3	4	1	2	3	4	1	2	3	4
1	No				No				No			
2	Maintenance				Maintenance				Maintenance			
3	76				50				50			
4	76				50	12			50	12		
5	76				12				12			
6	155				50				155			
7	155				100	50			155			
8	197	155			197	100	12		197	155		
9	197	155	20	12	197	100	50	12	197	155	20	12
10	400	197	20	12	400	197	50		400	197	20	12
11	400	197	155		400	197	155		400	197	155	
12	400	155	20	20	400	155	20		400	155	20	20
13	400	155	20	20	400	155	20		400	155	20	20
14	400	155			400	155			400	155		
15	400	197	76		400	197	76		400	197	76	
16	197	76	50		197	76	50		197	76	50	
17	197	76	50		197	76	50		197	76	50	
18	197				197				197			
19	No maintenance				No maintenance				No maintenance			
20	100				100				100			
21	100	50			100	20	12		100	50		
22	100	50			100	20	12		100	50		
23	No maintenance				No maintenance				No maintenance			
24	"				"				"			
25	"				"				"			
26	155	12			155				155	12		
27	155	100	50	12	155	100	50		155	100	50	12
28	155	100	50		155	100	50		155	100	50	
29	155	100			155	100			155	100		
30	76				76				76			
31	350	76	50		350	76	20	12	350	76	50	
32	350	76	50		350	76	20	12	350	76	50	
33	350	20	12		350	12			350	20	12	
34	350	76	20	12	350	76	12		350	76	20	12
35	400	350	76		400	350	76		400	350	76	
36	400	155	76		400	155	76	76	400	155	76	
37	400	155			400	155	76		400	155		
38	400	155	50	12	400	155	76	50	400	155	50	12
39	400	155	12		400	197	155	50	400	155	50	12
40	400	197			400	197	155		400	197	76	
41	197	100	50	12	197	155			197	100	76	
42	197	100	50	12	197	155	20		197	100	76	
43	197	100			155	20			197	100		
44-52	No maintenance				No maintenance				No maintenance			
	Plan 1 IEEE-RTS Maintenance Schedule using Risk Levelization Technique (Base Case) System Peak Demand = 2850 MW Planning Criterion 0.1 days/yr.				Plan 2 IEEE-RTS Maintenance Schedule using Risk Levelization Technique. System Peak Demand = 2850 MW. FOR of a 400 MW Unit = 1.5 x 0.12 Planning Criterion = 0.1 days/yr.				Plan 3 IEEE-RTS Maintenance Schedule using Risk Levelization Technique. System Peak Demand = 1.05 x 2850 MW Planning Criterion = = 0.1 days/yr			

An Analysis of Accidents with Casualities in the Chemical Industry Based on Historical Facts

ing. L.J.B. KOEHORST

TNO Division of Technology for Society
Department of Industrial Safety

1. Preface

On behalf of the Directorate-General of Labour, TNO carried out an analysis of accidents with casualties which occurred in the chemical industry.
From the databank FACTS, 700 accidents have been selected. The selection was focused on accidents during industrial activities with hazardous materials. The aim of the analysis was to find out how casualties occur and to give concrete suggestions and recommendations to improve safety during daily practice. The first results are presented in this paper. By the end of 1988 the full report of the analysis will be submitted to the Directorate-General of Labour, and most probable it will become available as a publication of the Directorate.

2. Scope of the analysis

The 700 accidents were selected from the databank FACTS.
The following selection criteria have been used:

- accidents with casualties;
- accidents with hazardous materials, which happened in the chemical industry;
- accidents that happened during the following activities;
  - processing;
  - storage;
  - industrial use and/or application;
  - transhipment to or from storage tanks.

Only accidents that were in line with all three criteria were selected.
Figure 1 represents the activities during which the accidents occurred.

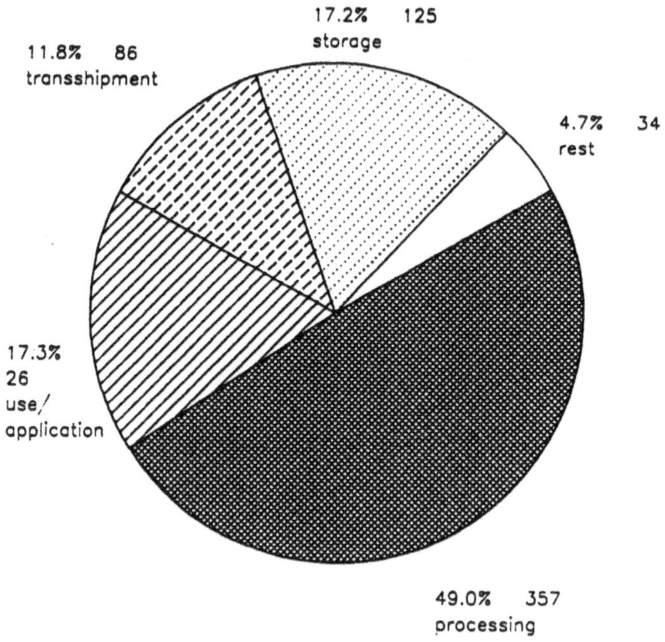

Fig.1. Accidents during several activities

3. The databank FACTS

FACTS is a database with technical information about accidents with hazardous materials that happened during all types of industrial activities (processing, storage, transhipment, transport, etc.). At this moment FACTS contains information of about 15,000 accidents that happened all over the world. Most information concerns accidents during the last 30 years.

The information stored in FACTS is derived from several sources such as:
- literature;
- periodicals;
- technical reports;
- environmental and labour inspectorates;
- industrial companies;
- fire brigades;
- police.

There will always be a discrepancy between the number or accidents that actually have happened and those that are recorded. Accidents having minor consequences may not have been recorded at all, while accidents with more consequences may be recorded incidentally. Only accidents where severe damage or danger is involved will be publicized, analysed and documented. The discrepancy between events that actually have occurred and those that have been recorded, is shown in figure 2.

The quality of the available information of recorded accidents is also related to their seriousness. The most serious accidents are also those of which good and detailed information is available.

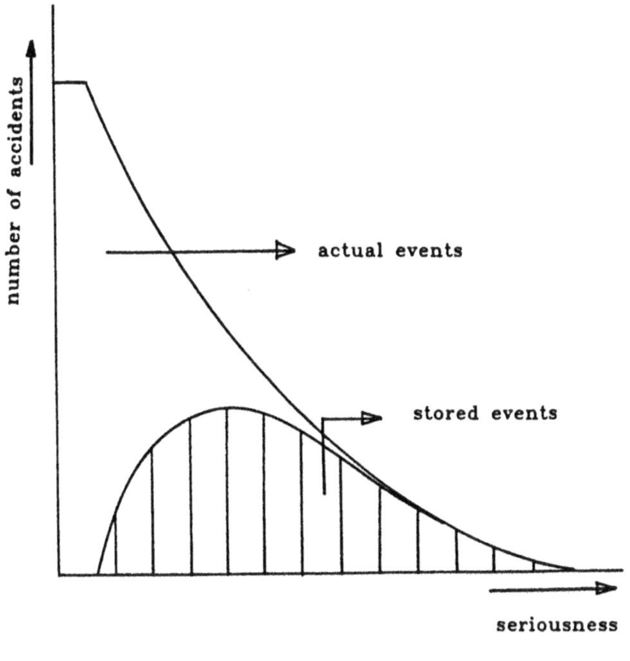

Fig.2. Comparison of actually occurred incidents and recorded incidents

4. Analysis

The selected accidents are divided into three categories:

- normal operation;
- maintenance;
- start-up/shut-down.

Each category represents one specific phase of the lifecycle of an installation or plant. During each phase certain activities will be carried out or chemicals and equipment will be used, which might be of influence on the cause and consequences of accidents. Each phase is analysed with respect to the following items:

- the course of the accident;
- the accident cause;
- type of involved equipment;
- human handling.

Figure 3 illustrates during which phase the accidents took place.

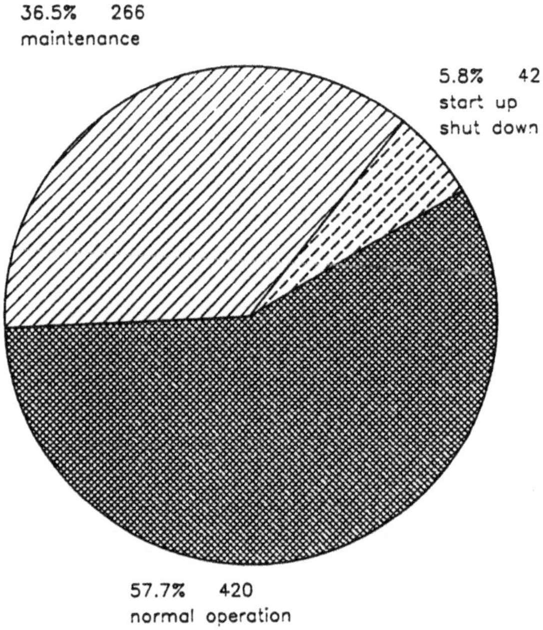

Fig.3. Classification of the selected accidents into system phases

It is obvious that most accidents took place during normal operation. Remarkable, however, is the great amount of accidents during maintenance and the very few accidents during start up and shut down procedures. With respect to the accidents during maintenance it should be noticed that maintenance time is only a small part compared to normal operation time. This indicates that the probability that accidents occur is relatively high during maintenance.

4.1 The course of accidents

During all three system phases a lot of dangerous situations were created by human actions. In order to describe the consequences of these actions and to indicate their relationship with other events that happen during an accident, the course of accidents has been analysed. The results are presented in a scheme. It will give an insight into the mechanism which leads to the release of chemicals. In the scheme the number placed before each event indicates how many times the event occurred. To keep the schemes readable some occurrences of the same type are replaced by a more general expression. For example,

wrong physical conditions: under-/overpressure;
under-/overheating;
overfilling.

technical failure:
- malfunctioning of equipment;
- broken equipment;
- wrong specifications of equipment.

In the figures 4, 5 and 6 phase an event-scheme is presented for each system phase. As might be expected a lot of accidents occur during human handling. This applies for all three system phases. In more detail, it can be concluded that:

* during normal operation a lot of accidents (23%) occur during unloading or pumping over of chemicals;
* 70% of the maintenance accidents occur during dismantlement of equipment (28%), welding (20%) and repair of equipment (20%);
* during start-up/shut-down, accidents often occur during manipulation of equipment (20%) and during dismantlement of equipment (19%).

All these different types of human action during which accidents occur can be described as simple human actions while people stay in close or direct contact with hazardous chemicals.

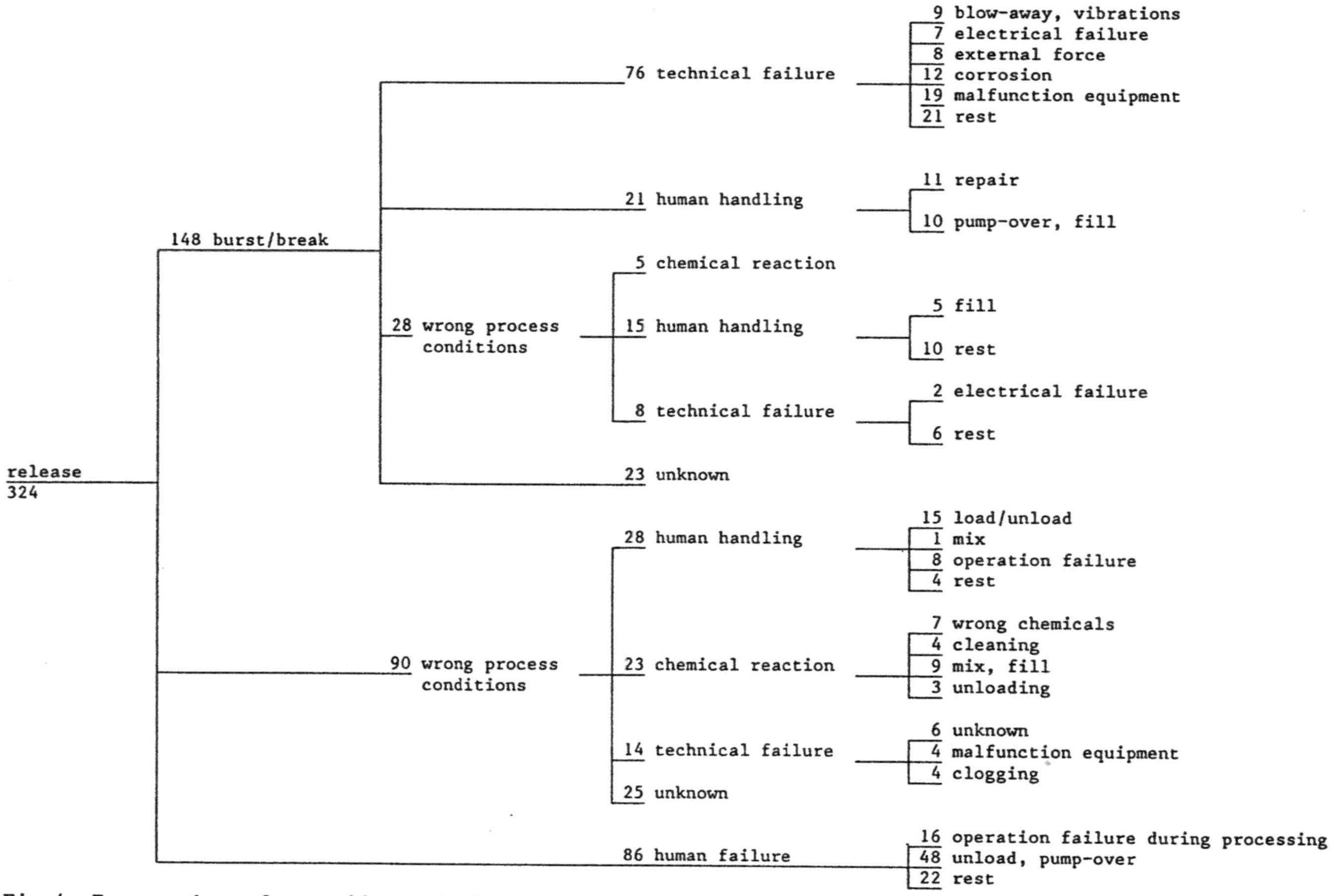

Fig.4. Event scheme for accidents during normal operation

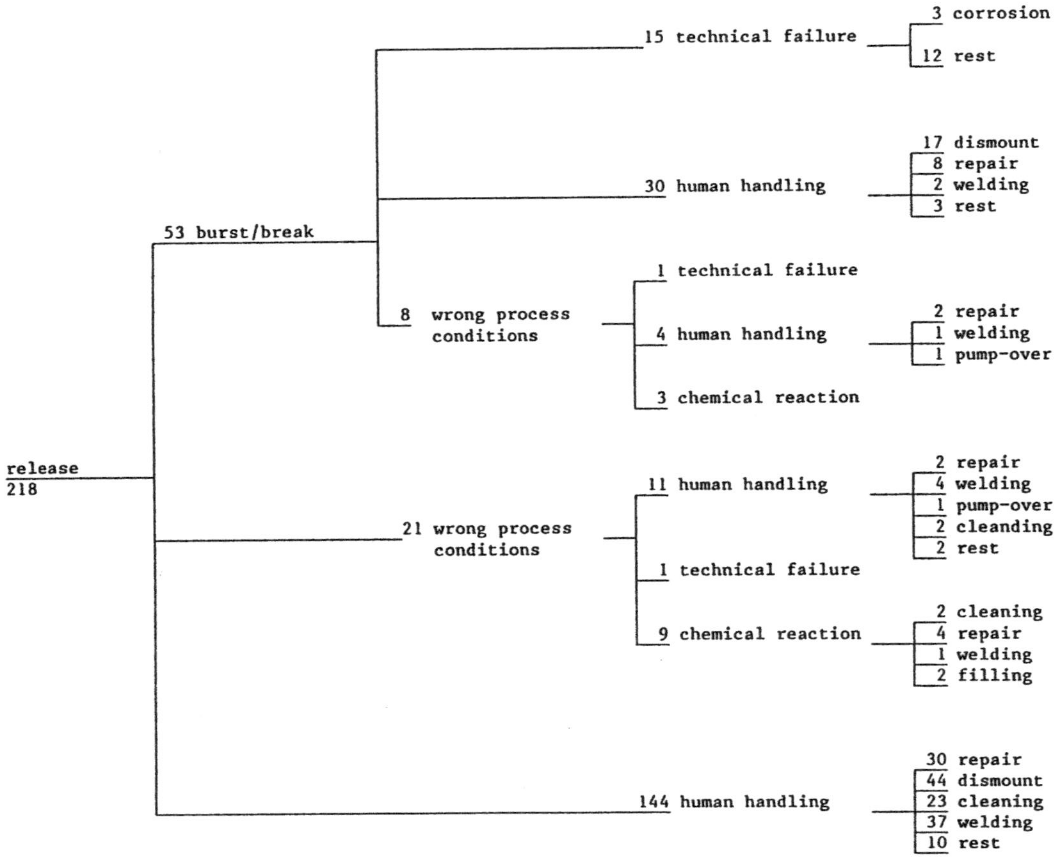

Fig.5. Event scheme for accidents during maintenace

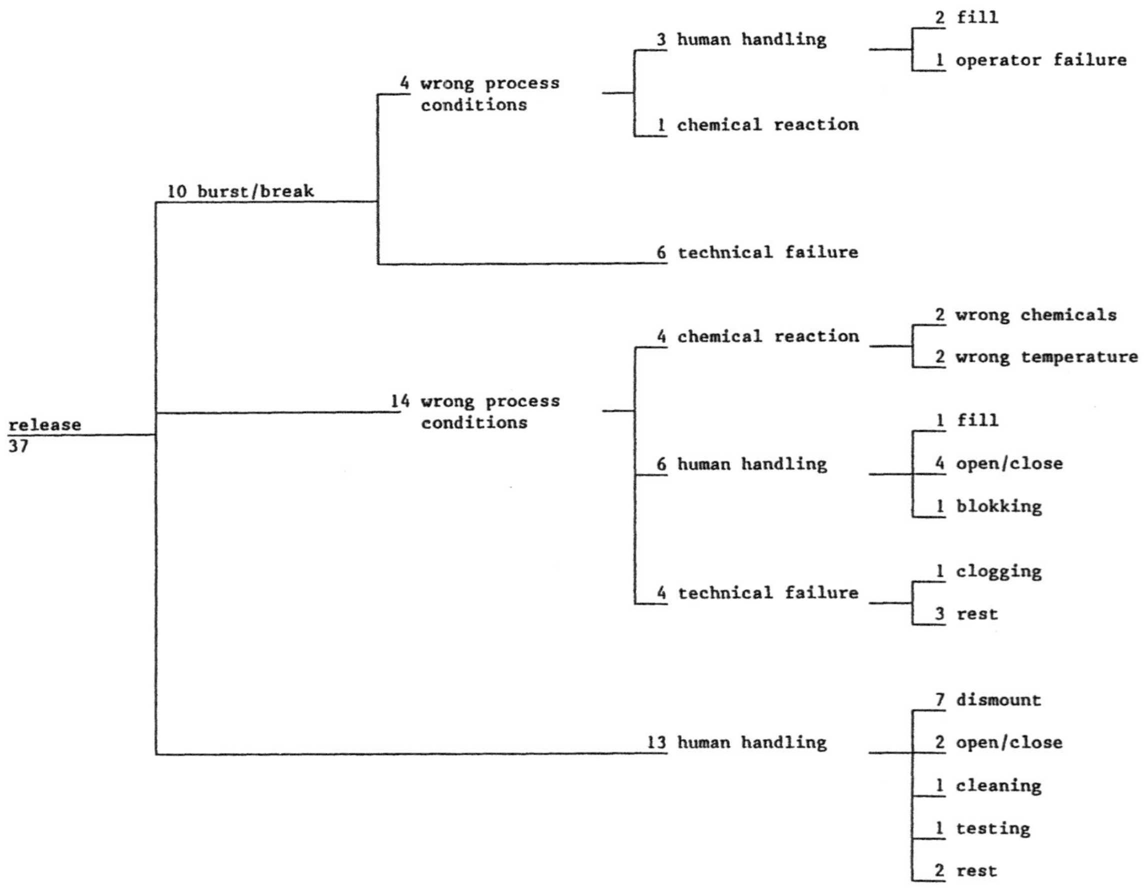

Fig.6. Event scheme for accidents during start up/shut down

Almost 80% of all selected accidents result into a release of chemicals. It is not possible to indicate which specific chemicals or groups of chemicals are mostly involved. Because most chemicals used in the chemical industry are highly inflammable, a release is often followed by a fire or explosion. Figure 7 presents the type of ignition source involved. Remarkable here is, that in 32% of the accidents during maintenance, welding was identified as the ignition source.

ignition sources	normal operation %	maintenance	start-up/ shut-down
hot spots	8	11	21
electrical sparks	24	3	16
static electricity	8	10	-
open fire	28	20	32
mechanical sparks	21	21	10
self-ignition	8	3	16
welding	3	32	5
total	100	100	100

Fig.7. Classification of ignition sources

For all three system phases the physical consequences of an accident are illustrated in figure 8. In about 30% of the accidents the consequences are limited to a release. In 60-70% a release is followed by a fire or explosion. Especially during normal operation the probability that a release will result into a fire is most likely.

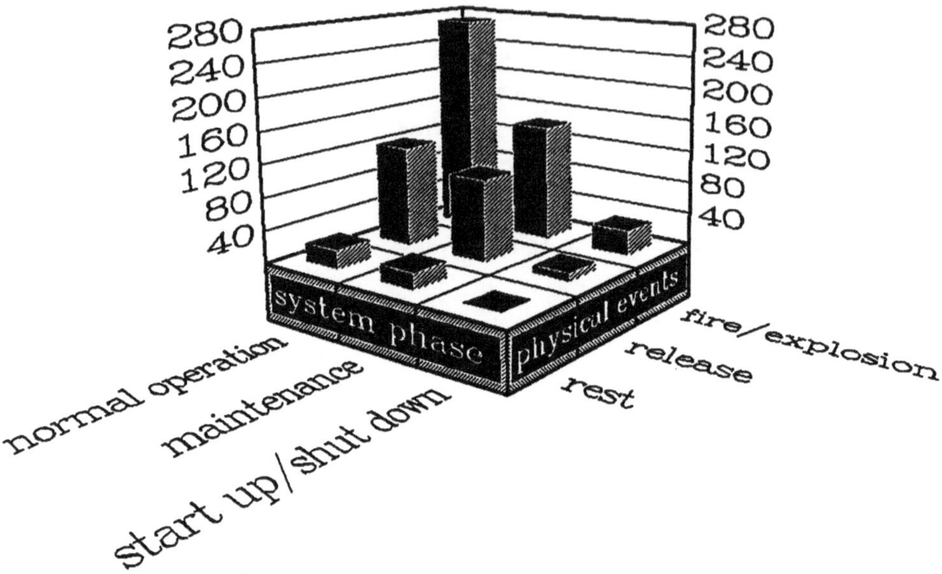

Fig.8. Physical consequences of accidents related to the system phases

4.2 Relation between involved equipment and human handling

With respect to the accidents during normal operation and during maintenance it has been investigated which type of equipment or installation is frequently involved in accidents. From tabel 1 in Annex 1 it can be concluded that during the most accidents the following types of equipment are involved:

- hoses involved in 95 accidents;
- lines involved in 85 accidents;
- valves involved in 152 accidents;
- drums involved in 82 accidents;
- cylinders involved in 58 accidents.

The above mentioned type of equipment are parts of the following installations:

- storage tank/vessel 256 times involved in accidents;
- reactors 72 times involved in accidents;
- transport equipment 134 times involved in accidents.

. In the tables 2, 3 and 4 of Annex 1 the above mentioned types of equipment and installation are viewed in relation to several human actions. From the event-schemes presented in paragraph 4.1 several human action, which are often involved in accidents, have been selected for this comparison.

. From the tables 2,3 and 4 it can be noticed that a lot of accidents happen in the following situations.

1. Normal operation:
   - handling of valves;
   - use of hoses during loading or filling;
   - filling of drums; regular accidents occur, caused by unclean drums or by using the wrong chemicals. In both situations unforeseen chemical reactions are generated;
   - filling of reactors and tanks; overfilling takes place due to inattention of the responsible persons or to failure of a level indicator.

2. Maintenance:
   - dismantling or disconnecting of lines, valves and hoses, connected to tanks or reactors;
   - cleaning of storage tanks;
   - use and repair of valves; wrong valves are open or closed causing the release of chemicals. Releases also occur from pressurized systems where valves need repair. In several cases the system was not depressurized before repair was started.

3. Start-up/shut-down.

   Although the available number of accidents that took place during start-up or shut-down is rather small, it can be noticed that most incidents were caused by simple human actions, such as handling of valves and lines.

5. Consequences to involved people

Taking in mind the fact that only accidents were selected where people got injured, figure 8 presents the number of accidents related to a cumulative number of injuries and fatalaties. A common rule is that the ratio between fatalities/injuries is about 1:30.

Figure 8 indicates a ratio of about 1:3 à 5. The reason for this discrepancy is the fact that for this analysis accidents are selected with more serious consequences. This was necessary because these accidents are mostly well documented, which is necessary for such an analysis.

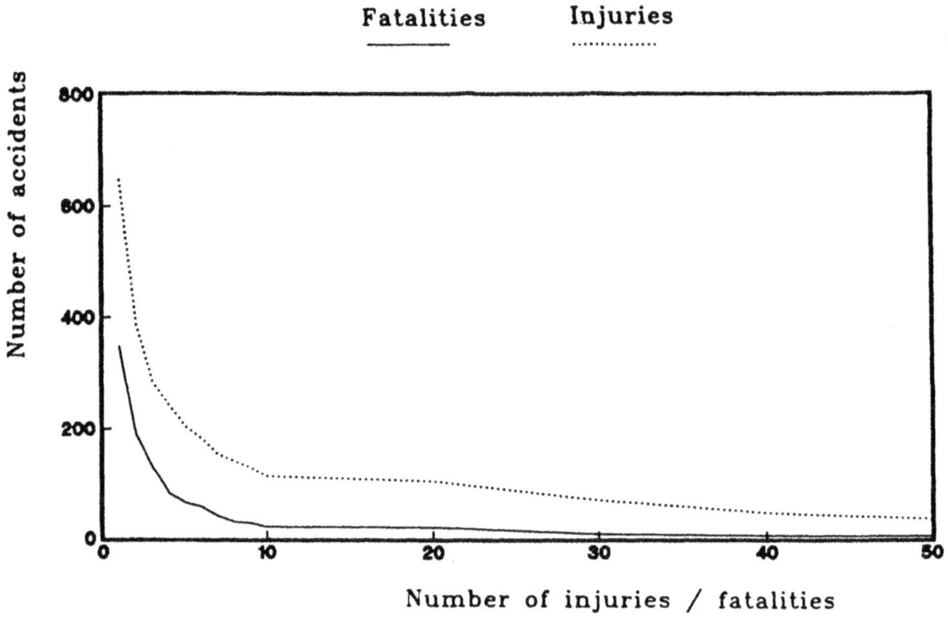

Fig.9. Cumulative number of fatalities/injuries.

Referring to the three system phases, figure 9 presents the cause of injury, related to the number of accidents. The figure indicates that during all system phases, fire is responsible for most of the injuries. During normal operation, also many injuries occurred due to contact with toxic compounds.

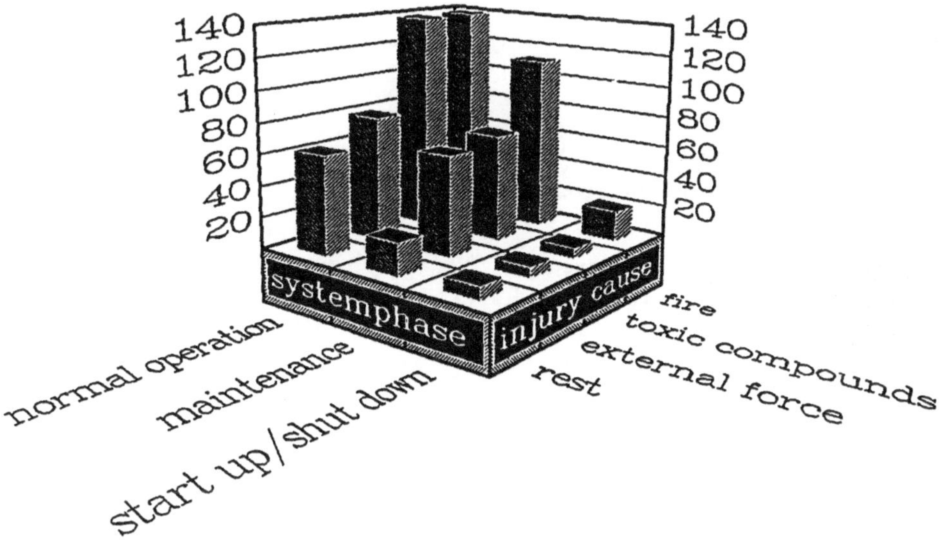

Fig.10. Injury cause related to the system phases

Further investigation showed that the numbers of people who were injured by fire or by toxic chemicals were about the same. Toxic chemicals can be present in process installations but can also be formed during a fire of non-toxic substances (plastics, synthetic material etc.). In about 10% of the accidents with a fire, toxic chemicals were formed by which people became injured.

6. Human role during accidents

From chapter 4 it become clear that regardless the system phase, human behaviour is of great influence on the cause and course of accidents. For this reason more specific attention was payed to this subject. In order to qualify the human role during accidents, the following classification has been made of human failure:

- operator failure;
- manipulation failure;
- installation failure;
- inspection failure;
- maintenance failure;
- assembling failure;
- organization failure.

A few examples of some of these failure types illustrate their meaning.
- operator failure:
  - wrong process adjustment;
  - wrong chemicals used;
  - injudicious use of an installation.
- manipulation failure;
  - wrong valves closed or opened;
  - dismantling of a line;
  - drop of drums, cans etc.
- organization failures;
  - activities executed in the wrong sequence;
  - unsafe or unclean surroundings;
  - wrong execution of another man's task

Table 5 presents for each system phase the share of the several failure types in percentages.

system phase / failure type	number of accidents in %		
	normal operation	maintenance	start-up/ shut-down
operator failure	34	28	29
manipulation failure	31	27	29
installation failure	12	1	10
inspection failure	3	6	4
maintenance failure	5	8	14
organization failure	15	30	14

Operator, manipulation and organization failures frequently cause accidents. Examples of such failures that frequently occur are:

* Operator failure:
  - wrong adjustment of an installation that deviates from the optimal setting;
  - the use of wrong combinations, quantities or type of chemical;
  - deviation of instructions or specifications, during loading or unloading. No earth wire was connected or the loading was done with too high a velocity, which causes a static electrical discharge.
* Manipulation failures:
  - replacement of valves, gaskets at pressurized or filled installations;
  - opening or closing the wrong valves; dismantling the wrong line or hose;
  - unpermitted smoking, unallowed entrance of an installation, fail to use personal protection equipment.
* organization failure:
  - accidents caused by a combination of activities which are executed at the same time e.g. dismantling a valve and testing a pump in the same system.
  - working in unsafe areas, e.g. vessels which have not been degased, welding in poor ventilated rooms:
  - fail to instruct contractors, or temporary employees.

7. Conclusions and recommendations

In the previous chapters several aspects of accidents with casualities have been analysed and where possible conclusions have been made.

The final conclusion of this analysis, is that most accidents are caused by failures during simple human actions. On the one side this conclusion may seem obvious because especially simple handlings occur very frequently. On the other side just because the high frequency of these handlings the people in charge are not (anymore) aware of the potential danger of chemicals and equipment they use.

Because it will never be possible to be 100% sure that these types of failure will be avoided, further investigation should be aimed at the improvement of safety in these situations. A possibility is the development of a working environment, where the consequence of failures will be limited to a minimum. In analogy to process installations which are inherently safe, this philosophy should lead to a inherently safe working environment.

8. References

1. Lees, F.P.: Loss Prevention in the process industries, 1980.

2. Bockholts, P. and Koehorst, L.J.B.: Accident analysis of vapour cloud explosions, 17th June 1987.

3. Bockholts, P.: Oorzaakanalyse, 1 november 1985.

4. Oorzakenboek (protoversie), DGA/TNO, 1987.

5. Health and Safety Executive. Dangerous Maintenance. ISBN 0 11 883957 8, 1987.

6. Gevaren van statische elektriciteit in de procesindustrie. Rapport van de stuurgroep Richtlijnen Veiligheid Procesindustrie, 1975.

ANNEX 1-1

Table 1

equipment	number of accidents	
	normal operation	maintenance
bots, nuts	8	15
drains	5	9
filters	6	4
fittings	11	10
fanges	4	21
hoses	40	54
lines	35	50
packings	9	6
pumps	20	23
valves	75	77
measure and controle devices	9	6
heat exchangers	7	17
compressors	4	5
drums	71	11
cylinders	41	17

installations	number of accidents	
	normal operation	maintenance
gasholder	1	7
reactors	50	22
tanks/vessels	150	106
transport equipment*	82	52
destillation units	10	3
flares	4	4
furnaces	10	4

* for example tankwagons, railwagons, vessels etc.

Table 2. Accidents during normal operation; human handling viewed in relation to type of equipment/installation

Human handling / Equipment/Installation	fill	load/ unload	collision	cleaning	pump over	actions: start, stop, open, close etc.	Total number of accidents
hoses	10	9	4	2	-	4	40
pipelines	2	-	5	3	1	6	35
valves	5	7	1	4	1	18	75
drums	25	8	1	1	3	-	71
transport equipment	11	15	6	1	2	6	82
reactors	12	-	-	1	1	9	50
storage tanks	34	11	2	4	5	23	150

Table 3. Accidents during maintenance; human handling viewed in relation to type of equipment/installation

Human handling Equipment/Installation	cleaning	actions start, stop, open, close etc.	dismount	fill	pump over	load / unload	repair	mount	mainte-nance	welding	Total number of accidents
hoses	6	10	21	6	2	3	9	5	5	6	54
pipelines	6	8	23	-	-	-	6	4	5	8	50
valves	7	23	33		2	1	18	8	2	8	77
drums	2	2	4	1	1	-	3		-		11
cylinders	-	3	4	1	-	-	-	3	1	2	17
transport equipment	5	5	8	2	2	5	3	-	1	8	52
reactors	2	7	10	2	-	-	5	1	3	2	22
storage tanks	27	7	21	7	4	4	14	2	7	15	106

Table 4. Accidents during start-up/shut-down; human handling viewed in relation to type of equipment/installation

Human handling Equipment/Installation	open/close	dismount	blocking	fill	Total number of accidents
valves	6	2	1	-	10
flanges	-	1	-	-	2
packing	-	-	-	-	2
lines	-	3	2	-	10
reactors	-	-	-	2	6
storage tanks	-	-	4	2	8

Reliability Follow-Up in the Pre-Production Process at Volvo Car Corporation

A. Wendel and O. Lindwall
Department of Quality Assurance
Product Engineering Division
Volvo Car Corporation
S-405 08 Göteborg, Sweden

Introduction

In the Product Engineering Division of Volvo Car Corporation,we have, since about 15 years, a special Quality Assurance Department which is responsible for the Reliability Programme of the division There are four major activities involved in this programme (fig.1 describes roughly how they are connected):

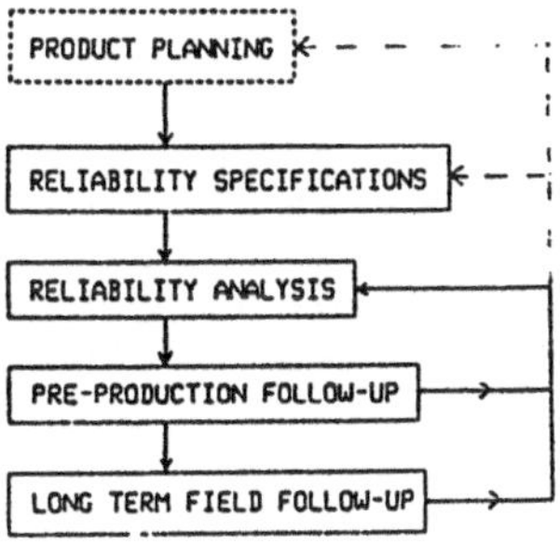

Fig.1. Connections between relibility activities

- Specification of reliability levels. General reliability demands from the Product Planning Department (e.g. "Best in class") are translated into technical terms (defaults per vehicle, L_{50} , probability of failure) by using competitor information from different marketing surveys, test results from motor magazines, own testing at VCC etc.
- Reliability analysis during the pre-production process to assure that the reliability requirements will be met. New designs and major design changes are analysed with techniques like FMEA and FTA.For this work, it is necessary to have field data and test results in order to make confident reliability assessments.
- The Long-term field follow-Up programme (LU) involves about 400 production cars per year model, from which repair data is collected and stored in a database. This gives us the possibility to measure the reliability level on the field, and see if it differs from the specifications. The LU programme is established since 10 years.

- Follow-up of pre-production test cars, the latest activity in the Reliability programme (started in 1986), which will be further described in this paper.

Together, all these activities make a powerful tool for us at VCC to assure that the final output (i.e. cars to customers) will meet Volvo standards.

Background and purpose

In today's very competitive automotive industry, it is necessary for the design engineering process to have a complete and accurate feedback system for failure incidents from all available sources.

Earlier, the reliability analysis work in the pre-production process has been based mainly upon the results from the LU system. The advantage of this system is that the failure data is very confident (compared with e.g. Warranty claim systems) and the very useful output facilities (Weibull plots, cumulated failure diagrams etc.), which make it possible to perform statistically relevant evaluations. The disadvantage with LU-data is that the "hardware" always is a couple of years old, and for some systems or components where major design changes recently have been made, therefore more or less inaccurate.

Results from the pre-production testing have, of course, always been considered in the reliability analyses, but due to the varying reporting standards and the difficulty to correlate the many different test methods, the pre-production failure data has been hard to use systematically.

Hence, we felt the absolute need for a follow-up system that would provide us with necessary test data, presented in a suitable way, for our reliability assessments.

Beside this, we also saw some other important areas where we could use a system like this:

- To "measure the temperature" of an entire project's reliability status, serving as a base for decisions to be taken by the project managment.

- General managment information.

- Feedback to designing and testing departments.

- Correlation studies between different types of testing.

- Other special case studies and further method development.

A complete reporting system including all repair data from the test cars, not only the specific test results from e.g. transmission or brake testing, would also give us a larger amount of failure data than before.

Pilot project

Before we went into "full scale" with the new follow-up routine, we decided to make a pilot study to see whether it was worth the effort to create a new system or not.

A rather new engine project was selected, old enough to have field data from production cars, but still new enough to have most of the pre-production test reports easy of access.

Based on the results from the different test reports, a relibility growth study was made, which showed how the project had evolved from the early phases of testing to the first years in production.
Though the result of the pilot study was encouraging, it was clear that if a project like this would succeed, there were some basic conditions that had to be fulfilled:

- The "raw"data input flow to the data processing group must be continous and without too much disturbance. Everyone involved in the failure data reporting must be informed of why and how they have to report every failure occurred.

- Full management support. The Product Engineering Division of VCC has more than 2000 employees, which makes a project of this size impossible to go through with, if the management doesn't support it.

- A database with suitable input and output routines.

The last problem was the easiest one to solve. We decided to work with data routines as similar as possible to our own department's LU system, which is very well functioning since about 10 years, and to use the same database. Only some smaller changes in the routines had to be made to adapt the system to pre-production test cars instead of "ordinary" production vehicles. Because of this, we choosed the project name LUPP (LU Pre-Production) for the new system (see fig.2)

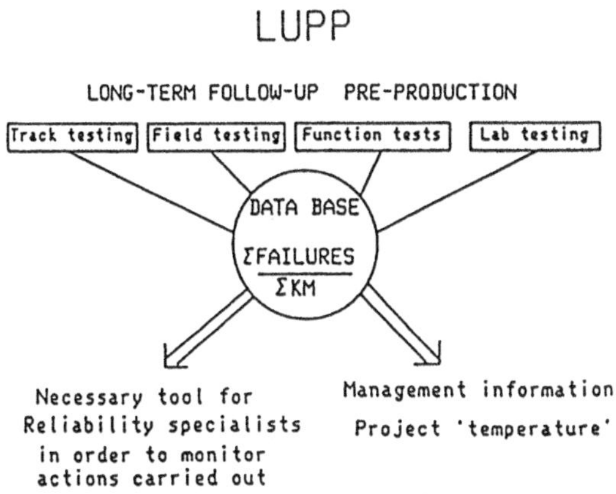

Fig.2. LUPP

To solve the remaining two problems mentioned above, we had to make a "marketing campaign" all over the Product Engineering Division, where the result of the pilot study was shown, and the LUPP- idea was presented. The project was approved by the management, so we decided to go ahead with it.

Reporting routines

The pilot study had shown us the importance of having clearly defined routines and instructions, so today when a project is decided to be involved in the LUPP-programme, the following procedure is used :

- Before the project cars are delivered, the testing departments and the QA department together pick suitable follow-up objects with a "sufficient" expected mileage.
  It is also decided who is going to be responsible forthe reporting of each car, and how often he has to report (usually once a month).

- When a car is delivered, it is equipped with a reporting journal and a special mark.

- In the reporting journal there is to be noted:
  - -Description of failures occured (with corresponding mileage)
  - -What caused the failure (if known)
  - -All repairs and adjustments
  - -Exchanged parts
  - -Date and mileage of repair

- Copies of the reporting journal are then sent to the QA- department for further data processing.

Some testing activities already have a databased reporting system. In these cases we can transfer files directly to our database, and do not have to go through all of the routines above.

Data Processing

The failure/repair information and the vehicle data has to be thoroughly checked and prepared before storing in the database.

- Vehicle data
  Each vehicle is specified with codes that shows:
  - -Vehicle project
  - -Year model
  - -Identification number
  - -Delivery date
  - -Specification of engine, transmission, chassi, body etc.
  - -Options and special equipment
  - -Typė of testing (high speed, climate, corrosion etc)
  - -Marketing area

- Failure data

  The failure and repair data from the reporting journal is prepared to fit a special reporting form, almost like the one used for the production car follow-up (see fig.3).

REPAIR REPORT

VOLVO
Volvo Car Corporation

Date: 88-10-15
Signature: [signature]

Importer (distributor)	Dealer

CT (1-3)	Date of service/repair Year (4-5)	Month (6-7)	Day (8-9)	Type (10-12)	Chassis No (13-19)	Km or miles (U.S.A.) (20-26)	Reg No
S47	88	09	22	760		48950	NOX 735

Notes

	Sub-group (27-30)	(31-32)	Defect (33-34)	Pos. (35-36)	Action (37-38)	Fail Class (39)	Part No (40-50)	(51-57)	(58-64)	Warranty (90) Yes	(91) No
1	1730										
2	2386	55	18		21	1					
3	2520	00	05	40	11	1					
4	8454	50	82	11	41	3	3389501				
5											
6											
7											
8											

Descriptions:

1. 50000 km service
2. Rough idle adjusted
3. Exhaust system checked. OK!
4. L.F. Sill-moulding replaced

* (65-89)

* Do not fill in (completed by VCC)!
Description: only necessary for information not included in attached work-order

Fig.3. Repair report

Here we fill in:

- -Vehicle identification
- -Date
- -Mileage
- -Function group code
- -Defect code
- -Action code (replaced, adjusted etc)
- -Position code (right, left, upper etc)
- -Part number
- -Severity code
- -Short failure description

The fuction group code occupies six digits, and is a way to describe parts, according to a standardized system used all over Volvo.

Finally, the vehicle and failure data is stored in the database.

Computer output routines

The data output used for presentation and further evaluation can be in many different forms. Both detailed lists and diagram routines are available.

The most useful diagrams are:

- -Weibull plots
- -Cumulated frequency diagrams, showing failure per vehicle as a function of mileage
- -Pareto diagrams showing most common problem areas

The most common lists are:

- -Lists showing all failures per function group
- -Complete failure record per vehicle

For evaluation of special vehicle types, failure causes, function groups, part etc, any parameter of the data or failure data set can be selected or excluded.

The excellent output facilities of this computer system is one of the main reasons why we choosed it for the pre-production follow-up project.

The project today and in the future

Today (October 1988) our system involves reliability data from 300 pre-production cars, with a total mileage of 8,5 * 10^6 km, in different stages of the product development process.

The results are used by the reliability engineers in the analysis work, and there is a great demand for the project status reports from the project managers.
Some of the results are presented below. Fig.4 describes how the defect levels of engines can vary depending on development stage, and fig.5 shows a Weibull diagram for defects on the same engine component in production (1) and in pre-production testing (2). Design changes were made in order to improve the reliability of the component, and they seem to have been successful.

The biggest problem today is that a lot of time is spent on manual data processing, which unfortunately makes it impossible to follow up as many projects as we maybe wanted.

We are therefore working on improving the data processing routines, so that we will be able to follow upp more projects, and spend more time analysing the material.

Another project that is being worked on, is to specify reliability requirements for the pre-production testing. The reliability level will, of course, be measured by the LUPP- system, and we are convinced that this will help us to build a car with the quality and reliability that the future customer demands!

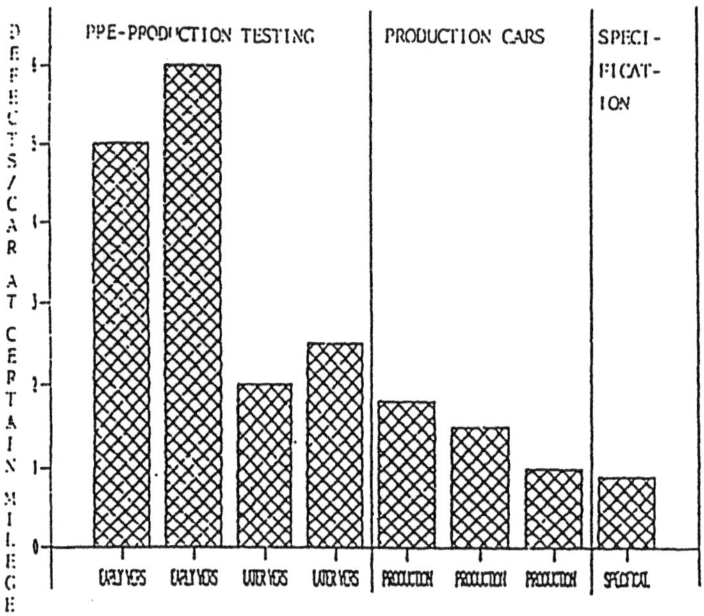

Fig.4. Defect levels of engines

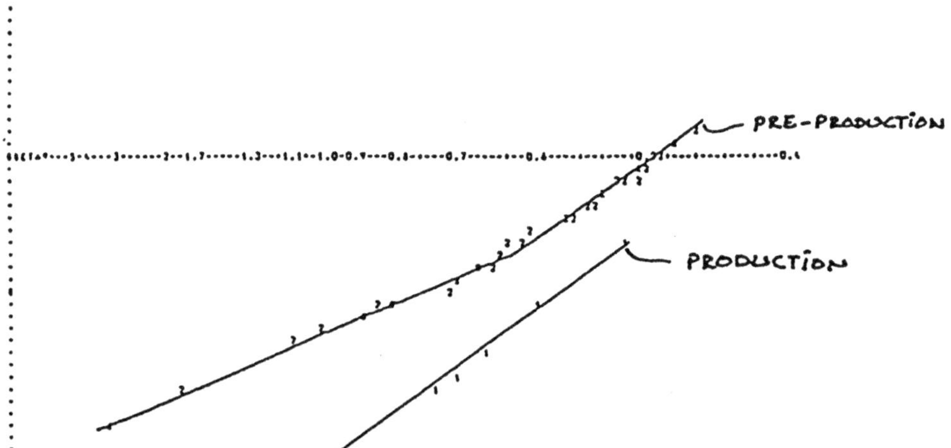

Fig.5. Weibull diagram based on LUPP results

Issues Associated with the Development of Nuclear Power Plant "Programm Effectiveness" Indicators for Regulatory Use [1]

E. LOIS

Office of Nuclear Regulatory Research

U.S. Nuclear Regulatory Commission

Washington D.C. 20555

J. WREATHALL

Science Applications International Corporation

2929 Kenny Road

Columbus, Ohio, 43221

J. FRAGOLA

Science Applications International Corporation

342 Madison Avenue

New York, New York, 10173

Abstract

This paper outlines several data issues related to the development of programmatic indicators for monitoring nuclear power plants (NPP) safety.

The programmatic indicators are used by management to monitor program effectiveness. The development of program effectiveness indicators for the use of regulatory organizations which are external to the "program" requires an entirely different approach than the development of indicators for program internal use. A regulatory agency is required to manage NPPs according to its objective, which is to protect public health and safety, and hence its indicators should reflect this objective. The NPPs, on the other hand, exist for productivity and profitability, and, therefore, their indicators should reflect primarily those goals. A safe plant may not be an economic plant, and the reverse is also not necessarily true; profitability and safety do not coincide at all times. This paper outlines methods for developing programmatic indicators for regulatory use. It is based on work sponsored by the US Nuclear Regulatory Commission (NRC).

The methods were developed within the context of one specific US NPP activity, the maintenance program. The method developed included frameworks for a systematic identification and evaluation of candidate indicators, and statistical techniques for indicator validation. The paper presents preliminary conclusions drawn from the methods applied and the data analysis performed. From the methods applied, it is concluded that regulatory programmatic indicators should monitor data related to program outputs (i.e. observable events), rather than data related to program processes (i.e. inputs and throughputs), although it is recognized that process indicators may be more suitable for internal plant management use. As a result of the data analyses, some preliminary statistical correlations of plant safety to plant performance outputs have been identified; these relationships show the potential for developing leading indicators of safety, and for combining regulatory and NPP senior management program effectiveness indicators.

Background

Since January 1987, the United States Nuclear Regulatory Commission (US NRC) has established a Performance Indicator (PI) program in order to enhance its ability to monitor NPP safety performance[2]. The performance indicators (PIs) are periodic (quarterly) measurements of plant performance based on data that are monitored and evaluated by the NRC in order to make timely regulatory decisions about NPP safety. Currently, the NRC is using a set of indicators based on reportable events (scrams, forced outages, outage rate, Engineered Safety System actuations, safety system failures, and significant events). While these indicators were adopted as a result of the PI task group effort, the search for improved indicators of safety has continued. In fact, two research activities have been established: direct, risk-based indicator development, and indirect, programmatic indicator development. The thrust of the first activity is to develop indicators reflecting the current or recent safety status; it is based on safety plant performance data such as safety system failures and initiating events. The aim of the second activity is to develop leading indicators of safety based on measures of program safety effectiveness. Plant programs include maintenance, training, radiation protection and management effectiveness. The authors' work was directed at this second activity, and specifically toward the development of maintenance effectiveness indicators. Therefore, this paper concentrates on the data issues pertaining to the development of indirect, leading indicators of safety at NPP as identified during this maintenance-related work.

NPP Program Effectiveness Indicators for Regulatory Use

Ideally PIs appropriate for regulatory use should have the following attributes:[2]

- Reflect a range of plant performance;
- Be independent of each other;
- Represent worthy goals for the utilities;
- Be supported by data available to the regulatory agency on a timely basis;
- Be directed from data not susceptible to manipulation;
- Be validated for there relation to safety;
- Lead plant performance; and
- Be applicable to all plants.

The problem of developing risk-based indicators which have these attributes, while not trivial, is much easier than the development of programmatic indicators since the former are based on events that have been determined to challenge plant safety through plant performance safety evaluations. Programmatic indicators are supposed to measure the effectiveness of plant programs to secure safety. Hence, they are not directly tied to observable events. Moreover they should more directly reflect human rather than equipment performance, although it is recognized that the two can not be distinguished rigorously. Thus, the development of programmatic indicators which have the above attributes is far from a trivial task; the more so since the dependent variable "safety" is somewhat elusive and difficult to express in a rigorous, quantified, and consistent manner.

A program can be characterized by three different elements: The inputs, the throughputs or processes, and the outputs. The inputs consist of the "what" of the program; the program's plan, objectives, policies, procedures, and structure are included here. The processes consist of the "how" of the program; the work flow within departments or from one department to another and the way the work is actually accomplished in the program are included here. For example in the case of the maintenance program, the work requests which operations submit to maintenance, the management/staff interaction, and the actual work performed by the maintenance personnel characterize the throughputs or processes intended to accomplish the program's mission. The "outputs" consist of the "how well", i.e. the quality, of the program. In the case of the maintenance program, the quality is measured by outcomes such as equipment operational readiness or the number of plant challenges caused by maintenance personnel. The issue is, how can "program effectiveness" be measured from the specific perspective of regulatory use? Should it be measured by individual indicators for input, or throughput, or output, or should the program performance be considered as a continuum, and should its indicators thus reflect "effectiveness" in all three areas?

In reality, no implemented program will produce high-quality outcomes (and therefore be effective) if it is not well defined, well planned, and well implemented. In addition a program is a dynamic process: planning, policies, and procedures feed into the process; process quality shows up in the results (the outcomes). If the program has a working self improvement mechanism, then "outcomes" or experience will feed back to "inputs". Therefore, program effectiveness indicators should be capable of tracking each of the individual elements of program performance, and this is exactly the approach utilized by the NPP utilities and other industrial organizations when they wish to have a healthy, self-correcting system.

However, a regulatory body is faced with a more difficult problem. Based on the previous discussion, a program can be represented by a closed loop of input-process-output-experience-input, which is a dynamic self-correcting process. Hence, it has the potential capability to adjust its indicators, so that it looks good on paper without true program improvements.

The question is, what data can an external regulatory organization use to monitor program effectiveness in an objective and consistent manner? Furthermore, how can one ensure that these data serve the objective for which they are monitored, in this specific case safety? And would these data have the capability to create warning signals regarding future performance?

These questions will be addressed by presenting the methods employed by the authors in their development of maintenance effectiveness indicators. Maintenance has attracted NRC's attention as the most vital area of plant safety after operations. Because of this, a plethora of candidate maintenance indicators have been proposed by the NRC, the NRC's research teams, and representatives of the commercial NPP industry. The research effort documented here evaluated these developed candidate indicators for their regulatory usefulness, and determined a reduced set upon which a detailed data validation analysis was performed and from which conclusions were drawn. In order for this effort to proceed in a systematic and tractable fashion, conceptual analysis frameworks were developed. These frameworks, the reduced indicator set, a preliminary validation analysis, and the preliminary conclusions reached are described in the paragraphs which follow.

The Development of NPP Maintenance Effectiveness Indicators

1. Literature Search

The first step undertaken was the performance of a literature search of available material related to existing maintenance performance indicators, the influence of maintenance on significant safety events, and issues affecting maintenance effectiveness, such as human factors problems. The literature search was augmented by discussions with NRC, NASA, a commercial air carrier, and nuclear utility personnel in order to better assure that the search was both current and complete. This step led to the identification of candidate maintenance indicators and the development of conceptual frameworks.

2. Development of Frameworks

Conceptual frameworks were developed to provide a structure for the systematic identification and evaluation of indicators. During the course of the investigation three frameworks were developed: (1) a process model; (2) a classification of the attributes of maintenance which have a significant impact on maintenance effectiveness; and (3) a classification of human factors having a significant impact on maintenance effectiveness. These frameworks can be considered complementary in the following sense. The process model describes the organizational interactions between maintenance and the other functional groups within a power plant; the classification based on maintenance-significant attributes deals with effects of maintenance on equipment performance; the classification based on human factors issues deals with the impacts of human performance on maintenance performance. A summary of each framework follows.

The Process Model

The process model is a portrayal of the influences among the various functional units in a nuclear power plant.

The purpose of developing this framework was to examine potential organizational influences within an operating plant. Those existing between maintenance and operations, training, engineering support, and management are considered the most important for the following reasons: (1) poor interactions between maintenance and operations can directly lead to equipment failures and scrams due to errors in equipment selection (a frequent cause of inadvertent scrams). (2) Training leads to systematic changes in knowledge and in the philosophy of maintenance activities within the plant. Errors or strengths in this area will have an extensive overall influence. (3) Relations between maintenance and engineering support have a strong influence on the effectiveness of root-cause identification, the implementation of design changes and other improvements directed at eliminate them. (4) Management decisions strongly affect the resources and priorities of the maintenance function.

The process model was created with the intent of identifying "probe points" (that is, points at which monitoring should be undertaken). However, once the strongly interactive nature of the "feedback" loops inherent in the plant operational and maintenance processes was recognized, it was clear that indicators from within the process were unlikely to provide the relationship to overall performance required for practical NRC use. There are too many alternative ways of successfully arranging priorities and influences within an organization to be captured in an internal indicator program such as this (i.e., there are too many "degrees of freedom"). For example, the utilities are using as an indicator the ratio of preventive to total (preventive+corrective) maintenance. This is a very strong process indicator and, for plants that have a preventive maintenance program, it reflects the program's goals for balancing preventive and corrective maintenance. The question is, would this indicator be appropriate for US regulatory use? Not all US utilities have a clearly defined preventive maintenance program. In addition, preventive and corrective maintenance are differently defined in different plants; some plants define as preventive, maintenance which is performed on items identified through monitoring and testing, while others define as preventive every maintenance action except repair after catastrophic failures. If the regulatory agency decided to monitor this specific indicator, the monitoring would favor plants that use a broad definition of preventive maintenance, and therefore plants would only have to broaden their definition in order to improve their standing. They would look better by making only a paper change, with no actual change in their maintenance program. Hence, although this indicator appears on the surface to match the regulatory objectives (i.e. to encourage preventive maintenance) its value in practice is questionable.

For this reason, the process model was used as a way of classifying indicators as "output-" or "process-" related. Output indicators are indicators directly tied to program outputs, and hence to objectively observable events; these include wrong part events, the number of misalignment errors, and scram frequencies due to test and maintenance. Process indicators are associated with how maintenance is carried out within the organization. These include the backlog of preventive maintenance items, the fraction of work requests reviewed by the quality assurance program, and the amount of overtime worked.

Indicator Classification According to Frequency, Duration, and Quality.

Classification was used to express ways in which maintenance impacts safety directly. These direct impacts are the following: (1) Maintenance can detect safety equipment failures through testing or inspection. (2) Maintenance can ensure operability of safety systems by checks and tests to ensure correct alignment of systems. (3) Maintenance can improve the reliability of safety systems through the use of preventive maintenance to eliminate wearout or other time- or cycle-related failure modes. (4) Errors in maintenance or testing can lead to reactor scrams that result in demands on safety systems. (5) Equipment failures may occur in balance-of-plant systems that result in reactor scrams which, with preventive maintenance, could have been prevented.

In these cases, three aspects of maintenance determine its impact on safety. These are how often maintenance is performed, for how long maintenance is performed, and how well maintenance is performed. Each of these affects the availability of safety equipment and the scram frequency. Therefore, a classification scheme was developed based on these three variables.

Indicator Screening Based on Human-Factors Issues

The literature search included the effects of human factors. The review included the EPRI evaluation of human factors issues in maintenance (EPRI NP-1567, 1981) and NRC's evaluation of maintenance status, NUREG-1212 (1986). Other related sources reviewed include the evaluation of issues in maintenance personnel reliability, NUREG/CR-2669 (1983); the estimation of the risk impact of maintenance activities, NUREG/CR-3541 (1983); and the previous development of programmatic performance indicators, NUREG/CR-5241 (1988). This literature review led to the creation of a classification of human factors issues found important in carrying out maintenance effectively at the personnel level. The classification elements identified were (1) workload/fatigue, (2) communications, (3) training, and (4) supervision/administration.

Selection of a Short List of Candidate Indicators

A total of 78 possible indicators were proposed. They resulted from both the literature reviews and the review of the process model. Each of the identified indicators was evaluated for its ability to reflect the attributes and characteristics of maintenance as shown conceptually in the developed frameworks. The identified issues were whether the indicator considered (1) the outputs or the process of maintenance, or both; (2) the frequency, duration, and quality of maintenance; (3) the human factors issues associated with workload/fatigue, communications, training, and administration.

The product of this evaluation was a matrix which displayed the degree to which each indicator reflects these attributes. Subsequently, a preference scheme was developed from the matrix and applied to reduce the 78 indicators to a set that could be analyzed. This preference scheme can be summarized as follows: (1) Output indicators were preferred over process indicators because of their closer tie to safety. However some process indicators were included for comparison, and because some process indicators might provide problem signals early on. (2) The most important single attribute in the set of frequency/duration/ quality was considered to be quality, again because of its direct tie to safety. The second preference was for frequency of maintenance because patterns of repeated repair activities indicate a lack of systematic elimination of equipment problems. (3) From a human factors perspective, the most important influences on overall performance considered were communications and supervision/administration. Problems in these areas can have their impact extended to many individuals and systems in a short time-scale.

Based on the application of this preference scheme the list of indicators was reduced to the following short list of nine potential indicators:

Number of repeat maintenance items;
Number of realignment errors;
Number of wrong unit/wrong train events;
Number of inadvertent Engineered Safety Features (ESF) actuations;
Backlog of engineering change notices (ECNs) related to performance;
Mean time between repairs;
Scrams due to test & maintenance;
Wrong parts events;
Heat loss rate.

Data Validation Analyses

The purpose of the validation analysis was to investigate how well changes in the values of the indicators correlate to changes in safety performance, and how well they anticipate changes in safety performance. The validation process involved estimating the time relationships and degrees of association between the indicators and a measure of safety. The analysis supporting this process was performed using methods from the disciplines of systems modeling and statistics. The evaluation of the effectiveness of indicators allowed for extremes in their behavior, and the fact that performance might sometimes be optimum at some midrange value of an indicator (rather than "more is better"). For the behavior of these indicators to be properly gauged some ultimate measure of safety had to be used as a basis for comparison. Several alternative measures were considered as a basis for the objective reference measure; these included "significant event" occurrences, risk-oriented measures of scrams and system unavailabilities, and measures of unit and systems availability.

Results of Analysis

Four of the nine potential indicators were selected as showing the strongest potential correlation to the selected measures of safety. These were:

Gross heat rate (GHR)
ESF actuations from test & maintenance
Scrams from test & maintenance
Mean time to return to service (MTRS).

Of these, the scrams and the MTRS indicators only appeared valid in a few cases and were therefore only considered as secondary indicators.

Gross heat rate (GHR) should be more or less constant for constant reactor power, with small changes due to variations in system thermal performance. As electric output power changes, GHR will vary because the "overhead" thermal losses remain approximately fixed. Therefore associations between GHR and unit availability measures (outage rate, mean availability, etc.) are to be expected. However, associations with failure rates of standby equipment were also found; these require further investigation.

In addition to ESF actuation indicator being correlated to safety, it also appears to be correlated to the way in which different plants respond to the events described in the original data. Those plants which showed infrequent and declining ESF actuations described the associated plant response in terms of equipment fixes to prevent the events from recurring; the plants which had both more frequent and recurring events described their responses in terms of "disciplining" or "counseling" the technicians involved. These different responses appear to indicate strongly differing attitudes within the plant's maintenance function, which have a strong influence on operational behavior. This indicator is being evaluated further to confirm its logical basis and to determine how good is its applicability.

Conclusions

First, it should be pointed out that the preliminary results discussed have not yet been reviewed and therefore it is not known whether they will be approved or adopted by the US NRC.

Bearing this in mind, the present work on maintenance effectiveness indicators shows the benefit to be obtained by external monitoring organizations from basing their monitoring on program outcomes rather than inputs or internal processes. It is believed that this output-oriented focus is a significant finding, and one which should contribute to the direction to be taken in the development of program effectiveness indicators for regulatory purposes. This focus automatically eliminates a large number of candidate indicators previously considered by experts as worthy of exploration, which in turn eliminates the significant data collection and data analysis effort that would otherwise have to be performed for proving or disproving their usefulness. Furthermore, the study also tentatively proves that higher-level output indicators such as gross heat rate are the most appropriate indicators for regulatory purposes. In particular, gross heat rate has additional desirable features. First, it is not intrusive to the program. This is very important, because when an external controlling mechanism such as regulation is involved in the program's processes it can have the negative effect of decreasing the amount of responsibility and creativity included in the program it seeks to control. Secondly, higher level indicators such as heat rate serve the utilities' goals as well as the regulatory goals, since high thermal efficiency is often translated into high profits.

The results of this research imply that the issues which are related to the development of indicators of NPP maintenance program effectiveness for regulatory use (which are for more complicated than the issues which must be addressed when indicators are developed for the program's internal use) are better understood if systematic methods are employed. Furthermore, the results show that high level programmatic indicators, i.e. indicators of overall plant performance, seem to be more useful for regulatory program trending than lower-level program-activity-specific indicators because the latter are more subject to manipulation by the licensee. Finally it appears possible for NRC to use indicators which couple the dual goals of safety and economics, and thus provide greater incentive to the industry to monitor these selected safety indicators and work for their improvement.

These preliminary conclusions are expected to be verified by performing further statistical analyses with additional data and additional qualitative analyses of plants' historical safety performance. Final results will be ready by early spring of 1989.

References

[1] The views expressed in this paper do not necessarily reflect those of the U.S. Nuclear Regulatory Commission.

[2] Commission Paper, "Performance Indicators", SECY-86-317, U.S. Nuclear Regulatory Commission, October 28, 1986.

[3] Wreathall, J. et al, "SAIC Draft Report on the Development of Maintenance Indicators by a Systematic Process", prepared for the US NRC, Nov. 1988.

The Role of a Disturbance Analysis and Surveillance System for the Operation and Management of Complex Plants

S. Messina, R.Galvagni -
Snamprogetti, Safety and
Reliability Dept. -
San Donato (Milan)
Italy

Abstract

The correctness of human action, especially in a complex situation, has very often proved to be a critical point, just like the correctness and unambiguousness of the information which is made available. This consideration led to the development of Disturbance Analysis and Surveillance Systems aimed at assisting the operator in decision-making, having regard to the operation or management of plants.
Unfortunately, the D.A.S.S's developed so far are not capable of foreseeing the potential dynamic behaviour of transients or accidents under surveillance. The studies conducted by Snamprogetti have led to the conclusion that it may be possible to achieve D.A.S.S.'s able to give at the same time:

- information tools based on ergonomic criteria making it possible to know in advance the probable development and outcome of transients and accidents, hence the potential impact on the site and the environment;
- tools for integrated Reliability, Availability and Safety analyses;
- tools for computer-based collection and management of the data coming from the plant, arranged in a Reliability, Availability, Mantainability Data Bank.

INTRODUCTION

The analysis of the plant made on the basis of codified guidelines and extremely accurate and systematic procedures has been recently imposed by the safety regulations in force in the most advanced countries.In time, the application of such procedures, which from the initial systematic analysis of the sources of hazard passed progressively to a real "Safety Analysis" and recently to the more exhaustive "Risk Analysis", has increasingly pointed out how the processing and systematic collection of information is of paramount importance. Indeed, a critical examination of the modes of operation of the plants and of their characteristics makes it possible to rearrange the information already available and obtain new data in view of a prevention policy which is proving more and more reliable in comparison with the less prudent method of intervention after the accident. Of course,the necessary condition is that the results of the analysis are correct, precise and not only of a qualitative nature; they should also be obtained in a reasonable time and at a reasonable cost. Furthermore, it is important that such results are not taken as a mere static acquisition; in fact, continuous updating linked to the changing of the situations and the progress of experience should be possible. Now, the problem arises of how to make the results of the analyses accessible and easy to read. Of course, there are many possibilities: from the production of conventional reports to the use of computerized means to guarantee the traceability and intelligibility of information.

Passing from the need to supply the necessary information on the plant - both for training purposes and for the correct interpretation of the conditions of the plant - to the willingness to assist the operator by means of systems able to process such information in real time is quite easy. However, putting this willingness into practice is a very difficult task which has already been faced more or less successfully and requires some evaluations which are considered fundamental for the successful solution of this problem.

OBJECTIVES AND RELEVANT INFORMATION

The objectives leading to the design of a disturbance analysis system for use on the plant can be numerous and need a clarification.

The system may play the role of a sophisticated alarm board. The latter, besides pointing out alarms, analyses its logic to control one or more "Top Events".

It can be seen as a prevention system which, by measuring physical parameters and their correlation, is capable of informing in advance of the occurrence of possible anomalies in a more or less variable time.

It can have the function of a sophisticated collection of data on the events which are considered important. Then, the objective is to make available to the plant operator data and results relating to the optimization of maintenance and its management or to events to be considered in the preparation of specifications for the construction of components and in the training of the personnel working on the plant.

The system can be considered as a static element to be updated only in case of exceptional events or modifications. However, it is possible to envisage continuous updating on the basis of every new experience or of a progressive changing of the situation according to a procedure which makes it possible to develop the model on which the analysis is based by introducing the acquired knowledge and the changes in the situation.

Of course, the choice of the system objectives depends on specific requirements which can be met only if they have been taken into consideration in the initial analysis of the plant.And that is true not only as far as analysis information, but also and even more important as far as analysis methodology are considered.

As the information and its use in the preparation of a disturbance analysis system is linked to the methodology of the analysis, it is worth mentioning some of the most important among those available.

The analyst can choose between a deductive or inductive approach. In the former case, the analyst follows some procedures and schemes enabling him to guarantee the completeness and correctness of the model he has produced. In the analysis, he starts from the consequences and goes back gradually to the causes following an atemporal scheme marked by groups of mainly static causes. Of course, the information obtained from the application of this method can be satisfactory for the preparation of a disturbance analysis static system aiming at the control of a relatively large number of "Top Events", in view of the implementation of a sophisticated alarm system.

In the latter approach, the analyst follows an inductive logic that guides him in a far more effective way and allows him to obtain best and more reliable results. Although it has a more linear logic, the consequent method is more complex to put into practice and needs a more sophisticated technology. The method consists essentially in searching systematically for those cause-effect relations which seem to be "possible" and in combining them to form a complete series (universe) of possible stories. The main difficulty in the application of this method lies in the extraordinary "imagination" one should possess to determine all possible stories (although in a given context) as soon as the system reaches a certain degree of complexity and in the remarkable "memory" needed to retain all the stories and submit them to a critical revision. Nowadays, these difficulties can be overcome thanks to the use of computers: a radical change for the better (perhaps not yet neatly perceived) in the obtention of information from analysis and its systematic use.

The efforts made so far for the design of disturbance analysis systems far beyond the logic of alarm systems and intended to guide the operator in the prevention of given occurrences or "Top Events" demonstrated that only a cause-consequence process can support the operator in every situation that can be "anticipated". (It is worth mentioning the enlightening studies made by Nielsen and Taylor at Riso and the choices for the implementation of the DAS - disturbance analysis system - made by EPRI and GRS).

Furthermore, the need for the use of the computer contemplated by the inductive method leads to the development of methods of interaction between the analyst and the computer that can be extended immediately to the interaction between operation and model. This is necessary to ensure a dynamic and representative model through continuous updating over time.

DESIGN of Snamprogetti D.A.S.S.

Snamprogetti designed together with Agip and developed a series of computer-based instruments which, by using an inductive logic similar to the well-known logic of the decision tree, makes it possible to obtain through an interactive process a logic-probabilistic model of the system or plant under examination, to generate all possible sequences of events or stories (compatibly with the logic of the model) and to use easily and effectively all the information so obtained to make a synthesis of it. The logic-probabilistic model is an ideal tool making it possible to update the store of knowledge characterizing the analysis, rather than the analysis itself which is inherent in the model. Another peculiarity of the model is that it represents the origin of the universe of the expected sequences of events or expected plant stories which provides the best reference point or the necessary element for the correct interpretation and classification of the experience deriving from the events which have occurred.

Obviously it can therefore represent the tool for the generation of the whole series of expected events to be used to know the possible evolution of the events which are occurring.

Within this framework, the kernel of a disturbance analysis and surveillance system (D.A.S.S.) consists of a model from which it is possible to originate a bank of expected stories. Of course, there can be a certain number of models representing different "stories" which concern the trend of process parameters or the spreading of external accidental conditions on the basis of a process according to which a given event can be considered as a cause of new events. However, this does not change the essential points of what has been described so far and therefore only one model shall be referred to.

The bank of expected "stories" represents a reference point for making comparisons between the "stories" which are occurring. The results of these comparisons can be directly seen by the operator who will read the expected evolution ordered by decreasing risk or probability and the relevant intervention steps in order of priority. For making the comparisons, the D.A.S.S. system must necessarily receive from the field the information allowing the operator to recognize the evolution in progress (among those which are expected) or correlate it with those which are similar. Furthermore, with the instruments available today, the system makes it possible for the operator to request the anticipation of the trend over time of significant physical parameters and compare them with the measured values. This is possible because the analysis system already available can use automatically the stories of evolution over time of the situations or sequences of events, to obtain the trend over time of the parameters chosen by the operator. Whenever a deviation occurs, it is necessary to review the behaviour of the D.A.S.S. to correct, improve or confirm it. For this reason, the system calls for the presence of a memory capable of storing the real trend of events together with the reaction and "reasoning" of the D.A.S.S. It is also necessary to update over time the reliability/probability parameters of the model so that it can reflect the real conditions of the plant. This implies the availability of a bank of anomalies, faults and intervention steps on subsystems and components, useful for getting the information needed to direct Bayes's process for the updating of the probability estimates. This is the same bank which provides the bases for the optimization of maintenance, of the management of the spare parts warehouse, of modifications and revamping. The evaluations made so far - although necessarily synthesized - give an idea of the D.A.S.S.'s architecture as is currently conceived by Snamprogetti. Considering this architecture and the information on which it is founded, the objectives set forth in the following paragraph appear to be quite feasible.

OBJECTIVES AND REQUIRED CHARACTERISTICS OF THE D.A.S.S.

The function of support by the D.A.S.S. system to the operator requires on line information tools based on systematic ergonomic criteria and able in the same time to supply data on the plant to the necessary storage and management systems in a position to act as an effective Data Bank. The latter should be used for the optimization of operation and maintenance which in turn is based on the capacity of availability, maintainability and safety integrated analysis.

In short, the following targets should be reached:

- interface with the plant: the D.A.S.S. should receive the information on the state of the plant directly from the instrumentation and control systems or from the process computers;

- interface with the operator: the mode of presentation of the results of the D.A.S.S. analysis should be integrated with the information coming from the control room. This integration should use ergonomic criteria, namely allow the operator to obtain in a direct and easily interpretable way ALL and ONLY the information he needs in the particular situation;

- time of the analysis: the D.A.S.S. should operate in real time and present the results so as to allow the operator to take any corrective measures to cope with the evolution of disturbance. The operator must be enabled to read such measures directly, together with the diagnosis of the situation and the causes which have determined it;

- improvement of information: the D.A.S.S. should improve

the quality and contents of the information needed by the operator by reducing the number of alarms (which are secondary in relation to the actual conditions of the plant) so as provide the operator with univocal information and by eliminating the information which may mislead him;

- disturbance analysis: the D.A.S.S. must be in a position to analyse the disturbance autonomously and in real time working on the basis of a pre-established and stored plant model. The D.A.S.S. system must be able to determine the nature, causes and consequences of disturbance and the corrective action which may possibly be taken to control it;

- possibility of verification: the D.A.S.S. must be able, at the operator's request, to show the procedure it has gone through to come to the stated conclusions. Furthermore it must supply the evolution of the physical parameters or anomalies, as requested by the operator, either in real time or a posteriori, in order to verify the procedure used for the analysis and the conclusions therein contained;

- plant model: the D.A.S.S. must include the software necessary to develop and modify the plant model used, according to the acquired experience;

- response to the operator: the D.A.S.S. must enable the operator to perceive the responses from the plant, even the delayed ones, as promptly as a pilot receives a response from the airplane. A pilot, indeed, usually perceives immediately the effect of every action of him, while the operator normally has to wait a long time before he perceives the effects of his action (or lack of action). To avoid this, it is necessary to anticipate the behaviour of the process and simulate it so quickly as to obtain sufficiently prompt expectations for immediate use by the operator.

- data base and recording of anomalies: the D.A.S.S. must provide the recording of the trend over time of the plant situation, alarm situation, the anomalies found and the consequent variations of significant variables. The purpose is to ensure the full verification - in real time or a posteriori - of the diagnosis made and provide the Bank of Plant Anomalous Events with the information useful for the optimization of management and maintenance strategies, including the management of the spare parts warehouse.

CONCLUSIONS

As actual experimentation has proved the need - in complex situations or plants - for an increasingly high capability of man to act correctly and on the basis of prompt and correct information, disturbance analysis and surveillance systems are being developed, capable of assisting the operator and all those who run the plant in making proper decisions when facing the risks linked to operation and maintenance problems.

However, most of the disturbance analysis systems developed so far do not make it possible to take into consideration, as they should, the actual dynamics of disturbance through its logic-probabilistic and phenomenological simulation.

This limitation led to begin studies for the design of improved systems capable of coupling the logic-probabilistic simulation with the phenomenological simulation in real time. This constitutes a valuable means of analysis at the disposal of those who run the plant, a means which in a farm independent of the designer, places at the disposal of the operator all the designer knowledge, integrated with the knowledge of an experimental nature directly acquired from operation.

The studies made by Snamprogetti will lead in a relatively short time to the implementation of D.A.S.Ṡ. systems capable of supporting effectively the running of the plant both in the phase of direct control and in the phase of analysis and strategic programming of its operation, and providing at the same time:

- information tools based on strictly ergonomic criteria making it possible to know in advance the probable development and outcome of transients and accidents, hence the potential impact on the site and the environment;

- tools for integrated Reliability, Availability and Safety analyses;

- tools for computer-based collection and management of the data coming from the plant, arranged in a Reliability, Availability, Maintainability Data Bank.

Data Fusion Problems in Intelligent Data Banks Interface

S. A. SANDRI*, A. BESI**, D. DUBOIS*, G. MANCINI**, H. PRADE*, C. TESTEMALE*

* L.S.I., Univ. P. Sabatier, 118 Route de Narbonne, F-31062 Toulouse Cédex
** CEC-Joint Research Centre-Ispra Establishment, I-21020 Ispra (Varese)

Summary
The reported work has been motivated by different kinds of data fusion problems encountered in the design of an advanced interface for exploiting heterogeneous reliability parameter sources.The considered data fusion issues originate either in the aggregation of pieces of information obtained from different sources or in the treatment of fuzzy queries addressed to the interface.The aggregation problem is discussed both in the presence and the absence of confidence estimates for the pieces of information to be fused.The different combination problems are dealt with in the framework of possibility theory.

Introduction

A series of difficulties arise in the design of an intelligent interface to a system consisting of a group of data banks, whose objective is to supply - upon a unique request - the fusion of the results obtained from each data bank. First, problems come from the fact that different data banks may have distinct representation frameworks. Another class of problems is directly related to the combination process itself. Lastly if the system allows for fuzzy queries, i.e., which concern vaguely specified sets of objects or which deal with categories different from the ones used in the sources, then a specific treatment is required,which leads to another combination problem.

The translation of the common request into the query language of each data bank may not be straightforward when the objects and attributes are not classified in the same way in each of the data banks. This translation might even generate fuzzy queries, since an accurately specified attribute in a query language might be only "fuzzily" specified in another language. When the results issued by the data banks are not encoded in the same representation framework a model should be chosen , into which the results would be transformed, in order to provide the user - if not with a unique answer (not advisable in some cases) - at least with the means of easily comparing the results.

The combination problem consists of choosing, among the combining methods available within the considered representation model, the most appropriate one, taking into account such issues as the nature of the data banks, the dependency among them, the possible existence of strong conflict among their results, the existence or absence of information on the reliability of each data bank (with respect to the request) and the properties the combination should satisfy. For instance, the reliability assigned to a bank may reflect the fact that the size of the sample used in the statistical evaluation is too small or that some extrapolation technique has been used for producing the information.

In order to deal with fuzzy queries, either requested by the user or stemming from the interface , the system has to be able to approximate the fuzzy query in terms of a small number of ordinary queries. The derivation of those queries and the aggregation of their results constitute another class of data fusion problems.

All these problems are encountered here, in the design of an intelligent interface for a multisource interrogation system under development at the JRC-Center Ispra with the collaboration of the LSI Laboratory. First a brief presentation of the application is provided. Next the issue of translating the responses of each data bank into the common representation model used in the present application is addressed.Then combination is discussed with emphasis on the adopted representation model (possibility theory) and on our strategy for interacting with the user. Lastly we briefly discuss the combination problem generated by the use of fuzzy queries.

The Current Application

The Reliability Parameter Estimation System (RPES) under development at Ispra is a multisource interrogation system that furnishes reliability parameters in the specific case of Nuclear Power Plants (NPP).It is composed of :

1) A set of databases dealing with either historical data or literature collections, of which the most important are :
   - the Component Event Data Bank (CEDB) [1]
   - the IEEE Standard 500 -1984 [2]
2) An intelligent interface

CEDB

CEDB is a database consisting of a collection of failure events and operating times/cycles of NPP components, and a package of statistical analysis methods. Its query language permits the definition of an unbounded number of classes based on their engineering and operating characteristics, the accessing of all the information stored and the application of the statistical methods to selected sample events related to the defined class, on a specific failure mode. The result of a query consists of probability density distribution parameters, such as the mean and the standard deviation (among others), of failure rates and failure repair rates for a specified class. A session to the CEDB is a selection of a set of components followed or not by a statistical analysis. The query language is composed of 3 groups of commands : selection, statistical and display commands. The selection phase consists of :

1) successively refining a set of components on its engineering characteristics , performing unions and intersections of classes (e.g. number of phases of a motor, its power, etc...)
2) specifying the category of failure mode (failures on operation or on demand)
3) specifying a type of failure (e.g. sudden failure on operation)

Part 2 may be omitted, in which case, the statistical analysis will be made considering all possible failure mode types, for the specified class of components and failure mode category. The user can analyse the failure and repair data connected with the selected sets, by various statistical methods. Upon the choice of a method the system automatically collects the sample (times to failure, survivals) and computes the reliability parameters of interest. The available methods more frequently used for the estimation of the constant failure rate λ are:

1) Classical estimation of constant failure rate - It calculates the mean, the standard deviation and the 90% confidence interval of λ.

2) Bayesian estimation of constant failure rate - The user prompts the name of the prior distribution followed by its parameters, or furnishes directly an histogram and the system yields an updated distribution.

IEEE

The IEEE Data Base displays, for a number of classes of components hierarchically structured in a tree of tables,the value of failure rates referred to operating times or operating cycles/demands. These parameters are (λ_{low}, λ_{rec}, λ_{high}), the lowest, recommended and highest failure rate respectively and λ_{max} in some cases. These estimates are the result of aggregating the opinions of experts by the Delphi method. This data base is incomplete, only few of its tables are completely informed. To deal with this problem, a system was designed [3] which, given a class of components whose table is partially or completely uninformed, uses expert rules to find its associated failure rate, by means of extrapolating the failure rate value of analogous classes. This system is presently implemented with reference to the electric motor category.

RPES Intelligent Interface

The intelligent interface is presently under development. It should be capable of:

. allowing the user to query each database i with its own associated language l_i. Should the user wish, it should be able to perform the combination of the various results; the system should check beforehand, if the queries formulated in the different languages are compatible;

. transforming the uncertain results of different nature into the same mathematical model;

. allowing the user to make a request in a general language L, translating the request into each language l_i, accessing each base i and merging the results taking into account the reliability of each source assigned by the user;

. allowing the use of fuzzy attributes; these attributes may pertain to the class of components or to the considered failure mode;

. extracting parameter estimates for non-informed tables in the literature databases (already implemented for the IEEE using an extrapolation technique).

Making the Data Bank Responses Compatible

Possibility Theory [4,5] offers a convenient framework for representing uncertain parameter values specially in the case of subjective information.A possibility distribution π, is a mapping from referential scale W to $[0,1]$ such that $\pi(w) = 1$ for some w, with the following interpretation : π is the characteristic function of the fuzzy set of possible values of a parameter v; $\pi(w) = 0$ means that w is totally impossible as the value of v and $\pi(w) = 1$ means that nothing prevents w from being the value of v. The set $Co(\pi) = \{ w \, / \, \pi(w) = 1 \}$ is called the core of π and $Su(\pi) = \{ w \, / \, \pi(w) > 0 \}$ is called the support of π. The sets $\{w \, / \, \pi(w) \geq \alpha\}$ are called the level-cuts of π.

The result of a query to the IEEE data bank, namely the triple $(\lambda_{low}, \lambda_{rec}, \lambda_{high})$ is approximated by a possibility distribution having $[\lambda_{low}, \lambda_{high}]$ as support and $[\lambda_{rec} - \delta, \lambda_{rec} + \delta]$ as core, where δ is an expectation of the error produced by the Delphi aggregation method, supplied by an expert.

A more complex transformation had to be done in the case of the CEDB data bank. To transform a probability density distribution p into a possibility distribution π, we have to make sure that the consistency constraint

$$N(A) \leq P(A) \leq \Pi(A) \quad , \forall A \subseteq W \tag{1}$$

is satisfied [5], where $\Pi(A) = \sup\{ \pi(w) / w \in A\}$, $N(A) = 1 - \Pi(\text{not } A)$, and $P(A) = \int_A p(w)\,dw$. The set functions Π and N, are called respectively possibility and necessity measures. (1) expresses that the possibility and necessity measures can be viewed as upper and lower probabilities that approximate a probability measure. Possibility distribution π is said to be an optimal approximation of the probability measure P, if it satisfies (1) and there is no π' satysfying (1) such that, $\forall w, \pi'(w) \leq \pi(w)$, or equivalently, $\forall A$, $\Pi'(A) \leq \Pi(A)$.

It can be proved that there exists an infinite number of optimal approximations of a continuous probability density. A simple way of characterizing approximations in the sense of (1) is to select a reference value w_0 belonging to the support $[\underline{w}, \overline{w}]$ of p and a function f of $[\underline{w}, w_0] \rightarrow [w_0, \overline{w}]$ such that $f(w_0) = w_0$, $f(\underline{w}) = \overline{w}$ and $\forall w, w' \in [\underline{w}, w_0]$, $w > w' \rightarrow f(w) < f(w')$. A set of level-cuts $\{[w, f(w)], w \in [\underline{w}, w_0]\}$ is thus characterized. Then the possibility distribution π such that

$$\forall w \in [\underline{w}, w_0], \qquad \pi(w) = \pi(f(w)) = \int_{-\infty}^{w} p(x)dx + \int_{f(w)}^{+\infty} p(x)dx$$

in an optimal approximation of p when w_0 and f are fixed. For instance, the approximation using the median as the reference value w_0 and $f(w) = w'$ such that $q_{w'} = 1 - q_w$, where $q_w = P([-\infty, w])$ is the quantile corresponding to w (particularly $q_{w_0} = .5$) is of this form. Other approximations can be envisaged in practice choosing as reference value the mode or the mean value of the probability distribution.

Combination in the RPES

The combination of various pieces of information supposes that their respective representations can be put in a common framework. Besides, there are some properties that a combination law should possess such as symmetry, associativity, closure and idempotence [6]. Associativity is not necessarily required; a weaker property such as quasi-associativity is often sufficient [7]. Upon the use of a symmetric quasi-associative combination method, the order in which the bases are combined does not influence the final result.

The closure property ,which states that the result of the combination should be of the same nature as the operands, is a useful technical feature. Idempotence ($\pi_1 = \pi_2 \rightarrow \pi_{12} = \pi_1 = \pi_2$) is more debatable. It is incompatible with reinforcement properties in the presence of independent sources but it copes with redundant information.

However combination cannot be discussed simply in terms of desirable algebraic properties. First too many algebraic properties lead to sterile impossibility results that cannot solve any practical problem, or to unicity results that are delusive because restricting too much the range of combination attitudes. Second the semantics of the numbers to be combined, their meaning, also helps in choosing the proper combination operations. Namely, combination laws should be in agreement with the axiomatics to which degrees of uncertainty obey. The solution of a combination problem does not only reside in the choice of a suitable combination rule. Questions such as the reliability of the sources, the possible dependencies between sources and the existence of conflicts between the supplied pieces of information, should be also addressed.

Set-Theoretic View of Combination

The combination problem can be studied from a set-theoretic point of view [8]. Set-theoretic operations such as unions and intersections are indeed the basic consensus rules in set theory. Consider the case of two sources of information giving evidence about the value of some variable v (the reliability of a component, for instance) under the form of a set. Namely

Source 1 - $v \in A \subseteq W$

Source 2 - $v \in B \subseteq W$

The choice of a type of consensus rule is clearly a matter of context. There is no theory liable of prescribing a unique way of pooling these two pieces of information.The framework of possibility theory is a straightforward extension of the set-theoretic point of view in terms of fuzzy sets. In this framework there are three attitudes for combination which arise as basic ones from the study of fuzzy set-theoretic operations i.e. conjunction, disjunction, and trade-off, as explained in [8]:

conjunctive consensus

If the sources are completely reliable and properly interpreted then a reasonable rule is to conclude $v \in A \cap B$. Note that we should have $A \cap B \neq \emptyset$, otherwise it is

self-contradictory to claim that the sources are reliable and that they are correctly interpreted. Let $\pi_i(w)$ be the possibility of $v = w$ given by source i. The main conjunctive operations in the framework of possibility theory are :
$\pi_{12}(w) = \min(\pi_1(w), \pi_2(w))$, $\pi_{12}(w) = \pi_1(w) \cdot \pi_2(w)$ and
$\pi_{12}(w) = \max(0, \pi_1(w) + \pi_2(w) - 1)$.

disjunctive consensus
If the sources are not completely reliable but we have no information about their reliability then a reasonable attitude may be to conclude $v \in A \cup B$. It amounts to assuming that at least one of the sources tells the truth without specifying which one. The gain in confidence is counterbalanced by a loss in precision. The main combination operations of this type are : $\pi_{12}(w) = \max(\pi_1(w), \pi_2(w))$, $\pi_{12}(w) = \pi_1(w) + \pi_2(w) - \pi_1(w) \cdot \pi_2(w)$ and $\pi_{12}(w) = \min(1, \pi_1(w) + \pi_2(w))$.

trade-off
An intermediary attitude consists in considering that $A \cap B$ is a more plausible range for v than $A \cup B$ although one should not reject the latter values. Several kinds of trade-offs between $A \cap B$ and $A \cup B$ may be envisaged. For instance, under the assumption of equal reliability of the sources, one can consider a possibility distribution such as $\pi_{12}(w) = 1/2(\pi_1(w) + \pi_2(w))$. Here, combining evidence comes down to trading off uncertainty (pervading precise results) against imprecision (that may be uninformative).

The maximum and minimum operations are respectively limit cases of disjunctive and conjunctive attitudes. They can be justified on the basis of requirements such as idempotence and associativity (see details in [5]). Choosing between the minimum operation and the product in the case of agreeing sources is dictated by the existing links between the sources. The product operation models a reinforcement effect that may be intuitively satisfying when the sources are unrelated enough : i.e. when an alternative in W has a low degree of possibility according to each source, the resulting degree of possibility is still lower. When the information about independence is not available, the minimum rule appears to be more cautious, due to the idempotence property.

The set-theoretic point of view considers that the 2 sources contribute in the same way in the combination process. Alternatively, combination could be viewed as a dissymetric updating process. For instance, when the information issued by the IEEE database is used as the prior distribution in the Bayesian statistical inference method on CEDB [9], the pieces of information are considered to be combined in a dissymetric way and the system should perform no further combination. (Note that the Bayesian approach is sometimes the only way to exploit CEDB information; namely when no failure rates are recorded in the component history [10]).

Dealing with Conflicting Sources

Let $h(\pi)$ be the height of the possibility distribution π given by :

$$h(\pi) = \sup_{w \in W} \pi(w)$$

π is normalized if $h(\pi) = 1$, that is, there exists at least one element in W, considered as completely possible. When the combination law produces an unnormalized result in a given representation framework, the pieces of information are conflicting. This phenomenon occurs specially in the case of conjunctive consensus. If we are absolutely certain that the referential W is exhaustive and that the sources are fully reliable it is reasonable to renormalize the result, by dividing it by $h(\pi_{12})$. This yields :

$$\pi_{12}'(w) = \pi_{12}(w) / h(\pi_{12}) \qquad (2)$$

However, in this case we shall have a delusively precise result, the strongest belief being assigned to a value strongly supported by neither of the sources. (2) is discontinuous when π_{12} goes to 0 (total conflict). In [11] it is suggested to interpret the degree of conflict $h(\pi_{12})$ as a degree of ignorance about the combined result, by allocating this weight to the whole referential set W, in the framework of the theory of evidence [11]. In possibilility theory this is equivalent to adding the quantity $(1 - h(\pi_{12}))$ to the whole distribution yielding :

$$\pi_{12}'(w) = \pi_{12}(w) + 1 - h(\pi_{12}) \qquad (3)$$

This method of normalizing preserves continuity but has the disadvantage of creating an uncertainty level on the whole domain W. This actually means that none of the sources is completely reliable and any value in the domain has a non-zero possibility of occurring. We can then use a variant of (3) instead :

$$\begin{aligned} \pi_{12}'(w) &= \pi_{12}(w) + 1 - h(\pi_{12}) \text{ , if } w \in Su(\pi_1) \cup Su(\pi_2) \\ &= 0 \quad \text{otherwise} \end{aligned} \qquad (4)$$

In this case, we favor the support of the union without creating an uncertainty level. Normalization methods (2) , (3) and (4) do not alter π_{12} if there is no conflict.

Even if π_{12} is normalized, too weak an overlapping between π_1 and π_2 may still express some conflict between the sources, that cannot be measured by $h(\pi_{12})$. One way to measure such a conflict is to take the Jacquard index :

$$k = (|A \cap B|) \,/\, (|A \cup B|)$$

where
$$|A \cap B| = \int_W \min(\mu_A(w), \mu_B(w))\, dw$$

$$|A \cup B| = \int_W \max(\mu_A(w), \mu_B(w))\, dw$$

with $\pi_1 = \mu_A$ and $\pi_2 = \mu_B$.

Incompletely Reliable Sources

The various levels of reliability of the sources can be dealt with either by discounting the information provided by each source or by using a weighted combination law. The first approach supposes that we know the absolute reliability of each source in terms of the probability or the certainty (necessity) that the source tells the truth. The second method corresponds to the knowledge of the various levels of reliability in a relative way.

The technique known as "discounting" is borrowed from the theory of evidence [12]. In the case of possibility theory the possibility distribution π_i is discounted by a factor of $(1 - c_i)$ into a new possibility distribution π_i such that

$$\pi'_i(w) = (1 - c_i) \cdot \pi_i(w) + c_i$$

where c_i represents how much the user finds plausible that source i is wrong. After discounting each source by $(1 - c_i)$ we can perform the combination using a conjunctive rule such as

$$\begin{aligned}\pi_{12}(w) &= \min(\pi'_1(w), \pi'_2(w)) \\ &= \min((1 - c_1) \cdot \pi_1(w) + c_1, (1 - c_2) \cdot \pi_2(w) + c_2)\end{aligned} \quad (5)$$

In this combination method, if the cores of π_1 and π_2 intersect, then the core of the π_{12} will be this intersection ($\min(\pi_1(w), \pi_2(w)) = 1 \Leftrightarrow \pi_{12}(w) = 1$). However, this result is not idempotent, which is natural. Since the sources are not completely reliable, the real results are possibly not the ones issued by the sources and the agreement is a mere coincidence. (5) is not necessarily normalized, and we would have to proceed similarly to (2) or (3).

However, another discounting formula (see [13]) seems more natural in the restricted framework of possibility theory, namely

$$\forall w \in W, \pi_i'(w) = \max(\pi_i(w), c_i) \tag{6}$$

where c_i is the possibility that this information is not correct. The possibility discounting formula (6) expresses that either w is a possible value for v to the degree $\pi_i(w)$ or the source is wrong.

As explained in [5], the natural manner for introducing weights in a min combination operation is $\min_i \max(\pi_i, c_i)$ where π_i corresponds to the information provided by source i and $(1 - c_i)$ is its relative reliability with the scaling condition $\min_i c_i = 0$. For $c_i = 0$ we recover the min operation, for $c_i = 1$ the information is ignored. But the above expression can also be viewed as a min-combination of pieces of information which are discounted in the sense of (6), thus the two approaches coincide in that case.

The degrees of conflict h and k between the sources are important factors to be considered in the combination process. Indeed they can be used as discounting factors for the combined result.

Sources of Various Levels of Specificity

It is not always advisable to combine the results issued from several sources. An important instance of this situation is when one source gives a piece of information which is *specific* of the case under consideration while what the other source furnishes is derived from a *general* law (which is somewhat uncertain in the sense that it may have some exceptions). For instance, when one of the databases in our problem is not informed on the specific class of items addressed by the query and returns instead the value concerning a larger class. In that case the difference of origin between the two pieces of information can be taken into account in the following way using a possibilistic representation. Let π_1 correspond to the specific information and π_2 be the representation of the general information. Then the following combination law

$$\forall w \in W, \pi(w) = \min(\pi_1(w), \max(\pi_2(w), 1 - h(\pi_{12}))) \tag{7}$$

which is proposed in [14]. Using this formula the information supplied by the general law is not considered when it is not consistent with the specific information. This mode of combination consists of discounting the most general information by the degree of conflict with the most specific one.

Combination Strategy for RPES

In the following the two sources are the IEEE and CEDB data banks. Considering the existence or absence of information on the reliability of the results supplied by the sources upon a query, the following situations have been depicted.

The user :

I) considers both bases completely reliable.

II) believes that at least one of the results is reliable but does not know which one.

II-1) considers one of the bases as completely reliable and the other as completely unreliable.

II-2) considers one of the bases as completely reliable but does not know the reliability of the other.

III) can indicate the absolute reliability of each base.

IV) does not know the reliability of each base but believes that base 1 is relatively more reliable than base 2.

V) does not know anything about the reliability of the bases.

In the following, we show the solutions considered for the present project in each case. The properties satisfied by the methods are closure, idempotence, quasi-associativity and normalization .

I) both bases are completely reliable

Upon this choice, the system takes the minimum between the results, thus improving the precision of the final result.

$$\pi_{12}(w) = \min(\pi_1(w), \pi_2(w)) \tag{8}$$

Since dealing with completely reliable sources, we would expect them to have a big agreement. However a conflict may occur due to the following facts :

- The sources of knowledge are not reliable although the user believes they are

- The considered referential set is not exhaustive and the sources agree "outside" of it
- The sources do not speak about the same event

Here we do not consider the 2nd case, since the referential set W is considered to be exhaustive. We consider that the 3rd case does not occur either, since an adequation of vocabulary between the bases is assumed to have been done. Therefore the methods shown here refer to the 1st case, that is, we consider that although the bases are considered reliable by the user, they are not so in fact. To deal with this problem, fix a constant d, that will represent the amount of agreement considered satisfactory by the system. Should the bases agree, that is the Jacquard index $k \geq d$, simply deliver result (8) to the user. If however, $k < d$, we shall inform the user that the bases have a strong conflict and let him choose between :

a) Changing his opinion about the reliability of the bases. The system would restart the dialogue and option I) would no longer be available.

b) Accepting the result as it is. The system will normalize the result if it is necessary and deliver it to the user. When the cores of the distributions do not intersect, then result (8) will be normalized using (2).

II) the user believes that one of the bases is reliable but does not know which one

II-1) one of the bases is completely reliable and the other completely unreliable
In this case, the most reasonable combination method would be to take the maximum of the possibility distributions [8] , by making sure that the real answer lies within the result.

$$\pi_{12}(w) = \max(\pi_1(w), \pi_2(w)) \tag{9}$$

II-2) one of the bases is completely reliable but the user does not know the reliability of the other
Here we could choose to be pessimistic, considering that the other base is completely unreliable.We would thus fall into situation II-1) and consequently take the maximum as the combination method. If however we let ourselves be more optimistic, we could take the minimum, privileging the results on which both bases agree to the same degree. At the same time the values that at least one of the bases regard as possible can be preserved, by assigning a non-zero possibility to the support of the union . This would be done as follows :

- if the result is unnormalized, by normalizing it using (4).

- if the result is normalized, by modifying (4) using (1-k) instead of $(1 - h(\pi_{12}))$ as the discrepancy factor.

III) the user can indicate the reliability of each base
We would then use the combination methods (5) or (6). As soon as the reliability of the bases are known, it seems useless to use the degree of conflict which is made up.

IV) the user does not know the reliability of each base but believes that base 1 is relatively more reliable than base 2
Case IV) is not a particular case of III) though it looks like it. If we would have in III) $c_1 = .20$ and $c_2 = .60$ it would mean that the reliability of both bases are known and corresponds to $(1 - c_i)$ $i = 1, 2$. However if in situation IV) we have reasons to believe that base 1 is 3 times more reliable than base 2, we still do not know the reliability of either one of the bases.In practice, the user may not be able to assess the relative reliability of the bases and can only point out that base 1 is more reliable than base 2. In that case, the reasonable attitude is to conjunctively combine π_1 and π_2 only if they are not conflicting and discard π_2 in case of conflict. Formula (7) is particularly adapted to deal with this situation.

V) the user cannot tell the reliability of the bases
In this case, it will be up to the system to decide which is the best combining method. A reasonable approach is to permit ourselves to be optimistic as long as there is no conflict and take a cautious attitude should it appear. That would mean taking the intersection of the results should they have a big agreement, and the union should they disagree. One way to accomplish it would be to compare the Jacquard index k to threshold d, in order to choose between a purely conjunctive $(k \leq d)$or disjunctive aggregation method $(k>d)$. However, this method heavily relies upon the threshold d. A better way of realizing our approach would be to choose a trade-off combination method, like performing a convex combination between the intersection and the union of the results :

$$\pi_{12}(w) = k \cdot \min(\pi_1(w), \pi_2(w)) + (1 - k) \cdot \max(\pi_1(w), \pi_2(w)) \qquad (10)$$

It should be noted that if $k = 1$, i.e. the bases agree completely, (10) will reduce to the intersection and if $k = 0$ it will correspond to the union. This result will be unnormalized if the cores of π_1 and π_2 do not intersect. To normalize it, we proceed as shown in (4). The result will be near to the union, generally.

Fuzzy Queries

A fuzzy query to a database will occur whenever a value of an attribute is vaguely specified. More formally; let $q = (A^1, A^2, \dots, A^n)$ be the query describing a class of components C, where A^i represents the range of attribute a^i for that class as specified by the query. Query q will be fuzzy if at least one of the A^i's is a fuzzy set. Potential fuzzy queries for a reliability database are over attributes whose values vary on the real line, such as power, voltage, rotation speed.

Let us consider the elementary fuzzy query $q_C = (A)$ that describes a fuzzy class C of items x through the membership of a(x) to the fuzzy set A on the parameter scale W. The response of the data bank concerns the value of the failure rate $\lambda(C)$ of items in class C. When C is not fuzzy, it is supposed that $\lambda(C)$ can be obtained and is described by its possibility distribution on the failure rate scale S. In order to derive $\lambda(C)$ when C is a fuzzy class, the basic idea is to represent A through a weighted family of subsets of W. The most natural representation is in terms of the level-cuts A_α, $\alpha \in [0,1]$ of A; the membership function μ_A of the fuzzy set A is then expressed as proposed in [15]

$$\mu_A(w) = \int_0^1 \mu_{A\alpha}(w)\, d\alpha \tag{11}$$

If $\lambda(C_\alpha)$ is the (supposedly non-fuzzy) failure rate for $C_\alpha = \{ x \,/\, a(x) \in A_\alpha \}$, the failure rate $\lambda(C)$ is then defined as

$$\lambda(C) = \int_0^1 \lambda(C_\alpha)\, d\alpha \tag{12}$$

In practice only a finite set of α- cuts $A_1 \subseteq \dots \subseteq A_n$ need to be considered, with $A_i = A_{\alpha i}$ and $\alpha_1 = 1 \geq \alpha_2 \geq \alpha_n \geq \alpha_{n+1} = 0$. Then A is approximated by A(n) with $\mu_{A(n)}(w) = \sum_{w \in A_i} \alpha_i - \alpha_{i+1}$ as a finite counterpart of (11). (12) becomes the following combination formula

$$\lambda_n(C) = \sum_{i=1}^{n} \lambda(C_i)\,(\alpha_i - \alpha_{i+1}) \tag{13}$$

already proposed in [16], with $C_i = C_{\alpha i}$. When $\lambda(C_\alpha)$ is described by a possibility distribution, (11) is the integral of a fuzzy valued function [17], and (13) is simply computed by means of fuzzy interval analysis (see [5] for instance). This method looks well-suited for the CEDB database but may require some adaptation in the case of the IEEE database.

Conclusion

Possibility Theory offers a simple modelling framework for the description of expert knowledge about parameter values. Moreover, probabilistic knowledge as obtained from statistical data can be made compatible with possibilistic knowledge, since a possibility distribution can be used as the approximation of a probability density. One advantage of possibility theory is the availability of a variety of information combination modes that do not require too much knowledge about the reliability of the sources.

The type of interface procedure for the fusion of multiple-source data described in this paper is currently under implementation. Among the pending problems related to this topic, one is the integration of dissymmetric types of combination schemes, such as Bayesian inference, among the other symmetric modes of combination described here. Especially,it would be worthwhile to devise Bayesian-like updating strategies that allow for poor a priori knowledge under the form of possibility distributions [18]. Besides, cases where combination of information is not recommended because one source is far better than others should be investigated in greater detail, in the spirit of control strategies in expert systems involving uncertainty [19].

References

1. Besi, A. : Component Event Data Collection. In : Proc. of the Ispra course on Reliability Engineering, Sept 22-26, Madrid. Amsterdam : D. Reidel Publ. 1986.

2. IEEE guide to the collection and presentation of electrical,electronic and sensing components ; reliability data for nuclear-power generating stations. IEEE Std 500-1984, IEEE, New York 1984.

3. Besi A. ; Mancini G. ; Prade H. ; Testemale C. ; Arrazola I. ; Plainfossé A. : Dealing with uncertainty in a multi-source interrogation system . Proc 3rd IFAC/IFIP/IEA/IFORS Conference on Man-Machine Systems, Analysis, Design and Evaluation, Oulu, Finland (1988) 489-494 .

4. Zadeh L.A. : Fuzzy sets as a basis for a theory of possibility. Fuzzy Sets & Systems 1(1) (1978) 3-28 .

5. Dubois D.; Prade H.(with the collaboration of H.Farreny, R. Martin-Clouaire, C. Testemale) : Possibility Theory. An Approach to Computerized Processing of Uncertainty. New York : Plenum Press 1988.

6. Dubois D.; Prade H. : On the combination of evidence in various mathematical frameworks. Proc. NATO Advanced Research Workshop on Knowledge Based Robot Control, Bonas, France (1988). To be published by Springer-Verlag.

7. Yager R. R. Quasi-associative operations in the combination of evidence. Kybernetes 16 (1987) 37-41.

8. Dubois D.; Prade H. : Une approche ensembliste de la combinaison d'informations imprécises ou incertaines. Revue d'Intelligence Artificielle (Hermés, Paris) 1(4) (1987) 23-42.

9. Colombo A. G.; Jaarsma R. I. : Combination of Reliability parameters from different data sources. Proc 4th EuReData Conference ,Venice,Italy (1983) .

10. Balestreri S.; Besi A.; Carlesso S.; Colombo A. G.; Jaarsma R. J. : Use of CEDB for PSA. Proc. Int. Topical Conference on Probabilistic Safety Assessment and Risk Management, Zurich, Switzerland (1987).

11. Yager R. R.: On the relationships of methods of aggregation evidence in expert systems. Cybernetics and Systems 16 (1985) 1-21.

12. Shafer G.: A Mathematical Theory of Evidence. Princeton N. J. : Princeton University Press 1976.

13. Prade H. : Reasoning with fuzzy default values. Proc. of the 15th IEEE Inter. Symposium on Multiple-Valued Logic, Kingston, Ont. (1985) 191-197.

14. Dubois D. ; Prade H. : Default reasoning and possibility theory. Artificial Intelligence 35 (1988) 243-257.

15. Dubois D. ; Prade H.: Evaluation of fuzzy sets and fuzzy integrations: a synthetic discussion. Proc of the 2nd IFSA- EC and EURO- WG Workshop on Progress in Fuzzy Sets in Europe, Vienna, Austria (1988) 131-135 .

16. Dubois D. ; Jaulent M. C. : A general approach to parameter evaluation in fuzzy digital pictures. Pattern Recognition Letters 6(4) (1987) 251-259.

17. Dubois D. ; Prade H.: Towards fuzzy differential calculus. Part 1:Integration of fuzzy mappings . Fuzzy Sets & Systems 8 (1982) 1-17.

18. Dubois D.; Prade H. : Modelling uncertainty and inductive inference. Acta Psychologica, to appear.

19. Cohen P. : The control of reasoning under uncertainty. A discussion of some programs. The Knowledge Engineering Review 2 (1987) 5-25.

Measures of Dissimilarities for Contrasting Information Sources in Data Fusion

L. OLIVI*, R. ROTONDI**, F. RUGGERI**

* CEC-JRC Ispra (VA) - Italy
** CNR-IAMI V. Ampere 56, (MI) - Italy

Summary

Information content of data coming from a given source is modelled and formalized as a probability distribution and, as such, considered as a point in a function space, where a concept of distance can be introduced.
In such a space, Kullback-Leibler Information is a contrast function able to measure dissimilarities between probability distributions and, then, a practical index for clustering different information sources according to the quality of their content.

1. Introduction

Analysis and fusion of data on plants meet severe problems because of strong dissimilarities in the quality and nature of the information sources.
The case in study deals with a collection of reliability data /1/, the failure times concerning pumps located in different plants. The whole population of pumps cannot be considered as homogeneous, because some important differences are due to the modes of operation, the construction and working characteristics and their locations themselves. Nevertheless according to engineering judgement, one can subdivide the whole population in classes of items similar or quasi similar one each other. From the analyst point of view it is important to establish: wether the information on a given item is biased by the procedure in situ of data detection, or the behaviour of the item itself is affected by the mode of operation and maintenance.
The method proposed in this paper aims to answer the previous questions and to provide a technique of stratification according to similarity among groups of items.

Section 2 will provide the theoretical approach enabling each group of items to be "labelled" by a probability density function and the reciprocal dissimilarity, "the contrast", to be measured by Kullback-Leibler Information.

A typical example will be discussed in 3.

2. Probability model and contrast function

2.1 Predictive model

Within a homogeneous set of items, e.g. the pumps of the same type and use, observation data can be considered as realizations of events $A \in S$ of a same measurable space (Ω, S), being Ω the space of elementary events and S the associated σ-algebra. The behaviour of an item is described by the probability measure $P_i(A)$, which defines the probability of the event A to occur to a pump of the type under study when located in the ith site.

It is assumed that, for pumps of the same type, the distribution of the events A is the same, when tested in ideal laboratory conditions. Therefore the probability $P_i(A)$ does not vary with i, when location effects and environment disturbs are neglible, whilst sensitive local disturbs give rise to variations.

It will be considered that the probability measure $P_i(A)$ has the Radon-Nikodym derivatives with respect to the measure λ. It is expressed by:

$$g_i = \frac{dP_i(A)}{d\lambda}$$

Henceforth λ will be dealt with as the Lebesque measure. The density function $g_i(x)$ is the predictive model for the behaviour of a pump belonging to a given homogeneous class, when operating in the ith site, with respect to the observation time x.

2.2 Construction of $g_i(x)$

A sensible approach to construct g(x) is to apply a bayesian procedure. It enables all the information about the pumps of the same class to be exploited either from direct observations in different operation sites, or from reference data.

In addition, being the aim of this method to detect possible differences in the pump behaviours, exclusively produced by location effects, one needs a formal representation of its knowledge in a function space, where a concept of distance can be introduced. The usual parametric model of the classical approach is not adequate to this purpose.

It is assumed $f(x|\theta)$ as the probabilistic model of the failure time of the item under study; θ is a random variable with known probability density function $p(\theta;\underline{\gamma}_o)$ of parameter vector $\underline{\gamma}_o$.

Given the observation $\underline{X} \equiv (X_1 \ldots X_n)'$, from the ith site and the likelihood function $L(\theta|\underline{X})$, the conditional probability $p^{(i)}(\theta|\underline{X};\underline{\gamma}_o)$ is given by:

$$p^{(i)}(\theta|\underline{X};\underline{\gamma}_o) = p(\theta;\underline{\gamma}_i) \propto L(\theta|\underline{X})\; p(\theta;\underline{\gamma}_o) \tag{1}$$

$p(\theta;\underline{\gamma}_i)$ is the formal representation of the knowledge about the item as far as the surrounding effects of the ith location are concerned.

Therefore the predictive $g_i(x|\underline{X})$ is expressed by:

$$g_i(x|\underline{X}) = \int_\theta f(\underline{X}|\theta)\; p(\theta|\underline{X};\underline{\gamma}_o)\,d\theta \tag{2}$$

2.3 Definition of the references

In the previous section the $p(\theta;\underline{\gamma}_o)$ is considered as known. In the sense of this approach, where the condensed information about the item, when located on the ith site, is $p(\theta;\gamma_i)$, $p(\theta;\underline{\gamma}_o)$, plays the role of the condensed information on the item when working in reference conditions.

In the bayesian view, the $\underline{\gamma}_o$ vector would be obtained by expert elicitation, nevertheless in many cases this information is not always available and some inference is needed.

In other cases, e.g. /2/, the need to incorporate prior information in a typical probability parameter of sampling, has induced the analyst to study the role played by the prior with respect to the final inference.

In particular, the posterior robustness with respect to prior variations, contamination of "extraneous" unwanted information, relative importance of the prior and likelihood, e.g. case of greatest average domination of the likelihood and maximization of information gain, are concerned.

In the present case an analogous analysis has been performed. The estimate $\underline{\gamma}_o$ so obtained, produced results which, in the final cluster analysis, seemed not easy to be interpreted. Then, in a more practical way, a set of possible $\underline{\gamma}_o$ vectors has been chosen and tested with respect to the final result of clustering. It turned out that, the aggregation of dissimilarity measures does not much depend on the possible values of $\underline{\gamma}_o$, which, rather, are effective on the order at the magnitude of cluster ranges. As consequence some arbitrariness on the choice of $\underline{\gamma}_o$ has been left to the analyst.

2.4 Kullback-Leibler Information

Always with reference to (Ω, S), given two defined probability measures Q_1 and Q_2 and $q_1 = dQ_1/d\lambda$, $q_2 = dQ_2/d\lambda$ the respective Radon-Nikodym derivatives with respect to the measure λ, Kullback-Leibler Information is given by:

$$KL(Q_1,Q_2) = \int_{\Omega} q_1 \ln\left(\frac{q_1}{q_2}\right) d\lambda \qquad \text{for} \quad Q_1 \ll Q_2 \tag{3}$$

$$KL(Q_1,Q_2) = \infty \qquad \text{otherwise}$$

Kullback-Leiber Information /3/ is a contrast function /4/ and provides a relative measure of dissimilarity between two probability distributions and is, in practice, an effective index of dissimilarity between the information contents of the sources represented by Q_1 and Q_2.

2.5 Clustering criteria

The functions $p(\theta;\underline{\gamma}_i)$ and $g_i(x|\underline{X})$, for the item under examination, represent respectively, the shifting with respect to the initial conditions and, given this shifting, the predicted behaviour. In both cases Kullback-Leibler Information can be exploited and clustering criteria applied. In particular for a class of items one can compute:

$$KL(i,o) = KL\left\{P(\theta;\underline{\gamma}_i),\ P(\theta;\underline{\gamma}_o)\right\} \tag{4}$$

and then

$$\Delta KL(i,j) = |KL(i,o) - KL(j,o)|$$

The choice of $\underline{\gamma}_o$ plays the role of unit of measure, but does not much influence the clustering aggregation.

3. The case in study

3.1 Data Sources

Data extracted from CEDB /1/ have been treated by a unit of the JRC ATI Division, in the context of the Risk and Reliability Evaluation project. As to the methodology and algorithm used, an external support has been given by the Consiglio Nazionale delle Ricerche/IAMI (Milan).

For a preliminary methodology exercise, the analysis has been confined to groups of data already classified. Data collected in the same location have been subdivided according to engineering characteristics; $p(\theta|\underline{\gamma}_i)$ will represent the statistical identification label for each group.

3.2 Modelling and clustering

As far as the model of failure is concerned, as a first preliminary approach, the classical exponential model has been chosen:

$$f(\underline{X},\theta) = \theta\, e^{-\theta x} \qquad \theta > 0 \tag{5}$$

Where θ is a random variable with Gamma probability distribution of parameter $\underline{\gamma}_o \equiv (\alpha_o,\beta_o)'$:

$$p(\theta;\underline{\gamma}_o) = p(\theta;\alpha_o,\beta_o) = \frac{\beta_o^{\alpha_o}}{\Gamma(\alpha_o)}\ \theta^{\alpha_o-1}\ e^{-\beta_o\theta} \tag{6}$$

The (1) of 2.2 becomes:

$$p(\theta|\underline{X};\alpha_o,\beta_o) = p(\theta|\alpha_i,\beta_i) = \frac{\beta_i^{\alpha_i}}{\Gamma(\alpha_i)}\ \theta^{\alpha_i-1}\ e^{-\beta_i\theta} \tag{7}$$

where $\alpha_i = n + \alpha_o$ and $\beta_i = \sum_{j=1}^{n} X_j + \beta_o$.

For sake of simplicity the n index has not been labelled, even if it is depending up on the set of observations considered.

This model has the advantage to be flexible with respect to data and easy to be dealt with from computational point of view.

By introducing the (6) and the (7) in the (4), by integration one gets:

$$KL(i,o) = \ln\left\{\frac{\Gamma(\alpha_o)}{\Gamma(\alpha_i)}\left(\frac{\beta_i}{\beta_o}\right)^{\alpha_o}\right\} - \frac{\alpha_i\sum_{j=1}^{n} X_j}{\beta_i} + n\ \psi(\alpha_i) \tag{8}$$

where $\psi(z) = \frac{\Gamma'(z)}{\Gamma(z)}$ is the logarithmic derivative of $\Gamma(z)$.

If z is an integer larger than 1:

$$\psi(1) = -C, \qquad \psi(z) = \sum_{k=1}^{z-1} K^{-1} + \psi(1)$$

where C is Eulero-Mascheroni constant.

When z is a real number, it can be approximated by trunkating the asymptotic form:

$$\psi(z) \simeq \ln z - \frac{1}{2z} - \sum_{k=1}^{\infty} \frac{B_{2k}}{2kz^{2k}}$$

where B_l are the Bernoulli numbers /5/.

The algorithm for clustering Kullback-Leibler measures is very simple:

Step 0. Compute:

i) The symmetric matrix of the absolute differences:

$$A = [\Delta KL(i,j)] \qquad i,j = 1,\ldots,n;$$

ii) $$dmax = \max_{1 \leqslant i,j \leqslant n} [\Delta KL(i,j)]$$

iii) $$dlim = \frac{dmax}{n-1}$$

Let K=1

Step 1. Find:

$$\Delta(i^*,j^*) = \min_{1 \leqslant i,j \leqslant n} [\Delta KL(i,j)]$$

(we suppose $KL(i^*,0) \leqslant KL(j^*,0)$)
i^* and j^* belong to the same cluster.

Indicate with 1_K the observation set within distance dlim from j^*, having $KL(j^*,0) \leqslant KL(1_K,0)$. The observation 1_K belongs to the K cluster. Set $j^* = 1_K$ and go on until such a set exists. Repeat the procedure with respect to i^* with the condition $KL(1_K,0) \leqslant KL(i^*,0)$. Then set $\Delta(i,j) = dmax$ for all the pairs i,j of observation sets belonging to K cluster. Set $K = K + 1$. Go to step 1.

4. Results and conclusions

Each group of items homogeneous for location and engineering characteristics has been considered as the minimal "cell" of investigation.
The aim of the work was to verify wether the proposed method is able to detect differences amongst such "cells". To this end, Kullback-Leibler Information has been applied among appropriate probability distributions "cells labelling". By the algorithm of 3.2 the respective Kullback-Leibler indexes have been computed and plotted as in the figures.

It is very evident an important discrimination among "cells" belonging to different locations and, in some significant cases, a screening effect appears also between two "cells" of the same plant with different engineering characteristics.

It can be concluded that the proposed method is effective with respect to the specimen analysed and seems robust to the poornes of data.

A definitive answer on its practicality can be given only by a calibration exercise, where the statistical screening provided by the method can be constrasted with a well established engineering classification.

Aknowledgements

The authors are indebted to Mrs. S. Carlesso, Dr. Ing. A. Besi, Mr. S. Balestreri and Mr. R. Jaarsma, who gave their precious collaboration as far as the data treatment and interpretation are concerned.

References

1. Besi, A.: Characteristics and Use of the Component Event Data Bank, IAEA Technical Committee Meeting on Evaluation of Reliability Data Sources, Vienna, 1-5 February 1988.

2. Dyer, D. and Chion P.: An Information-Theoretic Approach to Incorporating Prior Information in Binomial Sampling. Comm. in Statistics. Theory and Methods 13(17), 2051-2083, 1984.

3. Kullback, S.: Information Theory and Statistics, John Wiley and Sons, N.Y. 1959.

4. Dacunha-Castelle, D. and Duflo, M.: Probabilités et Statistiques, Masson, Paris 1982.

5. Abramorwitz, M. and Stegun J.A. editors: Handbook of Mathematical Functions. Dower Publications Inc. N.Y. 1965.

	KL	cell	
1	1.512657	C - 1.1.1	
2	4.383929	C - 2.3.2	
3	5.293092	C - 2.3.1	
4	6.466325	B3 - 2.3.2	
5	6.678900	B1 - 2.3.2	
6	6.682478	B2 - 2.3.2	cluster 1
7	6.701658	B1 - 2.3.1	
8	6.979459	B3 - 2.3.1	
9	7.840930	A2 - 2.3.1	
10	7.897698	A1 - 2.3.1	cluster 2
11	8.120742	A1 - 2.3.2	
12	8.383580	A2 - 2.3.2	
13	9.046462	B2 - 2.3.1	cluster 5
14	9.412206	D4 - 2.2.1	
15	10.303431	D1 - 2.2.1	
16	10.558337	D3 - 2.2.1	cluster 3
17	10.998700	D2 - 2.2.1	
18	11.155740	D1 - 3.2.2	
19	11.739430	E - 2.2.2	
20	12.322330	D1 - 1.2.2	
21	12.511090	F - 2.2.2	cluster 4
22	12.982740	E - 1.2.1	

Fig. 1

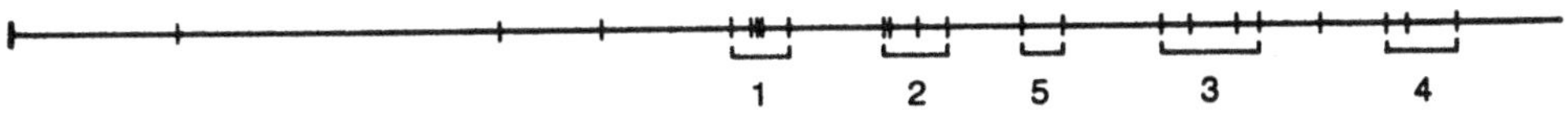

Figure 1 : Cluster analysis for all the considered cells. $\alpha_o = 3$, $\beta_o = 1,5$.

	KL	cell	
1	4.383929	C - 2.3.2	
2	6.466325	B3 - 2.3.2	cluster 1
3	6.678900	B1 - 2.3.2	cluster 1
4	6.682478	B2 - 2.3.2	cluster 1
5	8.120742	A1 - 2.3.2	cluster 2
6	8.383580	A2 - 2.3.2	cluster 2

Fig. 2

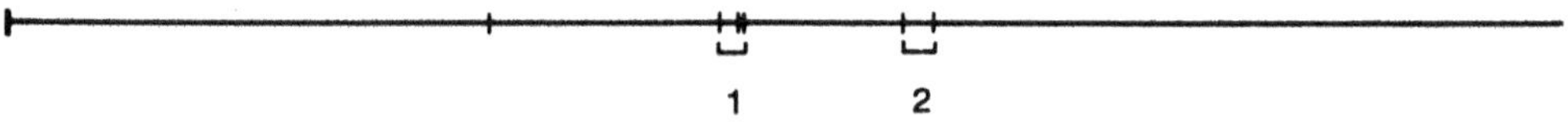

	KL	cell	
1	5.293092	C - 2.3.1	
2	6.701658	B1 - 2.3.1	cluster 2
3	6.979459	B3 - 2.3.1	cluster 2
4	7.840930	A2 - 2.3.1	cluster 1
5	7.897698	A1 - 2.3.1	cluster 1
6	9.046462	B2 - 2.3.1	

Fig. 3

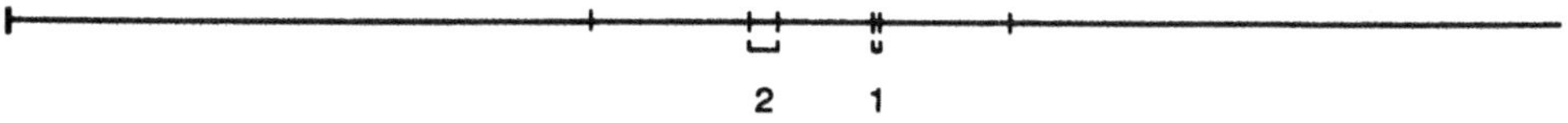

Figure 2,3 : Cluster analysis for cells with same eingineering characteristics. $\alpha_o = 3$, $\beta_o = 1,5$.

	KL	cell	
1	10.303431	D1 - 2.2.1	cluster 1
2	11.155740	D1 - 3.2.2	cluster 1
3	12.322330	D1 - 1.2.2	

Fig. 4

	KL	cell	
1	1.512657	C - 1.1.1	
2	4.383939	C - 2.3.2	cluster 1
3	5.293082	C - 2.3.1	cluster 1

Fig. 5

	KL	cell	
1	7.840930	A2 - 2.3.1	cluster 1
2	7.897698	A1 - 2.3.1	cluster 1
3	8.120742	A1 - 2.3.2	
4	8.383580	A2 - 2.3.2	

Fig. 6

Figure 4,5,6: Cluster analysis for cells with same location.
$\alpha_o = 3$, $\beta_o = 1,5$.

Computing Cumulative Measures in Reward Stochastic Processes by a Phase-Type Approximation

Andrea Bobbio
Istituto Elettrotecnico Nazionale Galileo Ferraris
Strada delle Cacce 91, 10135 Torino, Italy

Laura Roberti
Scolarship Associazione Tecnica dell'Automobile (ATA), Torino, Italy

Enrica Vaccarino
Centro Ricerche Fiat, Torino, Italy

Abstract

This paper illustrates new methodologies. that are under study for the analysis of complex degradable systems. The basic idea is to include into a single dependability model both the random variation of the system configuration in time, and the effective performance level of the system in each configuration. In order to characterize the system behaviour, cumulative measures are defined. The evaluation of the distribution function of these measures is a difficult task for the inherent computational complexity. This paper surveys the possibility of solving the above models by approximating the resulting non-markovian stochastic process by means of a suitably generated Markov chain.

1. Introduction

The classical reliability theory [1,30] is based on the assumption that each component, and the system as a whole, can be modelled by a binary variable representing two possible conditions: functional or non-functional. This assumption implies that the state space of the system can be univocally partitioned into two mutually exclusive subsets, one containing the up states, the other containing the down states. The classical reliability measures (*Reliability, Availability, Maintainability*) are defined on this binary partition of states, and reflect the system condition at a given point in time. For this reason we will refer to this measures as *instantaneous measures*.

The need of an accurate evaluation of the performance and reliability of highly reliable fault-tolerant systems has led to the definition of a new class of measures [15,5] that reflect system's operational characteristics over a finite interval rather then at a fixed time instant or in steady-state. We denote this measures as *cumulative* or *integral* measures. Typical integral measures are the interval availability [16,13], which is the fractional system up time in a given interval, delay or waiting time in queueing systems [23], or the total number of events (for example, failures and repairs) in a given interval.

The expected values of the integral measures can be computed by solving a linear system of differential equations derived from the transition rate matrix of the stochastic process [15,33]; thus, with a computational cost of the same order of the cost needed for the computation of the instantaneous measures. However, the knowledge of the expected values may not provide sufficiently accurate information about the system behavior in time, and in those cases the complete distribution function [31] is needed. The ability of evaluating the distribution function of cumulative measures is a recent development, and the computation of such distribution functions remains a difficult task [33,12,13,22,32,31].

Classical reliability modelling becomes inadequate when dealing with distributed degradable systems. The term distributed system defines a system made up of different resources, working independently, but cooperating through some communication network. The failure of one node or of one resource does not cause the system to fail, but reduces its working capacity so that the overall performance degrades. Typical distributed systems are computer networks, multiple processor systems, flexible manufacturing systems and, in general, any system whose configuration can be described by a network. The service or performance level and the reliability are both important in determining the ability of the system to deliver a given amount of useful work in a finite time period [2,25]. A model that takes into account combined performance and reliability measures is needed. The basic idea of this new modelling approach [33] consists in describing the degradation of the system in time by means of a stochastic process, called the ***structure-state process***, and by associating to each state of the structure-state process a positive real number representing the effective working capacity or performance level of the system in that state. The real variable is called the ***reward rate*** [19] associated to the state. The structure-state process together with the reward rates form the *Stochastic Reward Model (SRM)* [33,32].

2. Stochastic Reward Model (SRM)

In the present section we formally introduce the modelling technique based on SRM. Let $Z(t)$, $(t \geq 0)$ denote a stochastic process defined over a discrete and finite state space of cardinality n. $Z(t)$ is the structure-state process and represents the stochastic behavior of the system configuration in time. Let r_j $(j = 1, 2, \ldots, n)$ denote the reward rate associated to the structure-state j. With the above definition $r_j\, dt$ provides the amount of work performed by the system in the interval $t - t + dt$ given that the system was in state j at time t.

Two main different points of view have been assumed in the literature when dealing with SRM for degradable systems [22]. In the ***system oriented*** point of view the most significant measure is the the total amount of work done by the system in a finite interval; this measure is usually referred to in the literature as ***performability*** [2,25,24]. In the ***task oriented*** point of view the system is regarded as a server, and the emphasis of the analysis is on the ability of the system to process a task, whose execution requires a known amount of work (expressed in some convenient unit). The most significant measure is, in this case, the probability that a given task will be completed at a given time, taking into account degradation, failure and repair. The distribution of the completion time is affected by the way in which a change of state in the structure-state process interacts with the task in progress. We usually distinguish two different possibilities [14]:

- When a change of state occurs in the structure-state process, the system keeps memory of the work already done and the task in progress is resumed in the new state. This policy is called *preemptive resume (prs)*.
- When a change of state occurs in the structure-state process, the system cannot keep track of the past, and the work already done is lost; the task in progress must be restarted from scratch in the new state. This policy is called *preemptive repeat identical (pri)* if the repeated task has the same work requirement of the original preempted task, and *preemptive repeat different (prd)* if the work requirement is different, but sampled from the same distribution.

Different applications give rise to different interpretations of the structure-state process and of the reward structure superimposed on it. Usually, the reward rates represent the speed of the service, and the structure-state process the failure and repair of components; the time needed to accumulate a certain amount of reward will be the time to complete a task with that work requirement in a failure-prone environment. If the structure-state process is interpreted as modelling the arrival and departure of tasks in queueing systems and the reward rate represents the number of jobs in the queue, we can obtain the time averaged queue lenght distribution. By interpreting the structure state process as task arrival/departure and interpreting the reward rates as the portion of the server capacity allocated to a tagged task, the completion time distribution yields the response time distribution in the queueing system [21].

In application, for computational purposes, the structure-state process is usually assumed to be a continuous time, discrete state homogeneous Markov chain [11,20,22,32]. The extension to semi-Markov reward processes has been investigated in [21].

3. Measures of interest

If $Z(t)$ is a Markov process, the state probability vector $\underline{P}(t)$ may be computed by solving the standard Markov Equation

$$\dot{\underline{P}}(t) = \underline{P}(t)\,\mathbf{Q} \tag{1}$$

whose formal solution is

$$\underline{P}(t) = \underline{P}^0 exp(\mathbf{Q}t)$$

where $\mathbf{Q}$ is the transition rate matrix, and $\underline{P}^0$ the initial probability vector. The classical instantaneous measures, like reliability, availability and maintainability are directly derived from (1) by summing up the state probabilities over suitably defined partitions of the state space [1].

Define the vector

$$\underline{B}(t) = \int_0^t \underline{P}(z)\,dz \tag{2}$$

The $i-th$ entry of $\underline{B}(t)$ is the expected time spent by the process $Z(t)$ in state i during the interval $(0,t)$ [1]. By direct integration of equation (1), equation (2) takes the form:

$$\dot{\underline{B}}(t) = \underline{B}(t)\,\mathbf{Q} + \underline{P}^0 \tag{3}$$

with initial condition $\underline{B}(0) = \underline{0}$. The system of linear differential equations (3) can be solved with the same methods used for solving equation (1) [29], and thus with approximately the same computational load. The expected values of the integral measures are calculated directly from (3) [5] once the partition of the state space is identified.

Let us now introduce the most significant measures in the framework of SRM, when the reward rates are superimposed to the structure-state process. The instantaneous reward rate at time t is a random variable $f_r(t)$ defined as:

$$f_r(t) = r_i \qquad \text{if} \qquad Z(t) = i. \tag{4}$$

From (4) follows that the expected instantaneous reward rate at time t is given by:

$$E[f_r(t)] = \sum_i r_i \, Prob\{Z(t) = i\} = \sum_i r_i \, P_i(t) \tag{5}$$

The performability $Y(t)$ is the random variable defined as the total accumulated reward up to time t:

$$Y(t) = \int_0^t f_r(\tau)\, d\tau \tag{6}$$

From the above definition the expected performability can be computed as:

$$E[Y(t)] = E\,[\int_0^t f_r(\tau)\, d\tau\,] = \sum_i r_i \int_0^t P_i(\tau) d\tau = \sum_i r_i \, B_i(t) \tag{7}$$

so that (7) can be computed by solving (2). On the other hand, the distribution function $F_Y(t,x)$ of $Y(t)$ is defined as:

$$F_Y(x,t) = Prob\{\, Y(t) \leq x\} \tag{8}$$

Computing $F_Y(t,x)$ is a more difficult task [33].

Let us further suppose that the system must process a task whose execution requires W units of work. W is in general assumed to be a random variable with distribution $G(x)$. The degenerate case in which W is deterministic and the distribution $G(x)$ becomes the unit step function $U(W)$ located at $W = x$, can be considered as well. Given $W = x$, the task completion time $T(x)$ is the random variable representing the time to complete a task whose work requirement is x. The cumulative distribution function $F_T(t,x)$ is given by:

$$F_T(t,x) = Prob\{T(x) \leq t\} \tag{9}$$

The unconditional completion time T is characterized by the following distribution:

$$F_T(t) = Prob\{\, T \leq t\} = \int_0^\infty F_T(t,x)\, dG(x) \tag{10}$$

The distribution of the completion time, $F_T(x,t)$ is a very useful measure since it incorporates the effects of a random variation of the execution speed that results from degradation and reconfiguration in the structure-state process.

In order to completely characterize the completion time problem, the interactions of the task in progress with the system must be specified. If in $Z(t)$ a transition from state i to state j occurs at time t and the transition is of *prs* type, the task execution resumes in state j at the new rate r_j. If the transition is of *prd* type, the work already done is lost; the preempted task is restarted in the new state j with the reward rate r_j, and with a work requirement W resampled from the same distribution $G(x)$.

When the structure-state process $Z(t)$ is a discrete-state homogeneous Markov process, $F_T(t)$ in (10) has been expressed in [22] as two-dimensional Laplace transform equation. An algorithm for numerically inverting these equations has been presented in [31] when all the states are of *prs* type.

Assuming that structure-states are all of *prs* type, so that no loss of reward occurs, the distribution of the completion time is closely related to the distribution of the accumulated reward by means of the following relationship:

$$Prob\{Y(t) \leq x\} = Prob\{T(x) \geq t\} \tag{11}$$

In spite of the computational complexity it implies, the evaluation of the distribution function of the completion time is a foundamental step in system analysis.

4. Phase type approximation

Analytic approaches to the solution of even simple *SRM* become rapidly very complicated. Numerical techniques for the computation of the distribution function of the completion time (or of the performability) have been proposed in [11,20,31]; in [17] and [32] the first two moments of the distribution are computed by using the randomization [18,23] technique. The above mentioned methods have been developed under the hypothesis that all states are of *prs* type, and are not easily extensible to more general schemes of preemption policies.

A different and computationally more versatile approach has been proposed in [4].

This approach is based on the use of a family of distribution functions called Coxian [7] or Phase type (PH) distributions [26,10]. A probability distribution function $G(x)$ is said to be of Phase Type [26] if it is the distribution of the time to absorption of a continuous time homogeneous Markov chain with m transient states and one absorbing state; m is known as the order of the PH distribution. The PH distributions combine formal elegance with computational efficiency; moreover, the PH family contains a number of useful distributions often encountered in applied stochastic modelling, like: the Erlang distribution, the hyperexponential and, obviously, the exponential.

For what concerns the computation of the distribution of the completion time $F_T(t)$ (10) by resorting to the PH approximation technique, the main results are contained in [4]. It is shown that, if $Z(t)$ is a discrete state homogeneous Markov chain with n states, and the work requirement W has a PH distribution of order m, the distribution $F_T(t)$ (10) can be computed as the distribution of the time to absorption in a suitably derived Markov chain with $n\,m$ transient states and one absorbing state, under any mixture of preemption policies. In other words, if the work requirement of a given task W is a PH random variable of order m, the taks completion time is still a PH random variable of order nm.

The practical interest of this results lies in the following properties:

- $F_T(t)$ is computed by solving a linear system of differential equations of the type (1), over an expanded state space; thus, well established numerical techniques [29] can be employed. Moreover, it can be shown that the transition rate matrix of the expanded process is very sparse and has a highly structured configuration, so that specially designed integration routines can be implemented [28,27].

- Since it is possible to approximate with any prescribed accuracy a general distribution by means of a PH distribution by letting the order m increase to infinity [8,10], it is possible to conceive the above technique for approximating any completion time problem or performability problem by means of a Markov chain.

The case when W becomes a fixed constant rather than a randon variable can be considered as well. In this last case, however, other numerical techniques could be more effective [31].

- The generation of the expanded state space can be performed automatically from the knowledge of the transition rate matrix **Q** of the structure-state process and of the transition rate matrix of the PH distribution of the work requirement [4]. Both an analytical technique based on matrix algebra [4] and a graphical/simulative technique [9] are available.

The drawback of this method is the explosion of the state space in the expanded chain which is common to many methods based on state analysis; the only way to circumvent this difficulty is the development of efficient and powerful numerical tools for handling large sets of linear differential equations [6,3,29]. In this particular case the expanded process has a very specially structured transition rate matrix, and an "ad hoc" designed procedure can be imagined [28,27] for exploiting in a compact computer code the structural properties of the matrix.

4. Conclusion

We have surveyed new modelling techniques and measures for the analysis of complex systems whose behavior cannot be simply represented by a binary partition of the state space. A versatile numerical approach has been introduced, that is based on the approximation of a non-markovian process by means of a Markov process defined over an expanded state space. The application of this technique to large scale problems requires powerful numerical methods. Research is currently in progress in this area particularly for what concerns the possibility of saving computational load by exploiting the peculiar structure of the resulting transition rate matrix.

References

[1] R.E. Barlow and F. Proschan. *Statistical Theory of Reliability and Life Testing.* Holt, Rinehart and Winston, New York, 1975.

[2] M.D. Beaudry. Performance-related reliability measures for computing systems. *IEEE Transactions on Computers*, C-27:540–547, 1978.

[3] A. Bobbio and K.S. Trivedi. An aggregation technique for the transient analysis of stiff Markov chains. *IEEE Transactions on Computers*, C-35:803–814, 1986.

[4] A. Bobbio and K.S. Trivedi. *Computation of the distribution of the completion time when the work requirement is a PH random variable.* Technical Report, Accepted with revision on: Stochastic Models, 1988.

[5] A. Bobbio and K.S. Trivedi. *Computing cumulative measures of stiff Markov chains using aggregation.* Technical Report CS-42, Duke University Tech Report, 1987.

[6] P.J. Courtois. *Decomposability: Queueing and Computer System Applications.* Academic Press, New York, 1977.

[7] D.R. Cox. A use of complex probabilities in the theory of stochastic processes. *Proceedings of the Cambridge Phylosophical Society*, 51:313–319, 1955.

[8] D.R. Cox and H.D. Miller. *The theory of stochastic processes.* Chapman and Hall, London, 1965.

[9] A. Cumani. Esp - A package for the evaluation of stochastic Petri nets with phase-type distributed transition times. In *Proceedings International Workshop Timed Petri Nets*, IEEE Comp Soc Press no. 674, Torino (Italy), 1985.

[10] A. Cumani. On the canonical representation of homogeneous Markov processes modelling failure-time distributions. *Microelectronics and Reliability*, 22:583–602, 1982.

[11] L. Donatiello and B.R. Iyer. Analysis of a composite performance reliability measure for fault tolerant systems. *JACM*, 34:179–199, 1987.

[12] L. Donatiello and B.R. Iyer. Closed-form solution for system availability distribution. *IEEE Transactions on Reliability*, R-36:45–47, 1987.

[13] E. De Souza e Silva and R Gail. Calculating cumulative operational time distributions of repairable computer systems. *IEEE Transactions on Computers*, C-35:322–332, 1986.

[14] D.P. Gaver. A waiting line with interrupted service, including priorities. *Journal of the Royal Statistical Society*, B24:73–90, 1962.

[15] A. Goyal, S. Lavenberg, and K.S. Trivedi. Probabilistic modeling of computer system availability. *Annals of Operations Reserch*, 8:285–306, 1987.

[16] A. Goyal and A.N. Tantawi. A measure of guaranteed availability and its numerical evaluation. *IEEE Transactions on Computers*, C-37:25–32, 1988.

[17] W.K. Grassmann. Means and variances of time averages in markovian environment. *European J of Operations Research*, 31:132–139, 1987.

[18] D. Gross and D. Miller. The randomization technique as a modeling tool and solution procedure for transient Markov processes. *Operations Research*, 32:343–361, 1984.

[19] R.A. Howard. *Dynamic Probabilistic Systems, Volume II: Semi-Markov and Decision Processes.* John Wiley and Sons, New York, 1971.

[20] B.R. Iyer, L. Donatiello, and P. Heidelberger. Analysis of performability for stochastic models of fault-tolerant systems. *IEEE Transactions on Computers*, C-35:902–907, 1986.

[21] V.G. Kulkarni, V.F. Nicola, and K. Trivedi. The completion time of a job on a multi-mode system. *Advances in Applied Probability*, 19:932–954, 1987.

[22] V.G. Kulkarni, V.F. Nicola, and K. Trivedi. On modeling the performance and reliability of multi-mode computer systems. *The Journal of Systems and Software*, 6:175–183, 1986.

[23] B. Melamed and M. Yadin. Randomization procedure in the computation of cumulative-time distributions over discrete state Markov processes. *Operations Research*, 32:926–944, 1984.

[24] J.F. Meyer. Closed form solution of performability. *IEEE Transactions on Computers*, C-31:648–657, 1982.

[25] J.F. Meyer. On evaluating the performability of degradable systems. *IEEE Transactions on Computers*, C-29:720–731, 1980.

[26] M.F. Neuts. *Matrix Geometric Solutions in Stochastic Models.* John Hopkins University Press, Baltimore, 1981.

[27] B.D. Plateau, J.M. Fourneau, and K.H. Lee. A package for solving complex Markov models of parallel systems. In *Proceedings Fourth International Conference on Modelling Techniques and Tools for Computer Performance Evaluation*, pages 341–360, 1988.

[28] B.D. Plateau and S.K. Tripathi. Performance analysis of synchronization for two communicating processes. *Performance Evaluation*, 8:305–320, 1988.

[29] A. Reibman and K.S. Trivedi. Numerical transient analysis of Markov models. *Computers and Operations Research*, 15:19–36, 1988.

[30] M.L. Shooman. *Probabilistic reliability: an engineering approach.* Mc Graw Hill, 1968.

[31] R. Smith, K. Trivedi, and A.V. Ramesh. Performability analysis: measures, an algorithm and a case study. *IEEE Transactions on Computers*, C-37:406–417, 1988.

[32] U. Sumita, J.G. Shanthikumar, and Y. Masuda. Analysis of fault tolerant computer systems. *Microelectronics and Reliability*, 27:65–78, 1987.

[33] K. Trivedi, A. Reibman, and R. Smith. Transient analysis of Markov and Markov reward models. In P. Courtois G. Iazeolla and O. Boxma eds., editors, *Computer Performance and Reliability*, pages 535–545, North Holland, 1988.

The Use of Data to Identify Systems for Reliability Assessment Using Monte-Carlo Simulation Techniques

A M FEATHERSTONE and K GHAZVINI

Aston University
Aston Triangle, Birmingham, England

Abstract

Direct Monte-Carlo simulation offers enormous potential for the assessment of system reliability. The method suffers from serious run-time limitations (12), there are however far more complex Monte-Carlo techniques, known generically as VARIANCE REDUCTION TECHNIQUES (7), which can overcome these limitations. Such methods are however very difficult, for non-mathematicians, to implement and require expert knowledge of the systems under assessment to choose the appropriate technique. At Aston University we are working on the development of an expert system to use the optimal Monte-Carlo Variance Reduction Technique for the assessment of the reliability of a given system. The main thrust of the work at present involves the identification of problems and the assessment of which Variance Reduction Technique is most applicable. This paper sets out to explain what system attributes to collect data on and use in the identification of that system in respect of Variance Reduction Monte-Carlo methods of reliability assessment.

1. INTRODUCTION

In general terms reliability assessment is the application of techniques in the quantification of the reliability of a system. The most widely used techniques of quantification are analytical and include; Fault Tree Analysis[1], Event Tree Analysis[2] and Failure Modes and Effect Analysis[3]. These techniques can be applied manually or using various computer packages, for examples of such packages see references (4), (5) and (6).

Monte-Carlo simulation [7] is an alternative to the analytical methods since it is by nature iterative. Monte-Carlo simulation functions by observing random numbers chosen in such a way that they simulate the physical random processes of the system being assessed. The solution is inferred by the resultant behaviour and interaction of the random numbers.

Monte-Carlo simulation enables the reliability analyst when assessing a system to use, basically, the system engineering diagrams as the input. The method is then flexible enough to allow the analyst to add practicalities as and when they occur e.g. repair man problems [9], non-exponential failure and repair characteristics, unrevealed failures, producibility measurements [8], mission profile problems and stress factor interaction [9]. All these practicalities can be expressed in logic terms which are readily implemented on a computer simulation.

The Monte-Carlo simulation technique functions by observing the resultant interaction of random numbers. When the results "settle down"*, then a conclusive result is obtained. This failure to settle "quickly" is caused by one or both of the inherent limitations of the method [12].

THE LIMITATIONS EXPANDED

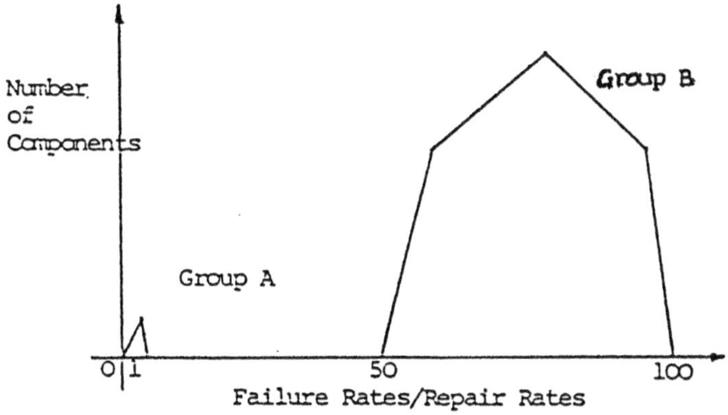

Fig. 1 A Distribution of System Failure/Repair Rates for a Given Example System

* The results "settle down" when a given variance (σ^2) is obtained for the spread of results.

Consider Figure 1, the bulk of system failure ** rates lie between 50 and 100 units, however a limited number of components have a failure rate of 1 unit which is far removed from the rest of the system characteristics. Let the failures associated with the low failure rate (Group A) have a "significant" effect on the system results. Note, if this is not so, then the problem to be described is inapplicable.

For the Monte-Carlo simulation technique to function a suitable sampling rate (time) has to be decided upon. The sampling rate required by Group A is however incompatible with that required by Group B. If a short step length is chosen to accommodate Group B, then the system result will appear to settle but will contain inconsistencies as Group A results sporadically occur and influence the system results. This is a description of the rare-event problem/dilemma. (Generically known as a STIFF PROBLEM [10].

The rare-event problem can also occur because of problem inconsistencies. Problem inconsistencies describes a wide range of situations but it can be explained in the following general terms, if a system result is required which is outside the scope of the component characteristics then a rare-event type situation occurs.

For an example of problem inconsistency, consider a fire alarm system, a component failure within such a system "may" only be revealed if a fire occurs, therefore it could remain dormant for a very long period (say years) whereas all other failures of the system and its components may have characteristics measured in hours and days. To accommodate the dormant failures in a Monte-Carlo simulation of the system would bring about the dilemma of sampling rates as described above.

** The characteristics described apply to both failure and repair rates

2. VARIANCE REDUCTION TECHNIQUES

The direct Monte-Carlo simulation technique attempts to model the problem under consideration as accurately as possible in a "true-to-life" manner. The sampling flows through the model and sampling probability distributions are chosen to reflect the "Real" situation as accurately as possible. Variance reduction techniques attempt to overcome the limitations, as described above, of the direct method by:

(a) Modifying the simulation procedure
(b) Utilising approximate or analytical information
(c) Studying the problem within a different context or abstract representation.

There are a large number of variance reduction techniques available, the major methods can be classified as follows:

(a) Sampling process modifiers

(i) IMPORTANCE SAMPLING
(ii) RUSSIAN ROULETTE AND SPLITTING
(iii) SYSTEMATIC SAMPLING
(iv) STRATIFIED SAMPLING

(b) Use of analytic information

(i) EXPECTED VALUES
(ii) STATISTICAL ESTIMATION
(iii) CORRELATED SAMPLING
(iv) HISTORY REANALYSIS
(v) CONTROL VARIATES
(vi) ANTITHETIC VARIATES
(vii) REGRESSION

(c) Specialised Techniques

(i) SEQUENTIAL SAMPLING
(ii) ADJOINT FORMULATION

(iii) TRANSFORMATIONS
(iv) ORTHONORMAL FUNCTIONS
(v) CONDITIONAL MONTE-CARLO

The sampling process modifying techniques attempt to employ more effective sampling techniques or attempt to modify the sampling distribution. Normally this involves altering the simulation such that the "rare-events" can be observed more frequently.

The techniques which incorporate analytic information into their methods, do so on the whole by "segmenting" the problem and dealing with each segment on the basis of whether it is more amenable to simulation or an analytic solution.

The specialised techniques, as the name implies, fall outside the catagories described above, some use a combination of the sampling modifiers and analytic information whereas others use completely different techniques.

Generally, Variance Reduction techniques can be aimed at reducing the variance of the estimate of only ONE parameter or aspect of the process being simulated. This is, however, often to the detriment of the overall effectiveness of the method in estimating other parameters. Thus it is of great importance when using such techniques to firstly determine all the results which are desirable from the simulation before searching for the techniques to apply to a given problem.

Variance Reduction can be viewed as a means of using known, usually qualitative, information about a system to be simulated in an explicit and quantitative manner. If very little, or nothing, is known about the system to be simulated variance reduction cannot be "easily" achieved. The variance reduction techniques have drawbacks as indicated, their effectiveness is dependant upon the users knowledge of the system under consideration, the more information that is available the more effective will be the variance reduction method. The method thus requires some "prior-knowledge" of the results, to obtain this prior-knowledge requires insight and intuition which can be obtained through experience and expertise. Thus, with the development of "expert computer systems" the use of such techniques to provide this knowledge for the variance reduction techniques is a logical step. An extension would then be given the type of system to be analysed and the "knowledge" available for that system and the information required from the simulation, what type of Variance Reduction technique should be employed?

3. REQUIRED DATA

As has been described above, the first piece of information required in making a decision about the type of Variance Reduction techniques to use is "What information is the simulation attempting to obtain?" Secondly one must consider the Variance Reduction techniques themselves and what system knowledge do they require and to what type of system are they best suited. A great deal of the work required to answer such questions has been carried out at the Oak Ridge National Laboratories in the U.S.A. As such, they have published a list of the Variance Reduction techniques and that information which must be considered when making a decision about which techniques to choose in ORNL-RSIC-38. Below is included an abstract from this list as an example:

Variance Reduction Technique	:		Sequential Sampling
Description of Technique	:		Development of estimates for parameters to be used in other variance reduction techniques by using information generated from previous sampling
Criteria for Application	:	(i)	Little or no information is available about system to be simulated
		(ii)	Other variance reduction techniques (e.g. importance, control variates etc.) are to be implemented.

Advantages : (i) Can be applied in most situations
(ii) Requires some prior knowledge
(iii) Can be used in conjunction with other techniques.

Disadvantages : (i) May require considerable initial calculation
(ii) Efficiency may be very low
(iii) In low probability problem the technique may be inapplicable
(iv) Not well developed

Comments on typical applications : Evaluation of multiple integrals, estimates of distribution parameters and some uses in lifetime studies.

All 16 variance reduction methods quoted in section (2) are examined in the same detail in ORNC-RSIC-38. This thus provides us with the basic information about the requred data. This information can than be structured to provide the expert system with the set of rules required to function.

4. THE EXPERT SYSTEM

Ferranti Computer Systems have provided the University with Inference ART (Automated Reasoning Tool) a software package for the development of Expert Systems.

In knowledge-based systems, the programmer devotes the majority of his/her efforts to developing the reservoir of facts, concepts, and rules (i.e. the knowledge) to be used, rather than organizing and proceduralizing it.

Since the capability of a system or program is directly related to the extent of its knowledge, this freedom from proceduralizing the knowlege makes development of more intelligent systems much simpler.

Knowledge-based programming permits the development of systems that would have been inadequate or impossible with traditional software engineering.

* Applications inherently limited by the sheer size of the knowledge base.
* Systems whose specifications cannot be fuly determined without operational experience and feedback.
* Systems with changing requirements and/or specifications.
* Programs that can react appropriately and intelligently to different situations. In effect, programs that tailor their own behaviour to address the problem at hand.

Conventional software engineering requires the program's function to be completely specified prior to code being written.

Where program requirements are dynamic, or formalization is difficult, conventional software engineering offers little.

The essential properties of knowledge-based systems and knowledge engineering, such as development by incremental refinement and rule-based knowledge representation, requires a special software support technology.

Conventional procedural programming languages and their satellite tools (editors, debuggers, version control packages, etc.) were not designed for knowledge engineering and are not appropriate.

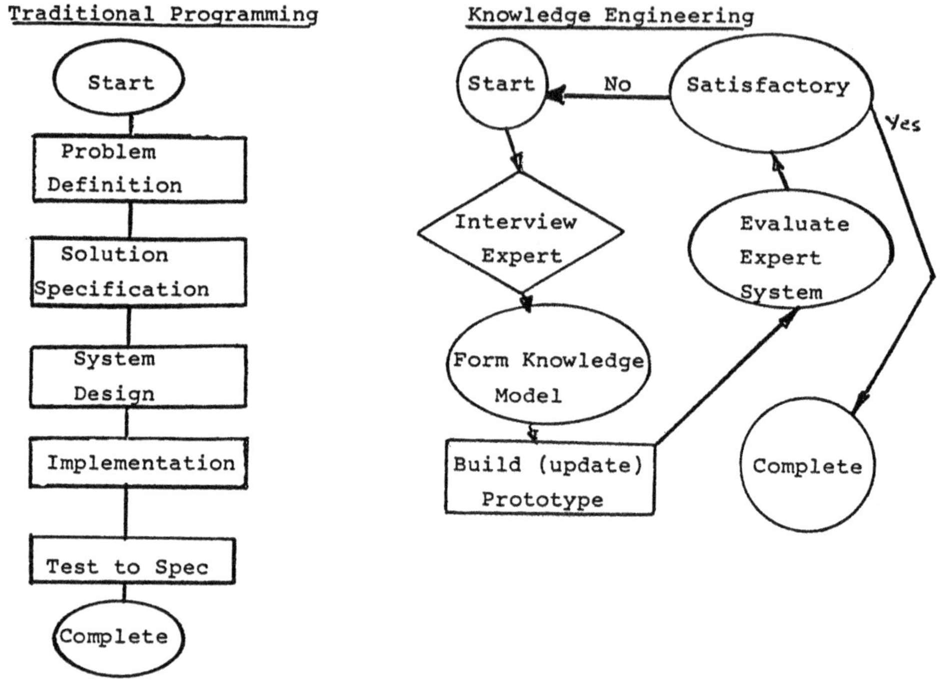

ART contains five major components:

* A Knowledge Representation Language in which rules, facts and concepts can be directly expressed.

* A Knowledge Base Computer that transforms rules, facts and concepts into automatically generated procedural code and indexed data bases.

* An Inference Engine that manages the reasoning process, i.e. the application of knowledge at runtime.

* The ART Interfaces Synthesis Tool, Artist^{TM}, which supports simple development of high quality graphic application interfaces for end-users. Artist provides an interactive graphics editor, command line and hierarchical menu specifications, mouse management, rule-based interface

animation, and automatic interface consistency management.

* The ART Studio™, which is a complete development environment for the knowledge engineer. The Studio supports rapid incremental development and detailed graphical access to the knowledge base and ongoing reasoning activities.

5. CONCLUSIONS

Monte-Carlo techniques have enormous potential in reliability assessment. However, the serious run-time limitations have caused severe problems in the past. To overcome these problems requires the use of complex, variance reduction techniques which require expert prior knowledge of the system to be simulated. The use of an expert system in this process is a logical step forward which if successful will enable the user to merely specify the problem, provide the available data and expect an answer from the optimum Monte-Carlo techniques. Be that a direct Monte-Carlo or Variance Reduction Monte-Carlo technique.

This paper describes what data will be required by the expert system in making such a decision. The decision will then be incorporated into a knowledge based expert system developed using the Ferranti ARTS software.

6. REFERENCES

1. J B Fussel
"A Formal Methodolgy for Fault Tree Construction
Nuc. Sci. and Eng. 52 (1973)

2. WASH 1400

3. MIL-STD-1629A
"Procedures for Performing a Failure Mode and Effect Analysis"
Dept. of Defence, Washington D.C.

4. W E Veseley and R E Narum
"PREP-KITT-Computer codes for Automatic Evaluation of a Fault"
IN-1349, August 1970

5. J R Taylor
"An Algorithm for Fault Tree Construction"
IEEE Transactions on Reliability, Vol. R31, No. 2, 1982

6. A Laviron, A Carnino and J C Manaranche
"ESCAF - A New and Cheap System for Complete Reliability Analyses and Computation"
IEEE Transactions on Reliability, p. 74-81, October 1982

7. J M Hammersley and D C Handscomb
"Monte-Carlo Methods"
Methuen and Co. Ltd. 1964

8. A M Featherstone and B K Daniels
"Simulation for Reliability and Availability Assessment"
4th International Conference on Reliability and Maintainability, Lannion, France 21-25 May 1984

9. N D Deans and D P Mann
"The Development of a New Hardware Reliability Simulator"
7th ARTS, Bradford, U.K. 1982

10. R E Shannon
"System Simulation, the Art and Science", Prentice-Hall, 1975

11. E J McGrath and D C Irving
"Techniques for Efficient Monte-Carlo Simulation"
ORNL-RSIC-38

12. A M Featherstone
"The Application of Monte-Carlo Computer Simulation Techniques in Reliability Assessment"
Proceedings of ASME Computers in Engineering Conference, Boston, U.S.A. August 1985

Data Analysis and Modelling in the Evaluation of Multiple System Organ Failure

P. Guadalupi [*$], R. Pizzi [*@], A. DeGaetano [*$], O. Chiara [*], C. Verna [&]

* Istituto di Chirurgia d'Urgenza - Universita' degli Studi di Milano - Ospedale Maggiore

@ Dipartimento di Scienze dell'Informazione - Universita' degli Studi di Milano

$ Maryland Institute for Emergency Medical Services Systems - Baltimore (MD)

& Istituto di Chirurgia d'Urgenza - Universita' di Torino

INTRODUCTION

At the Istituto di Chirurgia d'Urgenza of the University of Milan a computerized model with an original statistical algorithm has been developed to study the evolution of critically ill surgical patients with Multiple System Organ Failure (MSOF), defined as the insufficiency of two or more vital systems.
Aim of this paper is to show a method for evaluating the degree of disease of the organism through a representation of the human body as a system shared into nine vital subsystems. Its behaviour has been explained through a sequence of physiopathological states and has been displayed through the Markov chains formalism.
Starting from this representation, analyses and forecasts of system behaviour has been performed by processing the Markov model and its isomorphic stochastic Petri nets.

METHODS

The patient is represented as a system shared into nine independent subsystems:
- cardiovascular;
- respiratory;
- metabolic;
- immunologic;
- gastroenteric;
- neurologic;
- renal;
- coagulation;
- liver.

Each subsystem is defined by a set of variables selected because of their clinical significance. Variables deviations univocally establish a set of physiopathological states for each vital subsystem. Subystems and their variables are listed below.

ORGAN SYSTEM	DESCRIPTIVE VARIABLES
1)Gastrointestinal	motility hemorrhage serum amylase urine amylase
2)Immunology	white blood cells platelets core temperature
3)Coagulation	aPTT PT platelets FDP
4)Metabolism	oxygen consumption arterial bicarbonate Blood glucose arterio venous O_2 difference
5)Kidney	Blood urea nitrogen serum Creatinine Urine Output Urine Na^+ Urine K^+ serum K^+
6)Cardiovascular	Systolic arterial pressure Cardiac index Pulmonary wedge pressure
7)Gas exchange	Respiratory index Arterial PCO_2 Respiratory rate Inspiratory fraction of O_2 PEEP
8)Central nervous system	Glasgow coma scale
9)Liver	Bilirubin Glutammyc transpeptidase Transaminase Colinestherase aPTT PT Albumine

Criticall ill patients were admitted to our surgical ICU and monitored with Swan-Ganz catheter, arterial line, ECG, expired gases analyzer and with ventilatory parameters; besides we performed periodic off-line data recording, such as hemogas and other lab analysis.
Monitorings were recorded in our relational database devoted to ICU data storage, integration and processing (3).
Therefore data were immediately available by the procedure below described. Among these stored data, we selected the 34 variables before mentioned, chosen on the basis of our clinical experience with the purpose to quantify the grade of failure of each subsystem. In fact the choice of hemodynamic and laboratory clinical data and of their deviations have been derived from Apache II Severity Score (4) and MSOF Score System (5).

We implemented a set of production rules relating variables deviations with the physiopathological situation (6). Starting from these rules we implemented an original statistical bayesian procedure. The first step was to analyze our first 240 monitoring records (coming from 132 emergency patients), by creating a matrix W(i,j) whose elements are the a-priori probability (observed frequency) of transition from one state to another. The dispersion matrices D (i,j) of each state i, in comparison with the other ones j, were then computed for each subsystem; afterwards the program was conceived to compute the likelihoods for each transition concerning a new record of values Y , relative to the new patient whose evolution we were studying.

We employed a successive routine to obtain the probability for the patient of record Y to pass from the state established by Y to another state.

The result obtained using this model was to evidence the physiologic trajectories of the patient with the analysis of the sequence of successive records.

In fact the current probability matrix is transferred into a Markov chain (see the example below) which relates all the states by means of the probability of passing from a state to another one (7).

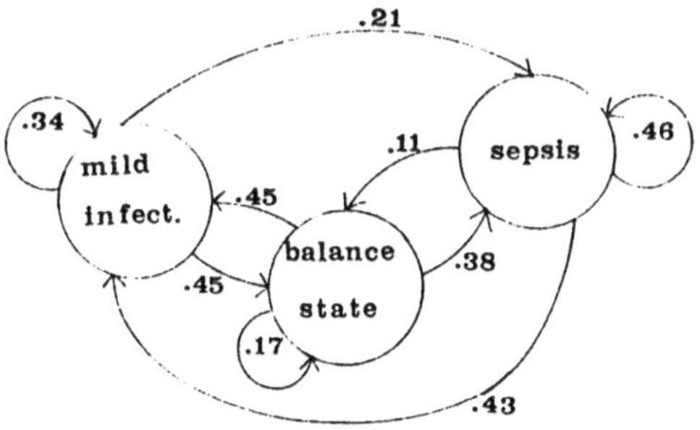

IMMUNOLOGIC SYSTEM - MARKOV CHAIN

The Markov formalism allows to obtain, starting from the array of the patient's probabilities at the time t , the most likely state for the patient at the time t+1, t+2,...t+n (i.e. after 1, 2, .., n monitoring steps).

We have realized that through a diagonal block matrix M, in which each block (one for each k=9 subsystems) represented the before mentioned probabilities for the patient of record Y to pass from the state j(t) to another state h(t+1).

If maxstate(k) is the maximum number of states for each subsystem k , the array v of the patient's probability of being in the state j at the time t can be expressed by v(t,Y). Then one can observe that

$$v(t=0) = \{0,0,\dots,1,0,\dots,0,\dots,1,0,\dots,0,0\}$$
$$= 0 \quad \forall\ j=1,\dots,p(k-1),p(k+1),\dots,maxstate(k), \quad \forall\ k$$
$$= 1 \quad \text{for } j = p(k) \quad \forall\ k$$

therefore it is possible to compute

$$v(t+1) = v(t) * M(t,Y) \qquad \forall\ t$$

It is also possible to predict the behavior of the whole system after n steps (i.e. at the time t+n) using the matrix W (i,j), which does not change varying the array v (such as Y) with the time. Following the same notation

$$v(t+n) = v(t) * W^n$$

Both in this case and in the previous one, the most likely state for the patient is that corresponding to the index at the time (t+n) of maximum element of v(t+n).
The record Y , once analyzed, is added to the knowledge base: this way the procedure is self-learning.

Actually this first approach is strictly deterministic, and passes over the relations among the subsystems. A lot of tools are today available to manage complex systems (such as the organism) in which relations among parts are the real feature of the system (8).
One of the most powerful tools of this kind representation are Petri nets (9). It can be demonstrated that any Markov chain is isomorphic to an appropriate Petri net.
So we have built the set of Petri nets isomorphic to the Markov chains concerning the nine subsystems (see the example below).

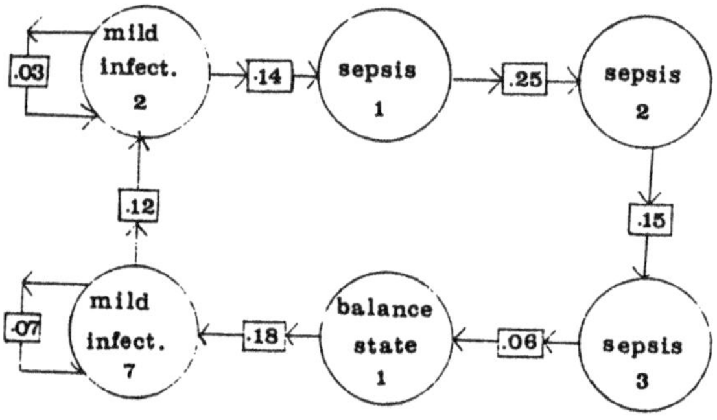

IMMUNOLOGIC SYSTEM PETRI NET

Using the Petri nets formalism we are now implementing the set of equations necessary to compute typical Petri net holistic objects, such as invariants and synchronic distances.
These objects will carry out some systemic information, such as:
- it will be possible to assess if a patient with a known physiopathological situation and a specific set of previous monitorings will follow a determined sequence of disfunctions or will reach a balanced condition.
- thereafter it will be possible to study the relationship among states suggesting possible associations and common ethiopathogenetic factor.

CONCLUSIONS

Aim of this work is to use clinical data and statistics to develop a dynamic knowledge representation which could support medical decision making. Comparing it with other A.I. tools as Expert Systems, our system allows not only knowledge extraction, but a real increase of information. This project can represent a useful tool to physicians concerned with evaluation and therapy of critically ill patient, mainly because modern monitoring devices allow an overload of information, sometimes strictly correlated, which need to be managed. Improving knowledge about the trend of a critically ill patient can therefore be useful in medical decision process.

REFERENCES

1. Carrico C.J. & coll.: MULTIPLE ORGAN FAILURE SYNDROME - INCIDENCE AND PROBLEMS. Arch.Surg. 121 (1986) 196-208.
2. Baker S.P. & coll.: THE INJURY SEVERITY SCORE: A METHOD FOR DESCRIBING PATIENTS WITH MULTIPLE INJURIES AND EVALUATING EMERGENCY CARE. J.Trauma 14 (1974) 187-196.
3. Bevilacqua G. & coll.: IL PAZIENTE DI TERAPIA INTENSIVA: DATABASE MODULARE PER LA GESTIONE E L'ELABORAZIONE DEI DATI RILEVATI. Urg.Chir.Comm. 10,2 (1987) 67-73.
4. Knaus W.A. & coll.: APACHE II: A SEVERITY OF DISEASE CLASSIFICATION SYSTEM. Crit.Care Med. 13 (1985) 818-829.
5. Goris R.J.A. & coll.: MULTIPLE ORGAN FAILURE - GENERALIZED AUTODESTRUCTIVE INFLAMMATION ? Arch.Surg. 120 (1985) 1109-1115.
6. Chiara O. & coll.: A PROSPECTIVE EVALUATION OF THE CRITICALLY ILL SURGICAL PATIENT BY MEANS OF PHYSIOLOGIC TRAJECTORIES. Acta of VIII Intern.Congr.of Emergency Surgery 1985.
7. Ajmone-Marsan M. & coll.: PERFORMANCE MODELS OF MULTIPROCESSOR SYSTEMS. Computer Systems Series, M.I.T. Press 1986.
8. Garfinkel D. : MODELLING TECHNOLOGIES AND TECHNIQUES: THE APPLICATION OF A.I. TO MODELLING PHYSIOLOGICAL SYSTEMS. Mathematics and computers in Biomedical Applications - Elsevier Science Publishers B.V. IMACS 1985.
9. Petri C.A.: INTERPRETATION OF NET THEORY . GMD Bonn, Internal Bericht, ISF-75-07, 1976.

Real-Time Operational Reliability

J. KNEZEVIC

School of Engineering Research Centre for Management of Industrial Reliability and Cost Effectiveness,
Univeristy of Exeter,
Exeter EX4 4QF, U.K.

Abstract

The classical approach to the determination of reliability characteristics is based on empirical data which represent operating time to the failure of several systems or components. Thus reliability assessment is a process which takes place after the failures have occurred. The main aim of this paper is to present a new approach which enables reliability to be assessed at real-time and reliability characteristics to be determined before failure takes place. The method used is based on the Relevant Condition Parameter reliability approach.

Introduction

The classical approach to the determination of reliability characteristics is based on the numerical values of the recorded operating times to the failure of the system/component under consideration. Hence, reliability characteristics are posterial because they can be determined only after failures have occurred.

The main aim of this paper is to show how reliability characteristics can be determined during the operation of the system in real-time, i.e. much before the failures actually occurred. The theoretical background used in this paper is the RCP approach to the determination of reliability characteristics proposed by Knezevic [1]. This approach has been successfully applied on several occasions Knezevic [2], Henshall [3], Knezevic[4], but in all cases the reliability characteristics were estimated after the tests or experiments.

The main intention here is to show how to obtain reliability characteristics during operation of the system in real-time, thus before failures have taken place.

RCP Reliability Approach

According to this approach reliability characteristics can be determined using the probability distribution of a relevant condition parameter, RCP, and its variation with time. This is a parameter which fully describes the condition of a system at every instant of operating time. For a system as a whole and its elements to be capable of functioning, their relevant condition parameter must lie between initial value, RCPin, and limiting value, RCPlim. In this case it was assumed that a failure of the system occurs as soon as the relevant condition parameter exceeds the limit. The life time of a system, T, is determined by the time when the relevant condition parameter reaches this limiting level.

As the change in condition has a completely random nature the relevant condition parameter which describes the condition of the system is a random variable whose probability density function at the instant of operating time t is denoted by f(RCP, t), as shown in figure 1.

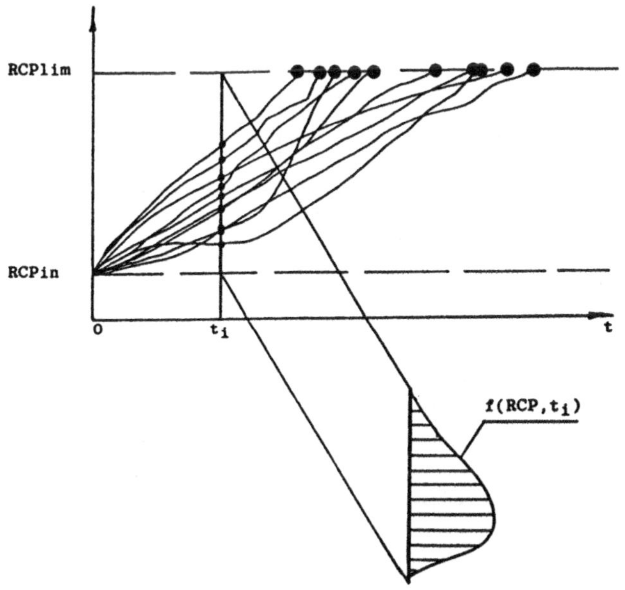

Figure 1. Change of relevant condition parameter during operation time and its probability density function at time t_i.

Real-time Operational Reliability

Applying the RCP reliability approach, at any instant of operating time the reliability of the system is equal to the probability of the relevant condition parameter being within the tolerance range expressed as:

$$R(t) = P(RCP, tk < RCPlim) = \int_{RCPin}^{RCPlim} f(RCP, tk)\, dRCP \qquad (1)$$

where f(RCP,tk) is the probability density function of the relevant condition parameter at the instant of operating time tk, as shown in figure 2.

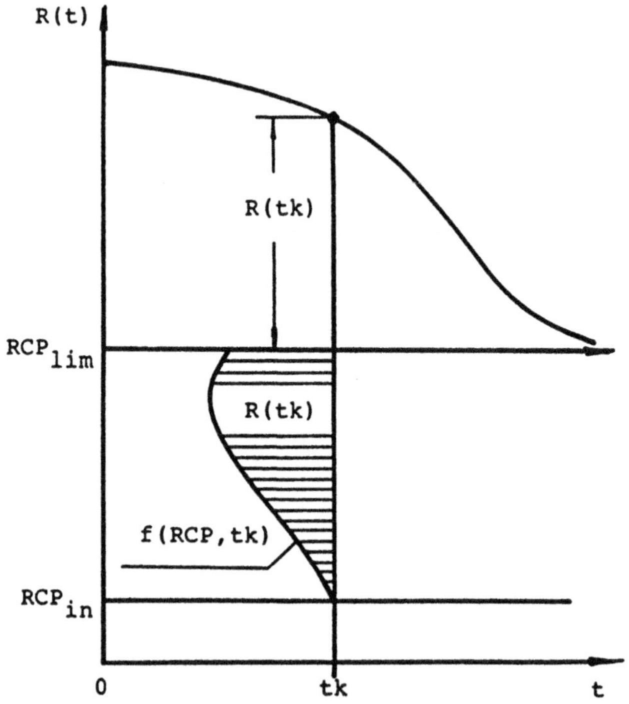

Figure 2. Relation between reliability function and probability distribution of relevant condition parameter at instant tk.

Predetermination of reliability characteristics

The approach presented in this paper enables reliability characteristics to be determined during operational life, prior to any failure.

The change of relevant condition parameter during operation is a random function of time which we shall denote as RCP(t). At every instant of operating time this function is defined by probability function and its distribution. There is an infinite number of such distributions, corresponding to the infinite number of possible instances of operating time. Hence the determination of these distributions is connected with many insoluble mathematical problems. The solution suggested by Knezevic [5] is related to the posterial determination of reliability characteristics. Whereas the solution proposed here is related to prior determination of reliability characteristics. During operation, at a very early stage, it is possible to find probability distribution of relevant condition parameter at several instances of time, f(RCP,tk), k=1,m. Knowing the probability distribution of relevant condition parameter in several instances of time, we still do not know the relation between them throughout operating time. On the other hand, by applying equation 1, it is possible to determine the probability of the relaible operation at these instances of time, as a result of which points (R(tk), k=1,m, can be obtained. If these points are plotted on appropriate probability paper the straight line fitted through them will determine the probability distribution of operating time to failure, and at the same time all necessary reliability characteristics can be determined. The graphical interpretation of the suggested approach is presented in Figure 3.

An Illustrative Example

The data used in [3] are from laboratory testing of six Double Edge Notched Tension specimens of AISI 316L stainless steel, which is commonly used in the power-generation processing industries. The length of creep crack is taken as the relevant condition parameter. The condition matrix required by the method proposed by Knezevic [5] consisted of the measured values of the length of crack measured every 10 hours during the 100 hours test. The limit value adopted was 2.5mm. The obtained reliability function is defined by the following expression.

$$R(t) = P(RCP,\ t<2500) = \exp(-t/96.6)^{42.6}) \qquad (2)$$

In order to apply the new approach data generated during the first 60 hours of the test were enough to obtain the following expression for

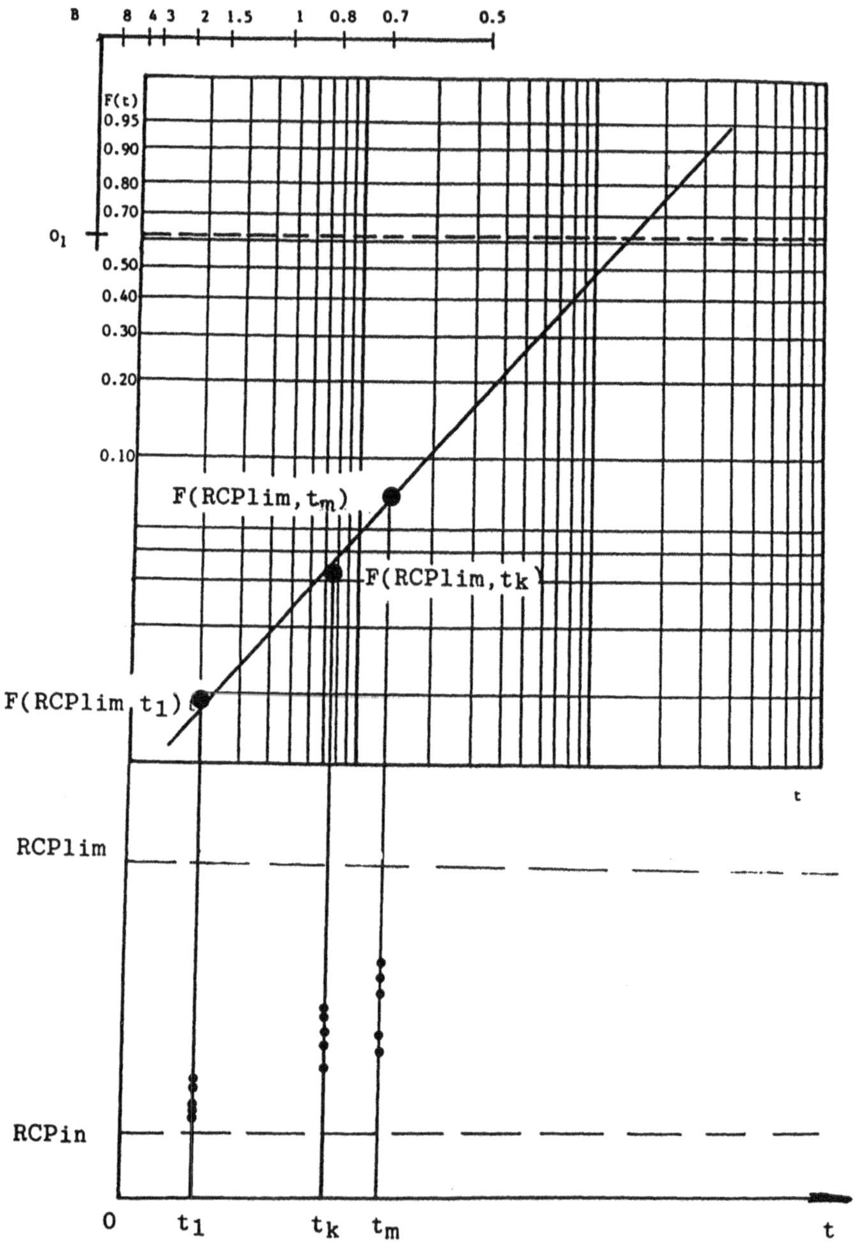

Figure 3. Graphical presentation of determination of reliability characteristics according to the approach suggested.

the reliability function:

$$R(t) = P(RCP,\ t < 2500) = \exp(-(t/98.01)^{41.5}) \qquad (3)$$

It is clear that equations 2 and 3 are very similar in spite of the fact that equation 3 was obtained 40 hours prior to equation 2, and that no failure occurred up to that instant of operation.

If the classical time-to-failure reliability approach was used the test would have lasted 190 hours which represents the time of sixth failure.

Conclusion

The advantage of this approach is the fact that reliability function can be determined at an early stage of the operation of the system, much before any failure occurs and also it can be used as a basis for decision-making related to the avoidance of the failure and the control of maintenance processes.

It is applicable to all engineering systems/components, but it is most likely that a majority of them will have a mechanical basis because they are subjected to processes with a gradual deterioration of material.

References

1. Knezevic, J., Condition Parameter Based Approach to Calculation of Reliability Characteristics, Reliability Engineering 19, pp. 29-39, Elsevier Applied Science Publishers Ltd., England, 1987.

2. Knezevic, J., Calculation of Reliability Characteristics in the Case of Corrosion Using the RCP-based Approach, Proc. 10th Advances in Reliability Technology Symposium, NCSR, University of Bradford and UMIST, Bradford, 6-8 April, 1988.

3. Henshall, L., Knezevic, J., Estimation of the Design Life for Notched Components Subjected to Creep Cracking Using a Condition Parameter Based Reliability Approach", Proc. of the Conference "Materials and Engineering Design", London, 9-13 May 1988.

4. Knezevic, J., On the Application of a Condition Parameter Based Reliability Approach to Automobile Maintenance", Proc. 4th International Conference on Automobile Reliability, Prague, June 1988.

5. Knezevic, J., Description on Continuous Change in Condition of Engineering Systems by Relevant Condition Parameter, Proc. International Symposium on "Risk Analysis in Environmental Impact Assessment", Milan November, 1987.

The Two-Sample Problem and its Practical Applications to Reliability Data Sets

G. Bohoris*
Department of Engineering Production
University of Birmingham

E. M. Aspinwall
Department of Engineering Production
University of Birmingham

D. M. Walley
British Rail

10 October 1988

Abstract

A number of non-parametric statistical tests, suitable for the solution of the two-sample problem with censored data are discussed. The tests presented are the Gehan's modification of the Wilcoxon test, the Logrank, and the modified Kolmogorov-Smirnov test. Their use in the reliability field is illustrated using British Rail data from both bogie ends and gearboxes. In each case two samples were drawn. With reference to the bogie ends, one sample was taken from one end of the car and the other sample from the other end. As to the gearboxes, a modified version was compared to the original. There was a statistically significant difference between bogie ends where as gearboxes were found to be more or less equally reliable.

Throughout this paper the results were obtained using FORTRAN routines developed in the Department of Engineering Production.

1 Introduction

A major problem often encountered in British Rail is to determine whether there is any difference in the life to failure distribution between two different build states of equipment. This occurs when assessing the effectiveness of experiments or modifications and when using Statistics to attempt to find the cause of in service problems.

Traditionally the approach to this problem has been to estimate lifetime distributions using Weibull analysis. The Weibull plots are compared visually and the parameters estimated using maximum likelihood techniques. The major disadvantage of this method is the inability to quantify the conclusions. Three years ago British Rail approached the Department of Engineering Production, University of Birmingham to study this problem. This paper summarises some of the work carried out on this project.

*Supported by the State Scholarship Foundation of Greece.

The above situation is certainly not confined to British Rail, a similar problem exists generally within the reliability field, i.e. on the basis of two samples from possibly different populations, can we conclude that the fitted lifetime distributions are the same ? The main complicating feature with regard to reliability life data as a whole is the presence of censored observations which rules out powerful, classical two - sample statistical procedures like the 't'-test. Censoring has the effect of increasing the variance of the data while decreasing the efficiency and accuracy of the analysis.

A large amount of work has been undertaken to overcome the complication implied by the presence of censored data in this area with applications confined to the medical field. A number of well known non-parametric tests have been upgraded and appeared in the literature as a modification or a generalisation of the original test, i.e. modified Wilcoxon test, modified Kolmogorov-Smirnov test, etc. All of these tests can be broadly categorised as either Rank Order statistics or Kolmogorov-Smirnov type tests.

The main difference between medical data and reliability data, however, is the percentage of censorings. In clinical trials, generally less than 50% of the samples are censored with the usual case to be around 10%. On the other hand in reliability the censorings can be as high as 95%.

2 Censored Data

Censoring occurs both in life testing and maintenance procedures as well as in medical trials and biological experiments. Life data generally contain running times of unfailed but replaced units. Similarly, clinical data contain treatment times of patients still alive at the time of the analysis. For such observations it is known that their lifetime exceeded a given value and such data are said to be censored on the right. Similarly, observations for which it is known that their lifetime preceded the first failure are said to be censored on the left.

Many types of data have censorings interspersed with the failures. This usually happens with field data where units go into service at different times and have different running times when are recorded. Such data are called progressively or multiply censored and are the most common in practice. The British Rail data used in this paper fall into this latter category.

3 A Review of the models used

Prior to discussing the tests used, it is necessary to define the notation used throughout the paper and the assumptions made to validate the procedures. The data consists of two random samples, one of size n and the other of size m. Their lifetimes will be denoted as follows :

$$\text{Sample 1}\ :\ t_{11}, t_{12}, t_{13}, \ldots, t_{1n}$$

$$\text{Sample 2}\ :\ t_{21}, t_{22}, t_{23}, \ldots, t_{2m}$$

Combining the observations of the two samples into one, results in a sample of size N=n+m and in k ordered distinct failure times written as :

$$t_1 < t_2 < t_3 < \ldots < t_k$$

Let also :

d_{1i}, d_{2i} be the number of failures in sample 1 and 2, respectively, at time t_i for $i = 1, \ldots, k$. If there are no failure ties, d_{1i} and d_{2i} are either equal to 0 or 1. When failure ties are present d_{1i} or d_{2i} is greater than 1, taking the value of the number of failures occurring at time t_i.

$d_i = d_{1i} + d_{2i}$

$d_0 = d_{10} = d_{20} = 0$, i.e. no failures at time zero.

e_{1i}, e_{2i} are the number of right censored observations in sample 1 and 2, respectively, that fall in the interval $[t_i, t_{i+1})$, for $i = 1, \ldots, k$.

$t_0 = 0$ and $t_{k+1} = +\infty$

$e_i = e_{1i} + e_{2i}$

n_{1i}, n_{2i} are used to denote the numbers of observations in sample 1 and 2, respectively, with survival times greater than or equal to t_i, for $i = 1, \ldots, k$.

$n_i = n_{1i} + n_{2i}$

In order to validate the procedures used, it must be assumed that

(a). Both samples are selected at random from their respective populations.

(b). Both samples consist of continuous, identically distributed random variables.

(c). There is independence within each sample (a result of randomness) and mutual independence between the two samples.

3.1 Rank Order Statistics

Several generalisations of the Wilcoxon statistic (Wilcoxon 1945) for the solution of the two-sample problem with censored data have been proposed. Among them the Gehan's test, the Peto and Peto's test, the Prentice's test all of which along with the Logrank test are discussed more fully in Bohoris and Aspinwall (1989).

In this paper only the Gehan's generalisation of the Wilcoxon test and the Logrank test are used. Both test can be expressed by the general linear rank statistics for censored data formula which is a weighted summation over the second sample of the observed number of failures d_{2i} minus the conditionally expected number of failures $n_{2i}d_i/n_i$:

$$W = \sum_{i=1}^{k} w_i \left(d_{2i} - \frac{n_{2i}d_i}{n_i} \right) \tag{1}$$

where the weight function w_i takes the value n_i in Gehan's modification of the Wilcoxon test and 1 in the Logrank test.

Varying the value of the weight function has the effect of increasing the sensitivity to survival differences occurring at specific points. For example, Gehan's test displays greater sensitivity to differences occurring earlier in the life distributions where as the Logrank test shows greater sensitivity to differences occurring late.

3.2 Kolmogorov-Smirnov Test

The classical Kolmogorov-Smirnov (KS) test statistic for complete data is based on the difference :

$$D = max \left[\, |\hat{F}_1(t_i) - \hat{F}_2(t_i)| \, \right] \tag{2}$$

for $i = 1, \ldots, k$, where $\hat{F}_j(t_i)$ is the cumulative distribution function of the jth sample.

Many authors including Barr and Davidson (Barr et al, 1973), Dufour and Maag (Dufour et al, 1978) and Chandra, Singpurwalla and Stephens (Chandra et al, 1981) have modified the critical values for the classical Kolmogorov-Smirnov statistic (Smirnov, 1939) to incorporate the presence of censorings in the two samples.

Fleming, O'Fallon and O'Brien (Fleming et al, 1980) modified the KS statistic for complete data and proved the increased efficiency of the new test using Monte-Carlo simulations. The new formulation was based on using the cumulative hazard estimator of the reliability function (Nelson, 1969) and the statistic is calculated through the difference :

$$d = max \left(\frac{\hat{R}_1(t_i) + \hat{R}_2(t_i)}{2} \left[\, |\hat{H}_1(t_i) - \hat{H}_2(t_i)| \, \right] \right) \quad (3)$$

for $i = 1, \ldots, k$, where $\hat{H}_j(t_i)$ and $\hat{R}_j(t_i)$ are respectively the cumulative hazard function and the reliability function of the i'th failure from the j'th population.

Having used the cumulative hazard estimator of the reliability function, a technique suitable when data are multiply censored, the KS test was further upgraded to take account of censored samples. Fleming et al (1980) laid down the formulation of this technique and provided a rather complex procedure for calculating the critical values.

The modified KS test is sensitive to any particular difference evident at least one point in time. It was developed to cope with situations described as 'crossing hazards' where the reliability distributions of the two samples cross (Figure 6, gearboxes). Under such circumstances the differences before and after the cross might cancel each other resulting in small statistics when using Linear Rank tests.

4 Bogie Ends

When undergoing a standard major overhaul after many years in service, cracks were found on the bogie frame. The part of the frame cracking was that which supported the gearbox mounting bracket.

Bogies are generally fitted to No. 1 End or No. 2 End of a power car and are slightly different to each other. A detailed data collection exercise was undertaken to assess whether there was any difference in the survivability of bogies fitted to the two ends using the above two-sample techniques.

The data was analysed so that events were generally grouped into failures or censorings, see Figure 1 and Table 1.

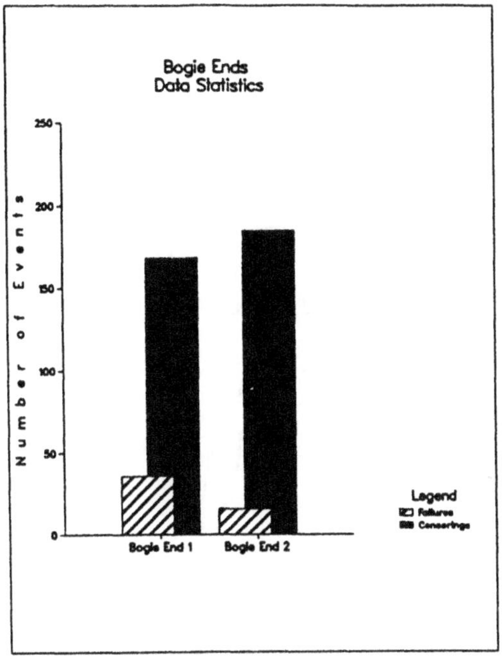

Figure 1: Bogie Ends - Data Statistics

	Failures	Censorings	Sample Size
Bogie End 1	36	168	204
Bogie End 2	16	185	201
Combined Sample	52	353	405

Table 1: Bogie Ends - Data Statistics

The survival analysis of the data was based on the following two assumptions :

(a). Bogies themselves were treated as non-repairable components (data included only times to first failure or replacement).

(b). Bogies' event times were assumed to be independent and identically distributed.

4.1 Survival Analysis

The data for the bogies fitted to No. 1 and No. 2 End were separated into two distinct data sets. Cumulative hazard estimates of the reliability function were calculated for each group.

Plots of the cumulative distribution functions and reliability functions (Figure 2) were superimposed. Visually the two curves indicated that the bogies fitted to No. 2 End were more reliable than the ones fitted to No. 1 End throughout their life range.

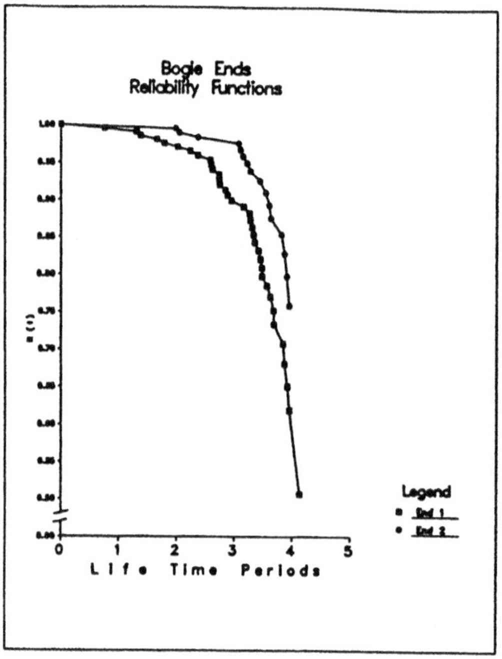

Figure 2: Bogie Ends - Reliability Functions

Statistical comparisons of the reliability functions of the two data sets were carried out using the Gehan's modification of the Wilcoxon test, the Logrank test and the modified Kolmogorov-Smirnov test. The results shown in Table 2 give the values of the test statistics and the corresponding probability (or P) values.

The probability or P-value associated with a test is the probability that the observed value, or greater, of the test statistic is obtained when the null hypothesis is true. It can also be described as the tail-end probability H_0, of the distribution of the statistic. It is used instead of the more classical text book approach of selecting a given significance value. When the P-value is small (typically less than 0.05) the null hypothesis that the two populations are the same, is rejected.

Test	Test Statistic	P - Value
Gehan-Wilcoxon	2.88310	0.0039
Logrank	2.76923	0.0056
Kolmogorov-Smirnov	1.14510	0.0545

Table 2: Bogie Ends

The Linear Rank methods gave high test statistics and therefore the null hypothesis, that the two samples came from identical populations, was rejected at least at the 5% level of significance. The modified KS test resulted in a relatively high P-value failing to reject the null hypothesis.

4.2 Conclusions

(a). Bogies fitted to No. 2 end are more reliable than the bogies fitted to No. 1 end.

(b). The two-sample tests validated our initial conclusions from the graph by showing that :

- (i). There was a significant difference in the Reliability of the two samples.
- (ii). The difference in the Reliability functions was more obvious in the early part of the bogie ends lifetimes. This is apparent since the Gehan's test which is more sensitive in survival differences occurring early in time, gave a slightly higher test statistic than for the Logrank test.
- (iii). The modified KS test, being conservative with heavily censored samples (82% and 92% respectively) containing a rather small number of failures, gave a higher P-value when compared with the other two.

5 Gearboxes

Problems have occurred on the gearbox transmitting the drive between the bogie mounted traction motor and the wheelset of one of British Rail's most intensively used passenger locomotives. The problems were divided between bearing failures and oil seal failures.

In an attempt to improve the reliability of these gearboxes (Type I gearbox) British Rail undertook a modification to both the bearing and the seals, costing approximately £500 per gearbox.

Unfortunately within 6 months of the first of the modified gearboxes (Type II gearbox) being fitted they started to fail.

A series of investigations was undertaken using Weibull analysis and it was found that population estimates of the lifetime distributions crossed over after 1.5 years with the modified gearboxes having a higher incidence of failure initially. Unfortunately due to other problems on the wheelsets, disc failures and tread damage, the average life of the wheelsets was approximately 1.25 years.

The graphical comparisons of the lifetime distributions indicated that the modification showed an overall improvement, but if the wheelsets lasted less than 1.25 years then it was not cost effective. Considerable difficulties arose in convincing the engineers that the modification was not cost effective since it was not possible to quantify the difference between the two distributions.

The data were, therefore, further studied using the above mentioned techniques for the solution of the two-sample problem with censored data.

5.1 Assumptions

Gearboxes themselves were treated as repairable systems. The survival analysis of the data was possible only if the data were independent and identically distributed (IID). Techniques used to justify the identically distributed (ID) assumption were :

(a). whether the Rate of Occurrence of Failures (ROCOF) was constant by plotting Cumulative Failures versus Cumulative usage.

(b). the Laplace Test for trend.

Initially, the Laplace test for trend for the Type I and Type II gearboxes gave test statistics of 0.67781 and 1.3288 and corresponding P-values of 0.24895 and 0.09196 respectively which were not significant at the 5% level.

Examining the data, it was found that there was a number of failure ties (equal mileage to failure) causing the relatively high value of the Laplace test statistic for the Type II gearbox. The ties occurred as a result of failures on the same date and on the same vehicle but on gearboxes fitted to different axles. Given that information was available for 197 vehicles, each having 4 gearboxes, fitted on all 4 axles, and that initially there were 85 and 35 failures for the

Type I and Type II gearboxes respectively, in a time span of two years, the simultaneous failures were considered as a not very likely alternative.

Looking at the way that data was recorded, it was found that the gearboxes were thoroughly inspected every 55 days. During inspection, gearboxes that did not meet certain criteria were reported to have failed on the inspection date. During this same period it was likely that a number of gearboxes might have failed but the possibility of more than one failure happening simultaneously was thought to be not very large.

It was assumed that if there was a way of detecting and recording a failure when it was actually happening, and simultaneously inspecting the condition of the remaining three gearboxes in the same vehicle, then possibly a preventive action, resulting in a censoring, would have been taken for gearboxes that were close to failure. If the assumption was true, then within any interval between two inspections, one or more mutually exclusive failures and a number of censorings could occur.

The above mentioned failure ties were removed for both gearboxes by considering as failures only one out of the many recorded on the same vehicle on the same date the rest being considered as censorings, making the assumption that failures did occur in different calendar times in between the inspection interval. Table 3 and Figure 3 show the resulting data for the two gearbox types.

	Failures	Censorings	Sample Size
Type I Gearbox	80	671	751
Type II Gearbox	32	381	413
Combined Sample	112	1052	1164

Table 3: Gearboxes, Full Data Set - Data Statistics

ROCOF plots were drawn (Figures 4 and 5) and Laplace tests for trend carried out again, this time giving test statistics of 0.5802 and 0.9735 and corresponding P-values of 0.28089 and 0.16515 respectively, which were not significant at the 5% level.

5.2 Survival Analysis

Having justified the identically distributed assumption, data was assumed to be independent. Cumulative hazard estimates of the reliability function were calculated for each group.

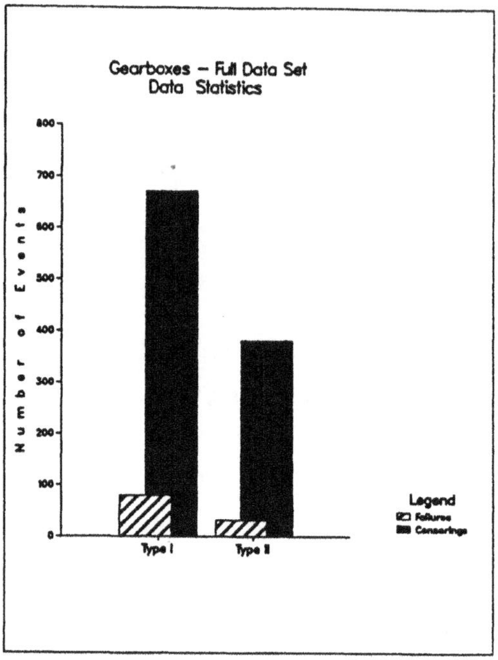

Figure 3: Gearboxes, Full Data Set - Data Statistics

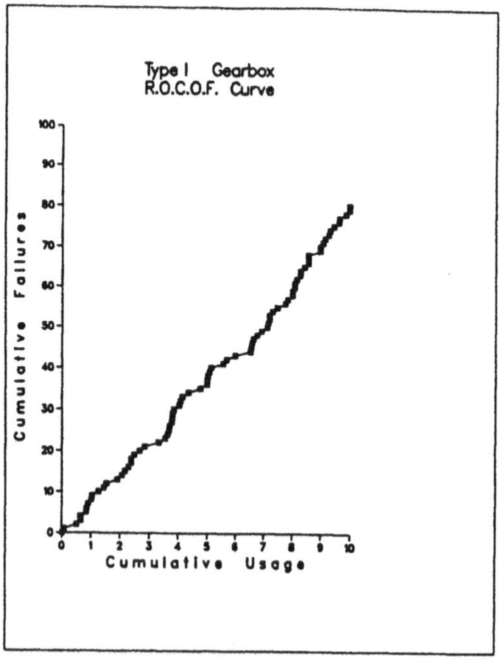

Figure 4: Type I Gearbox - R.O.C.O.F. Curve

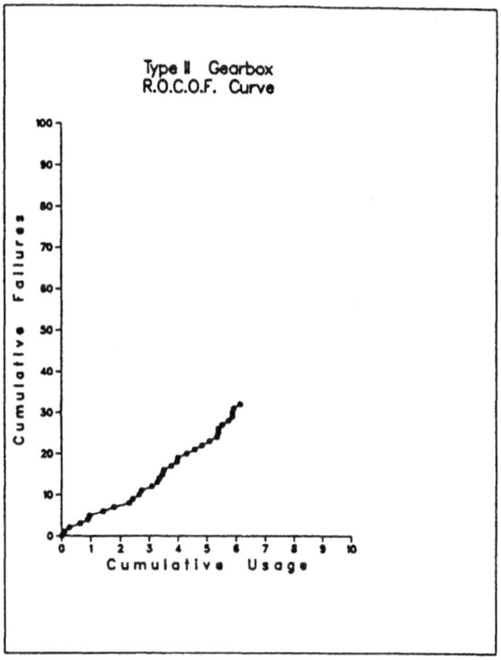

Figure 5: Type II Gearbox - R.O.C.O.F. Curve

Plots of the cumulative distribution functions and reliability functions (Figure 6) were superimposed. From the two graphs it was apparent that both gearboxes were more or less equally reliable, with the Type I gearbox being slightly more reliable in the first 1.5 years.

Statistical comparisons of the reliability functions of the two data sets were carried out using the same tests as for bogie ends (see Table 4).

Test	Test Statistic	P - Value
Gehan-Wilcoxon	0.55259	0.5805
Logrank	0.37791	0.7055
Kolmogorov-Smirnov	0.48900	0.4878

Table 4: Gearboxes, Full Data Set

All methods gave very high P-values and therefore the null hypothesis, that the two samples came from identical populations, was not rejected at the 5% level of significance.

The data was truncated using 1.5 years as the upper limit (i.e. all events with survival times

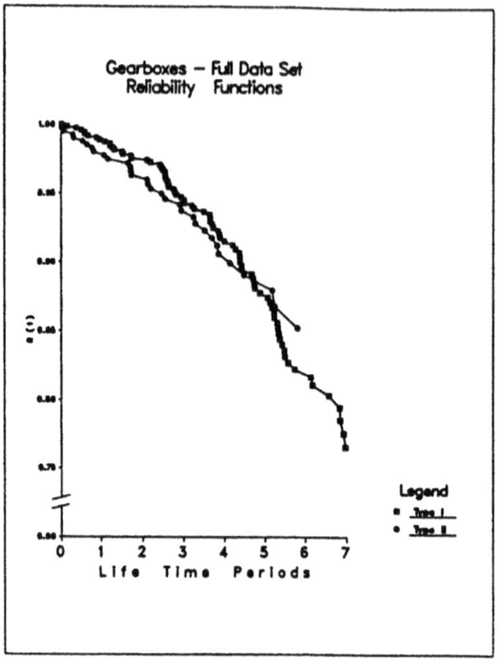

Figure 6: Gearboxes, Full Data Set - Reliability Functions

greater than 1.5 years were assumed to be censored, see Table 5). This particular value was chosen as the cut off point, because gearboxes are usually replaced after that amount of time.

	Failures	Censorings	Sample Size
Type I Gearbox	59	692	751
Type II Gearbox	29	384	413
Combined Sample	88	1076	1164

Table 5: Gearboxes, Truncated Data Set - Data Statistics

Estimates of the reliability function were calculated and plots were superimposed (Figure 7). All the two-sample tests (Table 6) resulted once more in high P-values not rejecting the null hypothesis at the 5% level of significance.

5.3 Conclusions

Based on the full data set the following inferences were made :

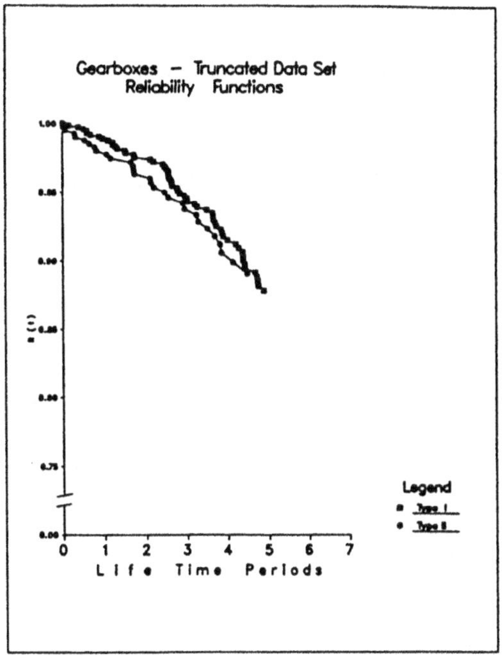

Figure 7: Gearboxes, Truncated Data Set - Reliability Functions

Test	Test Statistic	P - Value
Gehan-Wilcoxon	0.71369	0.4754
Logrank	0.23301	0.8158
Kolmogorov-Smirnov	0.04780	0.7728

Table 6: Gearboxes, Truncated Data Set

(a). The modified gearboxes are no more reliable than the original ones.

(b). All two-sample tests validated our initial conclusions from the graph by showing that :

(i). There was not a significant difference in the reliability of the two gearbox types.

(ii). Any slight difference in the reliability functions was more apparent in the early part of the gearboxes lifetimes. The Gehan's test which is more sensitive in survival differences occurring early in time, gave a higher test statistic than the Logrank test which is more sensitive to survival differences occurring late in time.

(iii). The modified KS test, being more efficient than the Linear Rank tests in the 'crossing hazards' situation encountered with the gearboxes, resulted in the smallest P-value.

With regard to the truncated data set, the difference in the reliability functions (the 'crossing hazards' situation being no longer applicable) existed only in the early part of the gearboxes lifetimes. The Gehan's test therefore, gave a lower P-value than the one obtained using the full data set. Similarly, the Logrank and the modified KS test gave higher P-values than the corresponding ones when the full data set was used.

6 Discussion

A number of two-sample statistical tests capable of dealing with multiply censored data have been discussed, using both non-repairable components and repairable systems. The tests enabled a formal hypothesis testing exercise to be undertaken with the end result being to reject or fail to reject the null hypothesis depending on the significance probability.

A very important part of the analysis was the superimposition of the reliability functions of the two samples. This graphical view of the differences helps to interpret the results of the statistical tests.

When dealing with repairable systems the independent and identically distributed assumption must always be validated. This is due to the fact that :

(a). the two-sample tests are based on the assumption that groups to be compared are independent samples from their respective populations and

(b). observations relating to a repairable system should be independent and identically distributed in order to estimation and plotting of any characteristic distributional function (i.e. reliability functions) related to those observations be possible.

The three tests described, allotted weightings to the two samples in different ways giving rise to large differences in the test statistics and therefore in the P-values as was the case with the gearboxes and with a number of other British Rail data sets analysed up to now. This exhibits the need for more information on the behaviour of the two-sample tests under a variety of circumstances, with data that are heavily censored and described by distributional forms and censoring mechanisms often met in the Reliability field. A Monte-Carlo simulation is currently being undertaken in the Department of Engineering Production, to assess the performance of the three techniques examined in this paper along with others, viz. Peto and Peto, Prentice and a further modification of the Wilcoxon test.

7 References

1. **ARMITAGE P.** (1959) The Comparison of Survival Curves, *Journal of the Royal Statistical Society*, Series A, Vol 122 Part III, pp 279 - 300.
2. **ASCHER H., FEINGOLD H.** (1984) *Repairable Systems Reliability*, Marcel Dekker, INC., New York.
3. **BARR D.R., DAVIDSON T.G.** (1973) A Kolmogorov - Smirnov test for censored samples, *Technometrics*, Vol 15 No 4, pp 739 - 757.
4. **BOHORIS G., ASPINWALL E.M.** (1989) Statistical comparison of Reliability data sets, *Reliability 1989, Conference Proceedings, Brighton, England, 14 - 16 June 1989*, pp 2B/3/1 - 2B/3/15.
5. **BRESLOW N.** (1970) A Generalized Kruskal - Wallis Test for Comparing K Samples subject to Unequal Patterns of Censorship, *Biometrika*, Vol 57 No 3, pp 579 - 594.
6. **BRESLOW N.** (1984) Comparison of Survival Curves, *Cancer Clinical Trials* by M.E. BRUCE, M.J. STAQUET and R.J. SYLVESTER, Chapter 22, pp 381 - 406, Oxford University Press, Oxford.
7. **CHANDRA M., SINGPURWALLA N.D., STEPHENS M.A.** (1981) Kolmogorov Statistics for tests of fit for the Extreme - Value and Weibull distributions, *Journal of the American Statistical Association*, Vol 76 No 375, pp 729 - 731.
8. **COX D.R., LEWIS P.A.W.** (1966) *The Statistical Analysis of Series of Events*, Metheun, London.
9. **DUFOUR R., MAAG U.R.** (1978) Distribution results for modified Kolmogorov - Smirnov statistics for truncated or censored samples, *Technometrics*, Vol 20 No 1, pp 29 - 32.
10. **EFRON B.** (1967) The Two Sample Problem with Censored Data, *Proceedings of the 5th Berkley Symposium on Mathematical Statistics and Probability* by J. Newman, Vol 4, pp 831 - 853, University of California Press, Berkeley.
11. **ELANDT - JOHNSON R.C., JOHNSON N.L.** (1980) *Survival Models and Data Analysis*, John Wiley and Sons, New York.

12. **FLEMING T.R., O'FALLON J.R., O'BRIEN C., HARRINGTON D.P.** (1980) Modified Kolmogorov - Smirnov Test Procedures with Application to Arbitrarily Right - Censored Data, *Biometrics*, Vol 36, pp 607 - 625.

13. **GEHAN E.A.** (1965a) A Generalized Wilcoxon Test for Comparing Singly - Censored Samples, *Biometrika*, Vol 52, pp 203 - 223.

14. **GEHAN E.A.** (1965b) A Generalized Two - Sample Wilcoxon Test for Doubly Censored Data, *Biometrika*, Vol 52, pp 650 - 653.

15. **GEHAN E.A.** (1969) Estimating Survival functions from the Life Table, *Journal of Chronic Diseases*, Vol 21, pp 629 - 644.

16. **GEHAN E.A., THOMAS D.G.** (1969) The Performance of some Two - Sample Tests in Small Groups With and Without Censoring, *Biometrika*, Vol 56, pp 127 - 132.

17. **GROSS A.J., CLARK V.A.** (1982) *Survival Distributions : Reliability Applications in the Biomedical Sciences*, John Wiley and Sons, New York.

18. **KAPLAN E.L., MEIER P.** (1958) Nonparametric Estimation from Incomplete observations, *Journal of the American Statistical Association*, Vol 53, pp 457 - 481.

19. **LATTA R.B.** (1981) A Monte Carlo Study of Some Two - Sample Rank Tests with Censored Data, *Journal of the American Statistical Association*, Theory and Methods Section, Vol 76 No 375, pp 713 - 719.

20. **LEE E.T.** (1980) *Statistical Methods for Survival Data Analysis*, Lifetime Learning Publications, Belmont, California.

21. **NELSON W.** (1969) Hazard Plotting for Incomplete Failure Data, *Journal of Quality Technology*, Vol 1 No 1, pp 27 - 52.

22. **NEWTON D.W., WALLEY D.M.** (1983) Reliability Analysis in Railway Engineering, *Proceedings of the 4th EUREDATA Conference*, Venice, Italy, March 23 - 25th 1983, Section 8, pp 1 - 25.

23. **PETO R., PETO J.** (1972) Asymptotically Efficient Rank Invariant Test Procedures, *Journal of the Royal Statistical Society*, Series A, Vol A135 Part 2, pp 185 - 207.

24. **SMIRNOV N.V.** (1939) Estimate of deviation between empirical distribution functions in two independent samples, (Russian) *Bulletin Moscow University*, 2:2, 3-16 (6.1, 6.2).

25. **TARONE R.E., WARE J.** (1977) On Distribution - free Tests for Equality of Survival Distributions, *Biometrika*, Vol 64 No 1, pp 156 - 160.

26. **WILCOXON F.** (1945) Individual Comparisons by Ranking Methods, *Biometrics*, Vol 1, pp 81 - 83.

Implementing Stochastic Filtering for Reliability Applications – The Arc Code

A. BESI*, C.A. CLAROTTI**, C.P. NICHELE*, L. PIEPSZOWNIK***

* CEC-JRC Ispra, Italy

** ENEA-TIB AQ, Casaccia Roma, Italy

*** Electricité de France, visiting scientist at the JRC Ispra

to be presented at the 6th EUREDATA Conference on Reliability Data Collection and Use in Risk and Availability Assessment, Siena, March 15-17, 1989.

Summary

A model for using field data for component reliability assessment is described. Field data (i.e. component histories consisting of maintenance operation descriptions, times to failure, survivals) are treated by stochastic filtering techniques, to "filter" the maintenance effect. The main aspects of the model, based on a Bayesian approach, and of its practical implementation in the ARC (Age Réel du Composant) are presented. Finally the results of an example of code validation are given.

1. Using field data contaminated by maintenance for component reliability assessment

In case of "reliable" components, such as some mechanical or electro-mechanical pieces of equipment important for plant availability and/or safety, life testing is practically impossible because of cost and time limitations. The more reliable a component is, the more expensive it is, the longer one must wait for observing a failure, the larger the sample to be put on test must be in order to observe a significant number of failures. As a consequence the duration and the cost of a life-testing-campaign would be prohibitive. For the purpose of inference on life lengths of "reliable" components one must then rely on field data, i.e. on data recorded during the components' operation in the plants where the components are installed. As a rule, a component installed in a plant is periodically maintained, whence field-data are contaminated due to maintenance, that is, loosely speaking, observed lives are longer than they would have been, had the maintenance-operations not been carried out. Because of the potential plant-to-plant-variability of the maintenance-policy, one cannot act as if the components were non-maintained. Doing this would be allowable if and only if the maintenance policy were the same all over the plants, so that maintenance could be considered simply as one of the operating conditions under which the reliability of the components is assessed.

Since pooling data from different plants is crucial to obtaining significant enough information on the components of interest, observed data must be cleaned up from the effect of maintenance. A method for doing this has been developed and a code named ARC implemented; for the method to become suitable for implementation in reliability-data-bank-practice a validation of the ARC, still in a preliminary version, is needed. This is being carried out in the frame of a cooperation involving EDF, ENEA and Ispra Establishment.

2. Definition of our goal

2.1 We have a file of maintained component hystories (our field data). For these components, assumed to have a continuos operating mode and to have been observed since the beginning of their operational life, we have maintenance operation descriptions, times to failure, survivals. These components are considered exchangeable and, given what we have observed, we want to assess the reliability of another component of the same family net of the effect of the maintenance. Suppose the component of interest fails according to a distribution of hazard rate $\lambda(s,\theta)$, with θ unknown (and may be vector) parameter. Following a Bayesian approach, let $\pi(\theta|H)$ be our posterior on θ, given the observation H, to assess the reliability of the component, we have to calculate the conditional expectation (conditional on H):

$$R(t) = \int \exp\left(- \int_0^t \lambda(s,\theta)ds\right) \pi(\theta|D)d\theta$$

where t is the mission time of the component in case of no maintenance.

In other words, reliability assessment in a Bayesian frame means: derive $\pi_o(\theta|H)$ and assess the component reliability according to the given formula /1/.

2.2 The implemented version of the code gives as output the component reliability (or its complement to one, the failure probability), as a function of the operating time t. The component is supposed to be new at the beginning of the mission and no maintenance is foreseen. The capability to compute the reliability versus time for a component which has already an history is an improvement planned for the near future. This improvement, easy to be realized from the model point of view, makes the code able to better meet the needs of plant operators interested in maintenance optimization.

3. The theoretical model

3.1 Aging and maintenance

3.1.1 If maintenance actions are foreseen for a component, it is implicitly assumed that the component is not exponential. An exponential component does not age; consequently, if it is "up" at time t, it is "as good as new" and there is no point in trying to make it better via a maintenance action. In view of the above, having decided to maintain a component tantamounts to assuming that it ages, i.e., in mathematical terms, the component-failure-time distribution is Increasing Failure Rate (IFR). The hazard rate $\lambda(s,\theta)$ is thus assumed to be increasing in time.

3.1.2 Maintenance is needed to compensate aging. A model of the maintenance effect is presented in /2/. One can think of a maintenance operation as "an attempt to reduce the age τ of the component". Modelling the effect of maintenance in terms of an age-reduction appear to be sensible.
A constant deterministic age reduction at each maintenance operation is not credible. Rather a maintenance operation can be regarded as an operation carried out with the aim, at most, of making the component "as good as new" (in the case of a complete overhaul of the component). In practice the achieved result is a random-age-reduction which ranges, according to a probability distribution, between 0 (no effect of maintenance) and τ (the component was τ time units old before maintenance, it is as good as new after maintenance).
Another important feature to be embedded in the model is the dependence of

maintenance-effectiveness on the age of the component which undergoes maintenance. There is no doubt that renewing a very "young" component is a much easier task than renewing a component which is almost at the end of its life. From a mathematical standpoint one can account for this dependence by considering the age reduction w and the component's age τ to be two dependent random variables so that each maintenance operation is characterized by the conditional probability density function:

$$g(w|\tau)$$

$g(w|\tau)\Delta w$ is the probability of reducing by an amount w the age of the component given that the component's age is τ.

3.2 Why Stochastic Filtering in Reliability?

The story of stochastic filtering in reliability began when field data came into the reliability business /2/.

Since field data are contaminated by maintenance, we must allow for the effect of maintenance /2/ when deriving our posterior on the unknown quantity of interest (typically the possibly vector parameter of a life distribution).

As a matter of fact, obtaining a posterior net of the effect of maintenance is possible by making use of just Bayes statistics under the following assumptions /3/, which appear reasonable on the basis of the previous discussions:

i. Components fail according to an IFR distribution (otherwise maintenance is point less);
ii. The j-th maintenance operation causes the component age to be reduced by a random quantity w_i distributed according the pdf $g(w|\tau)$.

Because of ii., component age becomes different from the calendar time, and then unknown to us, after the first maintenance operation. Inference is then a matter of deriving a pdf on the couple (θ,τ), rather on just θ.

Under i. and ii. Bayes theorem /3/ suffices for obtaining:

$$\pi_s(\theta,\tau|s < T_f) \propto \exp\left(-\int_0^{s-T_j} \lambda(\tau - s + T_j + x;\theta)\, dx\right) \cdot$$

$$\cdot\, \pi_{s'}(\theta,\tau - s + T_j \mid s' = T_j^+ < T_f),\ T_j < s < T_{j+1}$$

$$\pi_s(\theta,\tau| \, s = T_j^+ < T_f) = \int_\tau^{T_j} \pi_{s'}(\theta,r|s' = T_j^- < T_f)\, g(r - \tau\, |r)dr$$

$$\pi_s(\theta,\tau| \, s = T_f) \propto \lambda(\tau,\theta)\, \pi_{s'}(\theta,\tau|s'=T_f^-) \tag{1}$$

where:

$\pi_s(\cdot|\cdot)$ is our conditional pdf on θ and τ at time s, given what we observed up to s.

$\lambda(\cdot)$ is the hazard rate of the component life distribution.

T_f is component time to failure (a random variable).

T_j is time of j-th maintenance operation.

$g(\cdot|\tau)$ is the conditional pdf of the age reduction given that component age is τ.

The apices - and + mean "immediately earlier" and "immediately later" respectively.

The above shows that, contamination due to maintenance was just the opportunity for reliability and stochastic filtering to make friends. For the purpose of inference, stochastic filtering is not a compulsory choice (Bayes statistics being sufficient). On the contrary stochastic filtering is a compulsory choice (and this may be surprising) for the purpose of the numerical accuracy of inference /4/.

Two circumstances are crucial to the centrality of stochastic filtering as numerical tool:

* Once that our posterior $\pi(\theta|H)$ on θ has been derived (conditioning on H stands for conditioning on the observed history; in case of maintenance $\pi(\theta|H)$ is the marginal of $\pi(\theta,\tau|H)$ with respect to τ) a Bayesian is forbidden /1,5/ to estimate θ. Rather, if u is the safety goal of concern and $R(t_M|\theta)$ is component reliability at mission time given the value of

θ, the correct Bayesian use of $\pi(\theta|H)$ is to draw a decision as to the safety goal achievement via the comparison:

$$\int R(t_M|\theta)\ \pi(\theta|H)d\theta \overset{?}{>} u \tag{2}$$

** After data processing what is available to us is not our "true" posterior $\pi(\theta|H)$ but rather $\pi^{(n)}(\theta|H)$, a numerical approximation of its.

Of course

$$\pi^{(n)}(\theta|H) \rightarrow \pi(\theta|H)$$

but this unfortunately does not imply:

$$\int R(t_M|\theta)\ \pi^{(n)}(\theta|H)d\theta \rightarrow \int R(t_M|\theta)\,\pi\,(\theta|H)d\theta$$

which is what would be of interest to us, since we have to decide according to eq. (2). (Commutation between a limit and a linear operator always gives trouble in mathematics).

The very value of stochastic filtering lies in that it permits us to define a suitable metric space on the set of all pdf's whose domain is a discretization of (θ,τ), the metric of the space being such that a fine enough discretization of (θ,τ) entails:

$$\int R(t|\theta)d\theta \int q^{(r)}(\theta,\tau)d\tau \rightarrow \int R(t|\theta)d\theta \int \pi(\theta,\tau|H)d\tau$$

Details on how this can be done are in /4/; here only the sequential procedure for obtaining $q^{(r)}$ is reproduced.

- Components must be processed in sequence one after the other, the posterior we derived at the end of each component record being the prior for the next component in the file.

- Choose a discretization step size h for θ, τ and the calendar time, i.e. possible values for these quantities will be:

$$\theta = \theta_{min} + j_1 h$$
$$\tau = \tau_{min} + j_2 h$$
$$t = j_3 h$$

j_1, j_2, j_3 integer numbers.

- Start inference by setting:

$$q_o^{(n)}(\theta,\tau) = \begin{cases} \pi_o(\theta), \text{ if } \tau=0 \\ 0, \text{ otherwise} \end{cases}$$

where $\pi_o(\theta)$ is our prior on θ (components are assumed to be new at the beginning of the observation period).

- Latest news (i.e. what we observe in the time intervals) $(j_3h, (j_3+1)h)$ causes $q^{(n)}$ to be updated according to:

$$q^{(n)}_{j_3+1} = q^{(n)}_{j_3} \, E^{(n)}_{j_3+1} \, \pi^{(n)}_{j_3,j_3+1} \tag{3}$$

where:

$E^{(n)}_{j_{3+1}}$ is the following diagonal matrix:

$$E^{(n)}_{j_{3+1}} = \begin{matrix} \text{diag}\left\{\exp\,[(1-\lambda(\tau,\theta))h]\right\} & (j_3+1)h < T_f \\ \text{diag}\left\{\lambda(\tau,\theta)\,\exp\,[\lambda(1-(\tau,\theta))h]\right\} & (j_3+1)h = T_f \end{matrix}$$

$\pi^{(n)}_{j_3,j_3+1}$ is a time-shift matrix for all time-instants but T_j's (maintenance-operation-epochs), that is to say:

$$q^{(n)}_{j_3+1}(\theta,\tau) = \left[q^{(n)}_{j_3} \, E^{(n)}_{j_3+1}\right]_{\theta,\tau-h} \qquad \text{if} \qquad (j_3+1)h \neq T_j$$

if $(j_3+1)h = T_j$, then a random age reduction takes place according to $g(w|\tau)$.

- Iterate the above procedure by adopting a discretization step size h/2. Consider h to be adequate if the iteration yields no significant difference in calculating:

$$\int R(t|\theta)d\theta \int q^{(n)}(\theta,\tau)d\tau$$

(or better the discretization of this integral).

Once that a recursive procedure for deriving $q^{(n)}$ is available, the two major problems one has to face are:

1. The assessment of $g(w|\tau)$;
2. Programming, which is non trivial in that a large use of multi-dimen-dimensional arrays is needed.

These two problems will be discussed in the following sections.

4. Theoretical model implementation

We are going to examine some of the main problems met in the implementation of the described methodology of inference.

4.1 Adoption of a component-failure-time distribution

We are not at all certain whether the components fail according to a Weibull distribution (shape parameter β higher than 1) or may be rather according to a gamma distribution. /6/ suggests a way, coherent with the Bayesian approach, to select a component failure time distribution.
The Weibull distribution has been adopted in the implemented version of the program.

4.2 Settling the prior on $\theta(\beta,\eta)$

We could, for instance, adopt uniform priors for both β and η (the Weibull distribution shape parameter and scale parameter respectively) with limit values 1 and 4 for β (on the basis of our knowledge of the Weibull law application characteristics). But if experts are available, a more fruitful way to proceed is to exploit their knowledge.
They cannot be asked to give their informed opinion in terms of a pdf on

some probability distribution parameter θ which may be they do not understand throughoutly. Furthermore probability distribution parameters are fictional (see discussion in /7/), while in de Finetti's view probabilistic forecasts must concern observable quantities. A Bayesian way to proceed is to request experts to express their opinion on an observable quantity such as the component time-to-failure in absence of maintenance.

A reasonable hypothesis, in a Bayesian frame, is /8/: expert assessments of component-failure-times and observed failure times are exchangeable. According to this hypothesis, the prior π can be computed by using the Bayes formula as if expert-response were actual failure-times:

$$\pi(\beta,\eta) = \frac{\prod_{j=1}^{n} \left(\frac{t_j}{\eta}\right)^{\beta-1} \exp\left[-\left(\frac{t_j}{\eta}\right)^{\beta}\right]}{\iint d\beta d\eta \prod_{j=1}^{n} \left(\frac{t_j}{\eta}\right)^{\beta-1} \exp\left[-\left(\frac{t_j}{\eta}\right)^{\beta}\right]}$$

being $t_1,\ldots,t_j,\ldots,t_n$ the judgements of the n experts.

Note that the experts are asked, in this way, to express themselves on questions of the same nature as establishing times between maintenance operations.

4.3 Assessing the pdf of age reduction

Maintenance modelling is an essential part of the process. Its basic steps, each one based on expert judgement, are:

1. Subdivision of the component in n parts; each part is characterized by a typical aging process and a planned maintenance policy. Experience shows that a consensus between experts is easily reached on a suitable subdivision. Note that the complete failure of each part determines the failure of the component.
2. Assigning a weight to the age of each part in determining the age of the component (the expert will be aware of the fact that this weight is to be somehow connected with the frequency of planned maintenance of the part).

3. Assessing $g(w|\tau)$, which is maintenance operation specific, needs the following information relative to each maintenance operation:
   - Identifying the parts which have been maintenained (from the maintenance file).
   - The probability of renewal relative to each maintenained part (from expert consultation; see /6/ for the procedure of expert interrogation). The renewal can vary between 0 (maintenance cannot have a negative effect) and τ (part completely renewed).

   Further hypotheses for deriving $g(w|\tau)$ are:
   - Age of non maintained parts after maintenance is the same as it was before.
   - $g(w|\tau)$ is a function of just the ratio w/τ (judged to be reasonable by experts, in so far as components of concern were not too old). To ease computational aspects, a gamma law has been adopted in the programme.

5. Informatic aspects of the ARC code

The main issues to be kept within reasonable limits are:
- CPU time;
- the required memory.

CPU time "naturally" decreases as failures are observed; since the observation of a failure makes the posterior distribution shaper by pushing toward zero many entries of the θ, τ matrix.

Anyway one cannot rely on this "natural" decrease of CPU time and the latter must be kept small also by a suitable choice of the accuracy. The accuracy influences also the memory required for data processing.

A sensitivity study has then been performed for the sake of establishing a suitable balance between accuracy and memory/CPU-time requirements. The results of the study are summarized in the following examples.

We processed a sample consisting of observations on 16 components.

Tipically the observation on each component had lasted 98 months.

The number of observed failures was in total 21.

The components were assumed to fail according to a Weibull distribution. The generic vector parameter then specialized into the couple η, β, the scale and shape parameter respectively. Of course the data description above is not statistically sufficient in the case of a Weibull distribution; the only sufficient statistic being the whole set of failure and survival times. Nevertheless the data description we gave is enough for giving a rough idea of the amount of information needed for a given discretization on η, β, τ to be sufficient for getting an accurate enough assessment of component reliability.

In summary running the code with a discretization corresponding to a matrix η, β, τ of dimension 30x30x82 with respect to ranges such as $0 \leqslant \eta \leqslant 3$, $1 \leqslant \beta \leqslant 4$, $0 \leqslant \tau \leqslant 8.2$ years yielded no significant difference in the accuracy of reliability assessment in comparison of running the code with a discretization twice as fine as the above.

6. An example of the application of the code

6.1 An experimental application of the ARC code, to validate the code itself, has been carried out.

The set of components suitable for this application was selected on the basis of the following criteria:

- The components had performed a continuous function, in nuclear power plants;
- These components could be considered "exchangeable";
- The operator of the power plants (in which the components in question had operated) had a great interest in improving the adopted maintenance policy.

A set of 16 special steam valves, with reference to their pressure boundary function performed in the plants, was selected. A special inquiry was made to collect the full component-histories and the pieces of information to be given by the maintenance-experts, for the model application. The component was split by experts into 10 sub-components (ref. to 4.3). The procedure of this experimental code application consisted of the two following work steps:

1st step: The computer "jobs"

The 16 valve histories have been treated by the programme in a recursive way, one after the other (par. 3.2). The initialization of the process (par. 4.2) has been made using a uniform prior both for β and η. The posterior on β and η obtained as output after the treatment of each component history gathers the contributions to the parameters knowkedge due to the experimental evidence previously considered.

2nd step: CPU time analysis

A remarkable trend to a decrease of CPU time is observed along the process, going from one job to the following one. From an initial value of about 12 minutes, this time gradually reduces to less than 1 minute.

It is also to be noted the apparent influence of the integration step, which is to be indicated as an input parameter. This is assumed for the purpose of this exercise to be one month.

6.2 Result of the code application

As output, the code gives the component-failure probability of a component (of the considered family), supposed new at the beginning and in the absence of maintenance, as a function of its operating time (Tab. I).

Tab. I. Component failure probability versus operating time.

Operating time (month)	Failure probability
2	0.062025
4	0.148351
6	0.240342
8	0.330872
10	0.417479
12	0.497661
14	0.570025
16	0.635070
18	0.692417
20	0.742139
22	0.785331
24	0.822288

We note that the failure probability, is 0.5 after one year and 0.82 after two years.

According to the level of failure risk that the operator considers acceptable, he can identify the operating period after which the component has to undergo a maintenance intervention.

References

1. Clarotti, C.A.: "Failure Rate Estimation. A Dangerous Nonsense in a Bayesian View" Reliability Engineering and System Safety, Vol. 20, 1988, pp. 117-126.

2. Clarotti, C.A.; Koch, G.; Spizzichino, F.: "A Filtering Model for Bayesian Analysis of Failure Data Contaminated by Maintenance", to appear in Statistics & Probability Letters.

3. Clarotti, C.A.; Koch, G.; Spizzichino, F.: "Bayes Inference from Failure Data Contaminated Due to Maintenance", IEEE Transactions on Reliability, Vol. 34, No. 4, 1985 October, pp. 377-381.

4. Clarotti, C.A.; Pugno, G.F.; Runggaldier, W.J.: "Statistical Reliability Practice. From Sampling Theory to Stochastic Filtering", to appear in Reliability Engineering and System Safety.

5. Clarotti, C.A.; Runggaldier, W.J.: "On the True Value of a Failure Rate and Related Statistical Issues" to appear in Reliability Engineering and System Safety.

6. Besi, A.; Nichele, G.P.; Piepszownik, L.: "Implementing Stochastic filtering for reliability application", presented at the International Research Conference on Reliability, University of Missouri, Columbia, May 17-19, 1988.

7. Spizzichino, F.: "Symmetry Conditions on Opinion Assessment Leading to Time-Transformed Exponential Models", in Proceedings of the International School of Physics E. Fermi, Course CII, Accelerated Life Testing and Experts' Opinion in Reliability, C.A. Clarotti and D.V. Lindley Eds, North Holland, 1988, pp. 83-97.

8. Bologna, S.; Clarotti, C.A.: "Expert Opinions in Software Reliability Analysis", ididem, pp. 132-134.

A Development Flow Graph Method

Isa S. QAMBER and A.Z.KELLER

College of Engineering
University of Bahrain,
BAHRAIN

Postgraduate School of Ind.Tech.
University of Bradford,
ENGLAND

Summary

The paper demonstrates the use of graph theory to obtain steady-state behaviour of multi-state systems described by Markov models. A number of applications are given and compared with results obtained by other methods. The potential development of flow graph methods for further applications is also discussed.

Introduction

It is well known that engineers and others often wish to model ideas relating to new designs mathematically before investing large sums of money on hardware. In the conceptual stage, particularly in the synthesis of system design, it is desirable to obtain an understanding of how the various elements of the system will interact and how the overall system will perform. In putting systems together, the designers will wish to lay them out in a manner to obtain an idea of the overall operation of the system and to see how much equipment will be involved. In this way many potential mistakes can be eliminated at the beginning and improved system designs obtained. By developing and using models that simulate behaviour of the system and its components over the proposed lifetime, the designer can identify design deficiencies which can be removed. For these purposes it is suggested that graph theory is an appropriate tool [1].

In 1736, L Euler wrote the first known paper on graph theory [2], his interest around a problem occurring in Konigsberg, Germany. This is the so-called Konigsberg bridge problem.

One of the early uses made of graph theoretical concepts is illustrated by G Kirchhoff (1845,1847) in his celebrated study of the laws obeyed by potentials and currents in electrical networks. Since that time electrical engineers have increasingly used graph theory [2].

MASON'S RULE

Overall transmittance formulae are given by S J Mason [6,4,5]. These can be summarised as follows:

$$T = \frac{\Sigma T_n \Delta_n}{\Delta}$$

where: T is the transfer function

T_n is the transmittance of each forward path between a source and a sink node

Δ_n is the cofactor of T_n which is the determinant of the remaining subgraph when the path which produces T_n is removed.

Δ is the graph determinant given by

$$1 - \Sigma L_1 + \Sigma L_2 - \Sigma L_3 + \dots \qquad (1)$$

in which:

L_1 = the transmittance of each closed path in the graph

ΣL_1 = the sum of the transmittance of all closed paths in the graph

L_2 = the two non-touching loops transmittance product (when the loops do not have any common nodes)

ΣL_2 = the sum of the transmittance of the two non-touching loops transmittance

L_3 = the three non-touching loops transmittance product

ΣL_3 = the sum of the transmittance of the three non-touching loops transmittance.

Mason's original graphs are widely used in electrical networks analysis, control theory and in other different studies.

CALCULATION OF STEADY-STATE PROBABILITIES USING SIGNAL FLOW GRAPH METHOD

The following steps are recommended to calculate the steady-state probabilities:

1. Identify the characteristics of the system under study.
2. Draw a system model diagram and identify an appropriate source for the model.

Formulate the system of linear differential equations for the model.

4. Replace the source equation by the following equation:

$$\sum_{i=1}^{n} P_i = 1$$

where: n is the total number of states.

5. Draw the linear algebra signal-flow graph after writing down the system equations.

6. Using graph theory:

(i) identify in the signal flow graph all loops and paths and apply Mason's rule.

(ii) to follow this step a table as shown in the example given below is suggested. This table can simplify the identification of the common nodes in each of the loops. The set of non-touching paths can be obtained by preparing a table where there are as many columns as there are nodes in the set of closed loops. Each row of this table then corresponds to a particular closed-loop in the graph. An identifying symbol is then placed at the interaction of the row and column where a closed loop touches a node. Then multiply the gains of any two loops which do not have a common node. The process is then repeated for higher sets of loops (3,4 etc.). Again multiplying all individual loops again.

FLOW-GRAPH METHOD

In this section, applications of the flow-graph method are represented. A comparison made between this method and Billinton and Kumar [7,8] is given in the present paper.

A Two-State System Markov Model

A single repairable generator is considered for which the repair rate and failure rate are μ and λ respectively. Fig. 1a illustrates the state transition diagram for the generator. The two states in the figure are as follows:

State 1: the generator is operating

State 2: the generator is failed

Therefore, the generator system differential equations are as follows:

$$\left.\begin{aligned} \frac{dP_1}{dt} &= -\lambda P_1 + \mu P_2 \\ \frac{dP_2}{dt} &= \lambda P_1 - \mu P_2 \end{aligned}\right\} \qquad (2.)$$

The steady state probabilities for this simple Markov model can be obtained by solving the steady state equations using the flow graph method.

The above two equations are not independent, so one of the equations is replaced by the following equation:

$$P_1 + P_2 = 1$$

The system equations, when the first state is considered as a source, becomes:

$$P_1 = 1 - P_2 \qquad (3\ a)$$

$$P_2 = \frac{\lambda}{\mu} P_1 \qquad (3\ b)$$

For the two equations (3), the flow graph diagram is as shown in Fig. 1b.

From Fig. 1b, Table 1 is prepared as follows:

Table 1 Two-State Model (Nodes connections),
Node 1 is the source

Loop	Path	Nodes		Gain
		1	2	
L_1	1→2→1	x	x	$-\frac{\lambda}{\mu}$

From Table 1, one loop is obtained for the model of the present example. Then, equation (1) is applied to give:

$$\Delta = 1 - L_1 = 1 + \frac{\lambda}{\mu}$$

Therefore,

$$P_1 = \frac{1}{\Delta} = \frac{\mu}{\mu+\lambda}$$

$$P_2 = 1 - P_1 = \frac{\lambda}{\mu+\lambda}$$

Alternatively, node 2 can be considered as a source, and the previous steps are followed. Equation (3 b) is replaced by:

$$P_2 = 1 - P_1$$

where the system equations become:

$$\left.\begin{aligned} P_1 &= \frac{\mu}{\lambda} P_2 \\ P_2 &= 1 - P_1 \end{aligned}\right\} \qquad (4)$$

Therefore, the flow graph diagram is now as shown in Fig. 1c. The nodes connections are given in Table 2.

Table 2 Nodes connections (Two-state model)
Node 2 is the source

Loop	Path	Nodes		Gain
		1	2	
L_1	2→1→2	x	x	$-\frac{\mu}{\lambda}$

From Table 2, it is found that there is only one loop in Fig. 1c. To calculate Δ, equation (1) is again applied:

$$\Delta = 1 - L_1 = \frac{\mu+\lambda}{\lambda}$$

Therefore, the steady state probabilities are as before:

$$\left.\begin{aligned} P_2 &= \frac{\lambda}{\mu+\lambda} \\ P_1 &= \frac{\mu}{\lambda} P_2 \\ &= \frac{\mu}{\mu+\lambda} \end{aligned}\right\} \qquad (5\ )$$

It follows that one of the two nodes, 1 or 2, can be chosen as the source for the calculation of the steady state probabilities.

The generator availability and its unavailability are now given respectively as:

$$\left.\begin{aligned} A &= P_1 \\ U &= P_2 \end{aligned}\right\} \qquad (\ 6)$$

Although the above example is of a relatively trivial nature and the solutions (6) obtained easily using elementary algebra, it is of a limited value in illustrating the flow-graph methodology which remains the same irrespective of the size of the problem.

A Three-State Markov Model

A three-state Markov model of a system consisting of two identical components is now considered. It is assumed that at any time, the system is in one of three mutually exclusive states:

(i) both components are working

(ii) one of the components has failed and is being repaired

(iii) both components have failed and are being repaired simultaneously.

This model is based on the example given by Billinton and Kumar [36]. The equations describing this system are as follows:

$$\left.\begin{aligned} \frac{dP_1}{dt} &= -2\lambda P_1 + \mu P_2 \\ \frac{dP_2}{dt} &= 2\lambda P_1 - (\mu+\lambda) P_2 + 2\mu\, P_3 \\ \frac{dP_3}{dt} &= +\lambda P_2 - 2\mu P_3 \end{aligned}\right\} \qquad (7)$$

where λ and μ are the failure and repair rates respectively for a single component.

As before, the derivatives of state probabilities are set to zero to give the following equations:

$$\left.\begin{aligned} -2\lambda P_1 + \mu P_2 &= 0 \\ 2\lambda P_1 - (\mu+\lambda) P_2 + 2\mu P_3 &= 0 \\ +\lambda P_2 - 2\mu P_3 &= 0 \end{aligned}\right\} \qquad (8)$$

Since these three equations are not independent, the first equation is replaced by the following equation:

$$P_1 + P_2 + P_3 = 1$$

In the present, Fig. 2a shows the state–space diagram of the model.

The steady state solution can now be obtained following the same procedure a above. Then, the system equations become:

$$P_1 = 1 - P_2 - P_3 \qquad (9a)$$

$$P_2 = \frac{2\lambda}{(\mu+\lambda)} P_1 + \frac{2\mu}{(\mu+\lambda)} P_3 \qquad (9b)$$

$$P_3 = \frac{\lambda}{2\mu} P_2 \qquad (9c)$$

Fig. 2b represents the flow graph of the three–state model. Table 3 shows the nodes connections of the model.

Table 3 Three-State Model (Nodes connections)
Node 1 is the source

Loop	Path	Nodes			Gain
		1	2	3	
L_1	1→2→1	x	x		$-\frac{2\lambda}{(\mu+\lambda)}$
L_2	1→3→2→1	x	x	x	$-\frac{\lambda^2}{\mu(\mu+\lambda)}$
L_3	2→3→2		x	x	$\frac{\lambda}{(\mu+\lambda)}$

$$\Delta = 1 - \sum_{i=1}^{3} L_i = \frac{(\mu+\lambda)}{\mu}$$

The steady state probabilities are obtained as follows:

$$P_1 = \frac{1}{\Delta}(1 - L_3)$$

$$= \frac{\mu^2}{(\mu+\lambda)^2} \qquad (10a)$$

Equation (9c) and P_1 in (9b)

$$P_2 = \frac{2\mu\lambda}{(\mu+\lambda)^2} \qquad (10b)$$

P_2 in equation (9c):

$$P_3 = \frac{\lambda^2}{(\mu+\lambda)^2} \qquad (10c)$$

Since the above solution is independent of the choice of node as source, the same solution as equations (10) should be obtained where either of the other nodes are chosen as the source. This is now demonstrated.

Taking node 2 as a source instead of node 1 as a source, and following the previous steps to calculate the steady state solution, the second equation of (8) is replaced by

$$P_2 = 1 - P_1 - P_3$$

Then, the system equations are as follows.

$$P_1 = \frac{\mu}{2\lambda} P_2 \qquad (11a)$$

$$P_2 = 1 - P_1 - P_3 \qquad (11b)$$

$$- \quad \lambda \quad -$$

Fig. 2c represents the flow graph of the three state model and Table 4 represents the node connections for the model.

Table 4 Three-state model (Nodes connections), Node 2 is the source

Loop	Path	Nodes			Gain
		1	2	3	
L_1	1→2→1	x	x		$-\frac{\mu}{2\lambda}$
L_2	2→3→2		x	x	$-\frac{\lambda}{2\mu}$

$$\Delta = 1 - \sum_{i=1}^{2} L_i = \frac{(\mu+\lambda)^2}{2\mu\lambda}$$

$$P_2 = 1/\Delta = \frac{2\mu\lambda}{(\mu+\lambda)^2} \qquad (12a)$$

$$P_1 = \frac{\mu}{2\lambda}\left[\frac{2\mu\lambda}{(\mu+\lambda)^2}\right] = \frac{\mu^2}{(\mu+\lambda)^2} \qquad (12b)$$

$$P_3 = \frac{\lambda}{2\mu}\left[\frac{2\mu\lambda}{(\mu+\lambda)^2}\right] = \frac{\lambda^2}{(\mu+\lambda)^2} \qquad (12c)$$

It is obvious that when node 2 is considered as the source, equations (12) are similar to equations (10).

In a similar way, if node 3 is considered as the source, the system equations become:

$$\left.\begin{aligned} -2\lambda P_1 + \mu P_2 &= 0 \\ 2\lambda P_1 - (\mu+\lambda)P_2 + 2\mu P_3 &= 0 \\ P_1 + P_2 + P_3 &= 1 \end{aligned}\right\} \qquad (13)$$

The three equations (5.13) can be re-arranged as follows:

$$P_1 = \frac{\mu}{2\lambda} P_2 \qquad (14\ a)$$

$$P_2 = \frac{2\lambda}{(\mu+\lambda)} P_1 + \frac{2\mu}{(\mu+\lambda)} P_3 \qquad (14\ b)$$

$$P_3 = 1 - P_1 - P_2 \qquad (14\ c)$$

and from equations (14), the signal-flow graph can be drawn as shown in Fig. 2d.

Table 5 illustrates the common nodes between the loops.

Table 5 Three-State Model (Nodes connections)
Node 3 is the source

Loop	Path	Nodes			Gain
		1	2	3	
L_1	1→2→1	x	x		$\frac{\mu}{(\mu+\lambda)}$
L_2	1→2→3→1	x	x	x	$-\frac{\mu^2}{\lambda(\mu+\lambda)}$
L_3	2→3→2		x	x	$-\frac{2\mu}{(\mu+\lambda)}$

After the loops are specified, equation (..1.) is applied to calculate Δ. Therefore:

$$\Delta = 1 - (L_1 + L_2 + L_3) = \frac{\mu+\lambda}{\lambda}$$

Since the third equation of (13) is replaced, node 3 has been chosen as the source. The state probability for the third state is now given by:

$$P_3 = \frac{1}{\Delta}\ (1 - \text{sum of the loops which do not pass through node 3})$$

$$= \frac{1}{\Delta}\ (1 - L_1) = \frac{\lambda^2}{(\mu+\lambda)^2}$$

Substituting P_3 and equation 14a into equation 14c then gives

$$P_2 = \frac{2\mu\lambda}{(\mu+\lambda)^2}$$

and substituting P_2 into equation (14a) gives

$$P_1 = \frac{\mu}{2\lambda}\left(\frac{2\mu\lambda}{(\mu+\lambda)^2}\right) = \frac{\mu^2}{(\mu+\lambda)^2}$$

Therefore, the steady state probabilities are

$$\left.\begin{aligned} P_1 &= \frac{\mu^2}{(\mu+\lambda)^2} \\ P_2 &= \frac{2\mu\lambda}{(\mu+\lambda)^2} \\ P_3 &= \frac{\lambda^2}{(\mu+\lambda)^2} \end{aligned}\right\} \qquad (15)$$

It is seen that in choosing node 2 as the source, a simpler graph is obtained, as compared with other graphs. This simplification also reduces the amount of algebra involved.

The example is illustrative of the value to be placed on prior consideration as to which node should be chosen as a source before solution of the problem.

When the two identical components are connected in series, the first state is the up state of the system and the other two states are the down states of the system. Therefore, the system availability can be written as:

$$A = P_1$$

$$U = 1 - P_1$$

However, in the case that the two identical components are connected in parallel, the first and second states are the up states of the system, with the third state being the down state of the system. Therefore,

$$A = P_1 + P_2$$

The same model discussed above has been considered by Billinton and Kumar [7]. They also used graph theory to obtain the steady state probabilities for the model. After drawing the flow graph diagram (Fig. 3), the numerator term of the steady state probability of state (1) is computed, where state (1) is considered as a source and state (3) as a sink. In their solution, there is only one possible transition path, which is (1→2→3) and its gain is $2\mu^2$.

Similarly, in order to find the numerator term of the steady state probability of state (2) which can be considered as a source and state (3) as a sink, one possible transition path (2→3) can be considered which is 2μ.

In the case of state (3), if it is defined as the sink, there are two possible ombinations of non-touching loops:

(i) self-loops of node 1 (2λ) and node 2 ($\mu+\lambda$), where their gain is $(2\lambda)(\mu+\lambda)$;

(ii) loop between node 1 and node 2 ($+2\mu\lambda$), where its gain is ($+2\mu\lambda$).

It can be seen that the path gain is unity for both cases (i+ii) [7]. If gain (ii) is subtracted from (i), the following expression can be obtained:

$$(2\lambda)(\mu+\lambda)-2\mu\lambda = 2\mu\lambda + 2\lambda^2 - 2\mu\lambda$$

$$= 2\lambda^2$$

Table 6, given below, illustrates the numerator terms of Fig. 3.

Table 6 The numerator terms of Fig. 3

Sink	Source	Forward Path		Possible Combinations of Loops		$P_k\Delta_k$
		Graph	Gain (P_k)			
3	1	1→2→3	$2\mu^2$	-	1	$2\mu^2$
	2	2→3	2μ	1→1	2λ	$(2\mu)(2\lambda)$
	3	-	1	1→1 and 2→2	$(2\lambda)(\mu+\lambda)$	$2\lambda(\mu+\lambda)$ $-(2\mu\lambda)$
				1→2→1	$-(\mu)(2\lambda)$	

$$\Delta = \sum_k P_k \Delta_k = 2(\mu+\lambda)^2$$

They then apply the following formula to compute the steady state probabilities:

$$P_i = \frac{1}{\Delta} \sum P_k\Delta_k \qquad (16)$$

where P_k is the path gain of the kth forward path. If the sink and source are chosen to be the same, the forward path gain P_k is equal to unity [7]. This is because

$$\Delta_k = \sum_n (-1)^m \times \{\text{loop gain of a loop or gain product (in the case}$$

of more than one non-touching loop) of possible combination of non-touching loops. These loops should not touch the kth forward path}. In the cases in which there is no such possible combination, $\Delta_k=1$.

m is the total number of loops excluding self loops

n is the total number of all such possible combinations of non-touching loops as described above

k is the total number of all possible forward paths from source to sink

Therefore, the steady state probabilities can be computed using equation (16)

These are identical to those (10, 12, 15) obtained earlier using the method presented here.

It is seen that the flow-graph method presented here produces simplified graphs as compared with the procedure of Billinton and Kumar [7]; this is because all self loops are eliminated and that this graph simplification results in a significant reduction in the required algebraic manipulation.

A comparison of the difference between the Billinton and Kumar procedure and that presented here is given in Table 7.

Table 7 Comparison between Billinton & Kumar procedure and the proposed procedure

	Billinton & Kumar procedure	Proposed procedure
Source point	should be chosen and defined (any node can be chosen)	choose node that provides maximum simplification of graph
Sink point	should be chosen as any point	not applicable as sinks are not used
Self-loops	has self-loops	no self-loops
Transition rate matrix	should be obtained	should be obtained
Condition $\sum_{i=1}^{n} P_i = 1$	not used at all	used to replace source equation and simplify flow graph

Conclusions

The flow graph method which is based upon a graphical representation of the linear simultaneous equations governing the Markov model has the following advantages:

i. the procedure is general and can be applied to any Markov model

ii. the procedure can be applied manually

iii. the procedure is independent of matrix manipulations

iv. the choice of a particular node as the source can be critical, as an unsuitable choice of node can result in unnecessary additional manipulation.

v. procedures recommended in the present thesis require less manipulation on identical problems than those recommended by Billinton and Kumar.

vi. unlike sparse matrices usually required for complex systems which contain a large number of redundant zeroes, flow graph representations are free of redundant information.

vii. because flow graph methods to date have only been implemented manually, only problems of moderate size, i.e. of the order of say 10 states can be treated by this method.

REFERENCES

1. Lorens, C S, "Theory and Applications of Flow Graphs", DSc Report, MIT, 1956.
2. Harary, F, "Graph Theory and theoretical physics", Academic Press, London, 1967.
3. Buzacott, J A, "Finding the MTBF of repairable systems by reduction of the reliability diagram", Microelectronics & Reliability Vol. 6, pp. 105-112, 1967.
4. Dorf, R C, "Modern Control Systems", Addison Wesley Pub. Co., 1974.
5. D'Azze, J J & Houpis, C H, "Feedback control system analysis and synthesis", 1966.
6. O'Connor, P, "Practical Reliability Engineering", John Wiley & Sons, 1984.
7. Billinton, R & Kumar, S, "State availability analysis using graph theory", Proceedings of 8th ARTS, National Centre of System Reliability, University of Bradford.
8. Kumar, S & Billinton, R, "Graph theory concepts in frequency & availability, Vol. 34, No. 4, 1985.

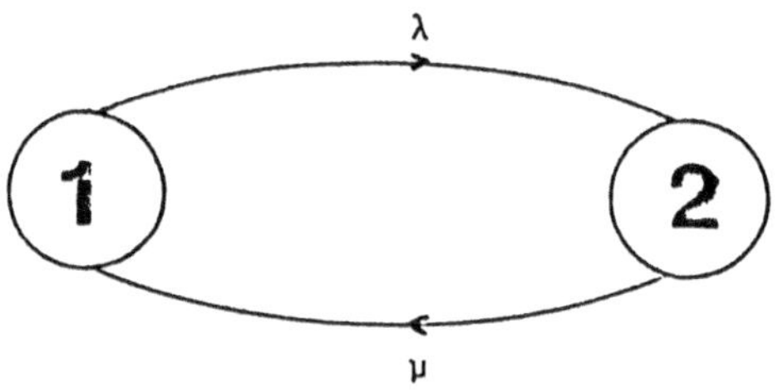

Fig. 1a State transition diagram for a generator

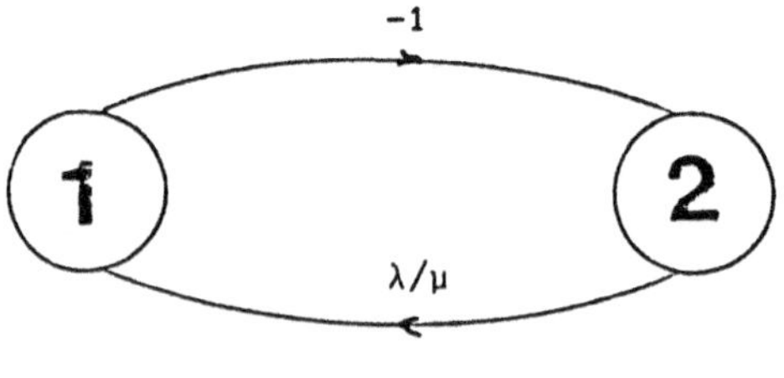

Fig. 1b Flow-graph diagram for the generator (Node 1 is the source)

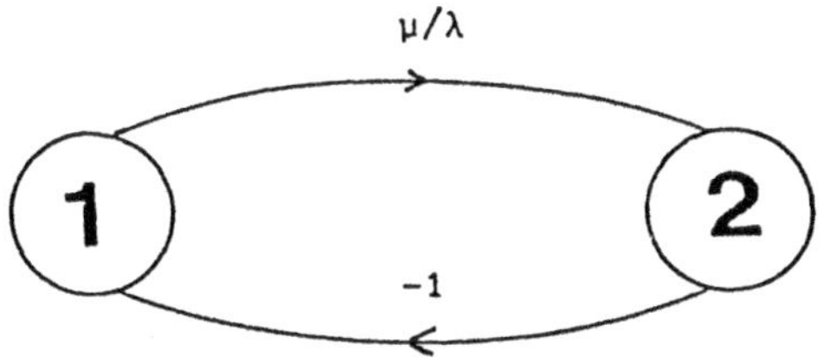

Fig. 1c Flow-graph diagram for the generator (Node 2 is the source)

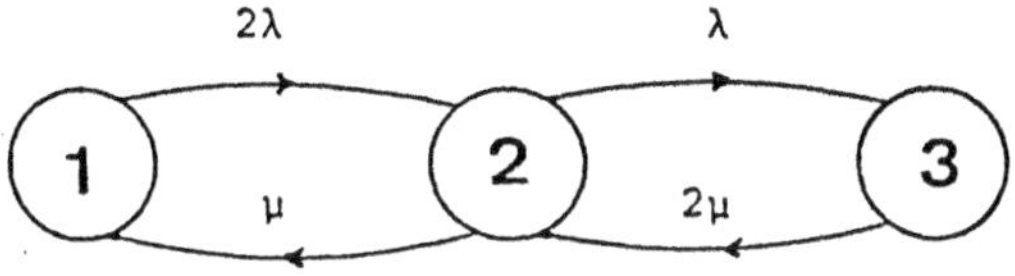

Fig. 2a State-space diagram of three-state model

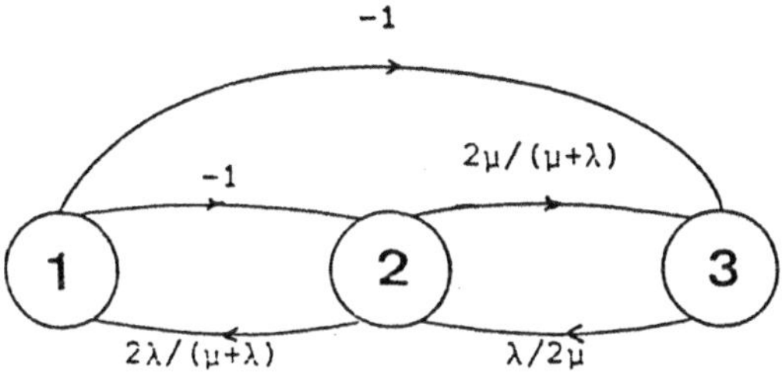

Fig. 2b Flow-graph diagram of the three-state model (Node 1 is the source)

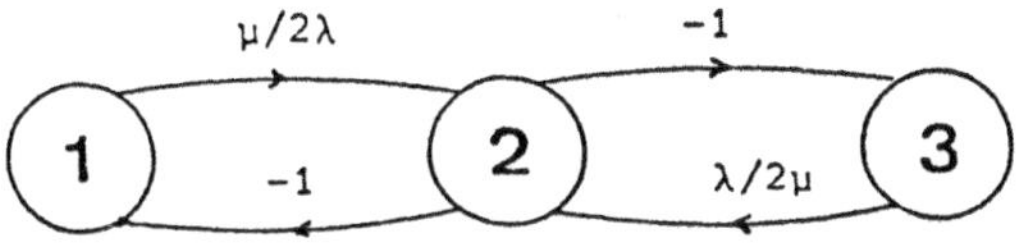

Fig. 2c Flow-graph diagram of the three-state model (Node 2 is the source)

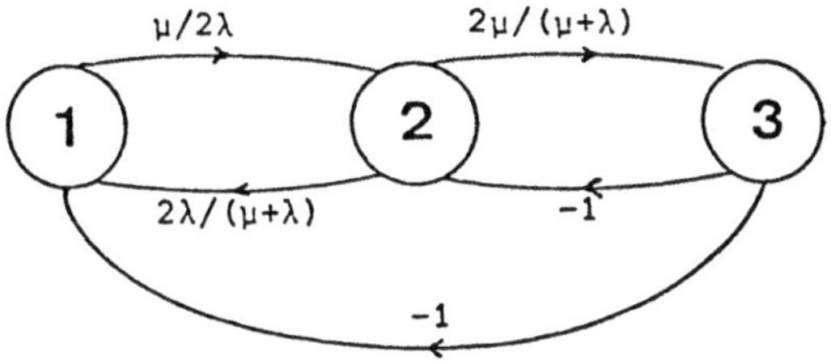

Fig. 2d Flow-graph diagram of the three-state model (Node 3 is the source)

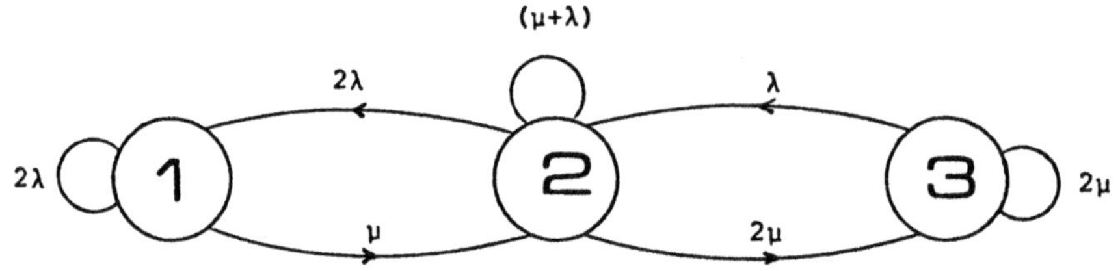

Fig. 3 Flow-graph diagram of the three-state system

KODABA – A Corrosion Data Information System

R. Leicht, G. Luthardt, R. Schönfeld, H. J. Wingender
NUKEM GmbH, P.O. Box 1313
D-8755 Alzenau, FRG

Abstract
The corrosion data base system KODABA is a shell to collect, document, and archive corrosion data from different fields. It provides fast and easy access to the stored data and offers comfortable retrieval options. Three dialogue languages are available for the user - English, French, and German. Different output options are offered to record the results of a query. It is written in CLIPPER- and FORTRAN-language and designed for use on IBM compatible personal computers operated under MS-DOS. Definitely KODABA is a helpful tool for people working in fields where corrosion is an important aspect, e.g. reliability engineering.

1 Introduction

The corrosion aspect is a very important point concerning the reliability and therefore also to the safety of components and installations where corrosive agents are treated or may occur, i.e. chemical and nuclear industry as well as the civil engineering field.

In the last 25 years studies with the aim to limit and to minimize corrosion damage were intensified because of the immense economic costs and the increasing account of the resulting safety and risk aspects. Great efforts have been made in basic research to find causalities and ways for efficient corrosion damage prevention. Considering the large number of important influencing parameters this was, and still is, a very difficult task. Current results are included in a variety of publications. Only a part of them are summarized and documented in publicly available data collections.

There is obviously a lack of good data collections and data information systems where published information on corrosion data is collected and can easily be retrieved and analyzed. As a consequence, much of the existing information is not available where it is required, and thus not taken into account in planning, construction and fabrication. Other difficulties are the use of nonstandard designations and definitions and the different levels of documenting conditions and results of corrosion tests. These facts suggest the idea of developing an information system for corrosion data, which contains factual data from basic research publications, and to propose a standard data collection scheme which also could be used by the authors of corrosion articles as a guideline to document main parameters of their corrosion tests.

2 Conception

The realization of the idea of a general information system shell for factual corrosion data was planned and performed in the following steps:

(1) Definition of requirements
(2) Analysis and evaluation of the important characteristics and terms which describe a corrosion system and the corrosion scenario within this system
(3) Analysis and evaluation of the data base structure
(4) Choice of the hard- and software environment
(5) Creation of a data base as a demonstration example
(6) Completion and optimization of user-adjusted support options for data input, retrieval, interpretation, documentation, explanation, and user guidance and help
(7) Generalisation

In the first step the following points were specified:
- availibility for a wide range of users (corrosion experts and engineers),
- user-friendliness,
- clear and simple structure,
- direct access, and
- low cost (investment and operation).

The objectives of the second and third step were to analyse and to evaluate those characteristics and terms which completely describe a corrosion system and the corrosion scenario within this system. With regard to develop a very general data base structure, which can consider a wide range of different corrosion aspects, a very common and thoroughly analyzed corrosion system such as "steel/acid" (stainless steel/nitric acid) was chosen. Based on this corrosion system a variety of corrosion studies were analyzed by corrosion experts to evaluate a general set of important corrosion parameters. The parameters were structured and classified into different groups and analyzed to find out what kind of data base would be appropriate - relational or hierarchical - and how it could be organized in a data base. A complete description of a corrosion data set must contain the following information:

- original reference,
- characteristics of the corrosive medium,
- characteristics of the material,
- characteristics and conditions of the corrosion experiment,
- corrosion results.

With regard to a clearly structured data base it seemed to be advantageous to organize the data in a relational data base system. The choice of hard- and software environment depends directly on the claims defined in the first step.

The system complying best with all the requirements discussed above is the Personal Computer. PC's are also widely used in companies, research institutions, and universities. The MS-DOS operating system for PC's is well known, as is the data base management system "dBase". The total investment costs for a complete PC-workstation including essential software amounts to less than 20.000 DM. Due to its graphical options, the user surface can be optimally designed. Furthermore, the requirements for interactive and cost-efficient operation as well as for low investment and operating costs are fully met. Thus, access is provided for a wide range of users.

The PC and the widely used data base management system "dBase III plus" and the related "CLIPPER"-compiler have been chosen to realize the corrosion data base.

According to the rules of information science, the total data set is subdivided into data subsets and and stored in separate data bank files. These are interlinked via so-called "external keys" of the relational data base management system. This concept provides

- high flexibility,
- minimal storage requirement,
- shortest access.

Finally the user friendliness of the data base system can be accomplished by features such as

- data input via standardized menus, screen masks, and use of window technique (e.g. to fade in thesauri),
- data output in form of data sheets, individual results, tables and diagrams,
- user-adjusted search options for different levels of expertise such as
  (1) assistance and guidance via an "intelligent" system of menus, providing information and help to less-experienced users (e.g. based on an expert system),

(2) retrieval according to the "Query by Example"- or "Thesaurus Windowing"-methods as an option for corrosion experts, and

(3) retrieval by means of a special query language for those users who know in detail the structure and contents of the data base and the data base management system.

3 Stage of development

General: KODABA consists of the two program modules KDBE and KDBR. The module KDBE is used for the creation of data base files (DBF) and for data input and editing. KDBR is the retrieval module. Access is provided to all data base files, and appropriate links between these files are established according to the requirements of a particular query. The data base and program structure of KODABA is shown in fig. 1.

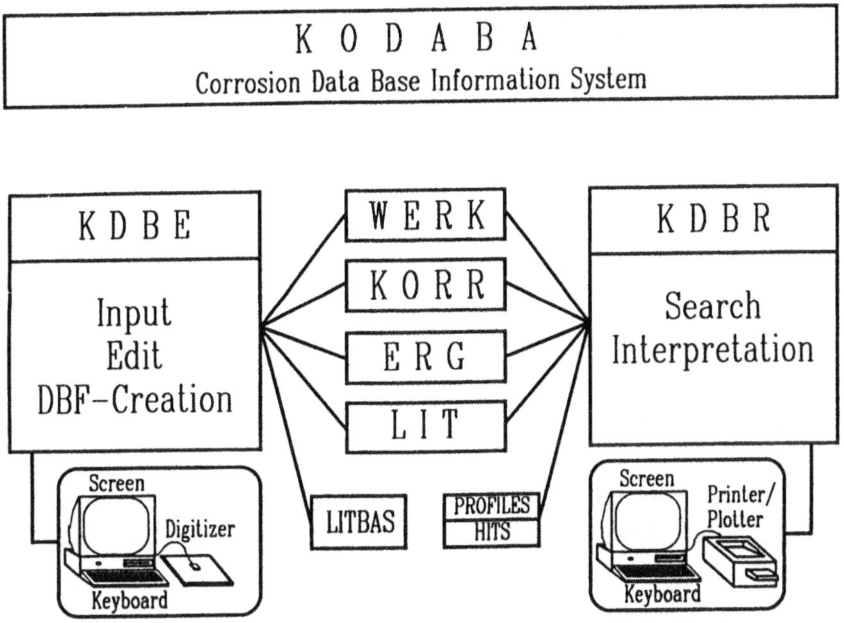

Figure 1: Data base and program structure of KODABA

<u>Storage of information:</u> As indicated above, the complete description of a corrosion data set is compiled in corresponding data sets and stored in the following four main DBF's:

LIT - original reference,
KORR - characteristics of the corrosion medium,
WERK - characteristics of the material,
ERG - characteristics and conditions of the corrosion test and results.

The particular parameters of each data set to be stored in the data bank were defined and selected in cooperation with corrosion experts. The main criterion is the importance of the information for a user. In addition, it might be necessary to provide information to the user which he did not explicitly ask for, but which is necessary to avoid misinterpretation of the data. There are three types of parameters:

- qualitative data (e.g. the description of material treatment or corrosion behavior),
- quantitative data (e.g. numerical values of the corrosion rate, acid molarity or impurities concentration),
- general data (e.g. material code number and name, or the title and author of a reference).

As far as possible, the qualitative data are stored in coded form to save storage space and reduce access time.

The quantitative results of corrosion tests are published as point values or tables with varied parameters, or as points or curves in figures.

The problem of data inhomogeneity is solved by organizing the results in standardized form in a separate sub-data base. This method allows to restore the original tables and to extract information for input to other programs, e.g. interpolation, fitting, or graphical output. The resulting structure of the data base and the linkage of the data base files is shown in fig. 2.

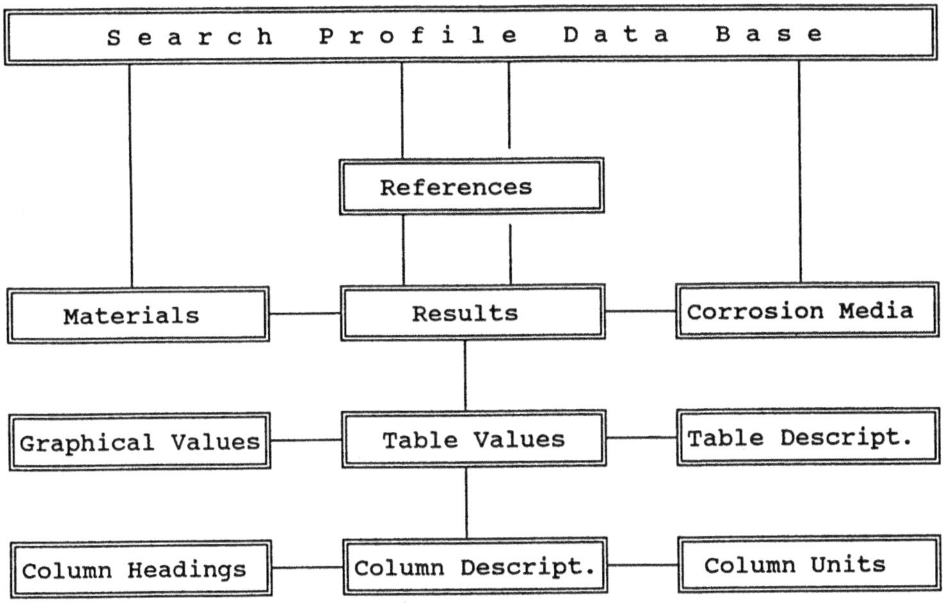

Figure 2: Data base structure of KODABA

<u>Data collection and input:</u> The transfer of information from original literature to the corrosion data base system requires in most cases reprocessing and adapting of the published data to the requirements of the data base. Besides this, care must be taken for identifying the very important information and data, and, in some cases, inconsistencies in the reports. After having analyzed a report, the data are recorded on appropriate data collection sheets. Certain information is to be stored in coded form, according to available coding tables.

Since numerical values are published in a variety of different tabular forms, it is not possible to introduce standard collection data sheets. Graphical data (points or curves in figures) are digitized by means of a graphics tablet.

Options of retrieval: The options of retrieval provided by a data base system very much influence its user-friendliness. Different user qualification and experience must be taken into account while realizing this point.

As already mentioned above, there are three different options for user-adjusted retrieval, according to different levels of users expertise:

(1) "Intelligent Assistent"

(2) "Query by Example" or "Thesaurus Windowing"

(3) "Special Query Language"

Only the "Intelligent Assistent" option is not yet finally realized. The options (2) "Query by Example" and "Thesaurus Windowing" are described in more detail in the following.

Running a retrieval in KODABA, the user first gets an overview of the actual contents and then has to specify a data base and the search field where he wants to define a search term. A thesaurus, i.e. an alphabetical list of all stored terms of a field and the corresponding frequencies, then appears in a window on the screen. The user can select a term by tagging it, modify it (for substring search) or immediately take it into the search data sheet.

Information which is stored in coded form in the data base, is provided decoded in the thesaurus window. Numerical search terms can directly be entered.

It is possible to define additional search terms in more than one data base. All search terms related to a given query form a "search profile" and are stored as a record in a search profile data base. During the search procedure, all actual search terms are linked via a logical "AND" condition. As a first result of a search, the number of hits in either of the 4 main data bases is displayed, and an overview of hits is given. The hits then can be selected for detailed output of information.

Generally, it is useful to run a query session by first specifying only one or a few search terms. Only if the number of hits is too high, more search terms should be introduced, and the search repeated. The history of such a stepwise retrieval is documented consecutively by the different search-profiles in the search-base.

KODABA offers the option to reactivate and, if desired, to modify previous search profile records. It is also possible to link previous search profile records with logical "AND" or "OR" conditions. The result of such a procedure is provided without a search by accordingly evaluating the existing sets of hits. At the end of a session, the profile data file can be saved for later use.

Data Output: The result of a retrieval is the set of all data records which meet the defined search conditions. These data are accessible to the user in the following steps and options:

- Overview of hits
  The first information provided to the user after completion of a retrieval is, for each of the 4 main data bases, the number of hits and the contents of selected fields. This information is displayed in one line per hit, such showing a number of hits in an overview.

- Detailed display of hits
  Selection of a particular data base and hit number in the overview mask shows the corresponding complete data sheet. It is possible to go to any data sheet (and table) of the 4 main data bases, and to any hit number, by simply using the cursor keys. In addition, the data sheets and tables can be printed.

- Graphics option

  This option offers the possibility to create diagrams of numerical results of one or, for immediate comparison, up to ten data sets. It is possible to draw data points and, if given, the related error bars, and to select a data range.

  Data to be taken from another table into an already specified diagram need only to have the same column units. Data points from different sources are represented by different symbols, and the relation between symbols and sources is also provided. The graphics option allows linear and/or logarithmic axis, grid lines, and linear or spline interpolation. Scaling is performed automatically. The coordinate descriptions are extracted from the column description of the associated table.

  Graphics output is available either on screen (monochrome or color) with the possibility to make hardcopies on a graphics printer, or, for better quality, on a Hewlett Packard compatible plotter (fig. 3).

<u>Contents of the Test Data Base:</u> The KODABA system has been filled with corrosion data for test purposes. Even if KODABA was originally designed for metal-acid-corrosion systems, it can be, and has been, used for a variety of other corrosion systems, e.g. metal/salt brine. The test data have been taken from the fast breeder fuel cycle field. Currently, the test data base contains about 200 different data sets on corrosion behavior from this field.

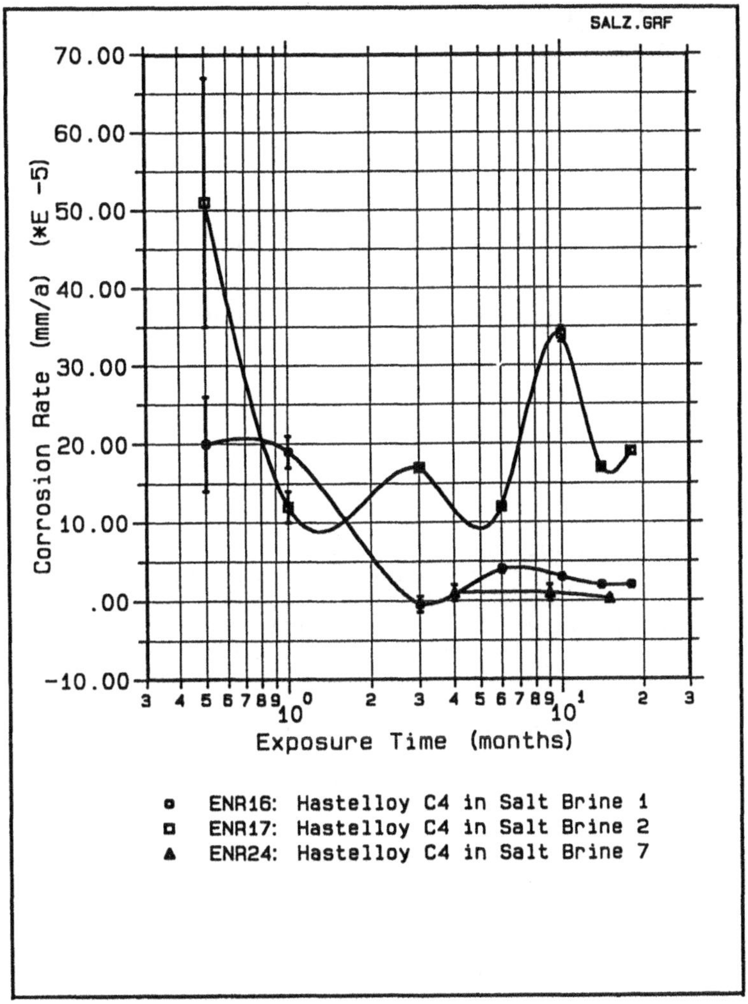

Figure 3 : Example of KODABA graphical output

Acknowledgements

The authors wish to thank the Bundesministerium für Forschung und Technologie (BMFT) for the financial support, the Projekt Wiederaufarbeitung und Abfallbehandlung (PWA) of the Kernforschungszentrum Karlsruhe for the friendly support, and our colleagues H.D. Baumgart, J. Bugiera, and V. Portscher for their contributions.

The Evaluation of Mechanical Resistance of Storage Tanks Exposed to Fire by Using Data Bank Information

Viviana Colombari, Maurizio Gilioli, Alfredo Verna

TEMA S.p.A.

via Medici del Vascello 26 Milano, Italy

Summary

The paper describes a mathematical procedure aimed at evaluating the heat flow and the transient variation of the wall temperature of a spherical vessel containing a liquified gas during exposure to thermal radiation. In fact in case of fire the vessel is subject to both overtemperatures and overpressures, which may lead to overheating and rupture of the wall.

The main objects of the study are to analyse the effect of dimension and duration of fire, different materials, insulation thickness and filling level in the vessel on the time needed to reach critical wall temperatures in order to help both designers and operators in the choice of suitable materials, the selection of optimal layout and the establishment of proper safety systems.

Introduction

Let us consider a spherical pressure vessel of internal radius r_1 [m] which contains m_L [kg] of a liquified gas; the wall is made of metal and an insulation is applied over the metallic surface, in order to reduce heat transfer between the atmosphere and the tank. r_2 [m] is defined as the distance of the interface metal-insulation from the center of the sphere, and r_3 [m] is the outer radius of the vessel (see fig. 1). The vessel is protected against overpressure due to either abnormal operative conditions or heat exposure by a pressure relief valve set to open at a pressure P_F [N/m^2] not in excess of design pressure.

If a spillage of a flammable liquid takes place close to the sphere, the liquid accumulates on the ground and, in case of ignition, a fire may occur.

As a result of fire, heat is added to the vessel, raising the wall temperature; and as a result of the increase in wall temperature, the temperature of the liquid in the vessel increases. The presence of the insulation introduces a resistance to the flow of heat to the wall and thus retards the heat flow to the liquid. Due to the wall temperature increase the integrity of the vessel may be compromised, as the strength of the material decreases. In addition the pressure and subsequently the stress in the wall rises, causing the pressure relief valve to lift and the evaporated liquid to leave the vessel. Further heat input causes the liquid to evaporate and the pressure not to rise further, on condition that the valve opens correctly and is able to discharge the total flow of vapour; otherwise the rise in pressure may burst the vessel.

In order to determine the time to failure of the vessel due both to overtemperature and overpressure it is necessary to calculate the temperature rise and distribution in the vessel wall. This calculation is complicated by a number of variables, such as:

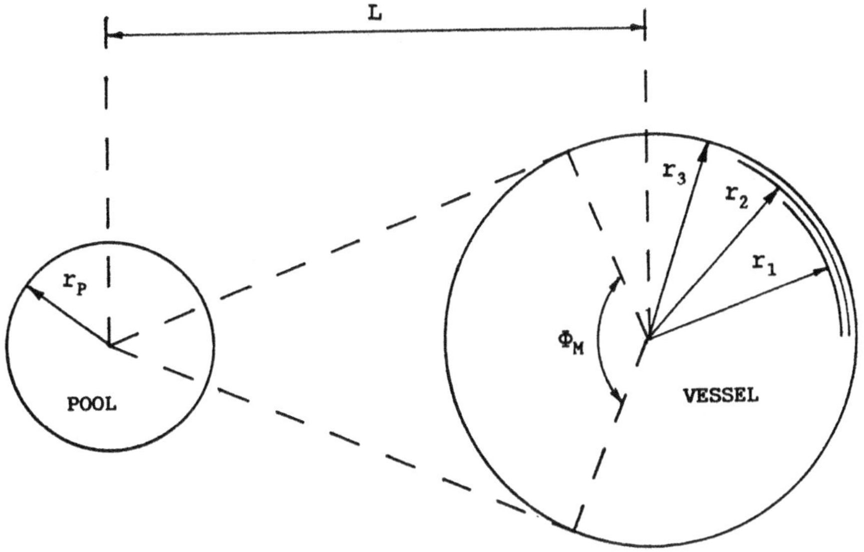

Figure 1: Schematic representation of the accident scenario

- the configuration of the pool of the flammable liquid;
- the distance from the vessel;
- the wind direction and speed;
- the type and thickness of metal and insulation materials;
- the type and quantity of substance into the vessel;
- the setting pressure of the relief valve, P_F.

The purpose of this study is to analyse the dependence of such factors on the speed of the system to reach hazardous conditions.

Procedure description

Let us refer to the case of an accidental release of a flammable liquid in proximity of the spherical vessel, which results in the formation of a circular pool of radius r_P [m] and, if ignited, eventually in a fire. Thermal radiation hazards from such a fire depend on a number of parameters, including the composition of the substance, the size, shape and duration of fire, the distance between the center of the pool and the center of the vessel, L [m], and the geometrical shape of the vessel itself.

In order to calculate the radiant flux $\dot{q}_0$ [W/m^2] received by a generic unit area of the vessel wall exposed to fire, the following expression is adopted:

$$\dot{q}_0 = \bar{F}\tau E \tag{1}$$

having assumed [1] that the radiative flux comes from a flame envelope shaped as a circular cylinder (hypothesis of solid flame model).

In (1) $\bar{F}$ represents the average fraction of energy radiated from the fire which is intercepted by the vessel, usually defined as the geometric view factor; τ is the atmospheric transmissivity; E [W/m^2] is the actual emissive power of the fire.

In order to evaluate $\bar{F}$, the parameter F_H has to be defined, as the element of the vessel closest to the fire, since it receives the highest rate of heat energy per unit area. F_H may be calculated by standard formulas [2]. The view factor decreases almost linearly with the solid angle Φ measured by the center of the sphere; it is zero for the elements located beyond the circular line Γ_M cut on the sphere by an imaginary cone which will have the vertex in the center of the fire and the lateral surface tangent to the spherical vessel (see fig. 1).

In such a way an angle Φ_M, identified by Γ_M, and the external surface of the vessel exposed to radiation from fire, A_3 [m^2], are defined. This analysis being limited to the cases in which the vessel is not enveloped by the flames, it is evident that A_3 cannot exceed the half of the exterior surface of the sphere. F_Φ represents the view factor for a generic element of the wall which lies on a circular line Γ_Φ at an angular position, Φ, with respect to the surface which connects the line with the center of the spherical vessel.

The average value of the view factor, $\overline{F}$, over the surface A_3 is determined through the following relation:

$$\overline{F} = \frac{F_H}{\Phi_M} \int_{\Phi=0}^{\Phi_M} \frac{G_1}{G_2} \, d\Phi \tag{2}$$

where G_1 and G_2 represent the analytical expressions for F_Φ and F_H [2].

As we have just examined, the region of the vessel exposed to fire receives heat as a result of thermal radiation only. At the time at which the fire starts the net heat transferred to the vessel is as shown in (1). Other initial conditions, thus relevant to time t_0, are that: the insulation outside surface temperature, $T_{3,0}$, the insulation inside surface temperature, $T_{2,0}$, the inside metal vessel wall temperature, $T_{1,0}$ and the liquid temperature, $T_{L,0}$ all equal the external environment temperature, T_A, so that at $t<0$ no heat is transferred from/to the atmosphere to/from the vessel.

As the insulation temperature exceeds T_A due to heating, the net heat to the sphere decreases, since the terms of cooling of the vessel - represented by the reradiative heat to the atmosphere and by both natural and forced convection - become effective.

Therefore at generic time t the expression for the heat flux is as follows:

$$\dot{q}_t = \dot{q}_0 - \varepsilon F_V \sigma (T_3^4 - T_A^4) - h(T_3 - T_A) \tag{3}$$

where ε is the emissivity of the insulation, F_V is the view factor between a generic element of the vessel and the atmosphere, σ [$5.7 \cdot 10^{-8}$ W/m^{2}K^4] is the Stefan-Boltzmann constant (in radiation) and h [J/m^{2}sK] is the convective heat transfer coefficient (total).

In order to estimate the maximum temperature rise of the vessel surface, equation (3) has to be solved in steady-state conditions, i.e at time t_∞, by imposing that the net heat transferred, $\dot{q}_\infty$, is zero; in such a way the equilibrium temperature, $T_{3,\infty}$, is evaluated. At this stage, however, nothing can be said about the time ,t_∞, required to reach such a temperature, since it strictly depends on the capability of the vessel to absorb heat.

In order to evaluate the distribution of temperature through the vessel as function of time, the non-stationary heat balance equation per volume unit of insulation layer is given:

$$\rho_s c_s \frac{\partial T_s}{\partial t} = \frac{k_s}{r^2} \frac{\partial}{\partial r} \left(r^2 \frac{\partial T_s}{\partial r} \right) \qquad (4)$$

where: ρ_s [kg/m^3] is the density, c_s [J/kg·K] the specific heat, k_s [J/m·s·K] the termal conductivity of insulation material, T_s is the temperature of a generic element of insulation located on the surface of radius r at time t. Both c_s and k_s may be expressed as polynomial functions of T, or roughly derived as the average values in the considered range of temperatures [3].

The partial differential equation (4) represents the variation of the heat flow passing through two imaginary spherical surfaces internal to the insulation, placed at a relative infinitesimal distance ∂r; such a difference corresponds to the amount of heat absorbed by the infinitesimal volume inside the two surfaces, and results in an increase of temperature ∂T.

Some boundary conditions have to be given in order to make equation (4) solvable; such conditions are relevant to the temperatures and heat flows at the outer border ($r=r_3$) of the space considered and are provided at t_0 and t_∞:

$$T_{3,0} = T_A \qquad (5)$$

$$k_s(\partial T_s/\partial r)_{3,0} = \bar{F}\tau E \qquad (6)$$

$$T_{3,\infty} = T_\infty \qquad (7)$$

$$k_s(\partial T_s/\partial r)_{3,\infty} = 0 \qquad (8)$$

while at generic time t the exchanged heat is:

$$k_s(\partial T_s/\partial r)_{3,t} = \dot{q}_t \qquad (9)$$

As for the insulation, the distribution of temperature within the metal is mathematically represented through a partial differential equation obtained by replacing, in (4), m_s, c_s, k_s and T_s with m_m, c_m, k_m and T_m respectively, which are the typical parameters for the vessel.

The boundary condition at the insulation-metal interface, i.e. at $r=r_2$, is as follows:

$$k_s(\partial T_s/\partial r)_{2,t} = k_m(\partial T_m/\partial r)_{2,t} \qquad (10)$$

the relation being valid at any time t.

Finally the heat flux into the liquid has to be evaluated, so as to determine the amount of heat which the liquid is able to absorb; this heat will be absorbed first as sensible heat of the liquid, and subsequently as latent heat of evaporation.

The heat exchange equation at the metal-liquid interface is given by:

$$k_m(\partial T_m/\partial r)_{1,t<t''} = h_i(T_3-T_L) \qquad (11)$$

where h_i [J/m^2sK] is the convective heat transfer coefficient and T_L is the temperature of the liquid.

As the radiative heat exchanged by the elements of the vessel adjacent to the vapour is little compared with convective heat, it is neglected. This assumption is conservative, as it gives a bigger value of temperature rise in the wall.

Therefore the heat exchange equation relevant to the metal-gas interface and valid at any time t is given by:

$$(\partial T_m/\partial r)_{1,t} = 0 \qquad (12)$$

Due to external heating the liquid temperature in the vessel grows for a time t', at which T_L equals the equilibrium temperature T_F of the substance at the set pressure of the relief valve, P_F. Up to this time the heat entering the vessel-liquid interface causes the temperature of the liquid to rise, as shown in the following equation:

$$h_i A_1 \cdot (T_1 - T_L) = m_L \cdot c_{pL} \cdot (\partial T_L / \partial t) \qquad (13)$$

where m_L and c_{pL} are the mass [kg] and the average specific heat [J/kgK] of the liquefied gas; A_1 [m^2] is the wetted area, i. e. the internal surface area of the vessel in contact with the liquid contents.

The solution of equation (13) over the temperature interval $T_{L,0} \div T_F$ allows the time t' at which the vapour is released through the relief valve to be evaluated.

The process of evaporation of the liquefied gas follows the heating stage; in the evaporation stage the heat is removed by the evaporating liquid according to the following equation:

$$h_b A_1 \cdot (T_1 - T_F) = H_v \cdot (\partial m_L / \partial t) \qquad (14)$$

where h_b [J/m^2sK] is the boiling heat transfer coefficient and H_v [J/kg] the latent heat of evaporation of the substance at temperature T_F.

The heat absorbed by the evaporating liquid is proportional to the wetted area A_1; contrary to the heating stage in equation (14) A_1 decreases with time, as the liquid level in the vessel decreases. The reduced capability of the system to remove heat results in a faster temperature increase in the vessel; therefore severe conditions of overtemperature in the material may be reached even before t", i.e. the time at which the last drop of liquid evaporates and leaves the vessel. This time is identified by the solution of equation (14).

For t>t" the equation (12) may be applied over the whole internal surface of the sphere; starting from this time the heat which enters the vessel only contributes to the wall temperature increase.

Stress analysis

Some general considerations are only given here, as a complete analysis of stress distribution would need the spatial variations in temperature and the correct boundary conditions to be known exactly.

Due to pressure and temperature increase as consequence of fire the vessel is subject to both pressure and thermal stresses. Tensile axial stress σ [N/m^2] caused by internal pressure may be easily calculated by treating the vessel as a thin-walled sphere, and is given by:

$$\sigma = \frac{P r_m}{2s} \tag{15}$$

where P [N/m^2] is the interior pressure; r_m [m] the mean radius of the vessel; s [m] the wall thickness.

The evaluation of thermal stress is more complex, as it depends by several factors: the distribution of temperature in the vessel; the restraint conditions; the liquid level in the vessel. It has been demonstrated [4] that the larger tensile stresses can result from the difference in wall temperature between the regions of the shell adjacent to the liquid and the regions of the shell adjacent to the vapour.

In order to estimate the time to failure of the vessel wall the decrease of strength of materials with temperature has also to be taken into consideration.

Conclusions

An analytical method has been proposed in order to foresee the behaviour of a vessel containing a liquified gas during exposure to thermal radiation.

The equilibrium vessel wall temperature and the time to reach steady-state conditions have been estimated; furthermore the distribution of temperature in the wall has been analysed as being function of time, thus enabling the evaluation of time needed to reach critical metal temperature. It has been proved that such a time depends both on design parameters and operating conditions.

References

[1] Mudan K. S.: "Thermal radiation hazards from hydrocarbon pool fires" Prog. Energy Combustion Science, 1984

[2] Rohsenow W.M., Hartnett J.P.: "Handbook of heat transfer" chap 15, McGraw-Hill, New York, 1973

[3] Perry R.H., Chilton C.H.: "Chemical Engineers Handbook" chap 3, McGraw-Hill, New York, 1973

[4] Johnson M.R. et al.: "Fire induced stresses in tanks containing liquefied gas" Pressure Vessels and Piping Conference, Miami Beach, 1974

Database of Mechanical Components in a Refinery. Structure and Operational Criteria for Inspection Planning an Reliability Evaluations

E. GALATOLA (*), G. SIMEONE (**), G.C. BELLO (*)

(*) ICARO Lodi (MI)
(**) AGIP RAFFINAZIONE (VE)

SUMMARY

The paper presents a recently established computerized database system for the collection and processing of mechanical inspection reports of a petroleum refinery. Major items under control are tanks, vessels, furnaces, pipes, heat exchangers and safety relief valves. The paper describes the type of data collected and their utilization, e.g. for determining life-expectations, strategic planning for the frequency of inspection, etc.

1. ACTUAL SITUATION

The testing and inspection department of the refinery has several items under control, e.g.:

- columns
- receivers
- separators
- degasers
- filters
- air-coolers
- reactors
- furnaces
- boilers
- snubbers
- heat exchangers
- coolers
- condensers
- tube bundles
- preheaters
- storage tanks
- safety relief valves
- pipe-lines
- hoses
- loading arms

The above mentioned items may be:

- active
- inactive
- exonerated
- demolished

For each item information is collected and periodically updated.

An example is shown in Fig 1, Fig 2 and Fig.3.

GENERAL EQUIPMENT CARD

PLANT	RC3
EQUIPMENT CODE	23-C-1
EQUIPMENT NAME	PRETREATING COLUMN
REGISTRATION N°	VE 9237
MANUFACTURE N°	085
PSV	P.4

OPERATIVE STATE : ACTIVE

TYPE : PRETREATING COLUMN

BUILDING FIRM : CANT.NAV. BREDA YEAR OF MANUF. 1971

CORROSION MARGIN 3÷6 CAPACITY (lt) 57300 NET WEIGHT (kg) 20608

PROJ. PRESS. (bar) 11.00 TEMP. (°C) F.248°C T.215°C

X RAY PHOTO : TOTAL ANNEALING : NO PRODUCT : GASOIL

HEAT INSULATION : NO METAL PACKING 18/8 ASBESTOS

NOTES

Fig.1 General information

PLANT : RC3 EQUIPMENT : 23-C-1 REGISTRATION N° VE 9237

MATERIALS CARD

BOTTOM AND SKIRT......:	Fe 42-1 UNI 5869
FLANGES...............:	ASTM A.105 Gr.I-A
...............:	Fe 42-1 UNI 5869
...............:	Aq.42 UNI 3986
PLATES................:	CARBON STEEL
PLATES VALVES.........:	AISI 410 S

Fig.2 Materials information

PLANT : RC3 EQUIPMENT : 23-C-1 REGISTRATION N° VE 9237

Pag.: 1 DATE: 01-01-82

MEAS.POINT	DESCRIPTION	ORIG. THICK.	MIN. THICK.
1	BOTTOM	16.7	10.7
2	PLATING I° MANHOLE	15.0	9.0
3	" 2° "	15.0	9.0
4	" 3° "	14.8	8.8
5	" 4° "	20.5	14.5
6	" 5° "	13.2	7.2
7	" 6° "	13.4	7.4
8	TOP	13.4	7.4
A	ø 20 " S.150 MANHOLE	12.7	6.7
B	ø 4 " S.150 160 INPUT	13.6	7.6
C	ø 8 " S.150 100 OUTPUT TOP	15.8	9.8
D	ø 10 " S.150 60 INPUT REBOILER	12.2	6.2
E	ø 10 " S.150 60 INPUT	13.0	7.0
F	ø 8 " S.150 80 INPUT	12.6	6.6
G	ø 3 " S.150 160 INPUT	11.0	5.0
H	ø 1½ " S.150 THERMOM.	15.3	9.6

Fig.3 Information about points of measurements

During the inspection the thicknesses are measured and written on the card; at the same time notes about the condition of the component (state, corrosion, forecast replacements, etc.) are reported.

Each item has a plan of inspection as follows; the type of controls to be made are shown in Fig. 4.

PLANT : RC3 EQUIPMENT : 23-C-1 REGISTRATION N° VE 9237

SITUATION OF INSPECTION

INSP. TYPE	LAST	NEXT	PLANN. INTERVAL
OPERATING TEST	01-01-88	01-01-89	12
COMPLETE CHECK	01-01-87	01-01-90	36
PSV SETTING	01-06-88	01-01-90	24
HYDRAULIC TEST	01-01-88	01-01-91	36
INTERNAL CONTROL	01-06-88	02-06-91	36

TYPE OF INT. CONTROL (U,M,L,E,R,T) UMP

U=ULTRASOUNDS M=MAGN. P=PEN.LIQ. E=EDDY CUR. R=RADIOG. T=THERMOVIS.

NOTES:

Fig.4 Plan of inspections per component

Using these cards, ticklers are produced to carry out the the internal controls and respect the law maturities.

2. INITIAL PURPOSES OF THE PROGRAM

To fulfill its aims, the testing and inspection department produces a considerable amount of paper.

In a refinery of this kind the number of different equipment is some hundreds, the items under control are thousands, there are several inspections per year per item and between ten to hundreds measurements per item.

The initial need, and the starting point for the work, is to collect all data and information in a computerized database.

Once data has organized on a computer, other features were made available.

2.1 The inventory of the equipment in the refinery has been made automatic, and this is particularly useful for the controls requested from ISPESL, the control and certification authority in Italy for pressure vessels.

2.2 The inspection reports are managed automatically and are easier to be found. In a report several information are written and qualitative judgements are formulated; in this way the retrieval phase is definitively simplified

2.3 The terms for the planned inspection, like operating and hydraulic tests, psv setting, complete check and the internal controls, are calculated from the computer and then printed, as in the example shown in Fig.5.

2.4 It´s possible to print individual reports for each equipment, in an extended or summarized form to know its history.

2.5 In each introduction, modification, deletion of information, congruency checks are made on the code, name, registration number, operative state, etc.

3. STRUCTURE OF THE PROGRAM

The program is organized to allow the following possibilities:

3.1 Creation of a card for a new equipment, with congruency checks on the code, name and registration number, as shown in the tables in Chapter 1.

3.2 Input of measurements data, inspection reports and notes for future controls.

3.3 Modification of a present card.

3.4 Reading of cards and former measurements.

3.5 Deletion of a card of a demolished equipment

3.6 Building the program to forecast internal controls, e.g. in Fig. 5:

PLANT : DP2											SHEET N° 1			
COMPONENTS		INTERNAL CONTROLS TERMS FOR THE YEAR 1989												
CODE	REG. N°	ICI	GEN	FEB	MAR	APR	MAG	GIU	LUG	AGO	SET	OTT	NOV	DIC
D/4	VE 1111				RM									
E/3	RM 1212				UM									
F/2B	GE 555									RE				
F/2C	GE 556									RE				

U=ULTRASOUNDS M=MAGN. P=PEN.LIQ. E=EDDY CUR. R=RADIOG. T=THERMOVIS.

Fig.5 Plan of internal controls per plant

3.7 Certification compulsory controls such as operating tests, complete checks, PSV settings and hydraulic tests, listed per year and per plant.

3.8 List of the controls per year per plant and per type of test.

3.9 List of equipment where tests were overdue, e.g. in Fig. 6:

PLANT RC3 : LIST OF EQUIPMENT
TESTS OVERDUE BETWEEN 01-07-88 AND 31-12-88

REGISTRATION N°	EQUIPMENT CODE	TYPE OF TERM
VE 9237	23-C-1	F S V
VE 111	D/3	F V
RM 1212	E/4	S M
GE 555	F/4B	F S

Fig.6 List of equipment that are overdue

4. FURTHER UTILIZATION

The collection of data as shown in Chapter 3 is a classical example of a computerized database of information previously gathered on sheet supports, but the utilization of the data may be more skillful.

Starting from this consideration more functions were added to the program:

4.1 Statistics of the corrosion rates are made.

The average corrosion rate is calculated:

- per item per plant
- per similar items in various plants
- per plant for all items.

The comparison of the corrosion rates points out critical points and critical conditions.

If we call:

S_{in} = Initial thickness

t = time

$C = \delta S/S_{in}/\delta t$ = corrosion rate

N_{ms} = number of measurements

the average corrosion rate will be:

$$C_{med} = \Sigma_i(C_i)/N_{ms}$$

4.2 The significance of the position and the number of measurement points is out-lined. If the corrosion of some point differs significantly from average corrosion and if the measurement is confirmed, the program will request further information and more investigation points in proximity.

The variance is defined:

$$\sigma^2 = \Sigma_i(C_i-C_{med})^2/N^2_{ms}$$

To evaluate the significance of the actual measurements points (location and quantity), the belongings of all the corrosion rates to the range $(0, C_{med}+2\sigma)$ is tested.

4.3 It's possible to make forecasts for the maximum operating time of the equipment:

$S_{i,lim}$ = minimum allowable thickness

$t_{i,lim}$ = maximum operating time

$t_{i,lim}/t = (S_{i,lim}-S_{i,in})/(S_i(t)-S_{i,in})$

$t_{lim} = \max(t_{i,lim})$

If t_{lim} precedes the date of the next inspection, a warning message is shown.

An example is presented in Fig.7.

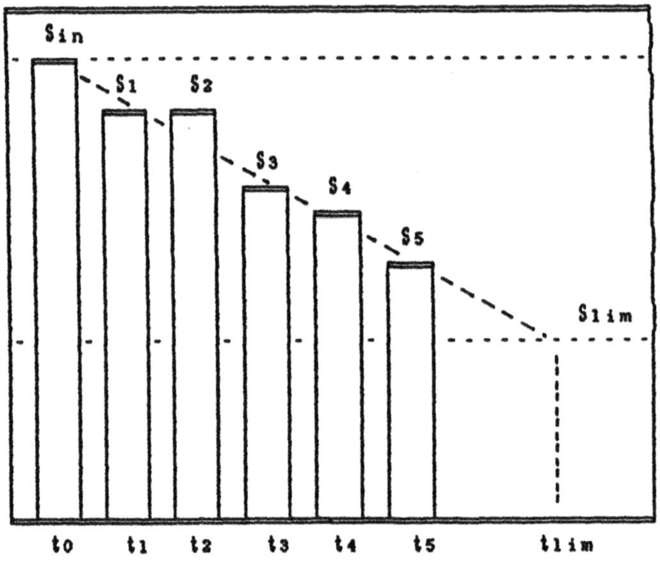

Fig.7 Maximum operating time per component

4.5 Some keywords are codified, so that it's possible to make statistics utilizing the notes of the inspection report; without reading every time all the reports.

Typically, the inquiries are done on:

- type of corrosive agent
- fluid involved
- type of corrosion
- operating time
- lifetime of the component

The statistics may be done:

- per item per plant
- per similar items in various plants
- per plant for all items.

4.6 At this moment attention was paid mainly to components with corrosion problems.

The reliability parameters are then calculated only for large failures; e.g. for an 8" pipe-line the possible failures are:

- significant leakage
- serious leakage
- catastrophic rupture.

It's possible to calculate the failure rate:

N = number of detected significant leakages

L = total length of 8" piping

T = operating time

fr = failure rate per unit of length

MTBF = mean time between failures

fr = L/N/T

MTBF =1/fr

5. FUTURE DEVELOPMENTS

5.1 The next need will be to transform the text database in a graphic database.

In this way each item will be identified also from its figure directly in the computer and the paper-dependency will be completely overcome; the modules where to write notes and enter measurements will be printed directly before inspections and the picture will be constantly up-dated.

5.2 Extending the analysis to other items, e.g. to actuators as valves, pumps, motors, switches etc., the introduced data will allow a more extended calculation of reliability parameters like the failure rate per operating mode and per type of failure, the mean time between failures, the mean time to repair, etc.

Applicability of First Order Reliability Methods A State of the Art

D. DIAMANTIDIS, G. FERRO, P. BAZZURRO

D'Appolonia S.p.A.

Summary

This paper introduces First-Order Reliability Methods (FORM) and presents in a condensed form examples from recent applications of FORM in structural analyses.

Introduction

The history of the first-order reliability methods is rather old. Ideas of using means and standard deviations to evaluate safety factors were proposed in the 1920's. The methodological background was available and was applied in civil engineering as a tool for uncertainty assessment in land surveying. For decades, however, the complexity of structural safety problems prevented any substancial development.

In the last 20 years there has been a rapid growth of academic interest in structural reliability theory and a growing acceptance by engineers of probability-based structural design, /1/, /2/. The development was encouraged worldwide by the work of safety code committees, who made clear the urgent need for some kind of rational basis for assessment of the safety factors specified in the building codes.

The theoretical background of component and system reliability is briefly described in this contribution. The benefits of utilizing FORM and the associated computer tools in structural analysis and design are illustrated on the basis of experience gained from case studies. The applicability of the methods to other industrial activities is finally emphasized.

Reliability of components

The computation of component reliability by applying FORM originates purely from second moment reliability methods, which are based on a simple representation of each basic variable by its mean value and standard deviation and by the covariances between each pair of variables. Recently, new methods have been developed to incorporate in FORM informations beyond the second moments, including marginal distributions, partial joint distributions and higher moments /3/. In these cases the multivariate distribution models have to be consistent with the available knowledge. As the quality of informations growths from second moments and covariances to perfect knowledge of multivariate distribution, the complexity of these methods increases as well, but not enough to prevent it from being used in practical applications.

The basic principles implied in FORM are summarized next:

- o Failure is defined by the limit state function separating the safe set and the failure set.
- o Random parameters influencing the limit state are modelled by a set of random variables and their joint distribution function.
- o The variables can be transformed by means of suitable transformations into a set of independent standard normal variables.
- o The limit state surface, now formulated in this new space, is approximated by its tangent hyperplane at the point of smallest distance to the origin, as illustrated in Figure 1.

The point of smallest distance is referred to as the design point or most likely failure point and is determined by an appropriate algorithm /4/. The probability of failure is then estimated by the probability content of the multinormal density function in the adverse domain defined by the approximating hyperplanes.

The influence of each random variable on the failure probability is expressed by the so-called sensitivity or weighting factors which give the coordinates of the design point.

Results obtained by adopting first-order reliability methods can be verified with the aid of Second Order Reliability Methods (SORM), which offer better estimates of the failure probability, especially for low values of the failure probability /6/. In SORM the limit state surface is approximated by a quadratic hyperparaboloid.

FORM can equally be used in cases where a target reliability is specified and some basic parameters must be selected to ensure that the reliability is not smaller than the target value.

In many practical problems the limit state equation is quite complex (e.g.: defined through finite element calculations), making computationally impossible to directly evaluate it at any iteration step in the FORM algorithm. This problem is overcome by using response surfaces (see /6/ for a general discussion) where the actual failure surface is approximated by a simple analytical form.

System reliability

System behaviour of structures arises from the presence of several structural elements, as well as from the possibility for each single element to fail in different modes.

From the structural analysis, a logical representation of the structural system can be derived, where component failures are logically related to the system failure in a suitable way, e.g. fault tree or event tree representation, as illustrated in Figure 2. This logical representation may be reduced, using the rules of elementary set algebra, either to a minimal cut set representation or to a minimal tie representation.

A critical review of implementation of first-order techniques in system reliability assessment is given in /7/. The practical application of those techniques requires explicit identifification of all possible failure modes of the structure.

Unfortunately, the practical application of FORM turns out to be a prohibitive task for most engineering structures which are in nature complex; therefore approximate approaches have been derived. The simple one consists in considering only a subset of all possible failure modes, selected according to a suitable algorithm in order to include the most likely failure modes. An upper bound of the system reliability can be thereby obtained.

A methodology for conducting a probabilistic seismic risk assessment at nuclear power plants has been presented in /8/. The practical application of the methodology based on FORM has been demonstrated by examining the effects of variability in soil-structure interaction parameters.

Implementation of FORM in structural design

Design of new structures

For structural concepts which are not simple extension of well known concepts, no records are available on which design can be based. In addition, the importance of failure modes and parameter uncertainties cannot be stated without the implementation of FORM. This is illustrated in the design of a subsea production system which is represented by composite shells consisting of two concentric steel skins with the annular space between the steel skins filled with concrete. Figure 3 illustrates the probabilility of failure and the corresponding reliability index as a function of the dimension of steel and concrete. Based on the obtained curves the design may be carried out such that target reliability requirements are fullfilled.

Development of reliability based codes

The main purpose of the safety requirements and regulations in structural codes is to ensure a socially acceptable reliability level and to implement this in a format which lends itself to practical applications. The choice of the safety level shall always be a compromise between theoretically desirable and convenient design.

Until now, safety requirements in technical standards have been mainly accomplished on the basis of tradition and accumulated experience. Occasionally, this can lead to a considerable non-uniformity of the safety level, with varying economic implications, in structural codes. Such a type of inconsistency in currently applied code frameworks is illustrated in an example dealing with the design of a platform in a damaged state (taken from /9/).

Safety checking of a platform at a damaged state is a special design case, which is specified as a 2000 year event. An analysis was performed /9/ to determine the return period consistent with:

- o i) current code requirements and corresponding safety factors;
- o ii) calibrated format based on a specified code space metric, which penalizes over- and under-design, and on FORM /9/.

The results illustrated in Figure 4 show that the return period covered by the current standards is not uniform and the corresponding format is not well calibrated. The format derived by applying FORM techniques is accurate since it implies return periods close to the target value.

Reliability analysis of existing structures

FORM provides a rational tool to measure and compare the relative change in the safety level of a structure to the original codified requirements. The methodology for updating of limit states and of random variables based on FORM has been described analytically in /10/. An example is illustrated here dealing with fatigue crack growth of a surface crack to a through - thickness crack.

The analysis has been performed for a tubular joint of a 12 years old platform. A stochastic model for fatigue crack growth has been applied which accounts for uncertainties in loading, initial defects, material parameters and in computation of stress intensity factors. The crack growth model has been combined with FORM to compute failure probabilities according to /11/.

Figure 5 illustrates the failure probability of two inspected joints A and B as a function of time, starting from the time of inspection. The crack growth parameters have been calibrated on the basis of the measured crack depth. The range of acceptable failure probabilities is also indicated in Figure 6. While the tubular joint A meets the specified requirements for a desired total life of 30 years, the tubular joint B is associated to a high (not acceptable) failure probability. The overall safety of the platform related to fatigue can be evaluated by combining the individual failure probabilities of the critical joints and performing a system reliability analysis.

Implementation of FORM to other domains

Although FORM have been mainly applied to civil engineering structures, they are equally applicable to other industrial activities such as transportation or chemical industry and to domains of investment planning or insurance policies. A main problem for the implementation of FORM seems to be the lack of sufficient exchange of knowledge between the scientists and the respective industry.

In /12/ FORM were applied to develope a conceptual basis for maximum loss prediction to be used in earthquake insurance policies, while in /13/ FORM were implemented as a decision tool for financial planning and investement strategies. Other example applications will probably follow in the near future.

Concluding remarks

The general statement resulting from the experience gained from FORM in structural analyses is an optimistic and promising one. The necessary analytical tools and the corresponding computer software exist. A significant progress has been also made in quantification of various types of uncertainties. The basic methodology has been briefly reviewed in this contribution together with main application domains. Concluding it can be stated that the first- (and asymptotic second-) order reliability methods (FORM):

- o have the potential for enhancing the quality of the system beeing engineered;
- o offer a unique tool for decision making in the face of uncertainty;
- o are applicable in areas ranging from structural design to financial planning.

REFERENCES

1. Madsen, H. O.; Kenk, S.; Lind, N. C.: Methods of structural safety. Prentice Hall: Englewood Cliff, NJ 1986.

2. Hohenbichler, M.; Gollwitzer, S.; Kruse, W.; Rackwitz, R.: New light on first- and second-order reliability methods. Structural Safety 4 (1987) 267-284.

3. Der Kiureghian, A.; Lin, P. L.: Structural reliability under incomplete probability information. Journal of Engineering Mechanics 112 (1986) 85-104.

4. Rackwitz, R.; Fiessler, B.: Structural reliability under combined load sequences. Computer and Structures 9 (1979) 489-494.

5. Breitung, K.: Asympototic approximations for multinormal integrals. J. Eng. Mech. 110 (1984) 357-366.

6. Myers, P. H.: Response surface methodology. Allyn and Bacon Inc., Boston, MA 1971.

7. Ditlevsen, O.; Bjerager, P.: Methods of structural system reliability. Structural Safety 3 (1986) 195-230.

8. Michalopoulos, A. P.; Swane, I. C.; Manfredini, G. M.; Silvestri, E.; Constantopoulos, I. V.: Effects of variability in soil-structure interaction parameters on probabilistic seismic risk assessment. SMIRT VIII. Brussels, Belgium (1985).

9. Diamantidis, D.; Lereim, J.: Calibration of technical standards for marine structures. Reliability of Marine Structures Symposium. Arlington, VA (1987).

10. Diamantidis, D.: Reliability assessment of existing structures. Engineering Structures 9 (1987) 177-182.

11. Madsen, H. O.: Probabilistic and deterministic models for predicting damage accumulation due to time varying loading. DIALOG 5-82. Danish Engineering Academy. Lyngby, Denmark 1983.

12. Ziegler, A.; Diamantidis, D.: Maximum loss prediction for earthquake insurance. Reliability Enginnering 18 (1987) 61-69.

13. Madsen, H. O.: Application of FORM to financial planning. Panel Lecture at Reliability of Marine Structures Symposium. Arlington, VA (1987).

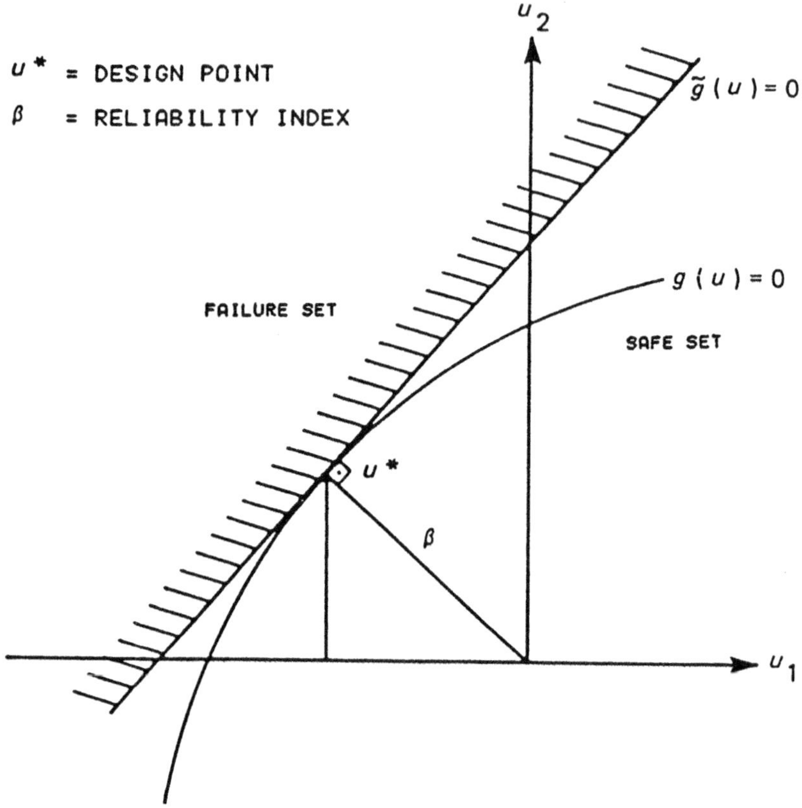

Fig. 1. Illustration of FORM in the two-dimensional space

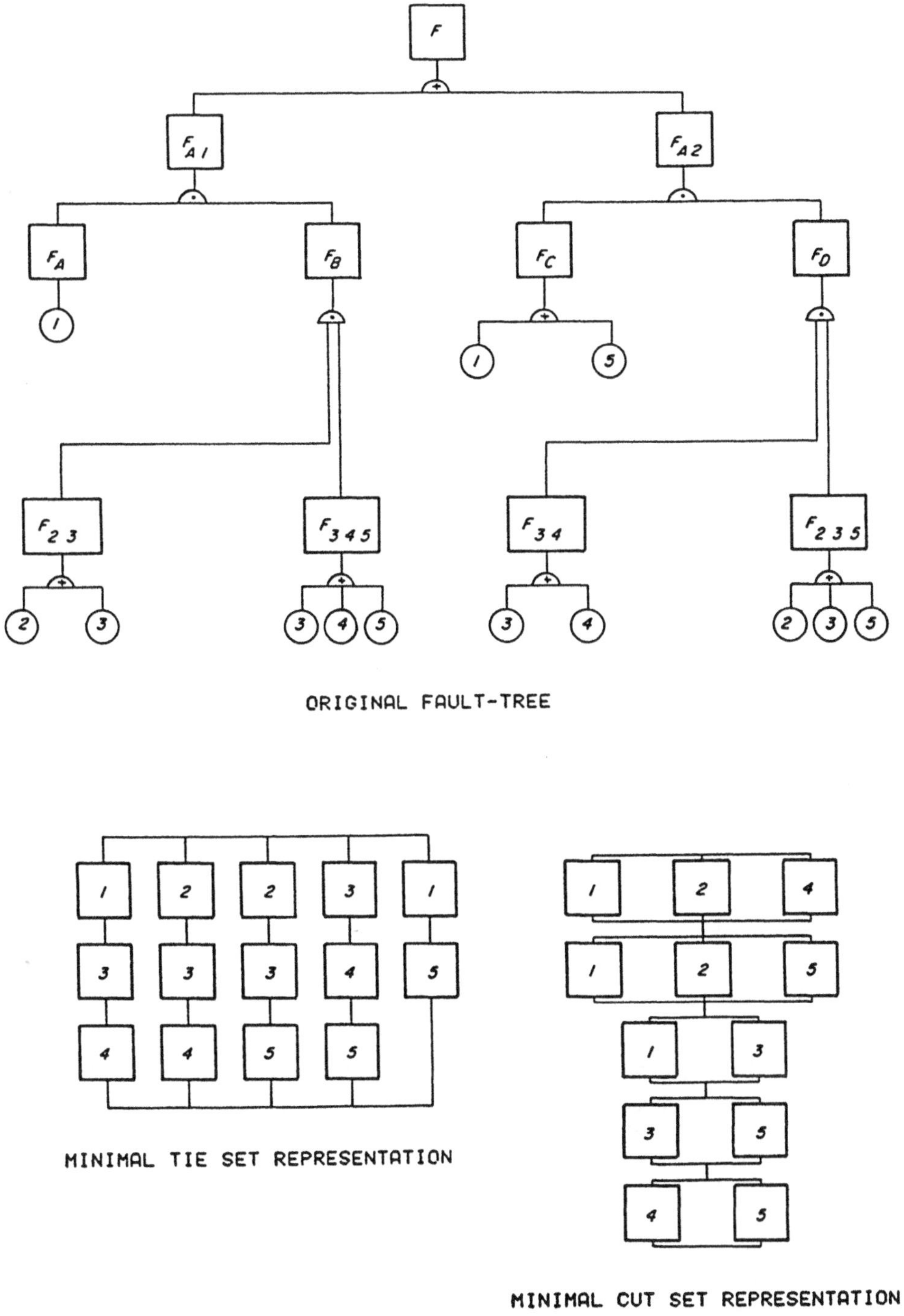

Fig. 2. Examplified fault tree, minimal tie set and minimal cut set representations

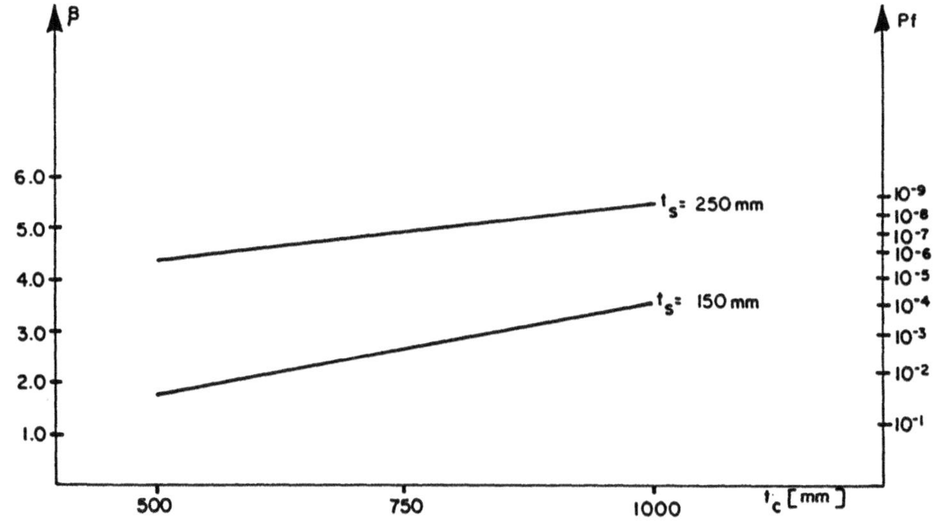

Fig. 3. Failure probability and corresponding reliability index for the subsea production system as a function of the steel thickness t_s and concrete thickness t_c

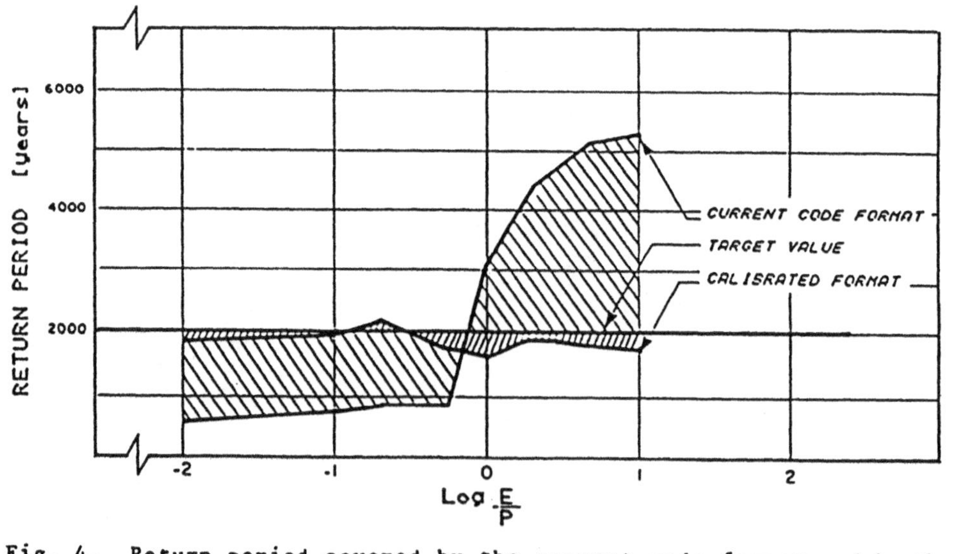

Fig. 4. Return period covered by the current code format and by the calibrated format as a function of the ratio environmental/permanent load

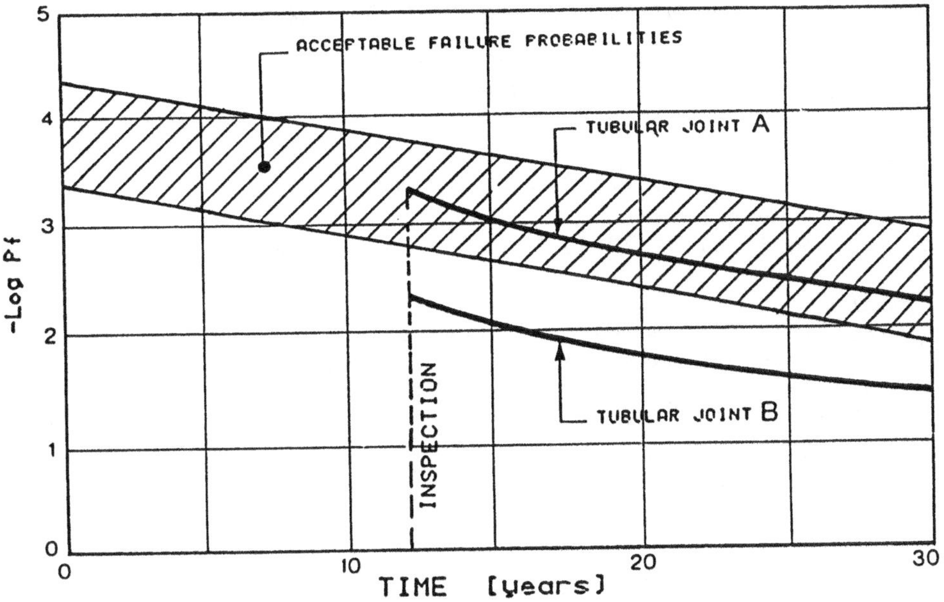

Fig. 5. Actual and acceptable failure probabilities for tubular joints A and B as a function of time

Estimate of Reliability Characteristics and Maintenance of Building Machines by Applying Statistics of Extreme Values

L.PAPIC
Engineering faculty, 32000 Cacak, Yugoslavia

P.DASIC
Engineering faculty, 37240 Trstenik, Yugoslavia

Summary

Statistics of extreme values, above all, has practical application with estimating extreme values of reliability characteristics and maintenance, both components and complex systems. The paper shows definition of extreme values, and according to asymptote distribution type I (Gumbel distribution) MINIMUM values of UP TIME and MAXIMUM values of DOWN TIME of machine stroke are determined as critical units of a bulldozer TG-50 manufactured in "14 Octobar" factory, Kruševac (Yugoslavia). Processing empirical data from exploitation is done according to a developed program packet in FORTRAN 77 program language.

Definition of extreme values

Distribution of extreme values is defined in the following way, |1,2|. Let the set $x_1,x_2,\dots,x_n$ represent the order of indipendent values of variable $x_k (k=1,2,\dots,n)$ in N samples of the same basic set. Then accidental values $Zn=\min\{x_1,x_2,\dots,n_n\}$ and $Wn=\max\{x_1,x_2,\dots,x_n\}$ represent extremes of separate samples (Fig.1).

Ordinal number $i=1,2,\dots,N$	1	2	...	i	...	N
Sample elements x_{ki} $(i=1,2,\dots,N)$	x_{k1} $(k=1,2,\dots,n)$	x_{k2} $(k=1,2,\dots,n)$	...	x_{ki} $(k=1,2,\dots,n)$	...	x_{kN} $(k=1,2,\dots,n)$
Sample minimum Zni $(i=1,2,\dots,N)$	$Zn_1=\min\{x_{k1}(k=1,\dots,n)\}$	$Zn_2=\min\{x_{k2}(k=1,\dots,n)\}$	...	$Zn_i=\min\{x_{ki}(k=1,\dots,n)\}$	...	$Zn_N=\min\{x_{kN}(k=1,\dots,n)\}$
Sample maximum Wni $(i=1,2,\dots,N)$	$Wn_1=\max\{x_{k1}(k=1,\dots,n)\}$	$Wn_2=\max\{x_{k2}(k=1,\dots,n)\}$	...	$Wn_i=\max\{x_{ki}(k=1,\dots,n)\}$	...	$Wn_N=\max\{x_{kN}(k=1,\dots,n)\}$

Fig.1. With considering definition of extreme values

Asymptote distribution of extreme values Zn and Wn may appear in several types, depending on the shape of distribution function of initial accidental variable $x_k (k=1,2,\dots,n)$ of a sufficiently big sample. Three types of asymptote distribution are known both for minimum and maximum values and depending on initial distribution. Particularly important are:

1. Asymptote distribution,type I for minimum values,
2. Asymptote distribution,type I for maximum values, and
3. Asymptote distribution,type III for minimum values.

Asymptote distribution type I is called distribution of extreme values type I, Gumbel distribution or symply a distribution of extreme values. Asimptote distribution of minimum values type III is called Weibull distribution.

In order to solve a problem of extreme values a general application methodology of Gumbel prognosis model is discussed, |4,5,6,7|.

Algorithm of forming Gumbel prognosis model of extreme values

Algorithm of Gumbell prognosis model of extreme values, shown on Fig.2 consists of following algorithm steps (procedure), |8|:

.1. Choice of prognosis model, on the basis of interactive dialogue user-computer: $m_p=1$ prognosis model of MINIMUM, $m_p=2$ prognosis model of MAXIMUM;

.2. Reading sample dimensions $n \geq 25$ and values of discrete accidental variable $x_k (k=1,2,...,n)$ of a sample according to accidental distribution;

.3. Arrangig values of accidental variable $x_k (k=1,2,...,n)$ of a sample according to decreasing set (distribution) if $m_p=1$ (at MINIMUM prognosis) or according to increasing set (distribution) if $m_p=2$ (at MAXIMUM prognosis);

.4. Forming tabular concept of determining empirical points in extreme and probability paper on the basis of accidental variable value $x_k (k=1,2,...,n)$ and probability $\emptyset_k(y)$, that is standardized variable $y_k (k=1,2,...,n)$;

.5. Estimating the center of scattering $\bar{x}$ and standard deviation σ_x of a sample;

.6. Adopting mean value $\bar{y}_n$ and standard deviation σ_n in function of sample size n, according to data placed in data set;

.7. Evaluation of value parameters $\frac{1}{\alpha}$ and q of theoretical straight line in extreme and probability paper;

.8. Forming theoretical straight line of the shape: $x=q-\frac{1}{\alpha}\cdot y$ (if $m_p=1$) or $x=q+\frac{1}{\alpha}\cdot y$ (if $m_p=2$);

.9. Forming a field of confidence of theoretical model for significance-level 5%, on the basis of mean square fault $\sigma(y_p)$ p-th values of standard variable, in function; y, $\emptyset(y)$ and n, according to data placed in data set;

.10. Determining a degree of agreement empirical and theoretical model. If empirical model is not sufficiently in agreement with correspodent theoretical one samples are made homogenous by the method of abandoning definite boundary empirical results (data) from a sample and by forming a sample $x_k (k=1,2,...,n)$,

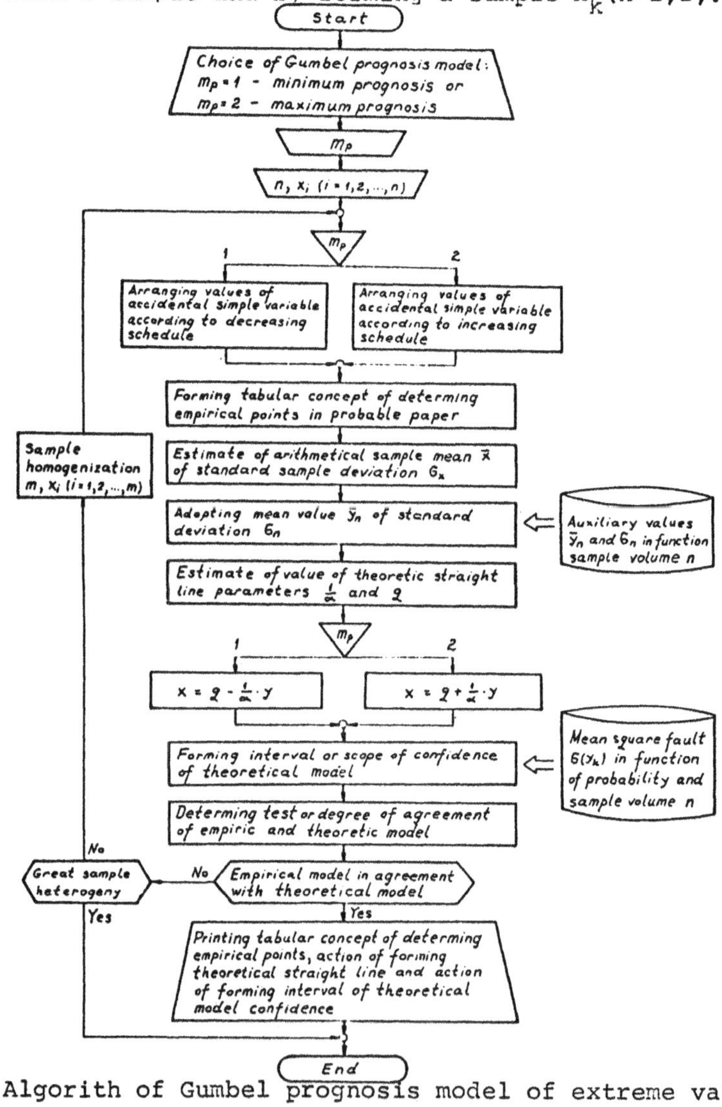

Fig.2. Algorith of Gumbel prognosis model of extreme values

so that $m<n$, and repeating a complete procedure. If after that empirical model is not sufficiently in agreement with theoretical one, the dimensions of the sample n are increasing. On the contrary, the following algorithm step is taken.

.11. Printing output processing results in the form of a concept of determining empirical points, theoretical straight line and the field of confidence of theoretical straight line in extreme probability paper.

Program realization of Gumbel prognosis model of extreme values

According to algorithm shown on Fig.2 a program packet is formed for Gumbel prognosis model of extreme values coded in FORTRAN 77 for compute system EI-Honeywell 6/95, designed on modulated principle. Program packet consists of main program, six general subprogram and two independent programs for forming datasets (one for auxiliary values $\bar{y}_n$ and σ_n and another one for mean square fault $\sigma(y_p)$, |9|. The main program controls six general subprograms, and contains just a choice of Gumbel prognosis model of extreme values on the basis of interactive dialogue user-computer, reading sample n dimensions and values of descrete accidental sample variable $x_k (k=1,2,\ldots,n)$ according to accidental schedule and calls for subprograms. Each general subprogram makes one independent unit, that is one mode, which is not bigger than one page of the list.

Estimate of up time minimum and down time maximum of building machines

Bulldozer TG-50 is a typical machine system represented by components and relation among them. Structural and functional link among components provides criterion function in correspondent exploiting conditions and in a planned time. The ability to perform criterion functions in that exploitation period is called bulldozer OPERATING ABILITY. Only until bulldozer is in its basic state UP TIME criterion function is provided. However, during the phase of real exploitation bulldozer is exposed to different influances according to the intensity, direction and character, what leads to deviation from the level of criterion function, that is, to reduction of bulldozer operating ability. Bulldozer is in the second basic state-DOWN TIME if the necessary interdependence of components in the system is disbalanced, as well as their characteristics and significance. In that case it is necessary to return the system in the state UP TIME by corrective maintenance or repair. Basic states, as a quantitative indicator of bulldozer TG-50 functional quality in the shape of time picture of states are given on Fig.3 |3|.

During real exploitation of bulldozer TG-50 manufactured in "14 October" Kruševac/Yugoslavia industry, machine stroke is identified as a critical unit. Time picture of the state in the shape

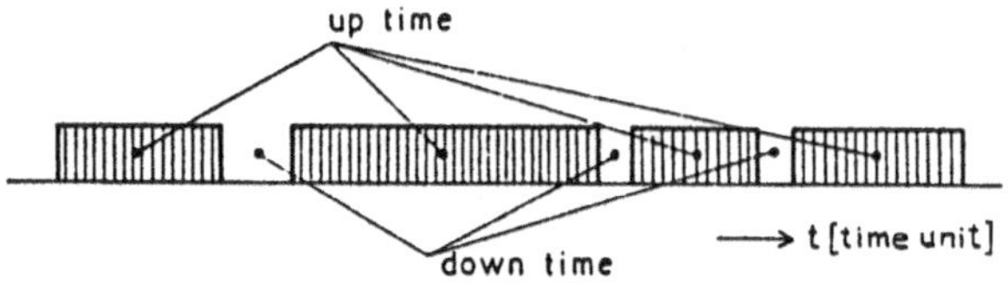

Fig.3. General time picture of system state

of alternating intervals UP TIME $|10^3 h|$; DOWN TIME $|h|$ of machine stroke (Table 1).

Table 1. Time picture of the bulldozer TG-50 machine stroke state

1	Up time $[10^3 h]$	1,314		2,001		1,803		0,787		1,758		2,115	
	Down time [h]		2,30		3,00		2,45		2,20		3,05		1,35
2	Up time $[10^3 h]$	2,234		1,587		2,636		3,795		1,731		1,012	
	Down time [h]		2,80		2,50		2,00		0,80		2,30		2,10
3	Up time $[10^3 h]$	2,173		2,203		3,701		2,880		2,533		1,631	
	Down time [h]		2,40		1,90		1,00		3,15		2,25		3,10
4	Up time $[10^3 h]$	2,574		3,452		2,422		3,153		3,678		1,421	
	Down time [h]		2,50		1,50		3,70		2,35		1,15		2,25
5	Up time $[10^3 h]$	2,391		2,672		4,250		1,299		1,347		2,022	
	Down time [h]		2,60		2,05		3,50		2,35		1,20		2,45
6	Up time $[10^3 h]$	1,552		1,847		2,905		2,521		1,153			
	Down time [h]		2,70		2,15		3,25		1,70		2,40		

Processing empirical data on the computer EI-Honeywell 6/95, according to developed program packet on FORTRAN 77, prognosis model of MINIMUM UP TIME of bulldozer machine stroke is formed (Fig.4). Empirical distribution with probability 95% corresponds to theoretical straight line in extreme and probability paper: X=2,6469-0,7449·Y according to this model it is possible to determine reliability of MINIMUM UP TIME realization to breakdown machine stroke and to define quantitatively a part of the sample where those conditions are not applied. Bulldozer TG-50 will be UP TIME with 90% of probability: 0,9706 moto h = 970,6 h, which does not apply to every tenth datum in the sample size N=35. The prognosis does not apply to any of 31 data in the examined sample.

Prognosis model of MAXIMUM DOWN TIME of bulldozer TG-50 (Fig.5), with 95% interval of confidence has the shape X=1,9735+0,6016·Y. According to this model, with probability of 90%, maximum time of corrective maintenance will be 3,327 h, it does not refer to the same quantitative sample part as well as with the model of minimum prognosis.

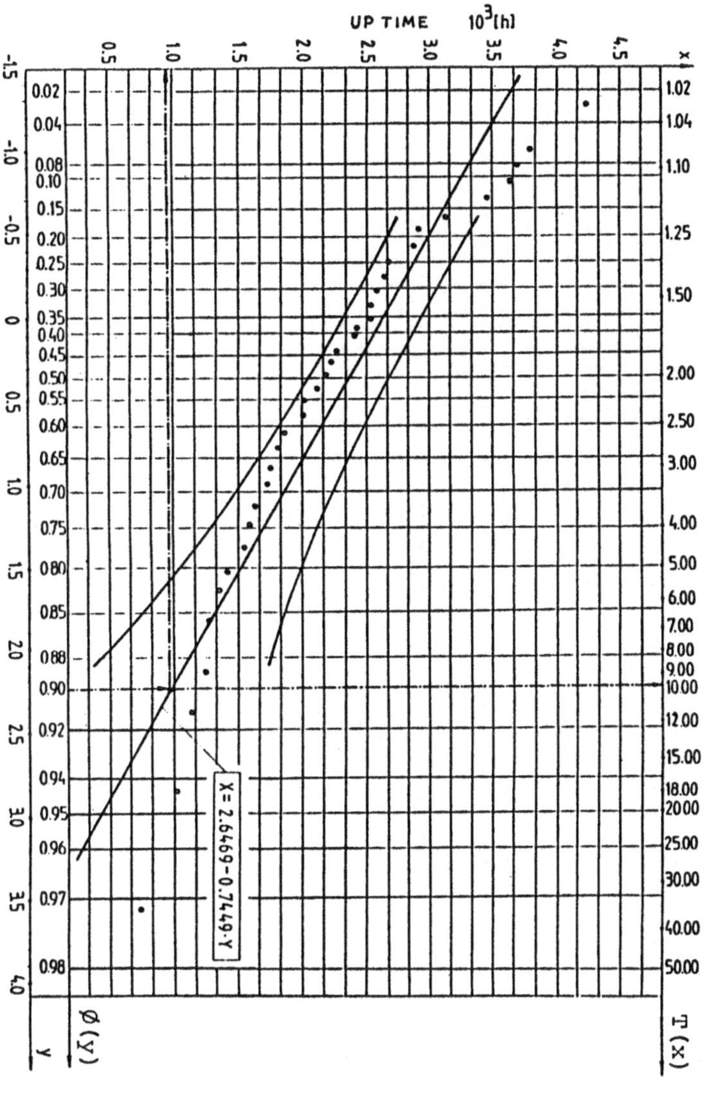

Fig.4. Variability dependence and MINIMUM UP TIME of bulldozer TG-50 machine stroke

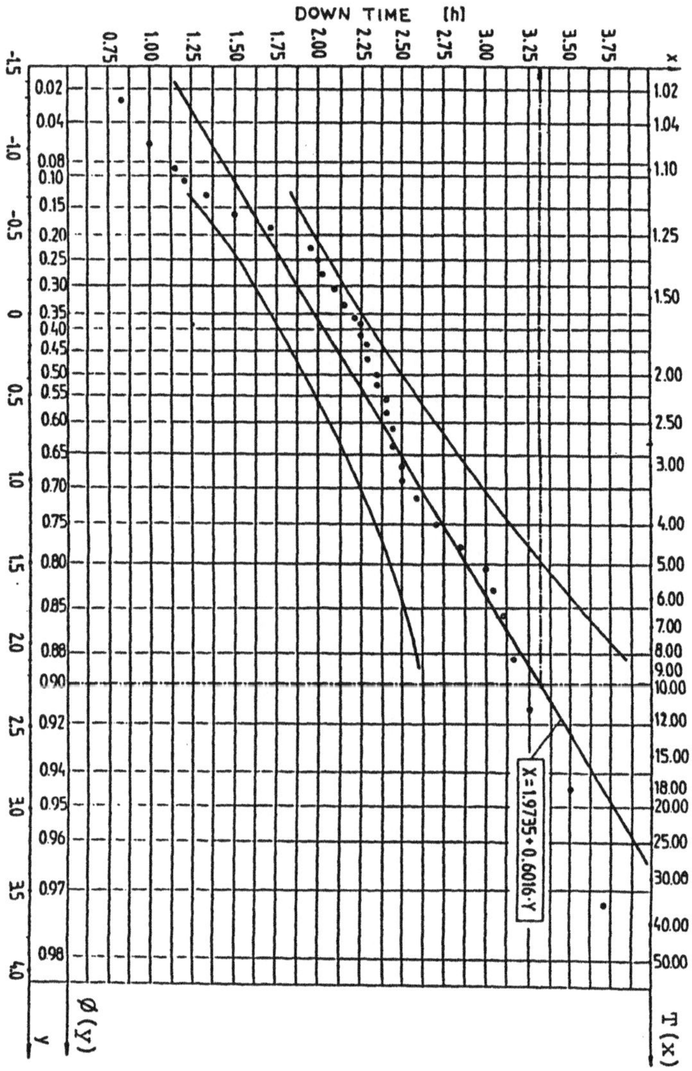

Fig.5. Variability dependence and MAXIMUM DOWN TIME of bulldozer TG-50 machine stroke

Conclusion

Estimate of reliability characteristics and maintenance of machine stroke as a critical subsystem of bulldozer TG-50 is done on the basis of time picture of state from exploitation.For the model of asimptote distribution of extreme values type I (Gumbel distribution) in program language FORTRAN 77 a program packet is formed which is used to process empirical data by computer. Obtained results make the prognosis of MINIMUM UP TIME values and MAXIMUM DOWN TIME values possible, with corresponding probability, as well as a quantitative part of a group of data the prognosis stands for. The created prognosis model makes objective limiting of continuous work and minimizing the time of corrective maintenance TG-50 bulldozer possible.

References

1. Gumbel,E.J.: Statistics of Extremes, Columbia University press, New York, 1962.
2. Galambos,J.: The Asymptotic Theory of Extreme Order Statistics; John Wiley and Sons, New York, 1978.
3. Todorovic J., Zelenovic D.: System efectivity in mechanical engineering (in Serbo-croat), Naučna knjiga, Belgrade, 1981.
4. Lisowski, Z.: Application of Gumbel theory while estimating rail vehicles reliability, Zagadnienia tarcia zažicia i smarowania No.10, Warshaw, 1972.
5. Lisowski,Z.: Freitting, pitting and spalling as wear forms of rail vehicles elements, Zagadnienia eksploatacijі maszyn No.4, Warshaw, 1972.
6. Papic,L.: Possibilities of applying Gumbel prognosis model of extreme values (in Serbo-croat), SYM-OP-IS ´87, Herceg Novi, 1987.
7. Papic,L.: Reliability analysis of complex systems by applying extreme values statistics (in Serbo-croat), SIPT ´88, Cavtat, 1988.
8. Dasic P., Papic L.: Computers application of Gumbel prognosis model of extreme values (in Serbo-corat), Računarstvo u nauci i obrazovanju No.3., Belgrade, 1988.
9. Dasic,P.: FORTRAN 77 in the field of production engineering, part II (in Serbo-croat), Kruševac, 1988.

VVF SIGEM: A Computerized System for the Emergencies Management

E. Marchionne[1] , M. Lanzino[2], M. Gilioli[3]

[1] Home Office - VVF Operative Center - Rome, Italy

[2] I.C.A.R.O. - Pisa, Lungarno Sonnino - Italy

[3] TEMA - Milan, via Medici del Vascello 26 - Italy

- ABSTRACT

In this paper it is described the general architecture of the Computerized system conceived by TEMA in collaboration with the Italian Fire Brigade Branch for the emergencies management.
The structure of the main Data Bank and the typologies of the interconnected simulation models are indicated as follow.

- VVF SIGEM

For a long time the Italian Fire Brigade Branch has had at is disposal at the Operative Centre of its General Management, an informative support able to coordinate, at the right moment, the Operative Force in case of disaster.
Such a system allows to determine at the moment of the accident

- The composition of the intervention teams all over the national territory.
- What the nearest special equipments necessary to the various type of intervention are, and where they are located.
- The estimated times necessary to reach the place of the accident efficaciously.

In order to give further development to the existing system and a qualified and specialized support to the aid activity carried on by the Branch, TEMA (a company of ENI group) has set up a procedure for the emergencies management consequent to accidents involving the presence of dangerous chemicals.

The system, called VVF SIGEM (Computerized System for the Emergencies Management) has been made operative first in relation to the accidental settings with accident or explosion, then to the accident caused by the release of toxic substances.

The VVF SIGEM software structure is mainly composed by data archives and by a whole of advanced simulation models.

The data archives are formed by a Substances Data Bank containing detailed information in relation to thousand chemical substances; for each product we have stated the danger characteristics, the indications for the measures in case of emergency, the suggestions for the first aid, the estimate of the precautionary areas for first evacuation and a complete survey of the physico-chemical data.

The access to the Data Bank is extremely flexible as the search of the concerned substance may be also effected through the commercial denomination or the possible synonymous, or through the chemical formula or the main international codification numbers (Kemler, CAS, CEE).

In Table 1 we find, by way of example, the cards which make up the information contained in the Data Bank for each of memorized substances.

We must underlined that, at the moment, the Data Bank contains nearly all the substances used in the main process, transport and storing activities.

The simulation models have been elaborated on the basis of the most advanced and up to date system of analyses and allow the evaluation of the dinamic and phisical consequences of the phenomena related to:

- fire
- explosion
- release of vapours and toxic and flammable gases in atmosphere

The Data Bank and the simulation models are interconnected, so that each model can directly withdraw, if it's wanted, the phisico-chemical parameters necessary to the simulation of the observed or simulated accident.

The start of the system is possible giving, by radio or telephone from the place of the accident, a description of the accidental setting with the essential information concerning micro-climate, shape and position of the vessel, substance and quantity involved, type of failure ecc.

The choice of the simulation model may be effected either directly by the operator of the Emergencies Room or automatically by the System: in such a case VVF SIGEM proposes the elaboration considered each time the most suitable on the basis of the fundamental input introduced in it.

In Table 2 we find the list of simulation models existing in the System.

The extremely reduced number of information (thank to the data of the Data Bank and the automatic choise System of the models) and the quickness of elaboration allow an immediate and continued assistance to the action of the Branch which operates on the site of the accident.

The output of the models are given under two forms:

- As an indication of the distances corresponding to the threshold of danger connected with the specific accidental typology (for example: thresholds for radiation, for over pressure ecc.).

- As planimetrical maps of the tendency of the phisical consequence related to the distance.

The former type of information gives important suggestions for the immediate emergency, while the latter may usefully support the intervention planning of the VVF equipment.

The information given by the system, rebrodcasted on the site of the accident, allow to adopt those safety precautions such as road blocks, evacuation of population, use of neutralising or estinguishing substances, organisation of the personnel intervention, protection of the exposed people.

The global architecture of the System represents therefore the most advanced study elaborated about Substances Data Banks and comprises the latest methodologies simulating the accidental events produced by the italian and international research.

The operative interconnection between Substances Data Bank and Simulating Models represent an important methodological innovation, able to bring to the best the execution times and the precision of data input.

Moreover it makes it possible the use of the System by personnel with simple operative knowledges and not appropriate scientific training.

Such characteristics put VVF SIGEM in a leading position among the existing system for the accident prevention , the emergencies management and the organisation of the interventions by the proper Branches.

Since VVF SIGEM has been started at the Operative Centre of VVF National Branch, there have been many alarms and interventions during which the informatic instrument has been able to supply a valid and opportune operative support, and it's natural to forsee a more and more generalised use of the System within the activities of the control and aid teams.

In such a prospective TEMA and the Ministry of the interior are arranging the further developments to be realized, concerning both the enlargement of the Substances Data Bank and the elaboration of new Simulation Models of particular accidents.

- CONCLUSIONS

The System described above represents something new which in its first operative phase has proved to be of a great importance as an on-line help in coming to operative decisions during emergencies management.

Moreover the easiness of use, along with the strength of simulation, allow an improvement in training of the personnel of the National or Local Branches put at the head of the defence of people from the consequence of accidents in industries and transport.

Table 1 : substance card printed by the System

CARD N.39

SUBSTANCE SAFETY CARD

IDENTIFICATION

SUBSTANCE	: PROPYL ALCOHOL
SYNONYMOUS OR CHEMICAL FORMULA	: PROPANOLO
KEMLER-ONU NUMBER	: 331274
CAS NUMBER	: 71238
CEE NUMBER	: 603003000

FIRE, EXPLOSION AND TOXICITY HAZARDS

FLAMMABILTY	: HIGH HAZARD
AUTO-IGNITION	: STABLE
TOXICITY	: MODERATE
REACTIVITY IN O_2	: SURE
REACTIVITY IN H_2O	: STABLE
SOLUBILITY IN H_2O	: HIGH

POTENTIAL RISKS

- TOXIC RELEASE	: NOT VERY DANGEROUS
- FIRE/EXPLOSION	: VERY PROBABLE

OPERATOR PRECAUTIONS IN CASE OF TOXIC RELEASE

- IN CASE OF A BIG RELEASE OF SUBSTANCE MAKE USE OF RESPIRATORY PROTECTIVE EQUIPMENT

- MAKE USE OF RUBBER GLOVES

Table 1 : Substance card printed by the System (cont.)

IMMEDIATE ACTIONS TO TAKE IN CASE OF FIRE

- CARBON DIOXIDE EXTINGUISHER
- POWDER EXTINGUISHER
- FOAM EXTINGUISHER

EMERGENCY ACTIONS TO TAKE IN CASE OF LEAKS

- STOP LEAKS IF IT IS POSSIBLE
- REMOVE AND SOAK THE RELEASED SUBSTANCE ON SAND OR OTHER INERT AGENT
- COOL THE VESSEL WITH WATER
- BLOCK ALL THE IGNITION SOURCES

EVACUATION DISTANCES IN CASE OF LIQUID PROPYL ALCOHOL LEAKED ON GROUND WITH WIND VELOCITY BETWEEN 2 AND 5 m/s

		SPREADING DIAMETER (m)			
DISTANCES OF		5	7	8.5	10
- IMMEDIATE DANGER	D	8.	10.	13.	17.
- ABSOLUTE SECURITY	A	78.	103.	128.	153.
	B	45.	81.	100.	110.

A : MAX. LEEWARD DISTANCE FROM POINT OF LEAK (m)

B : MAX. PERPENDICULAR DISTANCE FROM LEEWARD DISTANCE (m)

D : EQUIVALENT DIAMETER OF IMMEDIATE DANGER AREA (m)

Table 1 : substance card printed by the System (cont)

PHISICO CHEMICAL CHARACTERISTIC OF PROPYL ALCOHOL

- MOLECULAR WEIGHT		(Kg/Kmoli)	60.000
- CRITICAL TEMPERATURE		(°C)	264.000
- CRITICAL PRESSURE		(atm)	51.000
- LIQUID DENSITY AT	0 °C	(Kg/m^3)	822.000
AT THE BOILING TEMPERATURE		(Kg/m^3)	717.000
- GAS DENSITY AT	0 °C	(Kg/m^3)	2.680
AT THE BOILING TEMPERATURE		(Kg/m^3)	1.978
- LIQUID SPECIFIC HEAT AT	0 °C	(Kcal/Kg)	0.519
AT THE BOILING TEMPERATURE		(Kcal/Kg)	0.586
- HEAT OF VAPORIZATION AT	0 °C	(Kcal/Kg)	188.995
AT THE BOILING TEMPERATURE		(Kcal/Kg)	157.895
- VAPOR PRESSURE AT	0 °C	(atm)	0.010
- SPECIFIC HEAT RATIO		-	1.200
- SPECIFIC HEAT	0 °C	(Kcal/Kg °C)	0.531
- GAS VISCOSITY AT	0 °C	(Mpoises)	0.070
AT THE BOILING TEMPERATURE		(Mpoises)	0.095
- LIQUID VISCOSITY	0 °C	(Mpoises)	38.000
AT THE BOILING TEMPERATURE		(Mpoises)	5.000
- SURFACE TENSION AT	0 °C	(N/m)	0.032
AT THE BOILING TEMPERATURE		(N/m)	0.019
- GAS THERMAL CONDUCTIVITY AT	O °C	(W/m °C)	0.011
AT THE BOILING TEMPERATURE		(W/m °C)	0.018
- LIQUID THERMAL CONDUCTIVITY AT	0 °C	(W/m °C)	0.151
AT THE BOILING TEMPERATURE		(W/m °C)	0.142

Table 2 : VVF SIGEM Simulation Models

- CAMIN 1	:	Atmospheric dispersion of light gases and vapours (density lower than air) given out by continous and puntiform sources.
- CAMIN 2	:	High speed dispersion of neutral gases and vapours (as dense as air).
- HEAVY	:	Atmospheric dispersion of heavy gases and vapours (heavier than air).
- TANK 1	:	Outflow parameters calculation (flow, duration, pressure, temperature, density).
- EXPLO	:	Calculation of the overpressure peaks and areas of danger in case of unconfined explosion.
- FIREBALL	:	Thermal radiation effect consequent to fireball.
- POOLFIRE	:	Thermal radiation strength produced by a pool fire.
- TANKFIRE	:	Thermal radiation strength produced by a tank fire.
- JETFIRE	:	Thermal radiation strength produced by a jet fire.
- DANNI	:	Consequences coming from the catastrophic failure of vessels or other equipment with frowing of fragments.

Heavy Gas Dispersion

Mª Teresa Galvez
Labein
Dpto. Medio Ambiente
LABEIN

José Maria Renau
TEMA, S.A.

Introduction

Dispersion modelling for accidental releases has been a rapidly expanding and advancing field. One aspect which is causing growing concern is the hazard arising from an accidental release a with flammable or toxic gas. Knowledge of how such materials get dispersed in the atmosphere is important. The main objetive is to provide a good basis for the determination of hazardous zones in case of accidental releases. Predictive methods as used in air pollution problems may be considered adequated for various practical applications. Numerical modelling of heavy gas dispersion can be used. The mathematical techniques vary from simple box models to numerical solutions of turbulence equations over a three dimensional grid.
There are several sources of uncertainty in all dispersion models. In the heavy gas dispersion these incertainties are increased because this dispersion is subject to special physical phenomena, not included in the case os passive pollutants.

HEAVY GAS

A gas can be denser than air for a number of reasons: its molecular weight is greater than that of air (chlorine, propane, nitrogen dioxide, etc.), it is colder than the ambient air (LNG vapour, is a heavy gas because of its very low storage temperature), and/or it consist in part of a suspended aerosol of

liquid or solid particles (ammonia, when released under high pressure forms a cold cloud with part of the ammonia in liquid aerosol form). The density of the gas has a major influence on the dispersion of such a cloud. Considering an instantaneous release the first effect is that the gas spreads horizontally, or "slumps", with the height of the cloud initially decreasing, from an initial high cloud. Considerable dilution can be associated with this motion. The other major effect is due to the stable vertical density gradient at the top of the cloud, mixing, and thus reducing the dilution. On the other hand, owing to the slumping, the vertical mixing is ocurring over a larger area than would otherwise be the case.
In predicting the size and location of the hazardous region of concentration, it is important to consider the alteration in turbulent diffusion, gravity flow, and, for cold gases, heat flow from the ground.

Turbulent mixing

Turbulent mixing of any gaseous or particulate substance with air is primarily responsible for its dispersion. For a neutrally buoyant gas (conventional pollutants) the degree of turbulent mixing depends directly on wind speed and inversely on atmospheric stability. For a heavy gas there is an additional dependence; turbulent mixing decreases as the density difference between gas and air increases, leading to higher concentrations at any given downwind distance.

Gravity flow

Gravity flow is another phenomenon particular to heavy gas dispersion. Since the density of the gas is greater than that of air, the cloud tends to slump, spreading out horizontally while decreasing in height. This spreading makes the cloud wider, and can result in upwind motion of portions of the cloud and downwind motion at greater than wind speed. High values of wind speed and high levels of turbulence reduce the relative degree of gravity flow effects.

Gravity flow effects have a marked influence on concentration. The flattening of the cloud can raise or lower the concentration depending on position. The greater area of the cloud, due to flattening, causes more air to be mixed-in for a fixed level of turbulence, but the level of turbulence decreases with the reduction in cloud height.

Heat Flow

A phenomenon important to the dispersion of a cold heavy gas like LNG vapor is heat fow from the ground into the cloud. Heat flow lowers the cloud density, reducing the dense gas effects, and can provide the extra heat necessary to cause a LNG vapor cloud to become buoyant a long distance from the source.

In some accidental releases LNG may not evaporate as fast as it is spilled. Thus, understanding the phenomenology of the evaporation is required. Rapid phase transition (RPT) explosions can cause sudden releases of LNG vapor when the LNG es spilled on water, causing momentary extensions of the hazardous region, and are an additional hazard in themselves.
The existence and significance of these phenomena have been confirmed by quantitative measurements of their effects in some series of large-scale liquefied natural gas (LNG) spill experiments.
All afore-mentioned features and its physics basis must be incorporated into the numerical computer models used for predicting the evolution of heavy gas clouds.

Modelling

Several markedly different model types have been used for mathematical simulation of dense gas dispersion; they, usually, are placed into one of two categories: K-teory (eddy diffusivity) models and box models.

K-theory models numerically integrate suitably simplified equations of mass, momentum, and energy conservation, in either two-or-three-dimensional form. K-theory models can incorporate terrain effects and complex geometries, and describe wake effects around structures.

Box models assume that mass transfer occurs by entrainment across the density interfase of a cloud with an assumed shape (cylindrical, frequently) and that internal mixing is fast enough for the concentration within the cloud to be uniform. Air entrainment velocities are assumed to depend on turbulence models, density differences, and cloud speed.

The dynamics of the most essential state variables, over flat terrain, can be described by the box models. Possible refinements in some of the aproximations are obvious, but the design of a consistent and significantly more realistic model seems to introduce complications. These are associated with, for instance, the coupling between the release and entrainment processes, the coupling between the atmospheric flow and the gravity-induced flow, and the spatial variations of the gas cloud characteristics.

K-theory, three-dimensional grid models, offer the prospect of eventually calculating the interactions of dense-gas cloud with obstacles and terrain, but, also there are a number of difficulties. These relate to doubts about the level of turbulence modeling required. The modeling of turbulence and the corresponding dispersion constitutes a complex field of concurrent and complementary, approaches.

Uncertainties in the Models

In practice, models are based on aproximate descriptions, or parameterizations, of physical laws; they employ aproximate solutions of the governing equations; and they use values for the independent parameters that are derived from imperfect data.

In box models the requirements of input data are less stringent than for numerical models. As a direct consequence, the predictions and results of these models are less detailed and sometimes insufficient for the problem in question.
The principal parameter of interest in risk analysis is the distance to the point where the cloud is no longer ignitable or toxic. Of particular interest is the predicted extent of a flammable cloud at any time following a spill. At present, the dependencies of the predictions of different models on, for example, size of spill or windspeed are not in agreement. Thus some models show a steady increase of dispersion distance with windspeed while other show windspeed to decrease dispersion distance, to have no effect on it, or to cause it to go through a maximun as the wind speed changes.
Downwind dispersion distance is not the only parameter that can be calculated; the widin of the flammable or toxic cloud can be relevant, as well as the concentration as a function of time at a fixed location, and the distribution within the cloud. The box models provide only average concentrations.
In K-theory models there are an additional problem. In the construction of a turbulence model, which retains the influence of buoyancy, a choice must be made about the level at which closure will be made. Higher-order closure methods increase the generality with which the dynamics of the turbulence is described, however this approach also leads to a proliferation of unknown constants.

CONCLUSIONS

The predictions, even of a perfect model, cannot be expected to agree with observations at all times and locations, because there are some vagueness, not only, in the physical phenomena of the heavy gas dispersion, but also in the numerical approach used.

Software Data Collection and the Software Data Library

P. COMER

National Centre for Systems Reliability, UKAEA,
Wigshaw Lane, Culcheth, Warrington, UK, WA3 4NE

Summary

The Software Data Library Project began in 1985 with a one-year study of the problems involved in setting up a facility for the collection, storage, analysis, and dissemination of software engineering data. The second phase involved the design, development and administration of an appropriate database system. Data collection procedures were created, beginning with detailed definition of those entities about which data was to be collected and the metrics which applied to those entities. A data model was created which forms the basis of the design of both the database system and the data collection manuals produced by the project.

1 Background to the Software Data Library

The Software Data Library (SWDL) project was set up to create a UK facility for the storage, retrieval and analysis of Software Engineering data, operated for the benefit of the software community in general. The basic concept is that by collection of information about projects, such as the code and documentation generated (eg size, complexity, clarity), development environment and methods, occurrences of faults, effort and durations, it is possible to make systematic improvements in reliability, maintainability and productivity.

The project, sponsored by the Alvey Directorate of the Department of Trade and Industry, was envisaged as covering three phases. This paper reports upon the second phase.

The content of the three phases of the SWDL development may be summarised as follows:

Phase 1 This phase took place during 1985 and investigated the feasibility of setting up a SWDL, the interest of potential customers and its associated administration. An important activity was a survey of UK industry, whose results were to be used for Phase 2.

Phase 2 The prime objective was to set up an operational pilot SWDL, implementing the recommendations of Phase 1. In particular this comprised the creation of a computer database system, the collection of data, and development of data analysis and confidentiality procedures.

Phase 3 Phase 3 is a convenient nomenclature for whatever comes after Phase 2. It was not fully defined at the completion of Phase 2, but it was envisaged as a fully operational SWDL continuing to collect, store and analyse data, based upon the services and procedures developed in Phase 2.

The SWDL project was undertaken by a consortium of the following UK organisations:

National Centre for Systems Reliability (UKAEA)
STC (STC Technology Ltd and ICL)
NCC Ltd
GEC Software Ltd
Logica (Cambridge) Ltd

In Phase 1, there was also a sixth partner, Systems Designers Ltd.

Phase 2 of the project was sub-divided into a number of tasks. The following outlines the main functions and objectives of the tasks:

SYS Development of the System and Services
To develop the database and associated software to support data collection and verification, and the analysis tasks.

COLL Data Collection and Verification
The development of the data model and data collection techniques. The collection of a sufficient quantity and variety of high-quality software engineering data from a variety of projects. To provide, within a timescale of 1 month, analysis of a given project for statistical and software engineering anomalies. To provide feedback to the data provider on possible problem areas in the project, identified from the data, and helping to ensure that the data contained in the SWDL represents an accurate record of the project.

ANL Data Analysis
Tasked with comparison of projects within the database to correlate trends in productivity and reliability etc, with the use of particular methods, tool types or development environments. Derived trends etc would be published in open literature.

ADMIN Administration of Services
To provide the services and procedures for the day-to-day running of the library, for data entry, security and maintenance. This task was responsible for the protection of sensitive data received from clients and for the development of customer services. All data was treated as confidential and "sanitised" before being released for further processing.

TT Technology Transfer
Aimed at ensuring public awareness of the value of collecting and using software metrics and generally to advertise the existence of the SWDL. This task was

responsible for introducing the SWDL to individual members of organisations who were potential data providers.

Having outlined the main tasks, the rest of the paper focusses on the software data collection activity of the SWDL project.

2 The Software Data Collection Task

The diagram shows how the process of collecting and handling data operated:

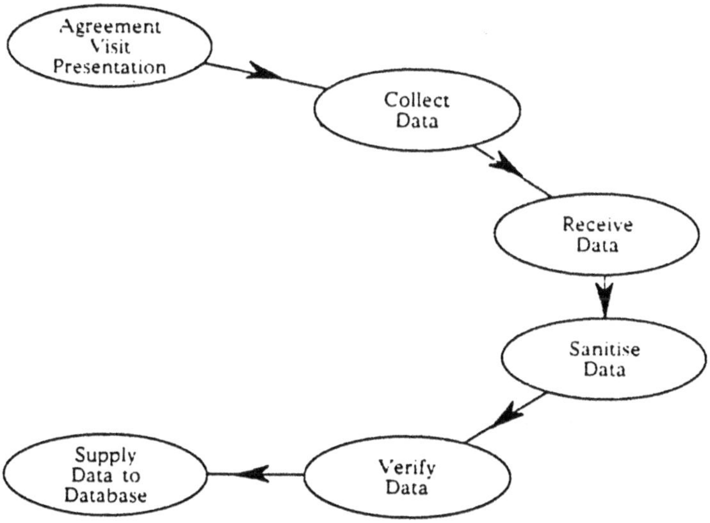

Fig.1. Collection Process

The first step in collecting data was to come to an agreement with an interested organisation as to how and when to collect, and as to what data should be collected. This step was mechanised in the SWDL by the Agreement Visit Presentation (AVP), which was treated as a follow-up to the Initial Contact Presentation (ICP).

Having made the agreement, data collection was started and, in due course, a dataset was passed to the SWDL, where it was sanitised and entered into the holding database pending verification. In this state it was treated as a distinct project, separate from and independent of any other SWDL data.

The dataset was then extracted from the database and checked for transcription errors and general plausibility, in conformance with scope and procedures laid down in the Verification Report Specification. The Verification Report, containing initial analyses and comment was passed back to the data provider (this being the only person in a position to confirm the correctness of the data and provide clarification of anomalies). On confirmation, the data could be transferred into the general database.

Two general categories of data were collected from each source:

1 The first to be collected was termed "Comparability Data". This is a description of the semantics of the data to be collected. It is of critical importance as the means by which it is possible for projects to collect data in their own terms and yet ensure comparability with data collected from other projects using different semantics. For example, if the size of a code module is provided, then comparability data would refer to lines or statements and whether blank lines or comments have been included.

2 The numerical data.
This provides the values of the data, measured against the provider's own "scales" as defined in the comparability data. any data form can refer to a specific version of the comparability data, so that the latter may evolve with time or be different for different parts of an organisation.

The following four sections deal with the stages involved in collecting a verified dataset, starting with the Agreement Visit Presentation to the prospective customer, through on-site collection to preparation and verification.

2.1 The Agreement Visit Presentation (AVP)

The AVP aimed to decide, actually during the meeting, as to "what" was to be collected. This was to be done on the basis of a detailed walkthrough of the SWDL data model. The model would be customised by deleting aspects not available or of interest to the organisation.

Documents distributed at the AVP were the text of the AVP itself, the Data Collection Manual and the SWDL Data Definitions.

2.2 Collection at Provider's Site

In every case a Collection Supervisor had to be nominated by the data provider who co-ordinated and supervised the work. This person was the primary liaison contact with the SWDL. Training in the data model, the data definitions, the use of the collection forms and collection techniques was given whenever possible.

2.3 Preparing Data for Verification

The initial step in validation was to put the data into the holding database, by transcribing from paper forms to formatted screens on the computer. This was done in conjunction with the encryption of identifiers.

Most of the entry was carried out by clerical personnel with little or no checking at this stage. Checking was then carried out on the data extracted from the holding database.

2.4 Data Verification

The objective of Data Verification was to ensure that, before data was put into the general database for comparison with other projects, the data was free from errors, was self-consistent and that it properly reflected the project from which it was derived.

A great deal of effort went into producing a specification for Verification Reports. This defined a standard set of statistical and software engineering analyses and the possible interpretations of a range of observations.

There was growing evidence that many datasets introduced their own special requirements for verification. This prevented "standard" analyses being performed and reduced the usefulness of the Verification Report Specification for these datasets.

The above is a brief introduction to the data collection activity that went on. Underlying this activity, and underlying the database developed to store the collected data, was a generic data model of the software development process.

3 The Data Model

When work began on identifying the data collection requirements it was quickly realised that a very general, highly flexible data model would be needed, because

- the SWDL needed to cope with data from a wide range of industrial software development environments,

- ongoing research into software reliability, quality, development practices etc, would generate new data requirements as it progressed.

This general data model that the SWDL used was begun in a sister project to the SWDL - the REQUEST project. Although SWDL has ended, the REQUEST project and the data model continues. The model consists of

- a definition of each entity for which data is collected (eg 'component', 'task')

- a definition of the relationships between the entities (eg a 'component' is modified by an 'operation' which may be due to a 'software engineering note')

- a list of attributes associated with each entity (eg a 'component' may have the number of lines of code counted, a 'task' may have the development method used recorded).

Figure 2 provides a top-level generic view of the data model:

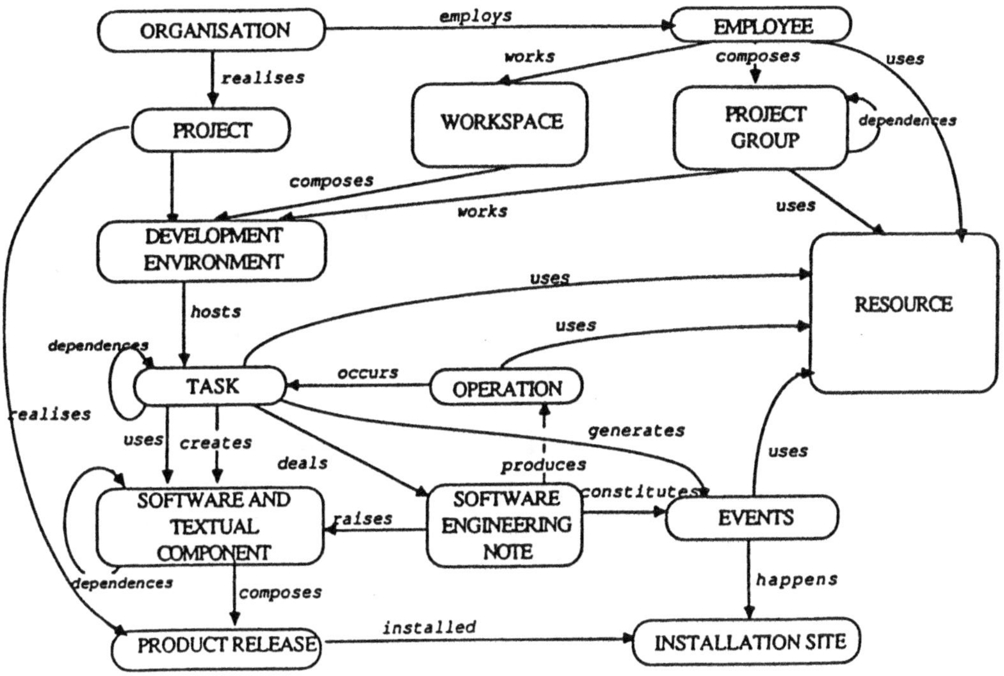

Fig.2. Data Model

Some of the boxes in Figure 2 represent entities, whilst others represent groups of entities, as explained below.

Organisation - a body which controls projects.

Project - an activity which develops product releases.

Employee - a person employed by an organisation.

Project Group - a group of employees.

Installation Site - the site at which a particular product release is being used.

Product Release	- a collection of software and textual components delivered to customers.
Task	- a planned software engineering process (eg design of a software component).
Operation	- an unplanned software engineering process (eg dealing with an incident).
Development Environment	- the environment in which software engineering tasks are carried out.
Workspace	- the physical environment in which an employee works.
Resource	- this comprises four entities recording usage of consumable resources (eg man hours) and non-consumable resources (eg men) by employees and project groups.
Software and Textual Component	- this comprises three entities recording data about items created by any of the software engineering processes involved in producing a product.
Software Engineering Note	- this comprises three entities recording data about incident reports, fault diagnoses or change requests.
Events	- this comprises three entities recording data about groups of incidents, faults or changes.

Having outlined the entities of the data model, the attributes which are recorded for each of these entities must be explained. These attributes are shown in Figure 3, together with the entities to which they apply.

The attributes (shown without underlining) appear in two ways in Figure 3. Attributes within parentheses actually refer to groups of individual metrics, whereas attributes without parentheses are themselves metrics. For example, the attributes of 'component' include (Quality) and Complexity. Here, (Quality) represents the individual metrics Reliability, Maintainability, Reusability, Extendability, Efficiency, Usability, Integrity and Generality - each of which is measured on a subjective 5-point scale - whilst Complexity is itself a metric, also measured on a 5-point scale.

The model which has been outlined above is deliberately a very general one and it is thus necessary to customise the model to the practice of any particular data provider, in order to make the task of data collection a practicable one. This customisation process can be thought of as a "pruning" of the "tree" shown in Figure 3. Because there is a strong correspondence between the data model and the data collection forms, it is a relatively straightforward task to customise the forms to the particular data collection activity by simply removing those forms or parts of forms which are not needed.

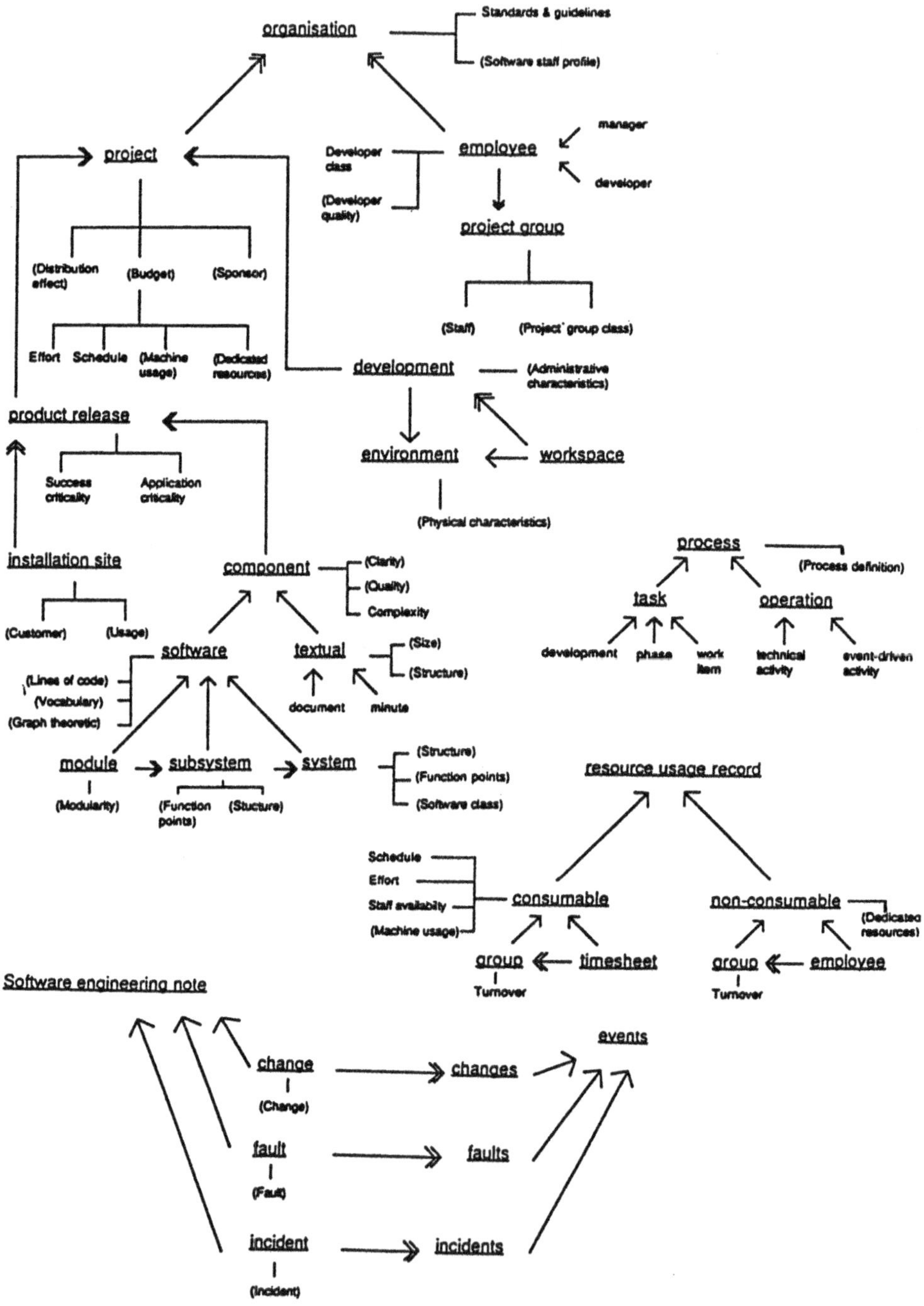

Fig.3. Entities and Attributes of the Data Model

4 Data Model Customisation and Comparability Data

As an example of customisation, a data provider might choose from the part of the data model shown in Figure 4a the simpler structure shown in Figure 4b. If the data provider designed all modules (not subsystems or systems) using flowcharts, had a compiler that recorded lines of code and a word processor that counted the number of words and pages in documentation, then this would be a natural choice.

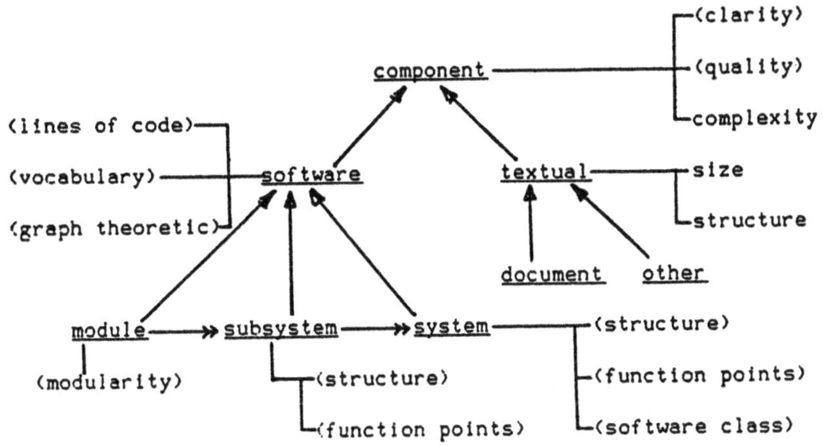

Fig.4a.

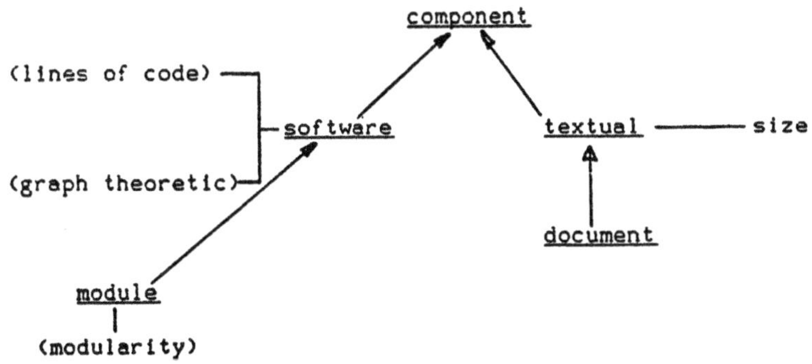

Fig 4b.

The process of customisation continues into the model attributes themselves. Metrics such as lines of code or effort often change their meaning slightly from one environment to another. One cannot expect a data provider whose compiler recognises a line of code in a certain way to replace it with one that uses some other 'standard' counting method.

A flexible data collection programme allows data providers to customise the definitions of attributes, recording the differences in a structured manner so that the data from any environments can be understood and compared. Recording these differences in a structured manner is called the collection of comparability data; there are three kinds of comparability data:

1 the definition of the unit being counted (which definition of a line of code is being used, are all incidents during the development process recorded or only serious ones, etc) or the axis of definition of an attribute (for level of skill this could refer to training and practical experience) - note that an axis definition can be multi-axial in nature,

2 the definition of the classification scheme being applied to the counted units (are only code lines counted, are code and comment lines counted together, etc) or the position of the turning-points (the points at which low becomes medium or medium becomes high) on an ordered scale,

The general classification schemes provided by the attribute definitions must be reduced to simpler classifications that are natural to the data provider's environment; by recording;

*any classes that are not collected
*any classes that are collected individually
*any classes that are collected but concatenated into one 'super-class',

3 textual information covers remarks that do not fit either of the two structured types above but are necessary for understanding the meaning of data collected.

Example.

The police are watching a road junction to catch cars speeding.

Attribute Name: Speeding Cars

Unit Definition:

a car is any wheeled vehicle

a car is any wheeled or tracked vehicle (tanks can speed too!)

a car is any wheeled vehicle - other than a police car (embarrassing to catch the Chief Constable speeding!)

Categories:

bubble cars	delete category (they can't speed)
two seaters saloons hatchbacks estates vans	merge into one category (they all have the same road restrictions)
lorries	keep as a separate category (they can have special road restrictions)

Textual:

The collection system is visual and shows no cars in fog.

5 Conclusions and Further Work

Among the major achievements of SWDL are:

- Development of a general data model of the software engineering process, incorporating the ability to compare data collected from projects carried out in diverse software engineering environments.

- Development of a database system based on this model, as a means of storing the data and providing the necessary services.

- Development of a data collection methodology, covering all aspects from initial contact with a potential data provider, to the verification of data using statistical and software engineering expertise.

- Enrolment of data providers from several software application areas.

From within the software data collection task came a large number of project deliverables:

- AVP
- Data Definitions
- Data Collecyion Manual
- Data Analysis Papers
- Comparability Data Definitions
- Comparability Data Collection Manual
- Regularisation Procedures
- Comparability Data Collection
- Verification Procedures

Many questions, including many we were not initially even aware of, were addressed and in general answered by the data collection task.

There are some important points to note, in terms of experience gained in collecting data:

- Organisations wanted to do their own in-house assessment before supplying data to SWDL.

- It was difficult to collect meaningful data retrospectively. Therefore, when collecting data, current projects were only considered if they were still at an early stage. New projects required time for a significant amount of data to become available.

- It was originally proposed to set up automated data collection via IPSE based projects. However, when project plans were reviewed after the first 12 months, resource for automating data collection in this way was seen to be lacking. It is considered that automation is essential to the health of software data collection in the long-term.

As to the future, the exploitation opportunities of the work that has been done are considerable, and extend beyond the scope of either (the now ended) SWDL, or the ongoing sister project REQUEST. There is a strong possibility of developing a generally available information service based upon the REQUEST and SWDL work which could be useful in many areas of software engineering. One example of this is the role that such a database could play in the area of certification, by assisting to identify the basis on which particular software should be certified.

Finally, it is appropriate to end this paper with a thank-you to all the people in SWDL, and REQUEST, for the hard work, determination and long hours that they have given to the work reported in this paper.

An Integrated Data Base and Reliability Assessment of Electrical Distribution Systems

R.N.ALLAN, T.Y.P.SO and D.K.GAZIDELLIS

Dept of Elect Eng and Electronics, UMIST, Manchester, England

Summary

This paper describes a reliability data base which has been developed as an integral part of a analysis package (RELNET) that evaluates the reliability of electrical distribution systems. It discusses the structure and file organisation of the data base, the management system which maintains the data base under central control and the interface between the data base and the evaluation routines. The concepts will be of interest in a wide variety of application areas.

INTRODUCTION

Power system engineers are responsible for the planning, design and operation of a complex system in which failures can cause effects that range from inconvenience and irritation to severe impact on society and on its environment. The impact of failures can be assessed by the use of quantitative reliability evaluation [1,2,3] which, in addition to providing a set of numerical indices, also indicates how the system may fail, the consequences of such failures and provide sufficient information to enable engineers and managers to relate the quality of their system to economics and capital investment.

In order to qualitatively and quantitatively evaluate the reliability of electrical distribution networks, UMIST have developed techniques which comprehensively model and evaluate the reliability of these networks. These techniques have been incorporated into a computer program (RELNET) using graphic/interactive computing methods. This was achieved in collaboration with the UK Electricity Supply Industry (ESI) and these evaluation techniques are already published [2,4-6].

The quantitative results of any reliability analysis are, at best, only as good as the input data. The data for distribution networks can be obtained by the UK ESI from their National Fault and Interruption Reporting System (NAFIRS). The problem of transferring this data efficiently and accurately to RELNET remains however. In order to overcome this problem, UMIST has recently extended their techniques by developing a reliability and load flow data bank which can act as an integral part of RELNET. It can also be used and managed as a completely separate entity. These objectives required three distinct but interrelated aspects, i.e. the development of:

(a) an appropriately structured data base and file organisation that incorporates easy, fast and direct access to any required data item.
(b) a data base management system (DBMS) which maintains the data bank under central control including input, retrieval, updating, deletion, searching and file re-organisation.
(c) an intelligent interface between the data bank and RELNET so that a distribution network under study can access and retrieve the appropriate and most up-to-date items.

The paper describes all three aspects. Although these details relate to a specific data bank and reliability evaluation program, the concepts and principles will be of considerable importance and interest to a wide set of application areas.

RELIABILITY ASSESSMENT OF DISTRIBUTION SYSTEMS

The models and techniques for evaluating the reliability of electrical distribution systems and incorporated in RELNET are based on both total and partial loss of continuity criteria. Five main reliability indices (expected failure rate, average outage duration, annual outage time, average load disconnected and average energy not supplied) for each load point in the system are evaluated. Both non-transferable and transferable load systems are analysed and all realistic failure modes including permanent outages, switching impact of active failures, temporary failures, stuck breaker conditions, failures due to overloads, effect of weather dependent failures and the effect of maintenance are incorporated. The restoration modes include repair, replacement, switching, reclosure and transfer actions. These techniques are described in detail in References [2,4-6].

In order to achieve flexibility, RELNET has been divided into seven distinct sections :

- Network diagram drawing and modification
- Input and edit of network numerical data.
- Input and edit of execution parameters.
- Input and edit of load duration curves.
- Storage and retrieval of networks.
- Full reliability analysis and results display.

One of the most tedious sections of RELNET, or any other computer program, is the inputting, editing and updating of the numerical data. This data management would be improved by integrating RELNET and its associated data into a powerful data-bank-centred reliability analysis program.

This approach and the related general concept will find application in a wide range of reliability evaluation packages.

CONCEPTS OF DATA BANK

A data bank is a comprehensive collection of interrelated data. Its primary task is to provide efficient storage and retrieval of data. It also serves as a well organised information centre which contains all the relevant data required by its users. A data bank, based on this concept, has been incorporated into RELNET. Sets of data identified by component type numbers are directly retrieved from the data bank and therefore the inputting of a large amount of data wich is tedious and error-prone is avoided. Without the data bank, each system network created in RELNET would need its own reliability and load flow data. This often leads to considerable data redundancy with resultant waste in computer storage. The use of the data bank can reduce the amount of redundant storage by sharing common data between system networks. Also the inconsistency inherent in maintaining and updating multiple copies of data is avoided and the data integrity can be preserved.

The data bank is designed so that it can be interfaced with RELNET or work on its own for data storage and management purposes. The choice is made during the use of RELNET via the interactive options displayed on the terminal screen. One of two options can be chosen on entry; 'RELNET' to attach to the analysis program or 'DATA BANK OF RELNET' to attach to the data bank. The selection is made by positioning the horizontal cursor over the required option and pressing the space bar. The data bank management covers the following types of data manipulations:

(a) The entry of new data
(b) the retrieval and display of stored data
(c) The updating of existing data
(d) The deletion of obsolete data
(e) The search of data according to manufacturer, operating voltage or rating of components
(f) The reorganisation of data files.

FILE ORGANISATION

The data storage for the data bank is accomplished by the use of files each of which is defined as a collection of similar records kept on computer external storage devices such as magnetic tapes and disks. The choice of a file organisation depends on the nature of operations to be performed on the data. It should allow flexible updating and deletion of individual records of the data bank without access to the remaining records of the file; otherwise the overall input/output performance will be downgraded. In the reliability analysis, it requires fast interactive access to the data bank and selective data retrievals. These are the crucial factors in considering the file organisation.

A direct file organisation was chosen. This avoids unnecessarily complicated file operations and makes the implementation simple and effective. A direct file can be visualised as a collection of fixed-length records. Each record is uniquely identified by a record number (an address) while each set of data is represented by a code number (a key) known as the component type number. A particular set of data can be selected by specifying the code number, which is translated into its corresponding record number. The record is then directly accessed through the data handling facilities.

In order to create efficient data management, individual direct files were created for each category of component, e.g. transformer, circuit breaker, cable, etc.

COMPONENTS AND DATA

The component categories stored in the data bank are all those used in RELNET: substations (busbars), transformers, circuit breakers, transmission lines (and cables), isolators and junctions. The data for each component category is subdivided into reliablity, load flow and type data.

The reliability data includes failure rates for each failure mode, scheduled maintenance data, repair times, switching times, reclosure times and stuck probability (for circuit breakers only). The load flow data includes operating voltages, ratings, tap data and impedances. The type data provides descriptive information about the components not used in the analysis but of interest to the data bank management and its users.

An example of this data, as displayed on the screen, is shown in Figure 1.

PASSWORDS

Passwords are used to protect the data bank from illegal access and modification. The master password is required to modify the information stored in the data bank. Hence, options 'EDIT THE DATA BANK', 'PACK THE DATA BANK' and 'CHANGE PASSWORDS' cannot be activated without the correct master password. Options 'PRINT THE DATA BANK' and 'BROWSE THE DATA BANK', which only allow read access to the data bank can be activated by either the master or user password. Either password may be changed provided the master password is known.

EDIT THE DATA BANK

The data of each component category is divided into three sections: load flow, reliability and type data.

After choosing the component and data categories, the first page of the relevant data table is displayed, as shown for instance in Figure 1. A maximum of eighteen data entries are displayed on each page. The conventional method is to copy all the data from the file into program arrays before displaying the data. However, the number of occasions for inspecting or modifying the whole file are relatively rare. The majority of data manipulations in the analysis is data searching and retrieval. Also, the direct file organisation allows efficient access and modification to individual data records. Therefore only eighteen entries are copied at a time from the data bank into data arrays. This technique saves both time and effort of copying the entire data file and reduces considerably the amount of memory allocated to data arrays during the execution of the program. Moreover, any subsequent changes of program array sizes due to the growth of the data bank can also be avoided.

The entry and modification of numerical data is performed in a simple interactive manner using the cursors to locate the starting point of data entry in each table. Data is then typed in free format and automatically formatted on redisplay.

Data can also be deleted and printed by using the same simple interactive procedure based on options and the two cursors.

BROWSE THE DATA BANK

There are many occasions when the data bank needs to be searched in order to identify components of the type required. This can be achieved using the "BROWSE THE DATA BANK" option. This option can be accessed from either the data bank management routines or the reliability analysis routines of RELNET. In both cases it permits components to be located according to some specified attributes. During the use of this option, the data can not be entered or modified, only searched.

The search attributes include manufacturer, voltage level and rating. After inputting one or more of these attributes, the data of the chosen component category is searched and only those components matching these attibutes are displayed on the screen. An example is shown in Figure 2 in which the attribute "manufacturer ABC" has been matched from the complete list shown in Figure 1.

HELP

An on-line help file is available which is divided into several sections. It can be obtained by choosing the option 'HELP' provided in each menu. Only the relevant part is displayed on the screen.

INTERFACING THE DATA BANK AND RELNET

The drawing routine establishes the system topology and simultaneously sets up two sets of arrays. The first set contains the topological information including co-ordinates, branch composition and component identification so that the system can be redrawn and the failure events of the system can be computationally deduced. The second set contains the numerical data which is entered manually or retrieved from the data bank.

Each component is identified by a type number. By a suitable choice of these numbers, a system in which some data (local) is inputted manually and other data (global) is retrieved from the data bank can be achieved. The ranges of the type of numbers are 1-10 for local data and 1001-2000 for global data. Numbers outside these two ranges are not accepted.

(a) Local Data. The data which is local to the system network is stored together with other system data such as graphical data in an unformatted sequential file. When a system is retrieved from the file, all the data is copied into program arrays. If it is a newly constructed system, the numerical data must be inputted manually.

(b) Global Data. The word global means that data can be shared among all system networks in RELNET. The only input data required for each component is the type number. The corresponding data will be directly retrieved from the data bank for either display or calculation. If the user is not sure what data or which type number should be used, an enquiry can be made into the data bank by browsing as described previously. The current quota of data types for each component category is 1000.

After specifying the type numbers, the failure, repair and load flow data of each type can be inspected. Unlike local data, global data cannot be modified (except in the data bank area) for the reason of data integrity. If the specified data entries do not exist in the data bank or have been deleted, the warning messages 'NO SUCH ENTRY IN THE DATA BANK' and 'DELETED' will be displayed accordingly.

CONCLUSIONS

This paper has described the concepts and features of a data bank which is incorporated into the reliability analysis computer program RELNET. The objective of the data bank is to integrate scattered files and data into a more cost-effective data management system. Data in separate files can be inconsistent, cause excessive storage and create problems in subsequent updating. Also users are able to refer to data items without having knowledge of record or file structure. A set of interactive data bank facilities has also been described. These facilities are designed to operate in a user-friendly environment in order to increase the operational efficiency and promote man-machine communication.

Although the techniques are described in relation to one particular program, the concepts are applicable to a wide range of other applications.

REFERENCES

1. R.Billinton, R.N.Allan. "Reliability evaluation of engineering systems: Concepts and techniques". Longman (London) and Plenum Publishing (New York), 1983.

2. R.Billinton, R.N.Allan. "Reliability evaluation of power systems". Plenum Publishing (New York), 1984.

3. R.Billinton, R.N.Allan. "Reliability assessment of large electric power systems". Kluwer Academic Publishers (Boston, USA), 1988.

4. R.N.Allan, M.F.De Oliveira. "Reliability modelling and evaluation of transmission and distribution systems". Proc IEE, 1977, 124, pp 535-541.

5. R.N.Allan, E.N.Dialynas, I.R.Homer. "Modelling and evaluating the reliability of distribution systems". IEEE Trans on Power Apparatus and Systems, 1979, PAS-98, pp 2181-2189.

6. R.N.Allan, B.Barazesh, S.Sumar. "Reliability evaluation of distribution systems using graphic-based interactive computational methods". IEEE Trans on Power Apparatus and Systems, 1982, PAS-101, pp 212-218.

TRANSFORMER DATA BANK : RELIABILITY DATA PAGE 1 OF 1

		FAILURE DATA								REPAIR DATA				
		NORMAL WEATHER			ADVERSE WEATHER					REPAIR TIME				
CODE	MANUF	TFR	AFR	TEMP	TFR	AFR	TEMP	MR	MT	MIN	MED	MAX	ST	RT
		(f/yr)	(f/yr)	(f/yr)	(f/yr)	(f/yr)	(f/yr)	(o/yr)	(hrs)	(hrs)	(hrs)	(hrs)	(hrs)	(hrs)
1001	ABC	0.010	0.010	0.05	0.010	0.010	0.05	0.50	96	24	168	720	1.0	0.083
1002	XYZ	0.015	0.015	0.05	0.015	0.015	0.05	0.50	96	24	168	720	1.0	0.083
1003	DUM	0.012	0.012	0.08	0.012	0.012	0.08	0.50	96	24	168	720	1.0	0.083
1004	XYZ	0.015	0.015	0.05	0.015	0.015	0.05	0.50	48	16	120	360	1.0	0.083
1005	ABC	0.020	0.020	0.10	0.020	0.020	0.10	0.50	48	12	120	360	1.0	0.083
1006	XYZ	0.025	0.025	0.10	0.025	0.025	0.10	0.50	48	16	120	360	1.0	0.083
1007	ANO	0.008	0.008	0.04	0.008	0.008	0.04	0.50	96	24	168	720	1.0	0.083

REDISPLAY
PREVIOUS PAGE
NEXT PAGE
TYPE DATA
LOAD FLOW DATA
HELP
RETURN

RELNET(vers 3.1) 20 OCT 1988

Figure 1 - Typical Data Bank Screen Display

TRANSFORMER DATA BANK : RELIABILITY DATA PAGE 1 OF 1

		FAILURE DATA								REPAIR DATA				
		NORMAL WEATHER			ADVERSE WEATHER					REPAIR TIME				
CODE	MANUF	TFR	AFR	TEMP	TFR	AFR	TEMP	MR	MT	MIN	MED	MAX	ST	RT
		(f/yr)	(f/yr)	(f/yr)	(f/yr)	(f/yr)	(f/yr)	(o/yr)	(hrs)	(hrs)	(hrs)	(hrs)	(hrs)	(hrs)
1001	ABC	0.010	0.010	0.05	0.010	0.010	0.05	0.50	96	24	168	720	1.0	0.083
1005	ABC	0.020	0.020	0.10	0.020	0.020	0.10	0.50	48	12	120	360	1.0	0.083

REDISPLAY
PREVIOUS PAGE
NEXT PAGE
TYPE DATA
LOAD FLOW DATA
HELP
RETURN

RELNET(vers 3.1) 20 OCT 1988

Figure 2 - Typical Screen Display After Browsing

Field Failure Data Collection and Analysis of Repairable Systems

C. KJÆRGAARD.

Danish Engineering Academy.
Electronics department.
Reliability Group.

SUMMARY.
During the last 4 years a research project concerning field failure data collection and analysis has taken place at the Danish Engineering Academy (DIA). Field failure data has been collected on two Danish electronic products.
In order to analyse the data it has been necessary to design a database to hold and manipulate the data.
This paper describes the level of information in the database and the work performed to handle incomplete datasets.

DATA COLLECTION.

In complicated systems such as aeroplanes, the collection of failure data is fairly easy because logbooks for the individual systems exist and qualified service personnel are constantly checking the systems to assure that they are safe to use, furthermore the logbooks keep's track on the use of the system in order to perform preventive repair of the system.

When dealing with small electronic systems produced in large numbers it is very difficult to perform a reliability analysis of field failure data due to the lack of feedback from the end user.

The problem is illustrated by a simple example:

Let us consider a domestic television set. What do we do when a failure occurs.
If the TV is in the guarantee period we would probably have the unit serviced at the place where it was purchased. The dealer would then charge the cost to the manufacturer and a dialogue between the dealer and the manufacturer would be established.

If the TV is outside of the guarantee period, a different situation exists. We would not necessarily go to the authorized dealer but instead look for the fastest or cheapest repair facility, in this case no information would be relayed to the manufacturer, because no benefit to the service shop or user could be obtained by doing so.
The conclusion of this example is that in order to set up a field failure data collection system the manufacturer must do following.

1) A communication path between the manufacturer and end user or repair facility must be established.

2) A format for this communication must be set up in order to assure that the necessary information is available for the reliability analyst.

It is obvious that the amount of information collected depends on the particular analysis one wants to perform on the dataset. In general though, it must be possible to track the life histories of the individual systems.

In order to do this there must be some kind of identification of the systems, subsystems and components.
The electronic systems that we have analysed can be described as shown in fig. (1).
This is the breakdown we have used for systems analysed at DIA.
A similar structure can be used on other systems.
The system used as an illustration in this paper has up to 25 boards, each with up to 400 components. The installation is identified by a user code, the boards by unique serial numbers, and the components by circuit reference positions in a parts list for each board type. The problem of tracking arises when a failed board is removed from the installation and another (new or renewed) board is inserted instead. The failed board is then returned to the manufacturer, repaired, tested and put into storage.

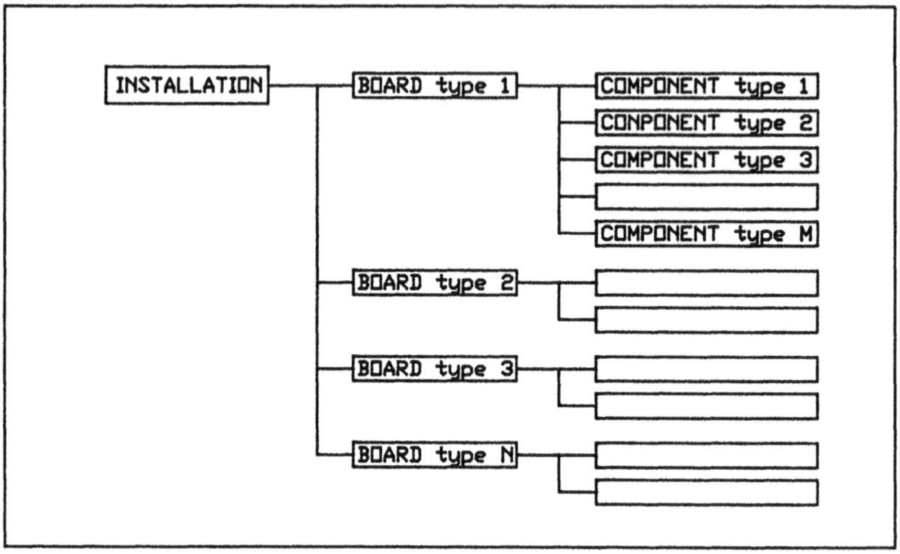

Fig. (1) Structure of an electronic installation.

This means that boards move around between different installations and the life history of a board becomes a mixture of time periods where the board was either operating or taken out.

Because of this it is not enough to report that a failure has occurred on a board, it is also necessary to report which board was used as a replacement instead of the failed one.

The report can then be divided into two categories with the following contents for each type of board:

Set-up reports

1) Serial number of board
2) Date of setup
3) User identification

Failure reports

1) Serialnumber
2) Date of failure
3) User identification
4) Description of failure and failed component

Apart from the user identification which is used for data validation purposes the above is the absolute minimum of information that should be collected to describe the individual board life histories.

The serial number is used to group events concerning a particular board and the setup and failure dates are used to produce the life histories for each board in a time scale of running time as shown below in fig. (2)

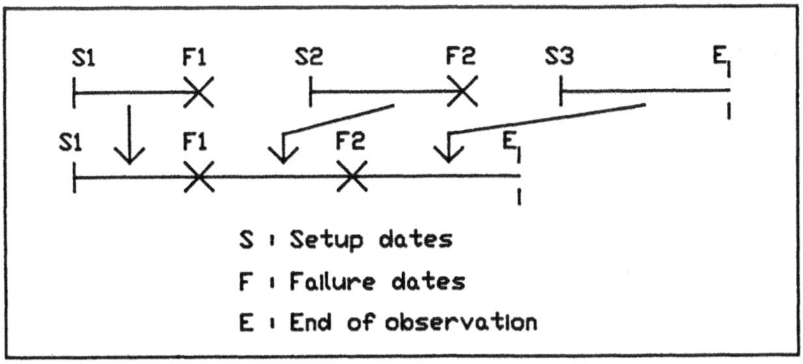

Fig. (2) Conversion of calender timescale to operating time.

This timescale is used to plot the mean cumulative number of failures pr. board as a function of time (M(t)) as shown by Rimestad et.al 1986 [1] .

If the failed components are described by their circuit reference position numbers it is possible to calculate the time to failure for the failed components, see fig. (2) for each component failure it is necessary to look back in the failure file to se if any earlier failures on the board (identified by serialnumber) has occurred at the same circuit reference position.
If not, the the time to failure for the component is equal to the time to failure for the board, if a previous failure is found on the same circuit reference position, the lifetime for the previous failure should be subtracted from the time to failure of the current component.

But not surprisingly we have found that there are often many holes in the life histories, due to missing reports both on failures and setup times.

As shown in fig (2) each life history is a mixture of running periods and repair periods placed in a sequence of S-F-S-F---. If either a setup record or a failure record is lost it can easily be detected.

However if a (S-F) or (F-S) combination is missing the sequence seems to be in order. Fortunately we can use the user identification to test if the failed board has been set up in the installation when it failed.

This test can be used on the whole sequence.

If it is true for all events in the board history, then the data on the board is accurate.

If not something is wrong and it is not possible to detect wether one or several reports are missing in the dataset.

DIABASE DESIGN.

We have now defined what information should be collected from the field and how it can be validated. The next step is to develop a database to store the collected data.

At DIA we have developed an integrated database and data analysis program called DIABASE. The aim of DIABASE is to design a small but effective field failure data analysis tool to be used on Personal Computers (PC's) under the operation system DOS 3.0 or higher. The program package DIABASE will be offered for sale to companies who wish to set up a field failure analysis system.

Due to the DOS operating system's limited memory access (640 KRam and 33MB on disk) it has been necessary to design the data file structure very carefully in order to store the necessary data, which often becomes quite voluminous.
In all programming problems it is good practice to structure the datasets and programs in a way, that reflects the real world and that is what has been done in DIABASE.

The data for each board type is contained in two files one describing the population and one describing the failures.

The population file contains records with the following information:

First start date
Serial number

The failure file contains records with the following information:

Serial number
Total operation time to failure
Operating time to failure since last repair
Failure date
Description of failure
User identification

The major strength of this structure is that every failure record contains the total operating time to failure in days. This means that groups of data can be taken out of the failure file and analysed without loosing the time information.

In principle an M(t) analysis can be carried out for each component type, and the sum of the curves would still match an M(t) analysis made of the total failure data set.

Apart from the two files just mentioned, provisions for adding a third file containing the parts list is made. This will be necessary if analysis of the generic type of the components is required. At DIA we have up to now concentrated on analysis at board level, using the circuit reference position as a description of the failed component.

To fit the actual failure reports into the DIABASE format prep-rocessing is necessary. The field failure reports contain failure dates and setup dates in a calendar time scale and this information is then translated into the DIABASE file format using the following method:

The data is sorted by serial number and by dates.
The first setup date for every board is put into the population file. The downtimes for the rest of the boards and the operation time to failure is calculated in days,
see fig. (2). At this point data is also validated and unusable information is removed.

During the translation of the datasets obtained from the Danish companies, we have found gaps in the life histories due to missing records.

To overcome this problem one of following three strategies can be used

1) Ignore datasets that do not have a correct life history in the analyses

2) Estimate the missing information and use all datasets

3) Develop statistical tools to work on incomplete datasets

Method 3 was out of the scope of this project although it is very interesting.
We started off using method 1, this gave good results,but unfortunately 60-70 % of the failure records were rejected by the analysis programs. The large number of rejects were due do the fact that a board with several correctly reported failures would be rejected if only one record were missing.
This was not satisfactory so we tried method 2.

The estimation was based upon the assumption that all failures on every board were recorded.
Using this assumption it can immediately be seen that only setup records are missing see fig. (3)

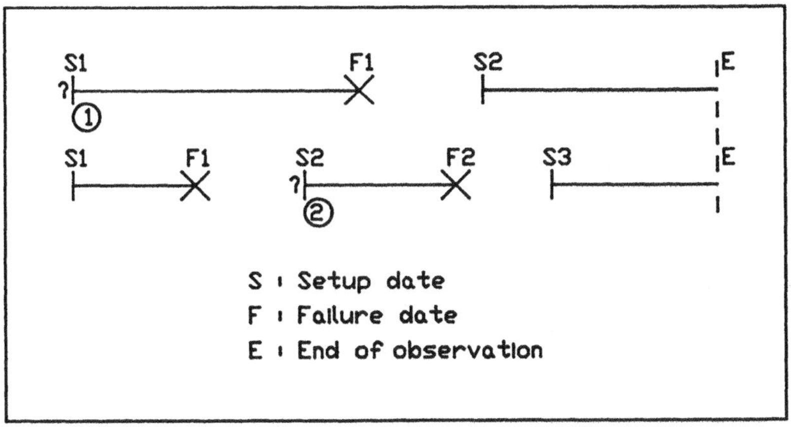

Fig. (3) Life history of systems with missing setup information.

In situation (1) the first setup date is missing.
A search using the user identification on failure F1 could be used to see if there were a first setup date recorded in the files for the other board types on the installation.

In situation (2) we know that a repair has taken place in the time interval between F1 and F2. An average repair time is then added to the failure date F1 to produce the missing setup date. If the average repairtime is too large to fit in the time interval, a percentage of the time interval F1 to F2 is used instead.
The various estimation methods are given a number and if an estimation has been used the failure record is marked with the number.
The following histogram, see fig.(4), shows the distribution of the estimates used on the total number of datasets from one Danish company. The letter N means that no estimation has been used.

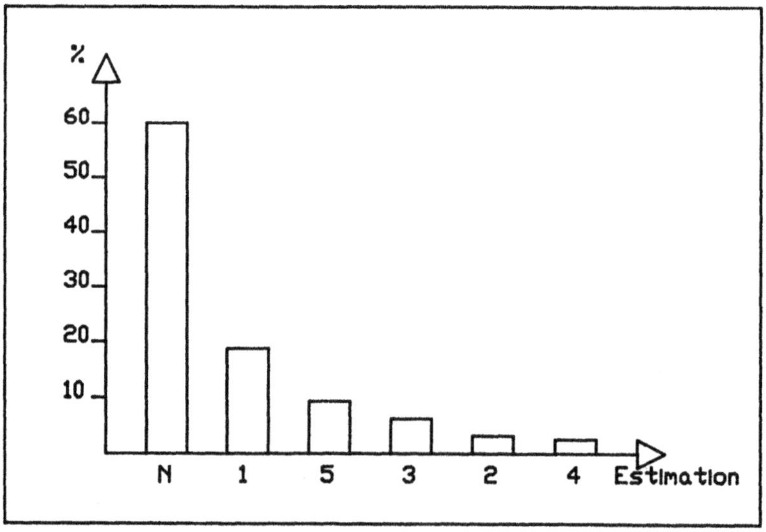

Fig. (4) Distribution of estimation types used in translation of data form a Danish company.

As it can be seen 60 % of the failure records are correctly reported which compared to the 60 to 70 % rejects of method (1) gives a far better result.

DIABASE STRUCTURE.

The DIABASE program is designed to be easy to use and friendly to the user.

With the program it is possible to examine and edit failure data, extract data, sort data , perform M(t), Weibull, Pareto analysis on selected datasets and plot the results on Weibull, log-log, lin-lin scales and histograms on the screen and/or a matrix printer.

The switch between subprograms is controlled by menu's in the bottom line of the screen using the principle that a choice of an item on the menu gives a new menu to select from.

Help facilities are present at every menu level, and the user can add to, modify, or translate the help messages to his own language if he posesses an simple text editor.

The hierachy of DIABASE is shown in fig (5).

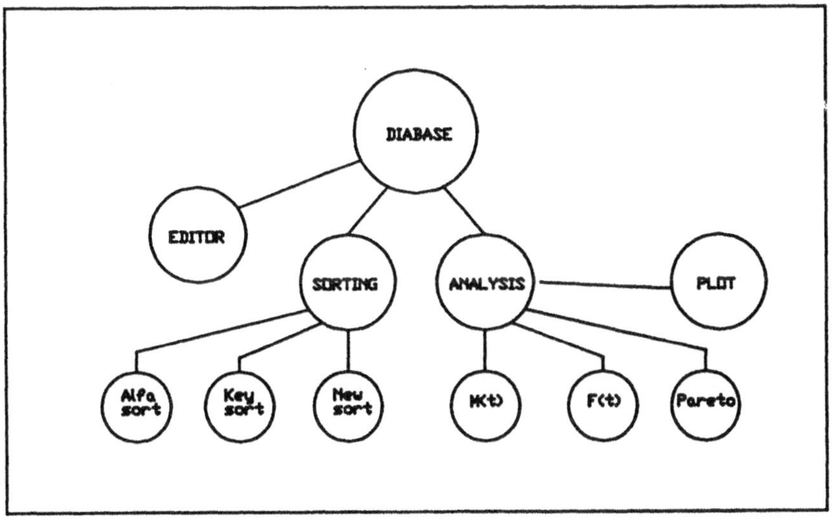

Fig. (5) DIABASE structure.

CONCLUSION.

The reliability group at DIA has many times heard the argument that the use of field failure data is not possible, beause the data is of too poor quality in some cases.
On the other hand we have found that if the right motivation is present in the company or organisation,then it is possible to obtain good field failure data.

It is especially important, that the the field service personnel are motivated to feed back realistic failure reports

It is then the task of the reliability analysist to develop models and analysis techniques that are representative and easily understood. One important tool in this work is the M(t) analysis briefly mentioned in the paper.

The DIABASE file formats developed at DIA has proved to be easy to work with. The information required is limited but but it is sufficient to make a realistic reliability analysis possible. We realise, of course that making estimations on missing datasets can lead to false results, but the advantage of doing so in the preprocessing of the data is looking only at a small subset of the data.

In the future work experiments will be performed with the preprocessing programs to determine the influence of various estimations on the M(t) analysis.

REFERENCES.

[1] RIMESTAD.L, KJÆRGAARD.C, MØLTOFT.J.
"Field failure data from repairable systems - A Danish case study"
RELCON Europe 1986.

Accelerated Test Procedure for Estimating Proportion of Early Failures of Metal Film Resistors

I. LIKAR and S. KOLENKO

Institute for Quality Testing and Metrology
Ljubljana, Tržaška 2, Yugoslavia

Abstract

The possibility of estimating the proportion of early failures using accelerated testing was investigated for metal film resistors. The aim of the investigation was to establish a test procedure equivalent to the standard endurance test. Thereby, a quick evaluation of a production lot would be possible to avoid long duration endurance tests on bad lots. In the paper a practical test procedure is presented together with accelerating test conditions which were determined by means of the Eyring mathematical model and on the basis of a series of accelerated tests performed together with the corresponding reference standard endurance tests on approximately 4600 test items. The proposed test procedure can be used as a simple tool for evaluating or/and comparing the reliability of individual lots in the manufacturing process as well as at the incoming inspection.

Introduction

Within the inspection procedures which are carried out during the qualification approval and quality assessment of electronic parts the endurance tests are most time-consuming. It is therefore in the great interest of the manufacturers to speed up the testing process or at least to get an approximate preliminary information on batch quality within a short time in order to avoid lenghty tests on bad lots. A simple accelerated life testing would also provide usefull information within a reasonable time frame for improvements and corrective actions in the manufacturing process.

This paper presents the results of an attempt to evaluate the quality of low-power metal film resistors in a short-duration test. The final objective of the investigation was to specify the operating and environmental conditions of an accelerated test that would be equivalent to the standard endurance test, e.g. such as prescribed in IEC Publications 115-1 and 115-2. According to inspection procedures and environmental conditions given in these publications, the resistors are to be subjected to a test of 1000 hours at ambient temperature of +70°C; the voltage shall be applied in cycles of 1.5 h on and 0.5 h off; this voltage shall be the rated voltage or the limiting voltage, whichever is the smaller; the sample size is 20 and the permitted number of defectives is 1 (group acceptance criterion); the item is considered defective if the change in resistance exceeds 1%. With the equivalent accelerated test the proportion of early failures should be assessed in a shorter time, e.g. within a shift (8 hours). The acceptance criterion should be equal or equivalent to the criterion prescribed for 1000 hour test (up to 5% proportion of failures).

All the experiments described in this paper were performed on the resistors of the manufacturer who ordered the investigation.

Initial estimation of accelerating conditions

A general study and a large scale of fundamental experiments concerning the accelerated life testing, which were previously carried out on passive electronic components, provided the guidelines for initially selecting the values of accelerating testing conditions within the appropriate scope. Three types of accelerated tests were conducted on metal film resistors in the course of this basic preliminary investigation: 1) tests with high voltage pulses at room temperature; 2) tests with different constant elevated voltages at room temperature; and 3) combined stressing tests at elevated (constant) temperature and voltage. The comparison of the cited accelerating techniques showed that the best effects could be achieved by combined voltage-temperature stressing. This mode accomplishes all three principal objectives imposed on accelerated testing: a) short testing times - from 0.3 h to 13 h; b) convenient values for temperatures and voltages, near rated values, which means that smaller risk exists to change the failure mechanisms; and c) well fitted mathematical model (for time-to-failure distribution as a function of combined electrical and temperature stress) that can be extrapolated to normal conditions of use.

The tests were conducted on metal film resistors of 0.33 W rated dissipation. The characteristics of the test samples were the following: rated resistance 301 kΩ, limiting voltage 250 V, isolation voltage 450 V, upper limiting temperature 155°C, temperature coefficient ±50 ppm/K and tolerance on rated resistance ±1%.

For description of the times-to-failure distribution as a function of loading, the Eyring mathematical model was chosen as it takes into account the influence of electrical and temperature stress as well as the interaction between them:

$$t = A.e^{b/kT}.e^{-S(c + d/kT)} \tag{1}$$

where t is the expected time-to-failure, A, b, c and d are constants (model parameters), T is the absolute temperature, S is a normalized stress (other than temperature, e.g. dissipation), and k is Boltzmann's constant.

The parameters of the "main-population model" were determined by fitting the equation (1) to the results of a series of accelerated tests . 13 accelerated tests were performed at the following temperatures and voltages:

	1	2	3*	4*	5	6*	7	8	9*	10*	11	12*	13	
Temp.:	155	155	155	175	175	175	175	175	175	200	200	200	200	(°C)
Volt.:	460	480	500	340	370	400	460	480	500	340	370	400	460	(V)

Remark: The results of tests denoted by asterix were later applied in defining the "early-failures model".

An automatic test system with capacity of 30 test specimens was formed around the HP 9836 Computer with Data Acquisition/Control Unit HP 3497 A and HP 3498 A Extender to control the testing and to evaluate the results.

The cumulative distributions of times-to-failure were plotted on Weibull paper. For the main populations, the failure points followed the straight lines with the slope $\beta>1$ in all 13 cases.

Thereby, the assumption of the Weibull distribution was confirmed for the tested resistors. For illustration, Fig. 1 shows the distributions of two test samples subjected to 340 V /175°C and 400 V /200°C.

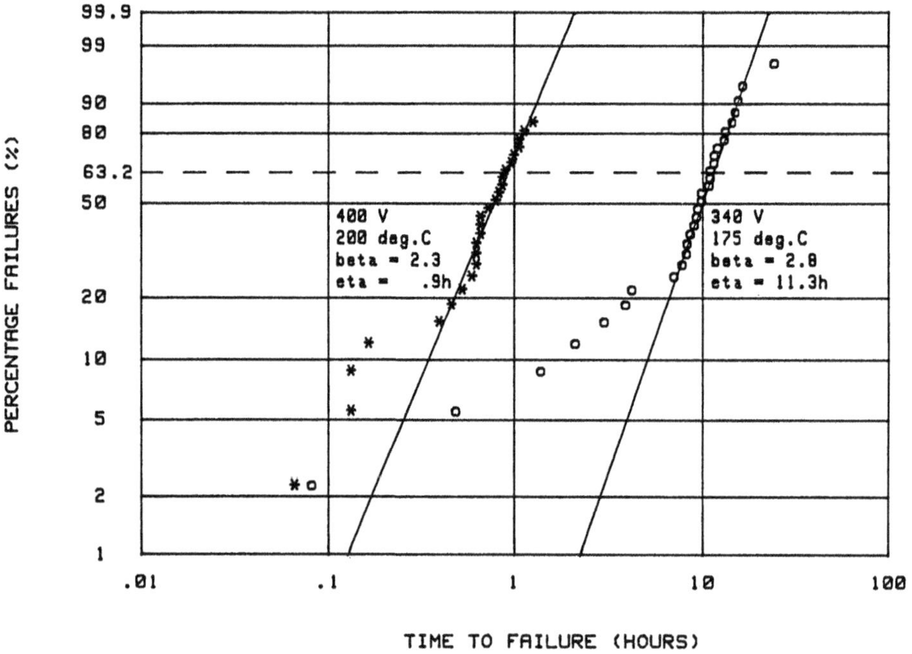

Fig. 1. Cumulative distribution of times-to-failure recorded in two accelerated tests on metal film resistors and plotted on Weibull probability paper

In the analysis and modelling, the early failures were neglected. The parameters of the "main population model" were consequently determined on the basis of the mean characteristic lifetimes (η), read-off from all 13 distributions which were accordingly re-drawn (without early failures). Graphical analysis and least-squares fitting were applied for determination of the parameters. For this purpose the model (1) is transformed into a more suitable expression:

$$\ln t = \ln A + b/kT - S(c + d/kT) \qquad (2)$$

This relation can be presented by two fields of straight lines if S and 1/T, respectively, are assumed to be constant. The pertaining equations, which evidently represent linear functions of 1/T and S, respectively, are namely:

$$\ln t = (b - S.d)/kT + \ln A - S.c \qquad S = \text{const.} \qquad (3)$$

and $$\ln t = -(c + d/kT).S + \ln A + b/kT \qquad T = \text{const.} \qquad (4)$$

The calculation of parameters is briefly explained later in description of "early-failures model" determination. The obtained parameters (b = 1.579 eV, c = -2.776, d = 0.172 eV and $A = 5.61 \times 10^{-16}$ h) determine a model that fits very well to the results of the accelerated tests as shown in Fig. 2.

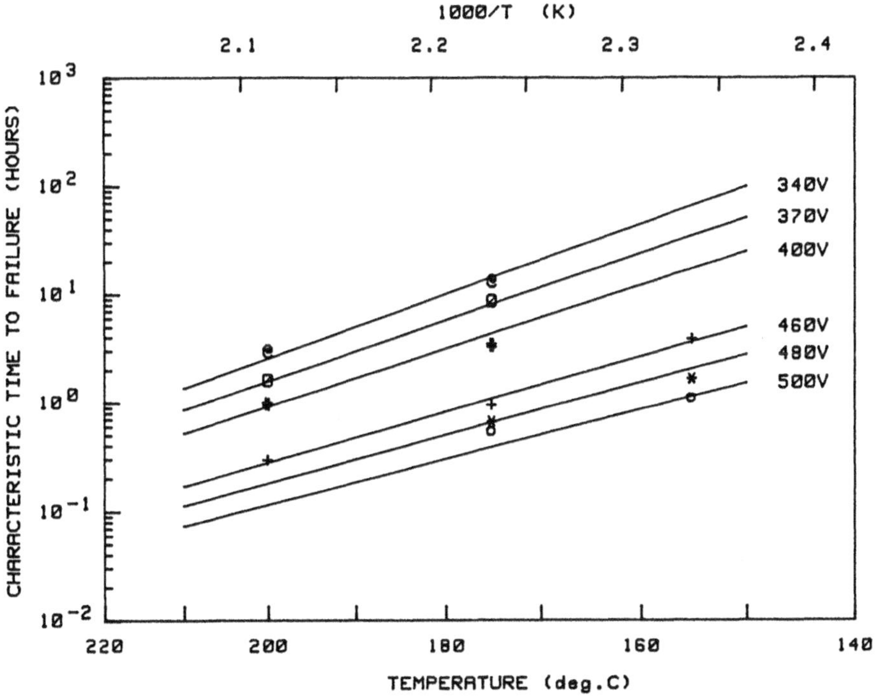

Fig. 2. Fitting of "main-population model" (straight lines) to results of accelerated tests (points)

Parameters of "early-failures model"

In the initial stage of the "equivalent test" investigation, the data on early failures recorded during the above basic experiments were applied as it can be expected that mainly early failures occur in standard 1000 hours endurance tests. Only those test samples were taken into consideration which exhibited a considerable amount of early failures. The pertaining straight lines were drawn through the early failure points using least-squares fitting, and the times to the 5th percentile of failures were read off. This is the proportion of failures equivalent to the reference percentage of failures (approx. 5%) observed in standard 1000-hours endurance tests performed on 3 samples of 1000 resistors from the same batch. The 5th-percentile-points are plotted as points 2 to 7 in Fig. 3.

As it is evident from Fig. 3, the selected distributions cover all three applied test temperatures (155°C, 175°C, and 200°C) and the minimal (340 V), a middle (400 V), and the maximal (500 V) voltage applied in accelerated tests. Each straight line in Fig. 3a is thereby defined by two points. The point 1, which is needed for the corresponding field of straight lines in Fig. 3b, has been read off from the line pertaining to 340 V in Fig. 3a. The lines were drawn through the points using the least-squares fitting, and their slopes (E_a and n, respectively) were calculated. The slope of the lines in Fig. 3a is denoted by E_a as it represents the activation energy according to Arrhenius law:

$$t = C.e^{E_a/kT} \qquad (5)$$

By comparing the expressions (1) and (5) we get the following relation between the activation energy and the model parameters b and d:

$$E_a = b - S.d \qquad (6)$$

In a similar way, the slope n of the lines according to equation (4) is related to parameters c and d:

$$n = c + d/kT \qquad (7)$$

The equations (6) and (7) express linear relationship between the slopes (E_a and n) and stress S and 1/T, respectively. The corresponding straight lines can be drawn as the best least-squares fit to the three known points for each line. The model parameters are calculated as the intersections on the ordinate (b and c) and the slopes (d). In this way, two values are obtained for d. We take the value that yields smaller spread of parameter A calculated for all observed points. A is finally calculated as the mean value of all parameters A pertaining to individual points.

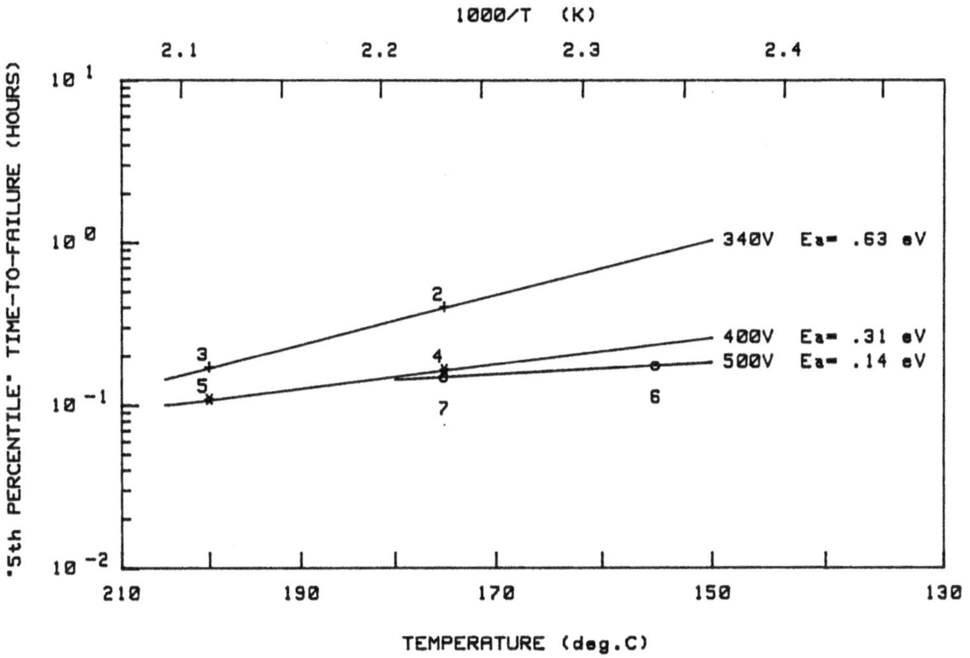

3a)

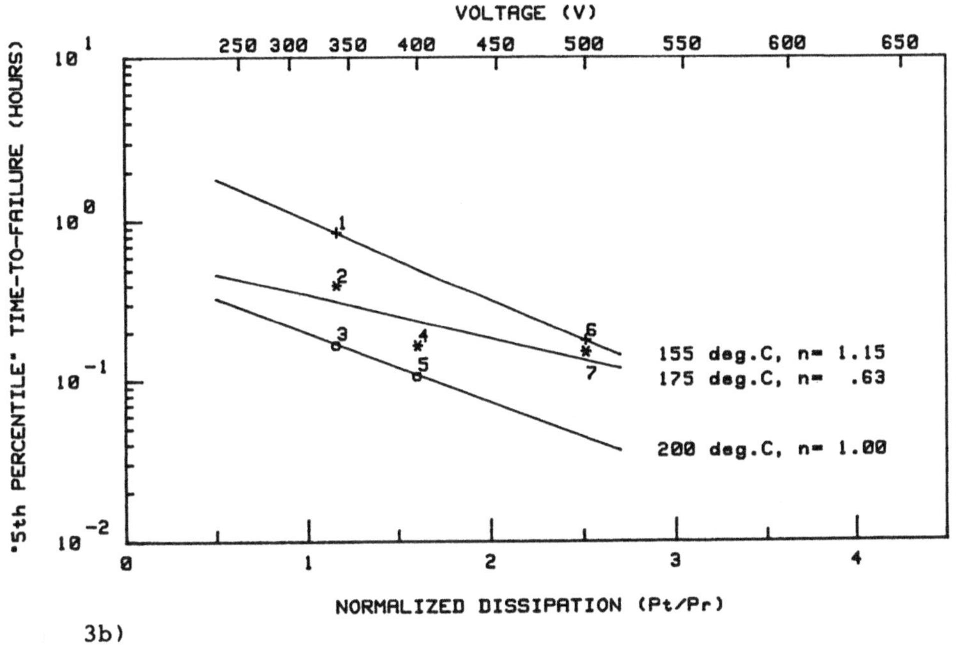

3b)

Fig. 3. "5th percentil" time-to-failure versus temperature and normalized dissipation, respectively, plotted on basis of selected accelerated tests

The "early-failures model" defined by the obtained parameters ($A = 2.26 \times 10^{-11}$ h, $b = 0.954$ eV, $c = -0.866$, $d = 0.070$ eV) is not so well fitted as was the "main-population model". The major cause for this lies in the fact that the tested samples exhibited a small proportion of early failures and that the times to failures were not measured with sufficient accuracy, as they were recorded at discrete (relatively long) sample times. The model was nevertheless found to be appropriate for initial estimation of accelerating test conditions. It is presented graphically in Fig. 4 by a field of straight lines for which the time to the 5th percentile of failures is constant. This presentation is suitable for reading-off the voltage-temperature stress required for attaining certain proportion of failures (corresponding to percentage of failures in 1000 hours in our case) in a predefined time.

Selecting the accelerating conditions

When selecting the values of accelerating conditions on the basis of the defined "early-failures model", we have taken into consideration the following requirements in order to define a simple inspection procedure:

- The inspection should be performed within the time of a shift. An appropriate time must also be foreseen for measurements at the beginning and at the end of the test as well as for recovery of test specimens at room conditions (e. g. 1 hour to 4 hours).
- The stress - temperature and dissipation - should be equal for all resistance values.

In addition, the following statements were considered concerning the extrapolation of the model and the level of the test voltage:

- Voltage-temperature stress should be selected so that the model would not be extrapolated too far beyond the values that were used for its determination.
- The test voltage should not greatly exceed the rated voltage or the limiting voltage (250 V).

With the respect to the above criterions, the following initial values were selected:

- Test duration: 2 to 5 hours
- Temperature: 130°C - 155°C
- Voltage: corresponding to 1 to 1.5 times rated dissipation

The range that corresponds to these conditions is shown as the encircled area in Fig. 4. As the rated dissipation is very convenient for electrical loading, we calculated the corresponding temperature for the test time of 4 hours. This temperature is equal to 137°C.

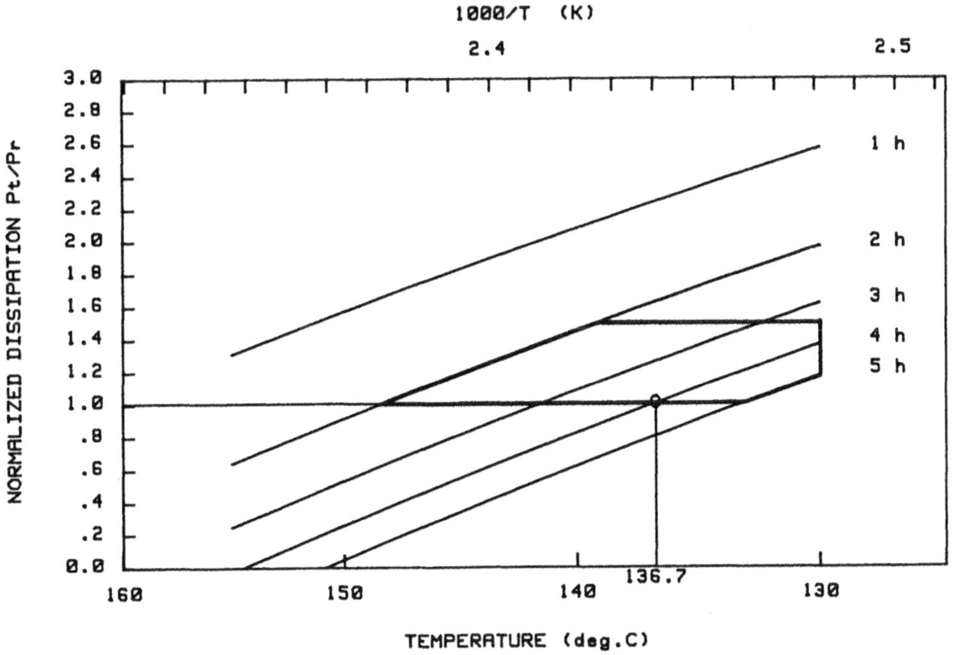

Fig. 4. "Early-failures model" presented by a field of straight lines for which the time to the 5th percentile of failures is constant

Accelerated tests at selected conditions

To confirm the model and the selected conditions, we performed about 30 accelerated tests at 137°C. The tests were conducted on subsamples of different resistor batches for which the reference data from standard 1000-hours endurance tests were available. For 12 characteristic samples the results of 4-hours accelerated tests and the reference data from corresponding standard endurance tests are compiled in Table 1.

Beside the proportion of failures (r/N), the mean change in resistance (ΔR/R) and the standard deviation (σ) of change in resistance were recorded in reference as well as in accelerated tests. This was necessary first of all because of small proportion of failures observed on most reference and accelerated tests. In this way, we introduced three criterions for judging the equivalency of test conditions on standard and accelerated tests, respectively.

The rated voltage was applied to the resistors with the resistance of up to 301 kΩ. The voltage was decreased below the rated value for 1 MΩ and 2.2 MΩ resistors because too great resistance changes (twice the reference values) were observed at the rated voltage (575 V for 1 MΩ). The resistance changes were comparable to reference values at 400 V.

Table 1. Results of some typical 4-hours accelerated tests at 137°C and comparative data of reference 1000-hours tests

	Reference tests			Accelerated tests			
Sample	r/N	ΔR/R(%)	σ(%)	r/N	ΔR/R(%)	σ(%)	Ut(V)
301 kΩ	83/1000	0.316	0.087	2/30	0.214	0.110	315
10 Ω	1/200	0.275	0.285	0/20	0.248	0.140	1.82
				0/20	0.233	0.151	1.82
249 kΩ	197/200	-	-	9/30	0.866	0.089	287
				10/34	0.848	0.088	287
249 kΩ	0/1000	0.237	0.043	0/29	0.247	0.053	287
				0/40	0.305	0.032	287
1 MΩ	6/1000	0.132	0.079	1/30	0.137	0.065	400
150 kΩ	0/7	0.500	0.070	0/40	0.339	0.073	222
1 MΩ	0/6	0.300	0.050	0/40	0.471	0.091	400
2.26 MΩ	1/20	0.040	0.010	1/10	0.069	0.014	400
				0/40	0.032	0.100	400

r/N ... proportion of failures
ΔR/R ... mean change in resistance
σ ... standard deviation
Ut ... applied voltage in accelerated tests

Discussion and conclusions

For majority of test samples the proportion of failures as well as the mean changes in resistance and the standard deviations of resistance changes observed on accelerated tests were equal to the reference values. Besides, it can be noted that the bad batch of 249 kΩ-resistors exhibited substantially greater resistance change - 3-times the value that was observed for good batches.

These statements lead to a conclusion that the proportion of failures as well as the change in resistance together with its standard deviation are the appropriate criterions for confirming the "early-failures model" as well as the selected accelerating conditions, and finally also for evaluating the quality of the produced batch. We also conclude that a short-duration test can be designed using the selected accelerating conditions. The corresponding voltage-resistance relation is shown in Fig. 5.

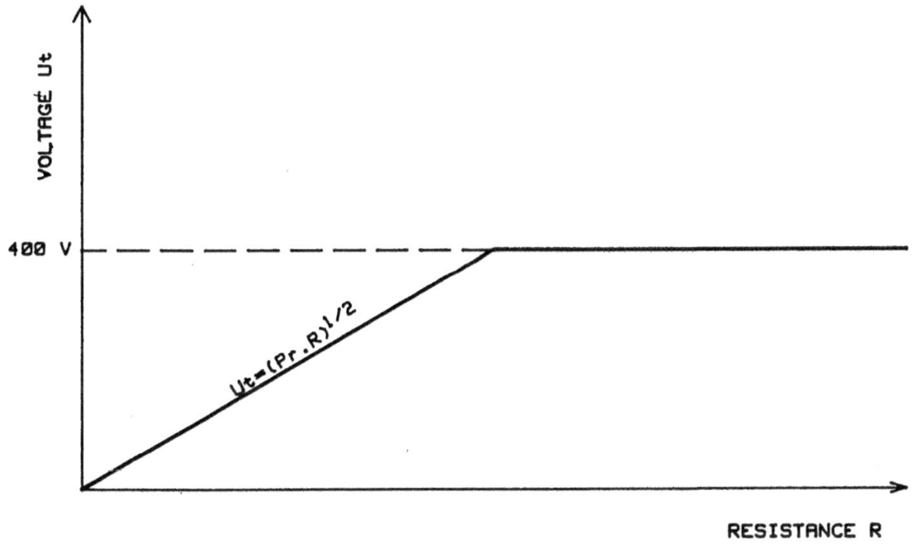

Fig. 5. Voltage-resistance relation for an accelerated test equivalent to the standard 1000-hour endurance test

In the following, we summarize the basic elements of an accelerated test which can be considered equivalent to the standard 1000-hours endurance test:

- Sample size: 20 specimens
- Test duration: 4 hours
- Ambient temperature: 137°C
- Applied voltage: constant direct voltage, corresponding to rated dissipation, or 400 V (whichever is the smaller)
- Time for recovery: more than 1 hour and less than 4 hours (at room conditions)
- Conditions for initial and end measurements: standard room conditions
- Failure criterion: resistance change $\Delta R/R > 1\%$
- Acceptance criterion: permitted defectives equal to 1

The testing and analysis approach applied in the investigation is estimated to be appropriate for further work, which could lead to effective collection and evaluation of reliability data obtained from accelerated testing and which could finally permit faster determination of reliability characterictics.

References

1. Institute for Quality Testing and Metrology, Ljubljana, Reports on Investigations "Development of Accelerated Life Testing", (1985/1986), and "Development of Screening Techniques for Assurance of Professional Quality in the Manufacturing Process of Electronic Parts", (1983/1984)

2. J. Nadrah, A. Pirih, I. Likar, "Accelerated Life Tests of Multilayer Ceramic Capacitors and Metal Film Resistors" Elektrotehniški vestnik, Ljubljana, 54 (1987)

3. IEC Publication 115-1 (1982), Fixed resistors for use in electronic equipment, Part 1: Generic Specification

4. IEC Publication 115-2 (1982), Fixed resistors for use in electronic equipment, Part 2: Sectional Specification: Fixed low-power non-wirewound resistors

Reliability Demonstration and Field Analysis of a Communications Controller

Dr U D PERERA
IBM UK limited
Havant Plant
P O Box 6
Havant
Hampshire PO9 1SA
ENGLAND

SUMMARY

This paper sets out to demonstrate the suitability of reliability growth models for reliability prediction during reliability demonstration and in the field for a communications controller.

The paper describes reliability demonstration tests, a failure review process, reliability prediction and follow up field reliability analysis for the cummunications controller. The reliability was demonstrated by simulating the product early life through two types of tests. One sample was run at ambient conditions and the other sample at accelerated conditions. In order to corroborate the demonstrated reliability, field failure data of a product sample was collated and analysed. A good correlation in reliability measures between the demonstrated and field observations was achieved.

In both test samples similar failure modes were observed. The acceleration factor of the stressed sample was estimated by modelling time to first failures using the Weibull model. This enabled combination of both test data. Through the 'test analyse and fix' process steered by a failure review board, a number of modifications were introduced to eliminate recurring failures. As a result, the demonstrated reliability of the product improved significantly. Weibull process model was applied to model the combined test data for reliability prediction.

The field performance of a sample of products was monitored for approximately one year. These systems also exhibited reliability growth. This was mainly due to natural screening. Weibull process and IBM models were applied to this data. Good agreement in reliability measures, demonstrated with field observations was observed.

The failure review board was seen as an effective vehicle to instigate corrective actions in the reliability demonstration program. The Weibull process and IBM models were suitable to model repairable systems failure data in the presence of reliability growth.

1.0 INTRODUCTION

In the development of complex electromechanical systems, the first prototypes manufactured are likely to contain various design and engineering deficiencies. To evaluate performance, reliability and latent early life problems so that corrective actions can be taken, these prototypes are typically subjected to a development and manufacturing test program. During this program product deficiencies are identified and corrective actions are introduced or initiated. Some of these corrective actions are introduced into the product throughout the design testing phase, while other corrective actions may be delayed and incorporated at the end of the test program. During this phase, the introduction of corrective actions results in improving the product reliability. This reliability growth process continues until the desired reliability is achieved. T The attainable reliability growth rate is directly related to a number of aspects such as, efficiency of the failure analysis, data recording, corrective actions and modelling.

This paper describes reliability demonstration of a communications controller. In order to identify early life failures and latent design deficiencies and to demonstrate the achievable reliability, two product samples were subjected to a 'test analyse and fix' (TAAF) process for up to three months. One sample was tested at ambient conditions while the other sample was subjected to an accelerated test condition. Also described is the failure review process and data modelling, in particular, reliability growth modelling. In addition, described is an analysis of field failure data of a product sample (120) monitored for approximately 18 months.

PRODUCT DESCRIPTION

The product in this study is one of a family of communication controllers designed to control the data flow between terminals and host processors in data communication networks. It can communicate with host processors or with other communication controllers. These are modular, programmable controllers and allow up to 60 line attachments, or a combination of lines and token ring networks.

RELIABILITY GROWTH

Reliability growth or reliability improvement with time is a common phenomenon observed for complex repairable electro-mechanical systems during the development and early life in the field. This phenomenon can be due to a single or a combination of aspects such as:

(a) Natural Screening

It is common for most system components, in particular, electronic components to have a small proportion of weak and/or 'infant mortality' type early life populations. These fail early during field usage and most likely replaced by good components. This in turn introduces reliability growth.

(b) Corrective Actions

The field performance of most systems is monitored and in order to prevent recurring or undesired failures often design modifications or preventative maintenance procedures are introduced through failure analysis. The effort put on to this process is a function of deviations in failure rate, repair costs and safety levels compared to expectations.

(c) Maintenance and Operator Learning

Often the operation of complex systems requires some sort of training. The acquisition of operating skills follow the familiar learning curve. This is usually a function of familiarity with similar systems, method of training, effectiveness of operating procedures etc. Hence, as the operator learns less operator related failures would be observed.

In a similar way, service engineers go through a learning process in familiarising system fault diagnostics. The learning process is a function of factors such as, system complexity, diagnostic capabilities/procedures and familiarity with similar systems.

RELIABILITY GROWTH MODELS

The failure times of a system is stated to follow a homogeneous process (with a constant intensity f) when during a small interval time (Δt), the probability of an event occurring ($f\Delta t$) is independent of how long the system has been operated prior to the beginning of the time interval. That is, the failure rate is constant and time between failures is exponentially distributed. for a system with failure times which follow a non homogeneous process, the intensity function is not constant, but varies with time. This process is applicable to a system showing reliability growth.

The earliest and most widely used reliability growth model is due to Duane {1}. Since, a number of alternative models have been proposed, based on different assumptions and hence it is likely to find a particular model to be more appropriate than other models for a given set of data.

In this paper Duane, Weibull process and IBM and models are outlined. Test data is applied to the Weibull process model while field data is applied to Weibull process and IBM models.

DUANE MODEL

Duane analysed reliability data of several systems in an effort to determine some consistent pattern for reliability changes during the development and design improvement phase. He observed a linear relationship between cumulative failure rate and the cumulative test time, when plotted on a log-log scale. He has shown that complex electromechanical and mechanical systems have remarkably similar behaviour of reliability improvement. Applications of this model are described in references {2,3}

Expressed mathematically, the Duane model assumes

$$f(0,t) = Kt^{-a} \quad \ldots \ldots \ldots \ldots \quad (1)$$

Taking logs gives:

$$\log f(0,t) = \log K - a \log t$$

where f (0,t) is cumulative failure rate and

K = a constant
t = Total test time
a = a measure of reliability growth or growth rate

$$\text{Now } f(0,t) = E\{r(0,t)\}/t$$

where E{r(0,t)} = expected number of failures over (0,t)

$$\text{Hence } E\{r(0,t)\} = Kt^{(1-a)} \quad \ldots \ldots \ldots \ldots \quad (2)$$

The instantaneous failure rate is given by differentiating the above equation:

i.e $f(t) = dE\{r(0,t)\}/dt$

$$= (1 - a) Ke^{-a} \quad \ldots\ldots\ldots\ldots \quad (3)$$

Again taking logs

$$\log f(t) = \log (1-a) + \log k - a \log t$$

Hence the plot of instantaneous failure rate f(t) verses time (t) on log-log paper lies also in a straight line of slope (-a), i.e. parallel to the cumulative failure rate time and displaced by the factor (1-a).

Combining (1) and (2) gives:

$$f(t) = (1-a) f(0,t) \quad \ldots\ldots\ldots\ldots \quad (4)$$

Also, since $f = 1/MTBF = 1/m$

$m(t) = 1/(1-a) \quad m(0,t)$

$m(t)$ = instantaneous MTBF

$m(0,t)$ = cumulative MTBF = $t/E\{r(0,t)\}$

WEIBULL PROCESS MODEL

Crow {4} observed that the Duane model for reliability growth is equivalent to a non homogeneous poisson process with a Weibull intensity function. He expressed the Duane model given in equation (1) above with a Weibull intensity function of the form:

$$f(t) = fb\, t^{b-1} \exp(-ft^{b}) \quad \ldots\ldots\ldots \quad (5)$$

where b is called the shape parameter and f is the scale parameter. This process is commonly called a Weibull process, and the instantaneous failure rate R(t) is given by:

$$R(t) = \frac{f(t)}{1-F(t)} = fb\, t^{b-1} \quad \ldots\ldots\ldots \quad (6)$$

where $F(t) = 1 - \exp(-ft^{b})$

The parameters K and a of the Duane model are related to the parameters b and f of the Weibull process by:

$$K = f$$
$$(1-a) = b$$

The Poisson process (homogeneous) corresponds to the case of b = 1.0, in which case the failure rate

$$R(t) = f$$

becomes a constant. This is the well-known case of the exponential distribution for the failure time.

NUMBER OF FAILURES IN AN INTERVAL

The number of failures expected in a time interval t1 to t2 (t2 > t1) is a random variable having the Poission distribution with mean given by:

$$\text{Expected number of failures} = f(t2^{b} - t1^{b}) \quad . . . . \quad (7)$$

ESTIMATION OF MODEL PARAMETERS

Method of estimating model parameters using the maximum likelihood method, which was applied in this study can be found in reference [5]. Also given in this reference is the method of estimating confidence intervals.

IBM MODEL

This model [6] is based on the following assumptions:

(a) Two types of failures are possible; random inherent failures occurring at constant rate f and non random or assignable cause failures due to design and manufacturing defects.

(b) At the beginning of testing the number of non random failures is unknown but is a fixed amount.

(c) If N(t) is the number of non random failures remaining at a time t. The rate of change of N(t) is proportiona
l
to N(t) at t.

That is,

$$d\,N(t)/dt = -BN(t) \qquad \ldots\ldots\ldots\ldots (8)$$

$$\text{hence} \qquad N(t) = A\,e^{-Bt} \qquad \ldots\ldots\ldots\ldots (9)$$

Defining V(t) as cumulative number of failures up to time t, then

$$V(t) = ft + A\,\{1 - \exp(-Bt)\} \qquad \ldots\ldots (10)$$

where
- V(t) - cumulative failures up to t
- f - constant failure rate
- A - total number of debugging failures
- B - a constant

A useful aspect of this model is the capability for assessing the fraction debugged, that is the percentage of non random failures that have been removed. The cumulative amount of debugging failures (non-random) removed at time t would be

$$N(0) - N(t) = A\,\{1 - \exp(-Bt)\} \qquad \ldots\ldots\ldots (11)$$

and hence the fraction (of K1 initial non random defects) removed by time t is

$$q = 1 - \exp(-Bt) \qquad \ldots\ldots\ldots\ldots (12)$$

Thus having estimated K2, the time at which 0.95 of the non-random defects have been removed by solving (12) for q = 0.95. That is:

$$t0.95 = \frac{-\ln 0.05}{B} \qquad \ldots\ldots\ldots\ldots (13)$$

A useful expression for planning a test programme.

TEST OBJECTIVES

The object of tests was to assure the product hardware reliability, excluding network related failures. This is due to the fact that during the test, product was not connected to a network, but powered on and running diagnostics continuously. During running diagnostics, the processor is scanning through the system hardware and in the event of a failure an error code is displayed. This code gives details about the problem, related components and probable causes of failure. In addition, in this operational mode, software failures are limited to system microcode errors. This excludes software failures that can be expected in normal operation at customer network environment. Moreover, in early products, the quality of software was at a low level and was upgraded continuously during the test program. Hence software failures were not included for data analysis. Due to these factors, the anticipated failure rate was a fraction of the expected product failure rate in the field.

There are several ways to test a sample of products in a limited period of time. These can be:

a) Test the whole sample at ambient conditions. This method is widely utilised for complex systems in TAAF programs, often tested for longer periods than three months. However, for the considered product, the total test time would not allow adequate test time to predict the field failure rate with a high level of confidence.

b) Test the whole sample at an accelerated environment. Although this can increase the equivalent ambient test time, there are a few uncertainties with this method. One is, how well the test simulates field typical failure modes and another is the method of estimating the acceleration factor of the applied stress. This type of tests are usually applied at component level when there is a known relationship between the applied stress and normal use conditions.

c) Split the whole sample into smaller samples and test each of these smaller samples at different test levels, for example at different temperatures. This is the ideal test for an accelerated test program aimed to estimate the life characteristics and the effect of environmental factors through global modelling. However, this is not the ideal test for a TAAF program. Furthermore, in order to estimate MTBF and the effect of environmental factors, at least three tests should be carried out, one sample at each temperature and each sample must be tested until a high proportion of the sample is failed.

d) Test two samples, one at ambient conditions and the other in an accelerated environment. This test enables an increase in total ambient equivalent test hours, provided similar failure modes are observed in both tests.

From the above considerations, the option adapted was (d), test two product samples, one at ambient conditions and the other in an accelerated environment.

The operation of the ambient sample was checked twice daily. In the event of a failure, product and failure information, such as, serial number, error code, time to failure, failure description, corrective action, etc were recorded in a failure report form. In most instances, it was necessary to carryout a failure analysis to obtain detailed information. Information from failure reports were used for failure review and data modelling.

The accelerated environment was thermal cycling between -40 and 60 deg C. The product was also powered on and off twice per cycle between the temperature range 10 to 40 deg C.

FAILURE REVIEW PROCESS

Prior to the test program a failure review board (FRB) was set up. This included personnel from various departments responsible for the product, quality, reliability, test and product engineers. The FRB was chaired by a quality manager and met weekly to review failures and the test status. FRB activities included:

a) Failure classification. This is to identify which failures should be considered as genuine failures and which should not. The latter category included software and test and operator induced failures.

b) Review of problem log (based on failure report forms), identification of problem owners, assignment of responsibilities for corrective actions and implementation. More emphasis was placed on unforeseen design weaknesses and safety related failures.

c) Review of ongoing failure rate, based on modelling of failure data. Also, assessment of which failures need corrective actions in order to achieve the desired failure rate by the end of the test period.

d) Review of the progress of both tests. This included activities such as, introduction of recovery plans to recover lost time due to shortage of products, loss of facilities, etc.

OBSERVED FAILURES

Details of failures observed in stress and ambient test samples are given in table 1.

UNIT/ASSEMBLY	QUANTITY	FAILURE DESCRIPTION/CORRECTIVE ACTION
STRESSED SAMPLE		
Power supply	2	Capacitor o/c - process improved
Power supply	1	Bridge rectifier-Modified to reduce leakage current
Power	1	Contactor - Process improved
Air flow sensor	2	Microcode modified
Logic card	1	Faulty module
AMBIENT SAMPLE		
Air flow sensor	2	Microcode modified
Logic card	1	o/c - modified
	1	Reseat
Display missing	1	Dry joint

TABLE 1 FAILURES OBSERVED DURING DEVELOPMENT TESTS

DATA ANALYSIS

From above section, it can be observed that failure modes of both samples is similar. Hence, in order to estimate the acceleration factor of the stressed sample compared to the ambient sample, time to first failure data of both samples was analysed using the Weibull model [7]. This is shown in fig 1. As would be expected, shape parameters of both plots are of the same order. However, they are displaced. The accelerating factor is the ratio of the characteristic life of the ambient to the stressed sample, estimated to be 17.5.

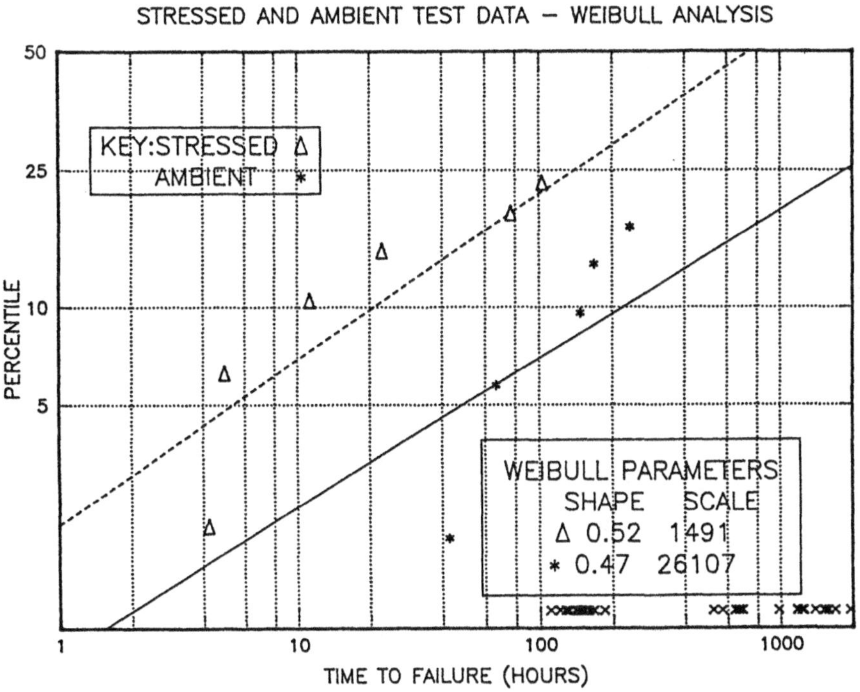

FIGURE 1. WEIBULL PLOTS OF STRESSED AND AMBIENT TEST SAMPLE DATAS

ESTIMATION OF THE COMBINED FAILURE RATE

In order to obtain the overall failure rate based on test data of both samples, time to failure and operating times of stressed data were multiplied by the accelerating factor and combined with the ambient sample. The overall failure rate was estimated by modelling this data to the Weibull process model described earlier. The goodness of fit of the model was tested using a Cramer-von Mises [8] statistic.

Estimated parameters of the Weibull process model:

Shape Parameter (b) = 0.55
Characteristic life (a) = 0.0734

Goodness of fit satisfied at 0.1 level of confidence

The shape parameter of 0.55 signifies the reliability improvement with time through the test analysis and fix phase, during which a number of modifications were introduced. At the end of the test programme a MTBF of 90 weeks was achieved. As the test environment was not simulating typical field conditions, only a fraction of this was expected in the field. The MTBF growth curve based on the above model is illustrated in figure 2.

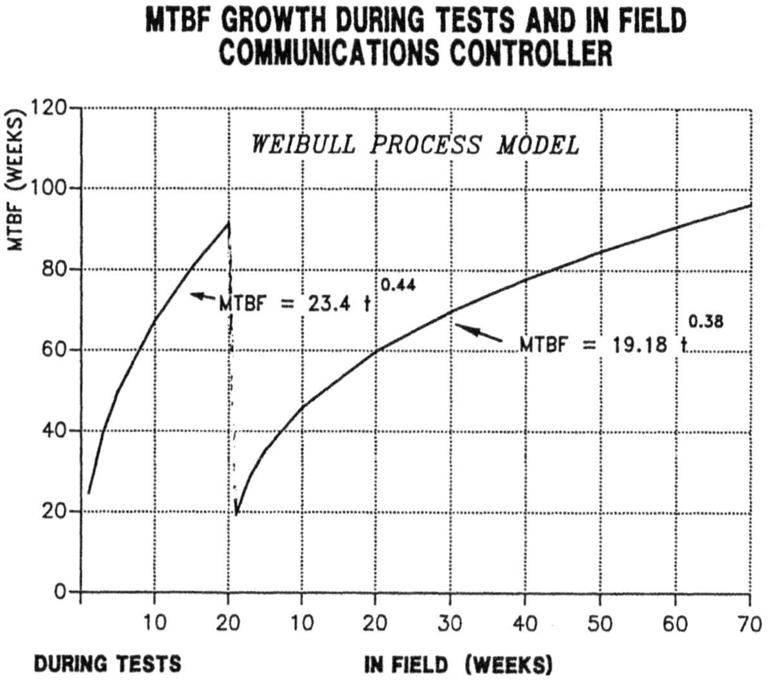

FIGURE 2 MTBF GROWTH DURING TESTS AND IN THE FIELD

ANALYSIS OF FIELD FAILURE DATA

In order to assess the field performance of the controller, the field failure data of a sample (120) of products was monitored. Data was from reported field repairs carried by service engineers. From this data it is possible to gather information regarding components replaced or assemblies repaired, repair category and calendar time to failure since installation.

The distribution of controller areas responsible for field failures is illustrated in figure 3. This illustrates two types of failures, internal system failures and those due to external cause in the network environment. Both types include a proportion of no fault found failures. The majority (70%) of failures are internal machine type. Logic cards are responsible for the majority of these. This is followed by cable/connector, switch/contactor and power related failures. Network/host systems and software are responsible for 8% and 22% of failures respectively.

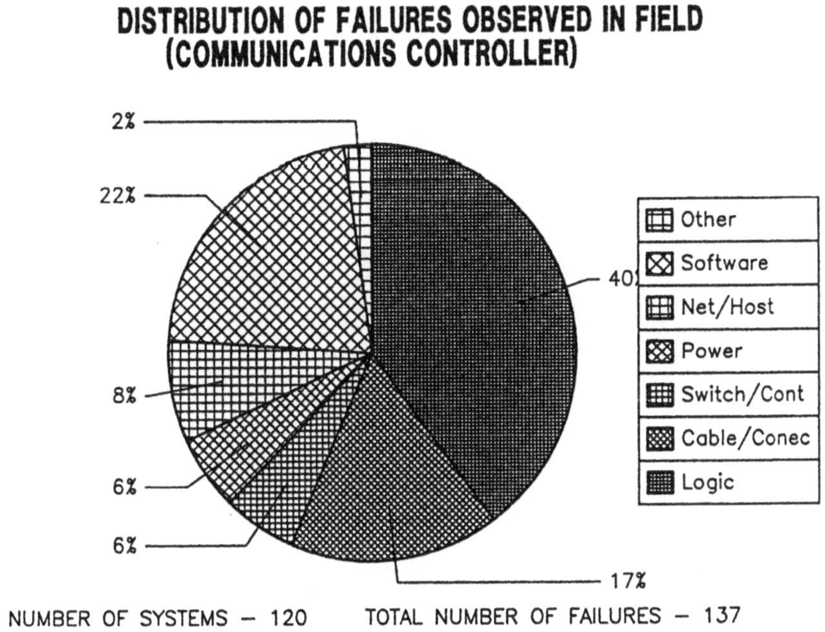

FIGURE 3 DISTRIBUTION OF AREAS RESPONSIBLE FOR FIELD FAILURES

APPLICATION OF GROWTH MODELS

As would be expected with field data, systems had varying field exposure times, from 70 to 82 weeks. Data was analysed using Weibull process and IBM growth models.

APPLICATION OF WEIBULL PROCESS MODEL

Estimated model parameters based on the maximum likelihood method for all (up to 82 weeks) time to failure data are:

Shape parameter = 0.62
Scale parameter = 0.084

The instantaneous field MTBF curve based on above parameters is shown in figure 2. It can be observed that MTBF improves from approximately 20 weeks after three weeks to 96 weeks after 70 weeks. The reliability growth is primarily due to the natural screening process; precipitation of latent early life failures. As would be expected, the growth rate during the test phase is higher than in the field.

APPLICATION OF IBM MODEL

Field sample field exposure time varied from 70 to 82 weeks. That is all systems have been in the field for at least 70 weeks. Due to restrictions in the estimation method, this model was applied to systems up to the 70 week period. Estimated parameters and related statistics are given in table 2. The distribution of observed and predicted cumulative failures based on the IBM model is illustrated in figure 4.

137 observations R-squared = 0.9933 Standard Error = 3.21

3 coefficients ADJ R-squard= 0.9932

CONSTANT	ESTIMATE	STD ERROR	T STAT	SIG LEVEL	0.95 CONF LIMITS LOWER	UPPER
A	1.6159	0.0243	66.42	0	1.568	1.663
B	20.56	0.9232	22.27	0	18.75	22.37
C	0.214	0.0159	13.47	0	0.183	0.245

TABLE 2 IBM MODEL CONSTANTS AND TEST STATISTICS

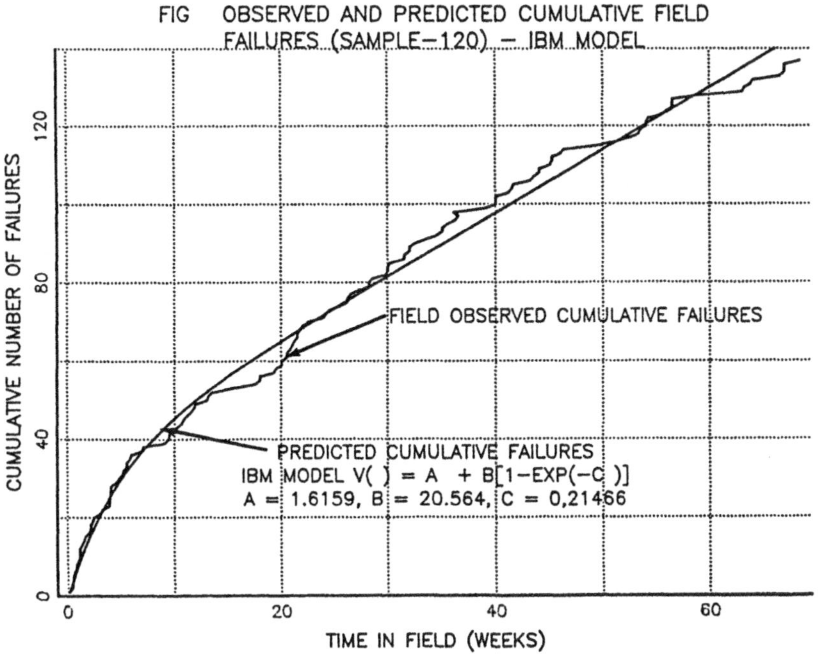

FIGURE 4 DISTRIBUTION OF OBSERVED AND PREDICTED CUMULATIVE FAILURES

A very good fit can be observed. From model parameters the following statements can be made:

1) The intrinsic failure rate is 0.01346 failures/week. MTBF based on this failure rate is 74.2 weeks and 95% confidence limits are 72.1 and 76.5 weeks.

2) The above MTBF is less than the Weibull process model projected MTBF of 96 weeks at 70 weeks, This is likely due to the different estimation methods and field exposure times.

3) Total non random failures removed by 70 weeks is 21. This is equivalent to 0.17 failures/system. The percentage of non random failures/system is 12.8%

4) 95% of non random failures have been removed by 14 weeks.

CONCLUSIONS

1) The TAAF program was successful in identification of latent defects and reliability demonstration.

2) The failure review board was an effective vehicle to instigate corrective actions.

3) The introduction of corrective actions resulted in a significant reliability growth. Weibull process model was suitable to model data in this environment.

4) The IBM model is useful model data in the presence of reliability growth. In particular, to estimate the intrinsic failure rate, remaining proportion of non random failures and time at which a specified proportion can be debugged.

5) The reliability of electromechanical systems improves with time. This is due to a combination of natural screening, operator and service personnel learning process and introduction of corrective actions.

REFERENCES

1. Duane, L.H.: Learning curve approach to reliability monitoring. IEEE transactions on Aerospace, Vol 2, No 2, (1964) 563-566.

2. Clarke, J.M.: No growth growth curves, IEEE proceedings 1979 annual reliability and maintainability symposium.

3. Bezat, at al: Growth modelling improves reliability prediction, IEEE proceedings 1975 annual reliability and maintainability symposium.

4. Crow, L H: Reliability analysis for complex repairable systems in : F Proshan and R J Serfing, Eds, Reliability and Biometry (SIAM Philadelphia, 1974) 379-410.

5. Crow L H: Confidence interval procedures for the Weibull process with applications to reliability growth, Technometrics, Vol 24, No 1, February 1982.

6. Rosner N: System analysis-non linear estimating techniques, Proceedings national symposium on reliability and quality control, 203-207, New York, 1961

7. O'Connar P D T: Practical Reliability Engineering, John Wiley and Sons, New York, 1985

8. Lawless J F: Statistical methods and methods for lifetime data, John Wiley and Sons, New York, 1982

Bayes Optimal Burn-In-Times

C. A. CLAROTTI

ENEA TIB-AQ
Casaccia, Rome

F. SPIZZICHINO

Dep. of Mathematics G. Castelnuovo.
University of Rome "La Sapienza"

Summary
The Bayesian definition of optimal burn in procerdure is given. The Bayesian approach to the burn-in problem is described in both the case of independent and in the case of dependent component failure times.

Introduction

Burn-in is a common practice in many branches of reliability engineering. In spite of this strong practical interest, the mathematical theory of burn-in has remained largely inexplored up to recently.

After the brilliant pioneering work by Proschan [1], almost nothing mathematical happened in the burn-in area till eighties (see the date of paper next to [1] in the references). As a matter of fact, at the critical epoch of transition from frequentist reliability to Bayesian reliability the only well established foundation of the burn-in theory was the bath-tube curve. This curve is a suitable guidance in classical statistics, but is of no use to a Bayesian who rather may be mislead by it.

For a correct Bayesian burn-in procedure to be obtainable the analyst must regard the burn-in problem as a decision problem. The decision problem of concern is: shall I stop the burn-in or not? And in the case of stopping, shall I judge the components to be suitable for the assigned mission or not?

The above decisions must be taken on the basis of the actual unconditional component life distribution. "Unconditional" means that the unknown parameters must have beeen integrated out. For instance, the exponentiality conditional on an unknown λ is not a sufficient condition for the burn-in to be pointless.

<u>The Case of Independent Failure-Times</u>

Let the possible decisions be

- a_o ≡ no burn-in, components are immediately put into operation;
- a_t ≡ a burn-in of duration t, components which survive the burn-in period are put into operation
- $a_{+\infty}$ ≡ a burn-in period of infinite duration, i.e. components are not put into operation since they are judged not to be suitable for that.

The first ingredient for settling a burn-in procedure is the superimposed cost structure.

In [4] the simple loss function below has been chosen as sample loss function

$$L(a_t, T) = \begin{cases} c_1 & \text{if} \quad T \leq t \\ C & \text{if} \quad t < T \leq t + T \\ -K & \text{if} \quad t + \tau < T \end{cases} \tag{1}$$

where:

- T is the failure time of the component of interest,
- t is the selected burn-in time
- τ is the duration of the mission that the component must accomplish after the burn-in.

$L(a_t, T)$ defined by eq. (1) is a reasonable approximation of costs incurred in many real problems of interest.Note that eq. (1) is understood to hold also for $t = +\infty$ (the components are not put into operation). In this case obviously, it can only be $T < t = +\infty$. The hypothesis underlying eq (1) is that the cost per unit time of burn in is negligeable with respect to the other costs. The negative cost corresponding to $T > t + \tau$ (mission accomplished) represents the gain in case of successful operation. Let us now come to the second ingredient of any decision problem in face of uncertainty, the probability model. We shall first consider the simplest probability model one can think of, that is the failure times of the components of interest are independent. Note that independence must be underdstood with respect to the unconditional life distribution Independence of component lives makes the many-component-burn-in-problem the same as the single-component-burn-in-problem.

Let F(t) be our unconditional component life distribution, and let $\Phi_s(t)$ be the expected cost following from a component of age s undergoing a burn-in period of duration t

$$\Phi_s(t) = c_1 - (C - c_1)\ \frac{\bar{F}(s + t)}{\bar{F}(s)} - (K + C)\ \frac{\bar{F}(s + t + \tau)}{\bar{F}(s)} \qquad (2)$$

In a Bayesian frame, setting the optimal burn-in time in the case of unconditional indepedence is just a matter of studying the function $\Phi_s(t)$.

Define (reasonably) the burn-in procedure a_t $(0<t<+\infty)$ to be convenient if it results

$$\Phi_0(t) < \min\ (\Phi_0(0),\ c_1)$$

which means that the expected cost of:

a burn-in time t, and then operation,

is smaller than the expected costs of

both : no burn-in, immediate operation,

and : no operation, the component is unsuitable

From this definition of convenient burn-in time it can be easily shown [4] that t being convenient implies:

$$\Phi_t(0) < \Phi_0(t) < \Phi_0(0) \qquad (3)$$

In words, eq(3) says that

a. going through the-burn in costs less then immediately starting operation $(\Phi_0(t) < \Phi_0(0))$

b. the expected cost after the burn-in (in the case that component survived) is smaller than the expected cost in case of no burn-in $(\Phi_t(0) < \Phi_0(0))$

Inequality in proposition b $(\Phi_t(0) < \Phi_0(0))$ is the rigorous mathematical counter-part of the well known, but vague, statement:

burn-in makes components better.

Let us now survey some other interesting results contained in [4]. Mathematical technicality will be skipped, the stress of this paper being on the practical impact of burn-in theory. In the frequentist age IFR distribution was synonym with no burn in. What about that, in the Bayesian age?

Proposition 1 in [4] states: if F is NBU then $\Phi_0(t) \geq \min(\Phi_0(0), c_1)$ and $\Phi_t(0) \geq \min(\Phi_0(0), c_1)$ for any t.

Under the NBU hypothesis for F, the burn-in is then not convenient. If F is IFR, it is also NBU, so is there nothing new in the age of Bayesian burn-in with respect to the age of the frequentist burn-in?

Not at all?

Remember that the distribution F in Poposition 1 is the unconditional distribution. It could be the case [5] that conditional on knowing the value of some "parameter" component life is IFR distributed, and yet since ths parameter is unknown to you, your unconditional distribution and the superimposed cost structure are such that the burn-in is convenient.

In the same line.
Suppose "you know for sure" that the component of interest is exponential, only the parameter λ is unknown to you .

Can you claim that the burn-in is unuseful given the conditional exponentiality of the component?

Before answering yes I can, have a look on the following results [4].

If F(t) has a density f(t) and the ratio

$$g(t) = \frac{f(t + \tau)}{f(t)}$$

is strictly increasing then

$$a_{+\infty} \text{ is optimal iff } \lim_{t \to +\infty} g(t) < \frac{C - c_1}{C + K}$$

$$a_0 \text{ is optimal iff } g(0) > \frac{C - c_1}{C + K}$$

$$a_{t^*} \quad (0 < t^* < \infty) \text{ is optimal iff } \quad g(t^*) = \frac{C - c_1}{C + K}$$

Note that in the case of an exponential component with unknown λ, $F(t)$ has a density and it is

$$g(t) = \frac{\int_0^{+\infty} \lambda\, e^{-\lambda t} e^{-\lambda \tau} \Pi(\lambda)\, d\lambda}{\int_0^{+\infty} \lambda\, e^{-\lambda t} \Pi(\lambda)\, d\lambda} \tag{4}$$

where $\Pi(\lambda)$ is the pdf which resumes our state of knowledge on λ.

The rhs of eq. (4) is increasing. Heuristically this can be proved as fallosws.

The pdf

$$\frac{\lambda\, e^{-\lambda t} \Pi(\lambda)}{\int_0^{+\infty} \lambda\, e^{-\lambda t}\, \Pi(\lambda)}$$

is the same as the posterior on λ we would have, were $\Pi(\lambda)$ our prior and did we observe a failure at time t.

The rhs of eq. (4) can then be regarded as our component reliability assessment after observating a failure at time t.

Obviously, the later the failure (the larger t), the larger the assessed value of component reliability.

That is, $g(t)$ is increasing. So, if a t^* exists such that

$$g(t^*) = \frac{C - c_1}{C + K}$$

the exponential component must undergo the burn-in.

Let us now complicate a bit the problem.

We are contemplating the possibility of N components undergoing burn-in before n of them be used for forming a k:n system (n<N). If during the burn-in more than N-n components will fail, which makes it impossible to form the k:n system, we have to pay c_1. The same cost will be incurred if we decide not to build the system, the components being unsuitable for that

If the burn-in duration is set at t, at least n components survive and the k:n system is built:

- an in-service failure of the system will cost us C,
- if the system survives its mission of duration τ, we will cash K.

Again, component lives are (unconditionally) independent.
Component life distribution is F(t).
This problem must be faced by the techniques of optimal stopping for Markov chains [6].

More precisely, the Markov chain of interest has state space (n-1, n, ..., N). (This is so because, the state n-1 is absorbing in that, reaching it makes it impossible to build the k:n system).

The cost of stopping the chain at time t is

$$c_t = \min(c_1, C - (C + K) \sum_{j=k}^{n} \binom{n}{j} [\frac{\bar{F}(t + \tau)}{\bar{F}(t)}]^j [1 - \frac{\bar{F}(t + \tau)}{\bar{F}(t)}]^{n-j} \quad (5)$$

First argument in the function min(.,.) is the cost we incur if we give up forming the system, and the second argument is the expected cost following from forming a k:n system from components of age t. At any time t, the only possible transition of the chain is the downwards transition which causes the number of alive components to decrease by one, which occurs at a rate

$$\lambda(t) = \frac{-\frac{d\bar{F}(t)}{dt}}{\bar{F}(t)}$$

In [6] the theory of optimal stopping is developed for continuous time-continuous-state-space and discrete-time Markov chains. Some work is being done [5] for adapting theorems in [6] to the case of interest, namely: continuous time - discrete-state-space Markov chains.

Hopefully, the result of this work will be to prove the seemingly reasonable conjecture:
Under the NWU (or at least DFR) hypothesis for $F(t)$, the cost structure discussed so far leads to the following optimal burn in procedure:
Delay the decision untill the time s at which only n out of the N original components are left. When only n components have remained on test then choose the optimal residual burn-in time by minimizing the expected cost function

$$\Phi_s(t) = c_1\left\{1 - \left[\frac{\bar{F}(t+s)}{\bar{F}(s)}\right]^n\right\} + C + \\ -(C+K)\left[\frac{\bar{F}(t+s)}{\bar{F}(s)}\right]^n \left\{\sum_{j=k}^{n} \binom{n}{j} \left(\frac{\bar{F}(t+s+\tau)}{\bar{F}(t+s)}\right)^j \cdot \left(1 - \frac{\bar{F}(t+s+\tau)}{\bar{F}(t+s)}\right)^{n-j}\right] \tag{6}$$

where:

- the first term on the rhs is the cost of giving up building the system due to at least one on-test failure of the remaining components;
- the rest of the rhs is the expected cost of forming a k:n system from components of age (s+t)

<u>The Case of Dependent Failure Times</u>

For learning to be possible, failure times of components under scrutiny for potential burn-in must exhibit some kind of statistical dependence.
In [7] component-lives are assumed to be exchangeable (This is more general than independence given a "parameter"). Let the available components be N and let $T(i)$ $i = 1,\dots, N$, be the time of ith earliest observed failure.
Each component-failure will yield a loss c_1 if it occurs on test and a loss C if it is an in-service failure. Each component which survives both the burn-in-time and the mission time will yield a reward K.

As usual an infinite burn-in time means: components are not suitable for operation.

Since due to statistical dependence, at any time s the behaviour of each component (its failing or surviving at s) is informative as to the future behaviour of all components still alive at s, the burn-in procedure has necessarely a sequential character as shown in the operational flow chart of fig.1.

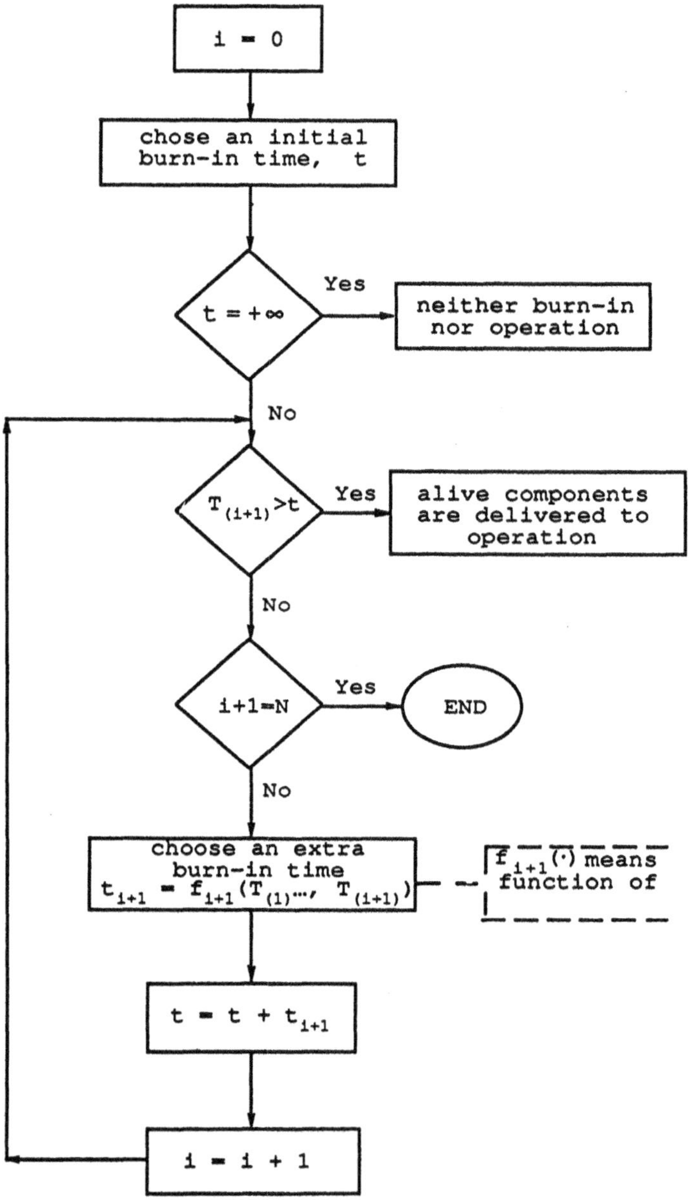

Fig. 1

The main results of [7] are:

- an existence theorem for the optimal sequential burn-in strategy;
- a procedure for calculating, at any time s and whatever the number and times of failures observed up to s, the cost of the optimal strategy.
- the rule for terminating the burn-in procedure which is simply: stop at the first time at which doing so will yield the cost of the optimal strategy.

References

1. Proschan, F.: Theoretical Explanation of Observed Decreasing Failure Rate. Technometrics, 5 n.3 (1963) 375-383.

2. Barlow, R.E.: A Bayesian Explanation of an Apparent Failure Rate Paradox. OOC-83-13 (1983).

3. Bergam, B.; Klefsjö, B: Burn-in Models and TTT - Transforms. Quality and Reliability Eng. Int. 1(1985) 125-130.

4. Clarotti, C.A.; Spizzichino, F.: Bayes Burn-in Procedures, Submitted for Publication in Probability in Eng. and Information Science.

5. Clarotti, C.A.; Spizzichino, F.: Optimal Stopping of Markov Chains in a Burn-in Frame, Working Paper, ENEA R-1-88 (1988).

6. Shiriaev, A.N.: Statistical Sequential Analysis. American Mathematical Society (1973).

7. Spizzichino, F.: Sequential Burn-in Procedures, Proc. of Int. Conf. on Rel.,Missouri Columbia (1988), to appear.

Review of the EuReDatA Project Groups' Activities and Results

A. AMENDOLA and T. LUISI

Commission of the European Communities
Joint Research Centre
Systems Engineering and Reliability Division
I-21020 Ispra (VA)

Summary

One of the most significant objectives of the EuReDatA association is the establishment of compatible standards for reliability data collection and analysis. This is being pursued by specialized project groups. By starting with the proposal of general reference classification schemes for component reliability and the issue of a guide for reliability data collection and analysis, the project groups' activity has moved in more recent times towards data analysis exercises and feasibility studies for data collection projects.

1. Introduction

Since the creation of the association, achievement of compatible standards and definitions for reliability data and agreement on methods for data authentication and validation are objectives which are being pursued by the set-up and the works of several ad-hoc project groups.

In fact, the establishment of common languages and standards is the necessary premise for any meaningful exchange of data among the members. These are, indeed, organizations from different countries and involved in industrial activities of very different kinds, where users needs are often not the same.

After the definition of general classification schemes concerning components reliability, and the issue of a guide for data collection and management, various components have been classified and problems connected with data exchange an electronics components and materials reliability have already been subjects of finalized projects.

Most of the results of these earlier project groups have already been the subject of a previous paper /1/. Some brief comments about them are however

still useful in order to link first activities and results with the more recent EuReDatA works. These are being directed towards proposals of criteria for collection of data an incidents and human reliability and exercises on data analysis aimed to assess corresponding methods and procedures.
Finally, feasibility of joint data collection exercises is being assessed as a preliminary step of a possible development of the association interests towards the implementation of some common data bases on selected components classes.

2. Data Collection and Analysis Procedures

Procedures for reliability data collection, management and processing have already been subjects of extensive studies within the association which, on the one hand, resulted in the issue of EuReDatA guidelines /2,3/ and on the other one in the diffusion of the experience gained by the Members through tutorial courses on Reliability Data Bases /4/.
The development of the guide is based on the experience that the members have acquired in their own previous activities. This experience was commonly discussed and written down in a Project Report /3/ which covers main topics like:

- Data Sources;
- Data Collection Procedures (Criteria for both Event Data and Historic Data Collection Schemes);
- Design of Data Collection Forms;
- Data Base Models and Managements;
- Updating and Checking Procedures;
- Statistical Data Processing in order to estimate failure rates, availability and utilization factors from event data;
- Uncertainties of Reliability Parameters.

After completion of this guide, the performance of a joint data analysis exercise was considered as a significant step to improve common knowledge on models and uncertainties linked with reliability data assessment. The exercise (of "Benchmark" type, since different teams analyse independently a same reference event data set) is extensively described in an other paper presented at this conference /5/. Benchmark exercises have already produced very valuable insights and results when performed on other reliability modelling topics /6,7,8/: indeed, owing to the intrinsic nature of any modelling and probabilistic prediction, a significant degree of subjectivity (assumptions, use of raw data and models, etc.) is introduced at any step into the analysis. Therefore, the comparison of studies independently performed by different experts results in a rather instructive picture of the different manners to approach the common problems; and it offers a unique way to assess employed methodologies and associated uncertainties.

The terms of reference for the project are extensively included in ref. /5/. Aim of the exercise is essentially the comparison of the methodology used by the various member organizations to estimate from raw event data reliability parameters like failure and repair rates, failure probabilities on demand and repair time distributions.

The common set of raw data concerns the operating experience of 450 pumps in various LWR nuclear power plants as recorded in the CEDB Data Bank operated at JRC, Ispra.

Participation in the exercise involves eleven member organizations from seven different countries and the Joint Research Centre, which is also coordinating the project.

Whereas a more detailed description of the intermediate results has to be found at the already quoted ref. /6/, at this stage of the exercise some relevant conclusions can already be drawn:

- The analysis of the raw data themselves has enlightened problems connected with completeness and correct interpretation of data sources;
- The first comparison of the approaches adopted has shown that very significant methodological differences exist in reliability data processing.

It is then confirmed that the issue of guidelines and procedures for reliability analysis can take a great advantage from work done in common projects of benchmark type. Therefore, after completion of the exercise, a natural objective of future EuReDatA activity should be the draft of guidelines for data validation and processing, as incorporating the lesson learned from the on going project.

3. Reference Classifications concerning Components Reliability

The criteria followed in establishing a general reference classification scheme for components reliability have already been described at ref. /1/ and in the EuReDatA project Report N. 1 /9/. Indeed, this report includes:

- Failure Classifications (concerning Failure Modes, Failure Descriptors and Causes of Failure);
- Common Classification Descriptors (concerning engineering descriptions, operational regimes, external and internal environmental influencing characteristics which were considered to be common throughout wide ranges of engineering components and systems);
- Components boundary, and engineering/operating characteristics.

Furthermore, the report has applied the proposed classification scheme to a first set of components, and, namely:

- Mechanical valves;
- Electric motors;
- Stand by emergency diesel-generators;
- Electro-mechanical actuators;
- Hydraulic actuators;
- Pneumatic actuators.

This classification has been assumed in many cases as a reference in designing raw data collection exercises. (A guide for attributing the corresponding codes is now being finalized by B. Stevens as a further contribution of NCRS (U.K.) to the association activities).

However, despite of the success of this first EuReDatA institutional effort, attempts for extension of this classification to further components have not yet given concrete results but as far as Automatic Fire and Gas Detectors (AFGD) are concerned.

AFGDs are devices of upmost significance for fire protection both on industrial activities and on civil buildings and facilities.

On the other side, spurious actuations often result on economic losses and losses of confidence in such systems. Thus, availability of reliable data on AFGDs behaviour is particularly important. Therefore these components were selected for a reference classification project, after a seminar held at

JRC, Ispra on the subject in 1984 /10/. The outcomes of the project are now being finalized in the EuReDatA Report N. 5 /11/.

Terms of reference for the project are the identification of a minimum set of attributes for characterizing AFGDs with respect to reliability data acquisition, as well as the improvement of existing failure classification schemes and boundary definition for such components.

Principal bases for the project group works have been:

- The experience gained an AFGDs reliability data collection with the OREDA project and included on the corresponding Handbook /12/;
- The general logical structure worked out within EuReDatA for components classification (see Fig. 1) /1,9/;
- The recommendations from the Ispra seminar /10/.

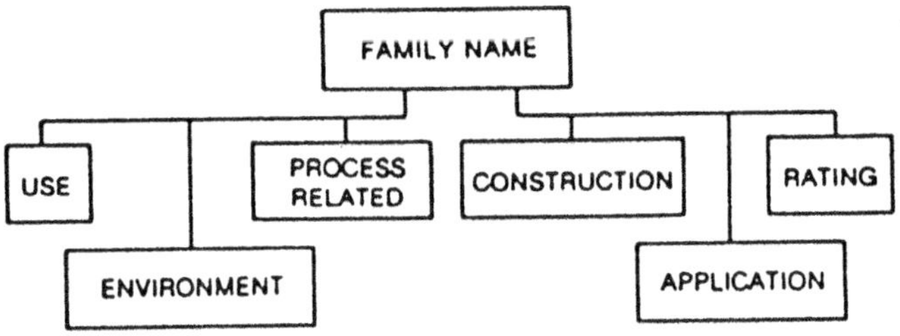

Fig. 1. Components' Classification Logical Structure.

The project report /11/ will cover the following items:

- Description of the most common types of fire and gas detectors;
- AFGD reference classification: the specialization of the general scheme in Fig. 1 for the AFGDs is represented in Fig. 2, which shows the hierarchical diagram of the attributes retained for characterizing the AFGDs;
- AFGDs boundary definition.

Automatic Fire and Gas Detectors (AFGD)

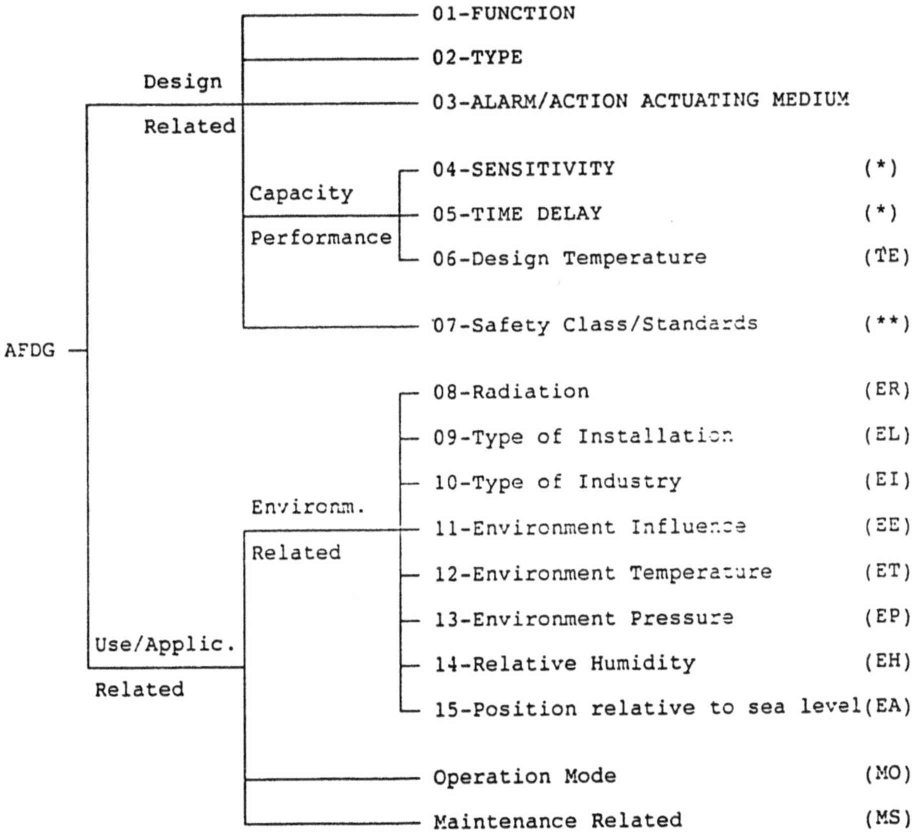

(*) Values must be established
(**) e.g. BS 5839 Part 1

Fig. 2. Proposed AFGDs Classification Scheme.

4. Data Collections on Incidents and Human Reliability

In addition to components performance data, other kinds of data relevant for safety reliability assessment have been subject of EuReDatA projects. Problems connected with Materials Reliability Data have been identified and suggestions for classifications and descriptive data collections have been laid down in Project Report N. 4 /13/.

On-going projects are being devoted to incident and Human Reliability Data Collections.

A project group on Incident Data (coordinated by P. Bocholtz TNO) is addressing problems connected with user needs and applications of incident data bases rather than problems of data structures, database management systems, retrieval methods etc. which are common to several other kinds of data bases. By keeping the focus of the investigation on risk identification, principal items being considered for drafting recommendations are:

- Definition of the scope of the data collection;
- Selection criteria for the incidents that should be considered for the collection according to the indended users (risk analysts, emergency planners, inspectorates or authorities);
- Reporting requirements, minimal information contents, application, limitations, uncertainties and incompletenesses.
- Data sources, practical experiences and examples.

The project group on Human Factors (coordinated by G. Mancini, JRC, Ispra) has started working by reviewing the main classification schemes for human reliability, like the well known taxonomy of human malfunctions established in a CSNI project of OECD /14/; the rather recent EG&G proposal /15/ more directed towards estimation of human failure rates to be inputed into the Nuclear Computerized Library for Assessing Reactor Risk (NUCLARR) system; the work still in progress at VTT /16/, etc. These preliminary investigations and previous experience shared by the participants in the project have resulted in a draft proposal for a reporting form for human errors which has to be seen in the prospective of an incident type (i.e. event description based) data bank.

Main information to be required involves:

- Free text for event description;
- Time of the event;
- Type of industry, process, plants;
- Modes of detection (as based on the CSNI taxonomy and completed according to the VTT indications);
- Plant state at event initiation (largely based on CSNI taxonomy);
- Free text for description of event course and actions taken after disturbance detection);
- System/Unit/Component affected;
- Human Error Description mostly based on CSNI cathegories, but including free formats for description of contributory factors concerning physical and human subsystems; specially control and safe devices, procedures, inspection etc.) information subsystem (training, instruction, coordination, communications, etc.) and management subsystems (fairly responsabilities, work organization and priorities etc.), and for giving indications of worst consequences possibly linked with the reported events.

Two major problem areas have been identified at the moment as worth of further investigation: a better identification and description of factors affecting performance, a proposal for which is being elaborated by Norsk Hydro, and the choice of systems classification schemes to identify units affected by the event. For these last problems the classification studies /17,18/ promoted by JRC for Nuclear and Chemical Plants are being considered as possible basic references.

A draft of the final report for the project is expected by spring '89. At this stage of work completion, it is already possible to remark how the participants are considering more appropriate to address the problem of human performance event collection quite differently from a direct approach to failure rates quantification; that is, in a way more similar to the CSNI proposal than to the NUCLEARR one.

5. Electronic Systems

Electronic components have been already the subject of a project group aimed to the identification of a minimum set of parameters in order to exchange data on their reliability /19/.

Recently EuReDatA has been joined by an increased number of members, whose major interest is with electronic components. The EuReDatA assembly has agreed on the proposal by J. Moltoft and L. Rimestad (Danish Engineering Accademy) concerning field reliability data for components in electronic systems with the following objectives:

1) The project group shall define and describe a structure, the type of content and the size of an European data base containing field reliability information on components used in electronic systems;
2) The project group shall identify the output from the data base and point out the corresponding methods of analysis;
3) The project group shall identify the problems involved in getting valid and dependable information into the data base and propose one or more organizational structures by which the problems can be solved.

For the first time a project group has been set up which is directed neither to produce collection guides on standards nor to assess methods and procedures for data analysis but to assess the feasibility of a European data base on a specific reliability topic.

This might really indicate an important change in the general policy of the association. Indeed, other proposals are being analyzed before a possible formulation of similar project groups for other kinds of components.

6. Conclusions

Since its foundation, the EuReDatA association has been active in producing guidelines for data collections and reference classifications to allow data bankers to collect and to exchange reliability data in a consistent way. Recent activities show an enlarged interest of EuReDatA not only with methods and procedures for data processing but also with special topics like incident and human event data; and, especially an important trend towards feasibility of European data collection exercises. This trend should find a favourable response from possible users: by however, these new activities should not slow down the important works which have been the early task of the association.

Production of standards for data collections should remain one of the major objective of the association.

Acknowledgements

The project groups have been made possible by the voluntary works of member organization and persons. Whereas the coordinators of the project groups have been already identified in the paper or in the references, it would be impossible to explicitly thank all other contributors. EuReDatA is really greatly indebted with all of them.

References

1. Luisi, T.: EuReDatA-Project Groups: An Overview of Objectives and Results. Proceedings of the 5th EuReDatA Conference: Reliability Data Collection and Use in Risk and Availability Assessment. Heidelberg, FRG, April 9-11, 1986 (H.J. Wingender ed.). Springer-Verlag - Berlin (1986) (pp. 47-56).

2. Stevens, B.: A Guide to Reliability Data Collection, Validation and Storage. Proceedings of the 5th EuReDatA Conference (Op. Cit. pp. 57-65).

3. EuReDatA Project Report No. 3: Guide to Reliability Data Collection and Management (B. Stevens ed.). CEC-JRC, Ispra - S.P./I.05.E3.86.20 (1986).

4. Amendola, A. and Keller, A.Z. (Eds.): Reliability Data Bases. Proceedings of the Ispra Course held at JRC, on 21-25 October 1985 in collaboration with EuReDatA - D. Reidel Publishing Company - Dordrecht (NL) 1987.

5. Besi, A. and Colombo, A.G.: Report on the on-going EuReDatA Benchmark Exercise on Data Analysis - Sixth EuReDatA Conference on Reliability Data Collection and Use in Risk and Availability Assessment, Siena, 15-17 March 1989.

6. Amendola, A.: Assessment of PSA Methods and Procedures: a Wide Scoping European Research Programme - Proceedings of the 5th EuReDatA Conference - (Op. Cit. pp. 273-280).

7. Amendola, A.: Uncertainties in Systems Reliability Modelling: Insight Gained through European Benchmark Exercises. Nucl. Eng. and Design 93, pp. 215-225.

8. Poucet, A. and Amendola, A.: State of the art in PSA reliability modelling as resulting from the international benchmark exercises project - NUCSAFE 88 Conference - Avignon, October 2-7, 1988.

9. EuReDatA Project Report No. 1: Reference Classification concerning Components' Reliability - (T. Luisi ed.) - CEC-JRC, Ispra - S.A./I.05.01. 83.02 (1983).

10. Proceedings of EuReDatA Seminar on Reliability of Automatic Fire and Gas Detection Systems. (A. Amendola and S.T. Peacock eds.) T.N. No. I.05.E3. 85.02, CEC-JRC Ispra (1985).

11. EuReDatA Project Report N. 5: Reference Classification Concerning AFGDs Reliability (T. Luisi ed.) to appear in 1989.

12. OREDA (Offshore Reliability Data) Handbook. First Issue (1984) - Veritee, P.O. Box 370, 1322 Hoevin, Oslo, Norway.

13. EuReDatA Project Report N. 4: Materials Reliability (G. Gavelli and A.L. Smith eds.) CEC-JRC, Ispra - S.P./I.05.E3.85.38 (1985).

14. Mancini, G.; Rasmussen, J. et. al.: "Classification Systems for Reporting Events Involving Human Malfunctions" EUR 7444 EN, (1981).

15. Gertman, D.I. and Gilmore W.E.: "Specification for the submission of raw human error probability data to the NUCLARR Data Cleaninghouse" EGG-REQ-7732 June 1987.

16. Pekka, Pyy: A method for human error identification in nuclear power plants. (Draft Report) VTT-(SF) Electrical Engineering Laboratory (1988).

17. Melis, M. et al.: Light water reactors Reference System Classification for the European Reliability Data System (ERDS), EUR Report No. 7905 EN, 1982.

18. Bello, G.C. et al.: Classification Scheme for Process Plants - 5th EuReDatA Conference, Heidelberg, 1986 (Op. Cit).

19. EuReDatA Project Report No. 2: Proposal for a Minimum Set of Parameters in Order to Exchange Reliability Data on Electronic Components (N. Garnier ed.) S.P./I.05.E3.85.25 CEC-JRC, Ispra (1985).